Mutant p53 and MDM2 in Cancer

Subcellular Biochemistry

Volume 85

Series editor

Robin Harris, Northumberland, UK

For further volumes:
http://www.springer.com/series/6515

Swati Palit Deb • Sumitra Deb
Editors

Mutant p53 and MDM2 in Cancer

Springer

Editors
Swati Palit Deb
Sumitra Deb
Department of Biochemistry and Molecular Biology, and the Massey Cancer Center
Virginia Commonwealth University
Richmond, VA, USA

ISSN 0306-0225
ISBN 978-94-017-9210-3 ISBN 978-94-017-9211-0 (eBook)
DOI 10.1007/978-94-017-9211-0
Springer Dordrecht Heidelberg New York London

Library of Congress Control Number: 2014946109

Printed on acid-free paper

Springer is part of Springer Science+Business Media (www.springer.com)

Preface

Since its discovery, p53 research has been the highlight of understanding the crucial role of oncogenes and tumor suppressors in regulated or deregulated cell growth. In the last decade, special attention was focused on gain of function mutations of p53 that can turn the tumor suppressor to an oncoprotein, as well as on abnormal expression of oncogenes such as MDM2 and MDMX that could inactivate the tumor suppressor function of p53 in the hope of devising cancer treatment. A wealth of information has emerged regarding what genes the gain-of-function mutants of p53 activate and how they induce oncogenesis, how these mutants are stabilized in cancer cells, how they respond to chemotherapy, and how interaction of p53 mutants with p53 family members may induce oncogenesis. Similarly there are exciting reports on how the oncoprotein MDM2, known to exist to control p53, can activate signaling pathways independent of p53 when overexpressed, and how MDMX is involved in the regulation of p53 by MDM2.

Mutant p53 and MDM2 in Cancer includes 19 chapters that discuss the activation of diverse oncogenic pathways consequent to p53 mutation and overexpression of MDM2 and MDMX and their splice variants. This book also includes chapters that discuss p53 mutation in hereditary cancer, response of cancers with p53 mutation to chemotherapy and radiation, structural aspects of mutant p53 that make it an oncoprotein and targeting of these structures for cancer therapy. The function of wild type p53 in response to stress and regulation of this function by MDM2 has also been included. Overall, this book provides an insight into the primary molecular events leading to oncogenesis consequent to p53 mutation and overexpression of MDM2. The information should be invaluable for beginning or experienced researchers, and even for future researchers opting to commit to cancer biology. To dissect the oncogenic functions of mutant p53 and MDM2, the book focuses primarily on human systems. Since a large volume of literature is available for the mouse models, perhaps it calls for a separate volume.

We thank Dr. Thijs van Vlijmen for giving us the opportunity of designing and editing the book. We also owe thanks to the staff of Springer Science and Business Media for their work in the completion of the book. We convey our sincerest thanks to the scientists who contributed the chapters for their insightful discussion. We are indebted to our graduate students for their untiring effort in every step of this work. We are particularly thankful to Isabella Pearsall for her help in communicating with authors during initiation and completion of the book. We also thank Catherine Vaughan for grammatical editing of chapters and Shilpa Singh for her support.

Richmond, VA, USA

Swati Palit Deb
Sumitra Deb

Contents

1 **p53 and Hereditary Cancer** 1
Diana Merino and David Malkin

2 **Alterations of p63 and p73 in Human Cancers** 17
Kazushi Inoue and Elizabeth A. Fry

3 **Cooperation of p53 Mutations with Other Oncogenic Alterations in Cancer** 41
Javier E. Girardini, Dawid Walerych, and Giannino Del Sal

4 **p53: Its Mutations and Their Impact on Transcription** 71
Catherine Vaughan, Isabella Pearsall, Andrew Yeudall, Swati Palit Deb, and Sumitra Deb

5 **Transcriptional Regulation by Mutant p53 and Oncogenesis** 91
Raffaela Santoro, Sabrina Strano, and Giovanni Blandino

6 **p53 Mutation in the Genesis of Metastasis** 105
W.A. Yeudall

7 **Structural Studies on Mechanisms to Activate Mutant p53** 119
Hector Viadiu, Gilberto Fronza, and Alberto Inga

8 **Mutant p53 and the Response to Chemotherapy and Radiation** 133
Leila Tchelebi, Hani Ashamalla, and Paul R. Graves

9 **The p53-Mdm2 Loop: A Critical Juncture of Stress Response** 161
Yaara Levav-Cohen, Zehavit Goldberg, Kah Hin Tan, Osnat Alsheich-Bartok, Valentina Zuckerman, Sue Haupt, and Ygal Haupt

10 **Mechanisms of Mutant p53 Stabilization in Cancer** 187
Rebecca A. Frum and Steven R. Grossman

11 Crosstalk Between Mdm2, p53 and HIF1-α: Distinct Responses to Oxygen Stress and Implications for Tumour Hypoxia 199
E. Douglas Robertson, Kostyantyn Semenchenko, and Bohdan Wasylyk

12 MDM2 Overexpression, Activation of Signaling Networks, and Cell Proliferation 215
Swati Palit Deb, Shilpa Singh, and Sumitra Deb

13 p53-Independent Effects of Mdm2 235
Stephen Bohlman and James J. Manfredi

14 Splice Variants of MDM2 in Oncogenesis 247
Melissa Rosso, Danielle E. Okoro, and Jill Bargonetti

15 Mdm2 and MdmX Involvement in Human Cancer 263
Steven J. Berberich

16 Targeting p53-MDM2-MDMX Loop for Cancer Therapy 281
Qi Zhang, Shelya X. Zeng, and Hua Lu

17 Involvement of p53 in the Repair of DNA Double Strand Breaks: Multifaceted Roles of p53 in Homologous Recombination Repair (HRR) and Non-Homologous End Joining (NHEJ) 321
Vijay Menon and Lawrence Povirk

18 The Role of Tumor Suppressor p53 in the Antioxidant Defense and Metabolism 337
Andrei V. Budanov

19 Lung Cancer Stem Cells, p53 Mutations and MDM2 359
Venkat Sundar Gadepalli, Swati Palit Deb, Sumitra Deb, and Raj R. Rao

Index 371

Chapter 1
p53 and Hereditary Cancer

Diana Merino and David Malkin

Abstract The roles of p53 as "guardian of the genome" are extensive, encompassing regulation of the cell cycle, DNA repair, apoptosis, cellular metabolism, and senescence - ultimately steering cells through a balance of death and proliferation. The majority of sporadic cancers exhibit loss of p53 activity due to mutations or deletions of TP53, and alterations in its signaling pathway. Germline *TP53* mutations have been identified in a group of families exhibiting a rare but highly penetrant familial cancer syndrome, called the Li-Fraumeni syndrome (LFS). Between 60–80% of 'classic' LFS families carry mutant *Trp53*. The most frequent cancers observed are premenopausal breast cancer, bone and soft-tissue sarcomas, adrenal cortical carcinomas, and brain tumors. Penetrance is nearly 100% by age 70. Although *TP53* is currently the only validated susceptibility locus recognized for LFS, recent studies have focused on the identification of genetic modifiers that may explain the wide phenotypic variability observed in LFS patients. Analyses of single nucleotide polymorphisms (SNPs), genome-wide copy number and telomere length have provided greater insight into the potential genetic modifiers of LFS. Moreover, the study of *Trp53* mutant heterozygous mouse models has elucidated novel functions of p53, and offers insight into the mechanisms governing tumorigenesis in LFS. The key findings outlined in this chapter provide an overview of the molecular basis of LFS and the role of p53 in this unique heritable cancer syndrome.

This work is supported in part by a grant from the Canadian Institutes of Health Research (DM). Ms. Merino is supported in part by a Vanier Scholarship of the Canadian Institutes of Health Research.

D. Merino, M.Sc. • D. Malkin, M.D. (✉)
Division of Hematology/Oncology, Program in Genetics and Genome Biology, The Hospital for Sick Children, 555 University Avenue, M5G 1X8 Toronto, ON, Canada

Departments of Pediatrics and Medical Biophysics, University of Toronto, Toronto, ON, Canada
e-mail: david.malkin@sickkids.ca

S.P. Deb and S. Deb (eds.), *Mutant p53 and MDM2 in Cancer*,
Subcellular Biochemistry 85, DOI 10.1007/978-94-017-9211-0_1,

Keywords p53 • Li-Fraumeni Syndrome • Cancer Predisposition • Cancer Genetics

One of the most notable tumor suppressor genes, *TP53*, encodes the transcription factor protein p53, a ubiquitous protein implicated in preservation of an intact genome. True to its name, the "guardian of the genome", p53 is a key suppressor of malignant transformation and somatic alterations are commonly observed in numerous cancers [56]. In response to cellular stress signals, p53 activates pathways that regulate the cell cycle, DNA repair, and apoptosis [23, 64] Additionally, p53 is also involved in regulating cellular senescence and metabolism, which have been shown to contribute to cancer progression [63, 77–79]. Due to the key role of p53 in restricting tumor initiation and progression, it is not surprising that numerous cancers have acquired mechanisms to inactivate p53 and/or its molecular pathway, thus bypassing the cell's innate tumor suppression system.

In vitro studies examining the status of p53 in numerous cancer cell lines and tumors determined that the activity of p53 was commonly lost due to gene mutations or deletions [5, 6, 19, 52, 80]. *Trp53* knockout mouse models demonstrated that although mice, for the most part, developed normally, they had an increased susceptibility to a variety of cancers, most frequently lymphomas, which developed earlier than in mice harboring wildtype p53 [24, 32, 36].

Similarly, in humans, germline *TP53* mutations are also associated with an increased cancer susceptibility and reduced age of onset than p53 wildtype carriers. The Li-Fraumeni syndrome (LFS; OMIM 151623), a rare but highly penetrant familial cancer syndrome, is characterized by germline *TP53* mutations inherited in an autosomal dominant manner [47]. Between 60 and 80 % of 'classic' LFS families carry a mutant p53. According to the International Agency for Research on Cancer *TP53* database (http://www-p53.iarc.fr/), germline mutations in 118 different codons have been reported within the 393 amino acid-long p53 coding region, demonstrating the wide variability of p53 alterations involved in this heritable cancer condition (Fig. 1.1). This genotypic variability is similarly reflected in the phenotypic variability LFS patients demonstrate in the number and type of cancer diagnoses, as well as their age at presentation. Although some correlations between genotype and phenotype have been identified in LFS, there are many more genetic modifiers to be identified and their effect on cancer risk remains to be fully defined. This chapter will highlight the role of germline p53 alterations in hereditary cancer, and outline key findings that refine our understanding on the molecular basis of LFS.

The Li-Fraumeni Syndrome

The Li-Fraumeni syndrome was first characterized by Drs. Frederick Li and Joseph Fraumeni in 1969 [40]. After the retrospective review of 280 medical charts and 418 death certificates of children diagnosed with rhabdomyosarcoma throughout the United States, an interesting familial cancer pattern was observed in that siblings

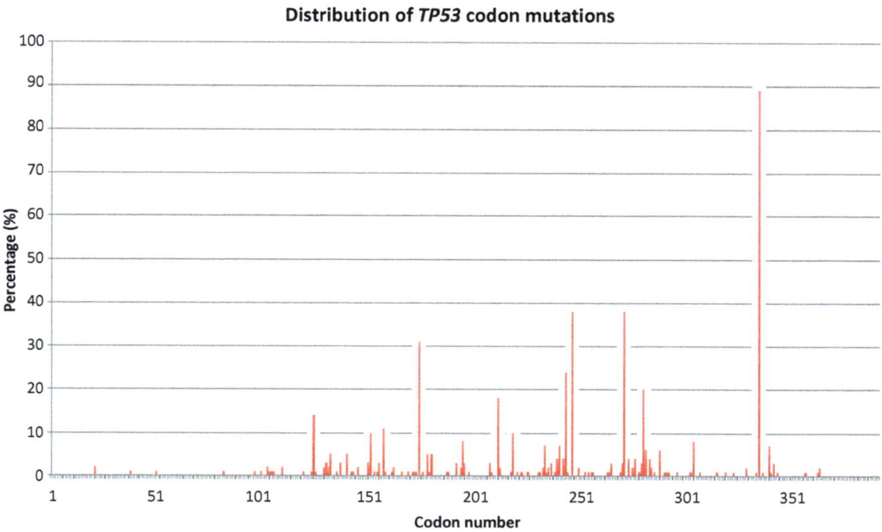

Fig. 1.1 Distribution of germline *TP53* mutations mapped to the coding sequence (Modified from IARC TP53 database, R16 November 2012)

and their biologically related cousins developed soft tissue sarcomas, while first- and second-degree adult relatives developed various forms of cancer [40]. These malignancies presented at higher frequencies in these families than expected by chance alone, and affected individuals had a tendency for developing multiple cancers during their lifetime. Age of onset varied between individuals; however, in all cases, specific cancer types appeared earlier than in the normal population. This observation was confirmed in larger population cohorts by Nichols et al. [51]. The inheritance pattern was autosomal dominant with affected relatives belonging to the same ancestral line, and at least one member being affected in each generation [40]. Numerous studies have reported on the types of cancer commonly observed in LFS families, and a list of LFS component tumors has been defined. The most frequent cancers in LFS are premenopausal breast cancer, bone and soft tissue sarcomas, adrenal cortical carcinoma (ACC), and brain tumors [40, 41, 44]. Less frequent cancers include choroid plexus carcinoma, gastric cancer, leukemia, lymphoma, melanoma, prostate and ovarian cancer, germ cell tumor, Wilms tumor, as well as colorectal and lung cancer[10, 51] (Fig. 1.2). Cancer penetrance in LFS patients is nearly 100 % by age 70, and females exhibit lower mean age of onset than males primarily, but not exclusively, due to the breast cancer incidence [45].

In 1990, Malkin and colleagues conducted a candidate gene approach in order to identify a genetic alteration that may account for the increased cancer risk in LFS families [47]. Because *TP53* was commonly inactivated in somatic cells of LFS component tumors occurring spontaneously, the mutation status of this tumor suppressor gene was studied. Germline mutations in *TP53* mapping to the

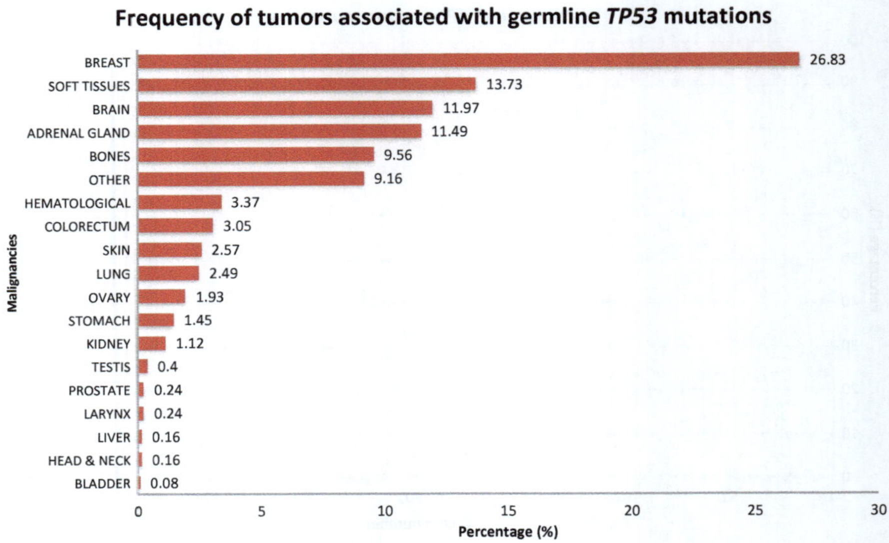

Fig. 1.2 Frequency of tumors associated with germline *TP53* mutations (LFS) n = 1,245 (Modified from IARC TP53 database, R16 November 2012)

DNA binding domain were identified in all five families analyzed. *TP53* mutation carriers in each family exhibited the same germline mutation, demonstrating autosomal dominant heritability, and almost all carriers were affected by cancer, while non-carriers were cancer-free, strongly suggesting its causal role in tumorigenesis [47]. Years of research have explored the role of mutant p53 in cancer susceptibility and its role in LFS. *TP53* is currently the only validated susceptibility locus recognized for LFS.

Definitions of the Li-Fraumeni Syndrome

The "classical" definition of the Li-Fraumeni syndrome was proposed in 1988. It is defined by a family in which the proband is diagnosed before 45 years of age, has a first degree relative developing cancer before 45 years of age, and another first or second degree relative with either any cancer diagnosed before 45 years of age, or a sarcoma at any age [41] (Table 1.1). Since its clinical definition, more than 500 families around the world have been diagnosed with LFS (IARC, http://iarc.p53/fr).

The clinical phenotype of LFS has evolved, and other less stringent definitions now exist. The Li-Fraumeni-like syndrome (LFS-L) is a term given to families for which the classical LFS definition does not apply. The Birch definition defines LFS-L families as those in which the proband is diagnosed with either any child-

Table 1.1 Current clinical definitions of the Li-Fraumeni syndrome

Classification	Description
Classic Li-Fraumeni Syndrome (LFS) [41]	Proband diagnosed with sarcoma before 45 years of age, and A first-degree relative with cancer before 45 years of age, and Another first- or second-degree relative with any cancer diagnosed under 45 years of age or with sarcoma at any age
Li-Fraumeni-like syndrome (LFS-L)	***Birch definition*** [9] Proband with any childhood cancer or sarcoma, brain tumor, or adrenocortical carcinoma diagnosed under 45 years of age, and A first- or second-degree relative with a typical LFS-related cancer (bone or soft-tissue sarcoma, breast cancer, brain tumor, leukemia, adrenocortical tumor, melanoma, prostate cancer) diagnosed at any age, and A first- or second-degree relative in the same genetic lineage with any cancer diagnosed under the age of 60 years ***Eeles definition*** [26] Two different tumors that are part of extended LFS in first- or second-degree relatives at any age (bone or soft-tissue sarcoma, breast cancer, brain tumor, leukemia, adrenocortical tumor, melanoma, prostate cancer)
Chompret criteria [20, 15, 69]	Proband with an LFS core tumor (sarcoma, brain tumor, breast cancer, or adrenocortical carcinoma) before age 46 years, and at least one first- or second- degree relative with an LFS core tumor (other than breast cancer if the proband has breast cancer) before age 56 years or with multiple primary tumors Or, a proband with multiple primary tumors, two of which belong to the LFS tumor spectrum (sarcoma, brain tumor, breast cancer, and/or adrenocortical carcinoma), with the initial cancer occurring before 46 years, regardless of family history A proband with adrenocortical carcinoma or choroid plexus carcinoma at any age of onset, regardless of family history

hood cancer or sarcoma, brain tumor or adrenocortical carcinoma before 45 years of age, and a first- or second-degree relative with an LFS component tumor at any age, and a first- or second-degree relative with any cancer before 60 years of age [9]. The Eeles definition defines LFS-L families as those in which two first- or second-degree relatives are diagnosed with LFS component tumors at any age (Table 1.1) [26]. The evolving definition of this familial syndrome takes into consideration recurrent phenotypes that, although not classified as LFS or LFS-L, would benefit from *TP53* testing. As such, the Chompret criteria [15, 20, 69] have led to creation of guidelines for *TP53* testing, and the classification of families in which: (1) the proband is diagnosed with an LFS component tumor before 46 years of age, and at least one first- or second-degree relative is diagnosed either with an LFS component tumor before 56 years of age, or with multiple primary tumors; or (2) the proband is diagnosed before the age of 46 with multiple primary tumors, two of which are LFS component tumors, regardless of family history; or (3) a proband is diagnosed with adrenocortical carcinoma or choroid plexus carcinoma at any age, regardless of family history (Table 1.1) [15, 20, 69].

The incidence of germline *TP53* mutations in "classical" LFS families approaches 70 %, 40 % in LFS-L families and 30 % in families defined by the Chompret criteria [45].

Although the clinical definitions of LFS are numerous and display variable degrees of rigor, they are effective in identifying families at a greater risk of developing cancer, and are useful as guidelines for patient diagnosis, management and care [62]. The evolution of these definitions, either by the inclusion of families not satisfying any of the current definitions, or the identification of novel molecular profiles associated to LFS, will refine our understanding of these cancer-prone families and will facilitate personalized approaches to patient management.

Genetic Alterations in the Li-Fraumeni Syndrome

TP53

Germline *TP53* mutations are the characteristic genetic aberrations observed in LFS, being found in 60–80 % of "classic" LFS families, 40 % of LFS-L, and 30 % of individuals meeting the revised Chompret criteria [45]. Of the somatic *TP53* alterations reported in the IARC p53 database, 73 % are missense mutations, 8.6 % splicing mutations, 8.1 % nonsense mutations and 6.1 % frameshift mutations [55]. Similarly, the majority of germline mutations in LFS families are missense (~75 %), and these alterations render the protein either non-functional, with a retained yet variable transactivation activity, or unaffected (functional) in 73 %, 20 % and 7 % of cases, respectively [54, 56].

The spectrum of germline mutations correlates with the spectrum observed in sporadic cancers that harbor somatic *TP53* mutations (Fig. 1.1). The majority of these are located in the DNA binding domain (DBD) and confined to small conserved regions [46, 56]. The most common germline or somatic mutations are found at codons 175, 245, 248, 273 and 282, although all codons exhibit at least one mutation [54, 74]. These hotspots correspond primarily to transition mutations at CpG sites [54]. In the germline, however, very few mutations have been found outside the coding region encompassing exons 5 through 8 [58]. Missense mutations in the DBD are generally associated with a higher incidence of breast and brain tumors, whereas missense mutations outside the DNA binding loops are common in adrenocortical carcinomas (ACC) [28]. The location of these mutations may clarify why ACCs exhibit such low penetrance compared to p53 mutant carrier families in whom no ACC has been diagnosed. Cancer penetrance in *TP53* mutation carriers is about 20 % before the age of 20 years, and greater than 90 % by the age of 70 [45].

A unique recurrent non-synonymous mutation at codon 337 (c.1010 G>A, genomic nucleotide number 17588), mapping to the oligomerization domain of p53 was initially identified in ACC patients of Southern Brazil [60]. The mutation, encoding a R337H amino acid substitution, has been identified in a large number of Brazilian families, and tumor frequency and spectrum of affected carriers shows an association with LFS/LFL [2]. Haplotyping analysis using 29 SNPs in 12 unrelated R337H carriers has revealed that this common mutation exhibits a founder effect [29]. Garritano and colleagues further mapped the origin of this founder mutation in which mutation carrier families were distributed along a commercial route commonly used by Portuguese merchants in the eighteenth and nineteenth centuries.

R337H families exhibit low penetrance in individuals under the age of 30 (20 %), compared to classic LFS families (50 %). However, penetrance increases with age and mirrors that of classic LFS families with a lifetime risk of cancer of 90 % [1].

Other Genes

The absence of detectable germline *TP53* mutations in several LFS families suggests that alternative mutations or alterations are involved in the etiology of this inherited condition. Several studies have conducted candidate gene and linkage analyses in order to identify other genetic alterations involved in LFS, although these findings have not been validated. Germline alterations of components of the p53 signaling pathway such as *BAX* [7], *BCL10* [67], *CDKN2A* [17, 57], *CHEK1* [73], *p63* [14], *PTEN* [16, 17] and *p73* [4] have not been documented. Germline mutations in *CHEK2*, a regulator of p53 activity, were identified in a few LFS and LFS-L families, however its role in the etiology of LFS has since been questioned as the missense and frameshift mutations found were shown to be polymorphisms or mutated in a duplicated exon of *CHEK2* [8, 66, 73]. Nonetheless, the *CHEK2* polymorphisms have been implicated with an increased risk of breast, lung, prostate and laryngeal cancers, suggesting a strong role of *CHEK2* in tumorigenesis [22, 50, 72]. Linkage analysis has drawn an association between LFS and chromosome 1q23, although no candidate gene has been identified [4]. Recently, Aury-Landas et al. conducted a genome-wide microarray analysis on a group of 64 individuals meeting the Chompret criteria and carrying wildtype *TP53*. Twenty novel copy number variants (CNV) were observed in 15 unrelated patients, spanning 49 genes [3]. Enrichment pathway analysis identified a subset of genes involved in chromatin packaging and remodeling. A heterozygous deletion in *KDM1A* (1p36), and duplication in *MTA3* (2p21), *TRRAP* (7q22) and *SIRT3* (11p15) were identified in patients diagnosed with brain tumors (medulloblastoma, oligodendroglioma, high grade glioma and astrocytoma, respectively). Although these findings are still to be validated in an independent cohort, they suggest an interesting link between chromatin remodeling genes and LFS, and specifically, a tissue-specific role of epigenetic alterations in brain tumor development within the LFS context.

Modifier Genes

The phenotypic variability of LFS may be explained by genetic and environmental factors, most of which remain elusive [12, 48]. To this date, it is not yet possible to ascertain the genetic modifiers that may define the age of onset or the cancer type developed by each individual. The role of non-genetic factors, such as epigenetic modifiers, on the phenotypic variability observed between affected family members is not fully understood complicating the creation of personalized therapeutic and surveillance approaches that may reduce patient mortality. Thus, it is imperative to refine the current understanding of genetic and epigenetic modifiers in LFS, which in addition to *TP53* mutations, alter the risk to multiple malignancies.

Correlative analyses of mutant p53 and cancer incidence have led to interesting discoveries. Compared to LFS families harbouring wildtype p53, mutant p53 LFS families have an increased prevalence of brain tumors, adrenocortical carcinomas (ACC), and a significantly lower age of onset for breast cancer [54]. Moreover, in all cases, *TP53* mutation carriers show an earlier age of onset than their sporadic tumor counterparts [54]. Functional studies have enhanced these findings and correlated degree of p53 activity with type of missense mutation, and age of onset. In particular, the mean age of onset in carriers of a functional mutation was higher than in carriers of a partially functional or non-functional mutation for several cancer types [56]. Non-functional or truncated p53 may arise as result of nonsense, frameshift, and splice mutations, which promote an earlier age of onset, particularly for brain tumors [54].

Polymorphisms in *TP53* and genes regulating the p53 pathway have also been shown to modify age of cancer onset. The polymorphism SNP 309 (T>G variation) of the mouse double minute 2 (*MDM2*), an ubiquitin ligase directly regulating p53 degradation, is a plausible candidate modifier in hereditary and sporadic p53 mutant tumors. [11–13, 49, 61]. The presence of the G allele at this locus has been associated with increased *MDM2* levels and an aberrant p53 pathway as demonstrated by the abrogation of DNA repair processes, increased mutation rates, and reduced apoptosis, which leads to faster and more frequent tumor formation [11]. On average, G-allele carriers developed cancers nine years before homozygous T allele carriers. In LFS patients diagnosed with soft tissue sarcoma, *MDM2* SNP 309 accounts for an age of onset difference of 12 years, while in patients with breast cancer, it accounts for a 10-year difference [11]. A correlation between the *MDM2* SNP 309 polymorphism and the occurrence of multiple tumors in LFS families was also observed, in which G-allele carriers exhibited a greater frequency of independent subsequent cancers [11].

Two *TP53* polymorphisms have been also associated with age of onset in LFS families. A polymorphism at codon 72 (Arg>Pro variation), located in exon 4 has been analyzed in the context of LFS, as well as sporadic cancer [13, 49]. In vitro studies have found that the Arg allele has an increased capacity to induce apoptosis in both p53 mutant and wildtype tumors, due to the efficient translocation into the mitochondria where it interacts with proapoptotic proteins, thereby leading to increased apoptosis (reviewed in [31]). Interestingly, in studying a Brazilian LFS population, Marcel et al. identified that there was significant excess of Arg-allele carriers and complete absence of homozygous Pro-allele carriers in families harboring a germline *TP53* mutation, compared to *TP53* wildtype carriers [49]. Furthermore, age of onset of Arg-allele carriers was lower than in Pro-allele carriers with a difference at first diagnosis ranging between 12.6 years in French LFS families [13] and 8.3 years in Brazilian LFS families [49].

Another *TP53* polymorphism was described as a 16 bp duplication in intron 3, named PIN3. The non-duplicated allele (A1) was associated with a significantly earlier age of onset in LFS families (28 vs. 47 years of age) [49]. Moreover, cancer occurrence before the age of 35 was only observed in homozygous non-duplicated (A1A1) PIN3 carriers. Further analyses have found that duplicated PIN3 allele carriers have an increased risk of sporadic and inherited breast cancer and colorectal cancer [21, 30], as well as increased risk of lymph node metastases in *TP53* mutation carriers [35].

A multivariate analysis of these polymorphisms demonstrated a cumulative effect on age of onset and cancer risk when both the *MDM2* SNP309 and p53 72Arg polymorphisms were expressed suggesting a synergistic interaction between these two polymorphic loci that alters the cancer phenotype of LFS patients [13, 27].

Anticipation, a pattern in which age at first diagnosis decreases, and/or symptom severity increases with successive generations, is a common characteristic of several inherited disorders, including dyskeratosis congenita (DC), Fanconi anemia (FA), ataxia-telangiectasia (AT), and LFS [70].

In LFS, genetic anticipation reduces the age of onset, increases the severity and the proportion of affected individuals of successive generations, supporting the hypothesis that additional genetic modifiers contribute to the variable clinical phenotype observed in affected family members. Although the molecular mechanisms of genetic anticipation are not fully understood, variability in telomere length has been identified to be a key contributor to the genetic anticipation observed in LFS. Telomere length analyzed in peripheral blood lymphocytes was significantly shorter in carriers of germline *TP53* mutations than in normal controls of matched age [71]. In both children and adults, telomeres were significantly shorter in affected *TP53* mutation carriers than in unaffected carriers and wildtype controls. Moreover, within each LFS family studied, telomere length was shorter in affected children than their non-affected sibling and their non-carrier (*TP53* wildtype) parent [68]. As well, telomere attrition was faster in *TP53* mutation carriers than controls [68].

The advent of high-resolution genome-wide approaches has refined the identification of genomic alterations in LFS, expanding on opportunities to explore novel potential mechanisms driving the variable clinical phenotype observed in LFS families. Shlien et al. conducted a high-throughput genotyping microarray analysis in which copy number variable (CNV) regions were assessed for 53 individuals from families harbouring germline *TP53* mutations. Compared to normal healthy controls, *TP53* mutation carriers exhibited a significant increase in CNV regions in their genome [65]. More than a reflection on the increased genomic instability of germline *TP53* mutation carriers, these CNVs encompassed various known cancer genes, suggesting a selection of cells carrying a unique mutator phenotype in LFS that increased cancer risk. It was also observed that an increased number of CNVs progressed in somatic cells of the tumor indicating that the CNV formation is a dynamic process that accompanies tumor progression [65]. Interestingly, the role of CNVs was also associated with an aggressive phenotype in the offspring, where CNVs from one parent were commonly inherited with a germline *TP53* mutation from the other parent [65]. The mechanisms by which CNVs are formed have not been fully understood, although studies have shown that non-allelic homologous recombination (NAHR) and microhomology-mediated events, such as microhomology-mediated break-induced replication (MMBIR) underlie these genomic changes [33, 34].

The phenotypic complexity of LFS may be in part due to the complex network and activity of p53, and its interaction with many different genetic modifiers, which are unique to each patient. Thus, refining the correlation between genetic modifiers and cancer predisposition phenotypes is imperative, as it will improve our understanding on this inherited condition and how to best provide personalized therapies.

Mouse Models and the Li-Fraumeni Syndrome

Genetically engineered mice have been crucial in the study of p53 function and the way organisms compensate for the loss of p53. The first mutant *Trp53* transgenic mouse was created in 1989, and was characterized by an increased incidence of early cancers of the bone, lung, and lymphatic system [39]. Later in 1992, the first *Trp53* knockout mouse was created, and although it developed normally, it was highly susceptible to the formation of a wide variety of spontaneous tumors, with lymphomas and sarcomas being the most frequently observed [24]. These mouse models, and more recently, mutant knock-in and inducible mouse strains, have elucidated the p53-induced mechanisms of apoptosis, senescence, cell cycle regulation, chromosomal stability, and abnormal p53 activity in tumor development (reviewed in [25]). The *Trp53* mutant heterozygous mouse models, however, have displayed the most accurate representation of LFS, as affected LFS individuals are invariably heterozygous for mutant p53.

Trp53 null heterozygous mice represent accurate models for LFS individuals whose *TP53* germline mutations are functionally null. The majority of *Trp53+/–* mice developed early tumors (by 18 months of age) which, considering the normal lifespan of the C57BL/6 mice (36 months), is in agreement with the 50 % incidence of cancers in LFS individuals by 30 years of age [32]. Moreover, *Trp53+/–* mice developed a wider tumor spectrum with an increased incidence of bone and soft-tissue sarcomas, and carcinomas than *Trp*53–/– mice. Although breast cancer is common in LFS families, this particular mouse model did not develop this cancer type due to the B57BL/6 strain's innate resistance to develop mammary carcinomas. To evaluate the effect of p53 loss in breast tumor formation, p53-null mice were back-crossed onto the BALB/c genetic background. Approximately 55 % of female BALB/c-p53+/– mice developed mammary carcinomas, demonstrating the effect of genetic background and different genetic modifiers on the development of different types of tumors in p53 null mice [37] .

Familial syndromes arising as a result of the loss of a tumor suppressor gene generally require the loss or inactivation of the second allele for a tumor to form. Some LFS families harbour functionally null *TP53* mutations, which require complete loss or inactivation of the remaining wildtype p53 allele for tumors to form. These tumors are characterized by an earlier age of onset and increased phenotypic severity. However, the majority of tumors in LFS families often retain a copy of the wildtype *TP53* allele [75]. Retention of the wildtype *TP53* allele is commonly observed in families harbouring missense mutations in the DNA-binding domain. These mutations have been shown to exhibit dominant-negative effects through the creation of heterotetramers with wildtype p53 proteins [18], thus inhibiting normal p53 activity.

Only about half of *Trp53+/–* affected mice exhibit loss of the remaining wildtype *Trp53* allele, which suggest than in mice, unlike humans, a reduction of p53 dosage is sufficient for tumor formation [76]. As with any disease model, the intricate differences between mice and humans need to be taken into consideration. These specifically include the mechanisms governing oncogenic transformation, which suggest that the requirement for disabling p53 in human and mouse cells are distinct [59]. Heterozygous p53 null mice that lose the remaining wildtype *Trp53* allele exhibit an earlier age of onset and more aggressive tumors than mice that retain the wildtype allele. Genetically engineered mice carrying a heterozygous germline missense mutation more often retain their *Trp53* wildtype allele much like humans and thus reflect a more accurate model of LFS [42]. The increased incidence of germline and somatic *TP53* missense mutations in LFS families and spontaneous tumors, respectively, indicates that in addition to the inactivation of *Trp53* these mutations may confer additional oncogenic properties. In order to study these alternative effects, the Jacks and Lozano laboratories developed mice harbouring heterozygous mutations in *Trp53* R172H, R270H, and R172P, corresponding to human hotspot mutations R175H, R273H, and R175P, respectively [38, 42, 53]. Interestingly, these LFS mouse models exhibit variable phenotypes consistent with LFS in humans. A model containing the *Trp53*^{R172H} mutation exhibited increased metastatic potential compared to *Trp53+/–* mice even though levels of the mutated p53 protein were as low as those observed in wildtype p53 mice due to a splicing abnormality [42]. These findings demonstrate that potent gain-of-function effects of missense *Trp53* mutations can significantly alter phenotype even when expressed at low levels. Moreover, *Trp53*$^{R172H/+}$ mice showed a significant increase in the number of carcinomas and a decrease in the number of lymphomas compared to *Trp53+/–* mice. The same mutation was studied on a different genetic background and similar results were observed [53]. In the 129S_4/S_v background, *Trp53*$^{R172H/+}$ mice exhibited highly metastatic tumors, a 2-fold difference in the incidence of osteosarcomas, and a slight increase in the incidence of carcinomas compared to *Trp53*$^{+/-}$ mice. Although both p53 mutant mouse models exhibit a loss of p53 activity, the phenotypes observed suggest that additional tumorigenic activities may be conferred by missense mutations.

Another model was created containing a heterozygous *Trp53*^{R270H} mutation in a $129S_4/S_v$ background [53]. Tumor burden was significantly increased, including increased incidence of carcinomas and hemangiomas compared to *Trp53*$^{+/-}$ mice. Similarly to the *Trp53*$^{R172H/+}$ mice, *Trp53*$^{R270H/+}$ mice exhibited increased tumor metastatic potential.

A mouse model carrying a mutation commonly found in spontaneous tumors, yet not identified in LFS, was developed in order to analyze the role of the apoptotic and cell cycle regulating functions of p53 in tumorigenesis [43]. R175P mutations in humans are completely defective in apoptosis signaling but still able to induce cell cycle arrest. Homozygous *Trp53*^{R172P} mice had a significantly better survival than *Trp53*$^{-/-}$ and exhibited decreased tumor burden, which included an escape of the early onset of thymic lymphomas commonly observed in *Trp53* null mice. These results indicated that p53-dependent apoptosis is not necessary to suppress the early onset of tumors, but rather, the chromosomal stability observed in the R172P mutant mice as a result of its intact cell cycle regulatory activity, is key in the suppression of tumorigenesis [43].

The variability in tumor spectra, and anti-neoplastic activity observed with different mutations suggests that different mutations may give rise to distinct phenotypes, although confirming these findings in humans may be challenging due to the inherent differences of every individual. Nonetheless, mouse models have been essential in studying the p53 activity and offering insight into the mechanisms governing tumorigenesis in LFS.

Although the role of p53 in cancer development has been investigated for decades, there is still much to be understood about the regulatory activity of this multifaceted tumor suppressor in tumorigenesis. Examining the activity of p53 in the LFS context will provide great insight into the mechanisms this protein utilizes to control irregular cell growth, and the regulatory partners that facilitate tumor suppression, which contribute to the complexity of the LFS phenotype.

References

1. Achatz M, Hainaut P, Ashton-Prolla P (2009) Highly prevalent TP53 mutation predisposing to many cancers in the Brazilian population: a case for newborn screening? Lancet Oncol 10:920–925
2. Achatz MIW, Olivier M, Le Calvez F et al (2007) The TP53 mutation, R337H, is associated with Li-Fraumeni and Li-Fraumeni-like syndromes in Brazilian families. Cancer Lett 245:96–102. doi:10.1016/j.canlet.2005.12.039
3. Aury-Landas J, Bougeard G, Castel H et al (2013) Germline copy number variation of genes involved in chromatin remodelling in families suggestive of Li-Fraumeni syndrome with brain tumours. Eur J Hum Genet 1–8. doi: 10.1038/ejhg.2013.68
4. Bachinski LL, Olufemi S-E, Zhou X et al (2005) Genetic mapping of a third Li-Fraumeni syndrome predisposition locus to human chromosome 1q23. Cancer Res 65:427–431

5. Baker S, Fearon E, Nigro J (1989) Chromosome 17 deletions and p53 gene mutations in colorectal carcinomas. Science 244:217–221
6. Baker S, Preisinger A, Jessup J (1990) p53 gene mutations occur in combination with 17p allelic deletions as late events in colorectal tumorigenesis. Cancer Res 50:7717–7722
7. Barlow J, Mous M, Wiley J et al (2004) Germline BAX alterations are infrequent in Li-Fraumeni syndrome. Cancer Epidemiol Biomarkers Prev 13:1403–1406
8. Bell DW, Varley JM, Szydlo TE et al (1999) Heterozygous germ line hCHK2 mutations in Li-Fraumeni syndrome. Science 286:2528–2531
9. Birch J, Hartley A, Tricker K, Prosser J (1994) Prevalence and diversity of constitutional mutations in the p53 gene among 21 Li-Fraumeni families. Cancer Res 54:1298–1304
10. Birch JM, Alston RD, McNally RJ et al (2001) Relative frequency and morphology of cancers in carriers of germline TP53 mutations. Oncogene 20:4621–4628. doi:10.1038/sj.onc.1204621
11. Bond GL, Hu W, Bond EE et al (2004) A single nucleotide polymorphism in the MDM2 promoter attenuates the p53 tumor suppressor pathway and accelerates tumor formation in humans. Cell 119:591–602
12. Bond GL, Levine AJ (2007) A single nucleotide polymorphism in the p53 pathway interacts with gender, environmental stresses and tumor genetics to influence cancer in humans. Oncogene 26:1317–1323
13. Bougeard G (2006) Impact of the MDM2 SNP309 and p53 Arg72Pro polymorphism on age of tumour onset in Li-Fraumeni syndrome. J Med Genet 43:531–533
14. Bougeard G, Limacher JM, Martin C et al (2001) Detection of 11 germline inactivating TP53 mutations and absence of TP63 and HCHK2 mutations in 17 French families with Li-Fraumeni or Li-Fraumeni-like syndrome. J Med Genet 38:253–257
15. Bougeard G, Sesboue R, Baert-Desurmont S et al (2008) Molecular basis of the Li-Fraumeni syndrome: an update from the French LFS families. J Med Genet 45:535–538
16. Brown LT, Sexsmith E, Malkin D (2000) Identification of a novel PTEN intronic deletion in Li-Fraumeni syndrome and its effect on RNA processing. Cancer Genet Cytogenet 123:65–68
17. Burt EC, McGown G, Thorncroft M et al (1999) Exclusion of the genes CDKN2 and PTEN as causative gene defects in Li-Fraumeni syndrome. Br J Cancer 80:9–10. doi:10.1038/sj.bjc.6690313
18. Chène P (1998) In vitro analysis of the dominant negative effect of p53 mutants. J Mol Biol 281:205–209. doi:10.1006/jmbi.1998.1897
19. Cheng J, Haas M (1990) Frequent mutations in the p53 tumor suppressor gene in human leukemia T-cell lines. Mol Cell Biol 10:5502–5509. doi:10.1128/MCB.10.10.5502.Updated
20. Chompret A, Abel A, Stoppa-Lyonnet D et al (2001) Sensitivity and predictive value of criteria for p53 germline mutation screening. J Med Genet 38:43–47
21. Costa S, Pinto D, Pereira D et al (2008) Importance of TP53 codon 72 and intron 3 duplication 16 bp polymorphisms in prediction of susceptibility on breast cancer. BMC Cancer 8:32
22. Cybulski C, Masojc B, Oszutowska D et al (2008) Constitutional CHEK2 mutations are associated with a decreased risk of lung and laryngeal cancers. Carcinogenesis 29:762–765. doi:10.1093/carcin/bgn044
23. Diller L, Kassel J, Nelson C (1990) p53 functions as a cell cycle control protein in osteosarcomas. Mol Cell Biol. doi: 10.1128/MCB.10.11.5772. Updated
24. Donehower L, Harvey M, Slagle B (1992) Mice deficient for p53 are developmentally normal but susceptible to spontaneous tumours. Nature 356(6366):215–221
25. Donehower L, Lozano G (2009) 20 years studying p53 functions in genetically engineered mice. Nat Rev Cancer 9:831–841
26. Eeles RA (1995) Germline mutations in the TP53 gene. Cancer Surv 25:101–124
27. Fang S, Krahe R, Lozano G et al (2010) Effects of MDM2, MDM4 and TP53 codon 72 polymorphisms on cancer risk in a cohort study of carriers of TP53 germline mutations. PLoS One 5:e10813. doi:10.1371/journal.pone.0010813
28. Figueiredo BC, Sandrini R, Zambetti GP et al (2006) Penetrance of adrenocortical tumours associated with the germline TP53 R337H mutation. J Med Genet 43:91–96. doi:10.1136/jmg.2004.030551

29. Garritano S, Gemignani F, Palmero EI et al (2010) Detailed haplotype analysis at the TP53 locus in p.R337H mutation carriers in the population of Southern Brazil: evidence for a founder effect. Hum Mutat 31:143–150
30. Gemignani F, Moreno V, Landi S et al (2004) A TP53 polymorphism is associated with increased risk of colorectal cancer and with reduced levels of TP53 mRNA. Oncogene 23:1954–1956. doi:10.1038/sj.onc.1207305
31. Grochola LF, Zeron-Medina J, Mériaux S, Bond GL (2010) Single-nucleotide polymorphisms in the p53 signaling pathway. Cold Spring Harb Perspect Biol 2:a001032
32. Harvey M, McArthur M, Montgomery C (1993) Spontaneous and carcinogen-induced tumorigenesis in p53-deficient mice. Nat Genet 5:225–229
33. Hastings PJ, Ira G, Lupski JR (2009) A microhomology-mediated break-induced replication model for the origin of human copy number variation. PLoS Genet 5:e1000327. doi:10.1371/journal.pgen.1000327
34. Hastings PJ, Lupski JR, Rosenberg SM, Ira G (2009) Mechanisms of change in gene copy number. Nat Rev Genet 10:551–564. doi:10.1038/nrg2593
35. Hrstka R, Beranek M, Klocova K et al (2009) Intronic polymorphisms in TP53 indicate lymph node metastasis in breast cancer. Hum Mutat 11951:1205–1211. doi:10.3892/or
36. Kemp C, Donehower L, Bradley A, Balmain A (1993) Reduction of p53 gene dosage does not increase initiation or promotion but enhances malignant progression of chemically induced skin tumors. Cell 74:813–822
37. Kuperwasser C, Hurlbut G (2000) Development of spontaneous mammary tumors in BALB/c p53 heterozygous mice a model for Li-Fraumeni syndrome. Am J Pathol 157:2151–2159
38. Lang GA, Iwakuma T, Suh Y-A et al (2004) Gain of function of a p53 hot spot mutation in a mouse model of Li-Fraumeni syndrome. Cell 119:861–872
39. Lavigueur A, Maltby V (1989) High incidence of lung, bone, and lymphoid tumors in transgenic mice overexpressing mutant alleles of the p53 oncogene. Mol Cell Biol 9:3982–3991. doi:10.1128/MCB.9.9.3982.Updated
40. Li F, Fraumeni J (1969) Soft-tissue sarcomas, breast cancer, and other neoplasms: a familial syndrome? Ann Intern Med 71(4):747–752
41. Li F, Fraumeni J, Mulvihill J (1988) A cancer family syndrome in twenty-four kindreds. Cancer Res 48(18):5358–5362
42. Liu G, McDonnell TJ, de Oca M, Luna R et al (2000) High metastatic potential in mice inheriting a targeted p53 missense mutation. Proc Natl Acad Sci U S A 97:4174–4179
43. Liu G, Parant JM, Lang G et al (2004) Chromosome stability, in the absence of apoptosis, is critical for suppression of tumorigenesis in Trp53 mutant mice. Nat Genet 36:63–68. doi:10.1038/ng1282
44. Lynch H, Mulcahy G, Harris R et al (1978) Genetic and pathologic findings in a kindred with hereditary sarcoma, breast cancer, brain tumors, leukemia, lung, laryngeal, and adrenal cortical carcinoma. Cancer 41:2055–2064
45. Mai PL, Malkin D, Garber JE et al (2012) Li-Fraumeni syndrome: report of a clinical research workshop and creation of a research consortium. Cancer Genet 205:479–487. doi:10.1016/j.cancergen.2012.06.008
46. Malkin D (2011) Li-fraumeni syndrome. Genes Cancer 2:475–484. doi:10.1177/1947601911413466
47. Malkin D, Li FP, Strong LC et al (1990) Germ line p53 mutations in a familial syndrome of breast cancer, sarcomas, and other neoplasms. Science 250:1233–1238
48. Marcel V, Hainaut P (2009) P53 isoforms – a conspiracy to kidnap P53 tumor suppressor activity? Cell Mol Life Sci 66:391–406. doi:10.1007/s00018-008-8336-3
49. Marcel V, Palmero EI, Falagan-Lotsch P et al (2009) TP53 PIN3 and MDM2 SNP309 polymorphisms as genetic modifiers in the Li-Fraumeni syndrome: impact on age at first diagnosis. J Med Genet 46:766–772. doi:10.1136/jmg.2009.066704
50. Meijers-Heijboer H, van den Ouweland A, Klijn J et al (2002) Low-penetrance susceptibility to breast cancer due to CHEK2(*)1100delC in noncarriers of BRCA1 or BRCA2 mutations. Nat Genet 31:55–59. doi:10.1038/ng879

51. Nichols K, Malkin D, Garber J (2001) Germ-line p53 mutations predispose to a wide spectrum of early-onset cancers. Cancer Epidemiol Biomarkers Prev 10:83–87
52. Nigro J, Baker S, Preisinger A (1989) Mutations in the p53 gene occur in diverse human tumor types. Nature 342:705–708
53. Olive KP, Tuveson DA, Ruhe ZC et al (2004) Mutant p53 gain of function in two mouse models of Li-Fraumeni syndrome. Cell 119:847–860. doi:10.1016/j.cell.2004.11.004
54. Olivier M, Goldgar DE, Sodha N et al (2003) Li-Fraumeni and related syndromes: correlation between tumor type, family structure, and TP53 genotype. Cancer Res 63:6643–6650
55. Petitjean A, Achatz MIW, Borresen-Dale AL et al (2007) TP53 mutations in human cancers: functional selection and impact on cancer prognosis and outcomes. Oncogene 26:2157–2165. doi:10.1038/sj.onc.1210302
56. Petitjean A, Mathe E, Kato S (2007) Impact of mutant p53 functional properties on TP53 mutation patterns and tumor phenotype: lessons from recent developments in the IARC TP53 database. Hum Mutat 28:622–629. doi:10.1002/humu
57. Portwine C (2000) Absence of germline p16INK4a alterations in p53 wild type Li-Fraumeni syndrome families. J Med Genet 37(e13):1–4. doi:10.1136/jmg.37.8.e13
58. Quesnel S, Verselis S, Portwine C et al (1999) p53 compound heterozygosity in a severely affected child with Li-Fraumeni syndrome. Oncogene 18:3970–3978. doi:10.1038/sj.onc.1202783
59. Rangarajan A, Weinberg RA (2003) Opinion: comparative biology of mouse versus human cells: modelling human cancer in mice. Nat Rev Cancer 3:952–959. doi:10.1038/nrc1235
60. Ribeiro R, Sandrini F (2001) An inherited p53 mutation that contributes in a tissue-specific manner to pediatric adrenal cortical carcinoma. Proc Natl Acad Sci 98:9330–9335. doi:10.1073/pnas.161479898
61. Ruijs MWG, Schmidt MK, Nevanlinna H et al (2007) The single-nucleotide polymorphism 309 in the MDM2 gene contributes to the Li–Fraumeni syndrome and related phenotypes. Eur J Hum Genet 15:110–114
62. Ruijs MWG, Verhoef S, Rookus MA et al (2010) TP53 germline mutation testing in 180 families suspected of Li-Fraumeni syndrome: mutation detection rate and relative frequency of cancers in different familial phenotypes. J Med Genet 47:421–428
63. Serrano M, Lin A, McCurrach M et al (1997) Oncogenic ras provokes premature cell senescence associated with accumulation of p53 and p16 INK4a. Cell 88:593–602
64. Shaw P, Bovey R, Tardy S et al (1992) Induction of apoptosis by wild-type p53 in a human colon tumor-derived cell line. Proc Natl Acad Sci 89:4495–4499. doi:10.1073/pnas.89.10.4495
65. Shlien A, Tabori U, Marshall CR et al (2008) Excessive genomic DNA copy number variation in the Li-Fraumeni cancer predisposition syndrome. Proc Natl Acad Sci U S A 105:11264–11269
66. Sodha N, Williams R, Mangion J et al (2000) Screening hCHK2 for mutations. Science 289:359
67. Stone J, Eeles RA, Sodha N et al (1999) Analysis of Li-Fraumeni syndrome and Li-Fraumeni-like families for germline mutations in Bcl10. Cancer Lett 147:181–185
68. Tabori U, Nanda S, Druker H et al (2007) Younger age of cancer initiation is associated with shorter telomere length in Li-Fraumeni syndrome. Cancer Res 67:1415–1418. doi:10.1158/0008-5472.CAN-06-3682
69. Tinat J, Bougeard G, Baert-Desurmont S et al (2009) 2009 version of the Chompret criteria for Li Fraumeni syndrome. J Clin Oncol 27:e108–e109. doi:10.1200/JCO.2009.22.7967
70. Trkova M, Hladikova M, Kasal P et al (2002) Is there anticipation in the age at onset of cancer in families with Li-Fraumeni syndrome? J Hum Genet 47:381–386. doi:10.1007/s100380200055
71. Trkova M, Prochazkova K, Krutilkova V et al (2007) Telomere length in peripheral blood cells of germline TP53 mutation carriers is shorter than that of normal individuals of corresponding age. Cancer 110:694–702. doi:10.1002/cncr.22834
72. Vahteristo P, Bartkova J, Eerola H et al (2002) A CHEK2 genetic variant contributing to a substantial fraction of familial breast cancer. Am J Hum Genet 71:432–438. doi:10.1086/341943
73. Vahteristo P, Tamminen A, Karvinen P et al (2001) p53, CHK 2, and CHK1 genes in Finnish families with Li-Fraumeni syndrome : further evidence of CHK2 in inherited cancer predisposition. Cancer Res 61:5718–5722

74. Varley JM, Attwooll C, White G et al (2001) Characterization of germline TP53 splicing mutations and their genetic and functional analysis. Oncogene 20:2647–2654. doi:10.1038/sj.onc.1204369
75. Varley JM, Thorncroft M, McGown G et al (1997) A detailed study of loss of heterozygosity on chromosome 17 in tumours from Li-Fraumeni patients carrying a mutation to the TP53 gene. Oncogene 14:865–871
76. Venkatachalam S, Shi YP, Jones SN et al (1998) Retention of wild-type p53 in tumors from p53 heterozygous mice: reduction of p53 dosage can promote cancer formation. EMBO J 17:4657–4667
77. Vousden KH, Ryan KM (2009) P53 and metabolism. Nat Rev Cancer 9:691–700. doi:10.1038/nrc2715
78. Wang P-Y, Ma W, Park J-Y et al (2013) Increased oxidative metabolism in the Li–Fraumeni syndrome. New Engl J Med 368:1027–1032. doi:10.1056/NEJMoa1214091
79. Wang Y, Blandino G, Oren M, Givol D (1998) Induced p53 expression in lung cancer cell line promotes cell senescence and differentially modifies the cytotoxicity of anti-cancer drugs. Oncogene 1299:1923–1930
80. Wolf D, Rotter V (1985) Major deletions in the gene encoding the p53 tumor antigen cause lack of p53 expression in HL-60 cells. Proc Natl Acad Sci 82:790–794

Chapter 2
Alterations of p63 and p73 in Human Cancers

Kazushi Inoue and Elizabeth A. Fry

Abstract *p53* and its related genes, *p63* and *p73* constitute the *p53* gene family. While *p53* is the most frequently mutated gene in human tumors, *p63* and *p73* are rarely mutated or deleted in cancers. Many studies have reported p63/p73 overexpression in human cancers while others showed that a loss of p63/p73 is associated with tumor progression and metastasis. Thus, whether p63 or p73 is a tumor suppressor gene or an oncogene has been a matter of debate. This controversy has been attributed to the existence of multiple splicing isoforms with distinct functions; the full-length TA isoform of p63 has structural and functional similarity to wild-type p53, whereas the ΔNp63 acts primarily in dominant-negative fashion against all family members of p53. Differential activities of TA and ΔN isoforms have been shown *in vivo* by creating isform-specific gene knockout mice. All p53, p63, p73 proteins bind to and activate target genes with p53-response elements; p63 also binds to distinct p63-response elements and regulate expression of specific target genes involved in skin, limb, and craniofacial development. Interestingly, several studies have shown that both p63 and p73 are involved in cellular response to cancer therapy and others have indicated that both of these molecules are required for p53-induced apoptosis, suggesting functional interplay among p53 family proteins. Consistent with these findings, aberrant splicing that result in ΔNp63 or ΔNp73 overexpression are frequently found in human cancers, and is associated with poor clinical outcomes of patients in the latter. Thus immunohistochemical staining of tumor specimen with ΔNp73-specific antibody might have diagnostic values in cancer clinics.

Keywords p63 • p73 • Splicing • Alteration • Overexpression • Knockout • Mouse • Cancer

---

K. Inoue (✉) • E.A. Fry
The Department of Pathology, Wake Forest University Health Sciences,
Medical Center Boulevard, Winston-Salem, NC 27157, USA
e-mail: kinoue2@triad.rr.com

S.P. Deb and S. Deb (eds.), *Mutant p53 and MDM2 in Cancer*,
Subcellular Biochemistry 85, DOI 10.1007/978-94-017-9211-0_2,

Introduction

The p53 tumor suppressor protein integrates endogenous and exogenous signals to modulate cell fate to stress and cellular environments [1–3]. Upon DNA damage or other cellular stresses, such as oxidative stress, hypoxia, carcinogen exposure, and oncogene overexpression, p53 becomes activated with increased levels. Then, p53 directs a variety of responses, including DNA repair, cell cycle arrest/senescence, apoptosis, and autophagy depending on the input signal and severity of the damage [4, 5]. The specific response depends on whether the damage can be repaired or is too serious that death of the cell is required to maintain tissue integrity. The genomic locus for *p53 (TP53)* is very frequently (~50 %) mutated in human cancers, which is associated with therapy resistance and poor prognosis of patients [6, 7]. Since p53 protects humans from damaged and life-threatening cells that may predispose to tumor development, recent research efforts have been made on reconstituting p53 function to effectively treat cancer patients [8, 9].

In the late 1990s, two other p53 family members, p73 and p63 were discovered [10, 11]. These three proteins, encoded by the *TP53*, *TP63*, and *TP73* genes (*Trp53, Trp63,* and *Trp73* in mice, respectively), are transcription factors that bind directly to DNA as tetramers, interact with other transcription factors and the transcription machinery, and together control the expression of genes involved in all aspects of life. It has now become clear that both p63 and p73 are involved in a broad spectrum of biological activities, such as cell proliferation, apoptosis, development, differentiation, senescence, and aging. In particular, p63 has emerged as a critical player in embryonic development, epithelial stem cell maintenance, and differentiation. Both p63 and p73 express as a variety of protein isoforms that originate from two different promoters and extensive gene splicing at the N- and C-termini [12, 13]. Moreover, the *p63* and *p73* genes encode a sterile alpha motif (SAM) domain at the C-terminus that is not found in p53. This domain is responsible for protein-protein interactions and is found in a diverse range of proteins that are involved in developmental regulation. In this chapter, we discuss the structure, splicing isoforms of p63 and p73 in normal and their distinct functions in tumor suppression/proliferation. We also explain their possible interaction with Mdm2 and MdmX. Whether these molecules are tumor suppressors or oncoproteins have been a hot topic of debate. Gene knockout studies will tell us the answer; since both of the genes have multiple splicing isoforms, we have put special interest on the phenotypes of splicing isoform – specific gene knockout mouse models. Finally, we summarize the mechanisms and frequencies for alterations of these genes in human cancers and their prognostic significance.

Structure of the p63 and p73 loci

Both *p63* and *p73* loci (*TP63, TP73*) generate mRNAs that produce multiple protein products resulting from use of distinct promoters and alternative mRNA splicing (Fig. 2.1) [10, 11]. Transcription of *p63* and *p73* occurs from two

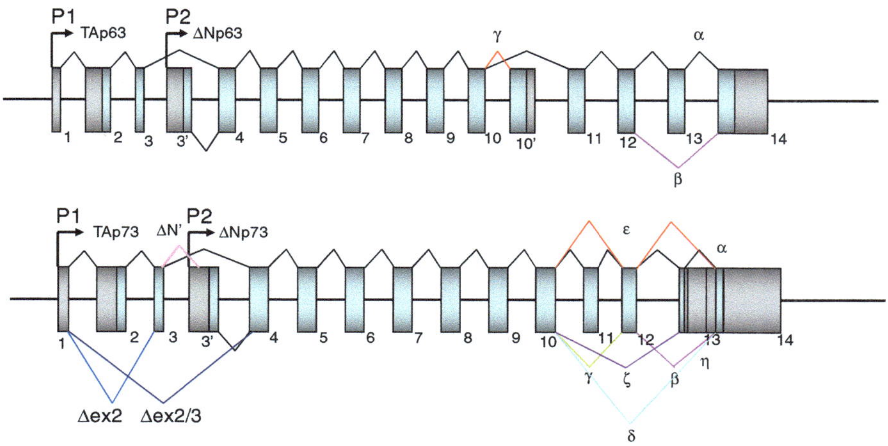

Fig. 2.1 Genomic structure of the *p63* and *p73* loci. Genomic structures of human *p63* and *p73* loci. *Numbered boxes* indicate exons, *black shading* denotes untranslated sequences, and *light blue shading* denotes coding regions. Distinct transcription start sites (P1 and P2) are indicated by *arrows*. N-terminal alternative splicing for p63 and p73 are indicated by *blue* and *light pink lines*, and C-terminal splicing events for these proteins are indicated by different *colored lines*

promoters: one upstream of exon 1 (P1) and the other located within intron 3 (P2). In both proteins, splicing isoforms transcribed from the P1 promoter have an N-terminal transactivation (TA) domain (i.e., TAp63 and TAp73), which is highly homologous to the TA domain of p53, whereas transcripts generated from the P2 promoter lack the N-terminal TA domain (39 amino acids; called ΔNp63 and ΔNp73, respectively; Fig. 2.1) [10, 11]. The unique structural differences for p63 and p73 are explained below.

p63

The structure of the genomic locus for p63 is shown in Fig. 2.1, upper panel [11]. Both mouse and human *p63* genes consist of 15 exons spanning around 210 kb and 270 kb, respectively, on the genome. The human version has been mapped to chromosome 3q27. The structures for the TAp53 protein, representative p63 isoform proteins are shown in Fig. 2.2 [12]. Wild-type TAp53 has an N-terminal transactivation domain (TA) for recruitment of core transcriptional factors, a central DNA-binding domain (DBD) for recognition of promoter sequences, an oligomerization domain (OD) for tetramerization, and a short basic stretch of 30 amino acids for regulation of transcriptional activity (Fig. 2.2, top panel). The *p63* gene encodes two alternatively spliced isoforms (TA, ΔN) with different ATG at the N-terminus with three alternatively-spliced C-terminal isoforms (α, β, γ), generating 6 different splicing isoforms,

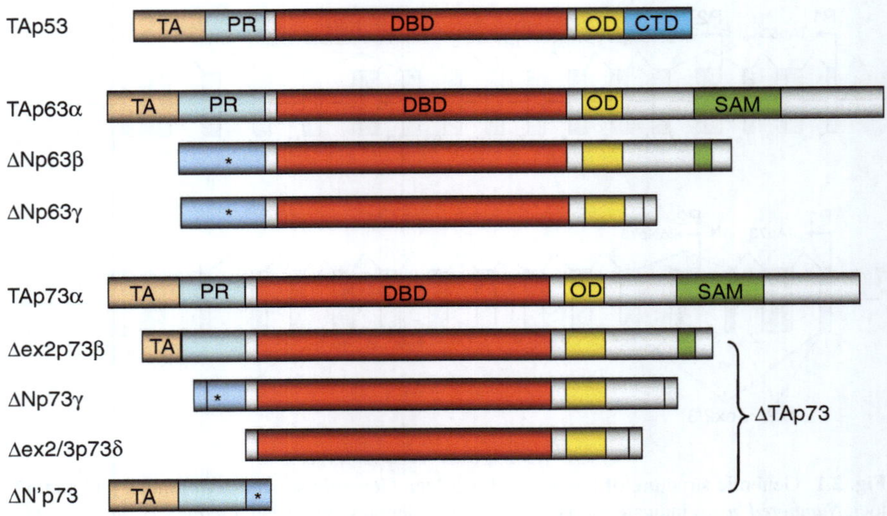

Fig. 2.2 Structure of the p53 family proteins. Protein domains of p53 family members. The transactivation (TA) domains shared by p53, TAp63, and TAp73 isoforms are shown in gold. The proline-rich domain (PR: *light blue*), DNA-binding domain (DBD: *red*), oligomerization domain (OD: *yellow*), carboxyl-terminal regulatory domain (CTD: *blue*), and sterile alpha motif (SAM: *green*) are shown in colors. The alpha isoforms of p63 and p73 possess a C-terminal SAM domain followed by a transactivational inhibitory domain (TID: *silver*). TAp63γ/TAp73γ isoforms most closely look like p53. N-terminally truncated ΔN isoforms for p63 possess the unique N-terminal sequence. p73 has four different isoforms at the N terminus (ΔN, ΔN', Δex2, Δex2/3) dependent on the usage of two different promoters and alternative splicing including exons 2 and 3. ΔN'p73 encodes a small protein having a unique sequence at the C- terminal end, but lacks the DNA-binding domain. * denotes the unique region encoded by exon 3' [12]

i.e., TAp63α, TAp63β, TAp63γ, ΔNp63α, ΔNp63β, ΔNp63γ (Figs. 2.1 and 2.2, middle panels). The p63α transcript has all 14 exons while the β transcript lacks exon 13. The γ transcript lacks exons 11–14 by splicing into a unique exon 10' (Fig. 2.1, top panel). The full-length TA isoform of p63 has structural and functional similarity to wild-type p53, whereas the ΔNp63 acts primarily in dominant-negative fashion against all family members of p53: p53, TAp63 and TAp73. Thus, it is generally assumed that *TAp63* is a tumor suppressor gene while *ΔNp63* is an oncogene. In addition, the C-terminus of p63 (and also p73) contains a sterile alpha motif (SAM) domain and a transcriptional inhibitory domain (TID) (Fig. 2.2). The SAM domains are small protein–protein interaction modules that are found in a wide variety of proteins, ranging from kinases and transcriptional regulators to cell surface receptors [14, 15]. The TID, an unstructured region C-terminal to the SAM domain, was shown to inhibit the transcriptional activity of p63 by interacting with the TA domain [16]. These two domains are not found in p53 (Fig. 2.2), suggesting unique functions for p63 and p73.

p73

Both mouse and human *p73* genes consist of 15 exons spanning around 80 kb on the genome. The structure of the genomic locus for *p73* is shown in Fig. 2.1, lower panel. The human version has been mapped to chromosome 1p36.33. The *p73* gene encodes 4 alternatively spliced isoforms (TA, Δex2, Δex2/3, ΔN) with distinct ATG at the N-terminus and 7 alternatively spliced isoforms at C-terminus (α, β, γ, δ, ε, ζ, and η; Fig. 2.1) [17, 18]. In addition, splicing-associated frameshifts yield unique C-terminal sequences for some p63 and p73 isoforms [17, 19]. This alternative splicing can generate 28 plus one (ΔN'; total 29) different splicing isoforms for p73. Of note, both ΔN and ΔN' isoforms have unique amino acids at exon 3' (Fig. 2.1). The difference in the N-terminal region contributes to different protein-protein interactions dependent on the isoform. The C-terminus of p73 has at least 7 splicing variants as shown in Fig. 2.1. The p73α transcript has all exons 1–14 while the β transcript lacks exon 13. The γ transcript lacks exon 11, the δ transcript lacks exons 11–13 (Fig. 2.1, lower panel). The ε isoform lacks 11 and 13, ζ lacks exons 11 and 12; η is close to α, but is different at exon 14 (Fig. 2.1). The TAp63γ and TAp73γ isoforms most closely resemble the full-length wild-type TAp53 (Fig. 2.2). In over-expression studies, TAp63γ has been shown to be as potent as p53 in transactivating target gene expression and apoptosis, whereas the most potent transcriptionally active p73 isoform reported is TAp73β [10, 11]. Since the ΔNp73 acts primarily in dominant-negative fashion against all family members of p53, it is generally accepted that *TAp73* is a tumor suppressor gene while *ΔNp73* is an oncogene.

Unique C-terminal Domains and Transcriptional Targets for p63 and p73

Both p63α and p73α isoforms also contain a protein–protein interaction domain known as sterile alpha motif (SAM) (Fig. 2.2). This is a globular domain composed of four α-helices and a small 310 helix. Although this motif is often found to mediate homodimerization with developmentally regulated proteins, the SAM domain does not contribute to homodimerization in p63 and p73 [20]. The SAM domains also appear to possess the ability to bind RNA. The post-SAM region known as the transactivational inhibitory domain (TID) has been identified in p63α and p73α isoforms [16]; Fig. 2.2). This region consisting of ~70 amino acids, which is absent in p53, has been proposed to inhibit transcription of both TAp63α and TAp73α through inter- or intra-molecular association with the TA domain [16]. Indeed, both of these proteins show decreased potency in transactivation and apoptosis induction as compared to other TA isoforms, and deletion of this region restored transactivating potential for both TAp73α and TAp63α [16, 21].

Since both p63 and p73 share strong structural, biochemical and biological homologies to p53, they bind specifically to conventional p53 response elements (p53RE: RRRCWWGYYY) and transactivate target genes such as *p21*^{Cip1}, *MDM2*, and *BAX*. In spite of their structural similarities between p53 and p63, the latter functions are greatly different from those of p53. The most striking difference is the apparent involvement of p63 in skin and limb development [22] (the details of phenotypes will be explained later). Global *p63* knockout mice that lack all splicing isoforms exhibit skin and limb defects as well as craniofacial abnormalities, but are not tumor prone. This is in contrast to *p53* knockout mice that develop normally, but are prone to develop various cancers from an early age, esp. thymic lymphomas and hemangiosarcomas [23, 24]. In humans, germ line mutations of *p53* cause Li-Fraumeni syndrome, in which affected individuals are very prone to cancer development [25–27]. These differences may be due to the differential regulation of target genes by p53 and p63. The p53 and p63 proteins can bind to two or more tandem repeats of RRRCWWGYYY (p53RE) or some other motifs and subsequently activate target gene expression. By using oligonucleotide expression microarray analysis and analyzing the promoters of p63-induced genes, Osada et al. [28] identified novel p63-specific response elements (p63REs) in the promoter regions of *EVPL* and *SMARCD3*. These p63REs exhibit characteristic differences from the canonical p53RE (RRRCWWGYYY) in both the core-binding element (CWWG) as well as the RRR and/or YYY sequences [28]. Their data indicate that p53 preferentially activates and binds to the RRRCATGYYY sequence, whereas p63 preferentially activates RRRCGTGYYY. Whereas EVPL protein is highly expressed in epithelial cells of the skin and pharynx in the *p63*$^{+/+}$ mouse, it is undetectable in these tissues in the *p63*$^{-/-}$ mouse. Thus p63 can regulate expression of specific target genes such as those involved in skin, limb, and craniofacial development by preferentially activating distinct p63-specific response elements [28]. Until now, a number of genes have been reported to be targets of p63 and p73, such as *REDD1* (regulation of reactive oxygen species), *JAG1*/*JAG2* (Notch signaling), *IL4R*, *ΔNp73*, and *AQP3* (glycerol and water transporter) [28–30]. Among these, *ΔNp73* is a splicing variant from the *p73* locus, suggesting its autoregulation [31]. In Notch signaling, Sasaki et al. found that the genes encoding ligands for the Notch receptors (*JAG1/2*), are up-regulated by p63 and p73 but not by p53 [30]. They identified a p63-binding site in the second intron of the *JAG1* gene, which could directly interact with p63 *in vivo* as demonstrated by chromatin immunoprecipitation. They also found a target of Notch signaling; HES-1 was up-regulated in Jurkat cells with high expression of Notch1 when co-cultured with p63-transfected cells, suggesting that p63 can trigger the Notch signaling pathway in neighboring cells. This suggests a potential molecular mechanism for the involvement of p63 in normal development [30]. Recently it was reported that BRCA1 activates the Notch pathway in breast cells by transcriptional upregulation of Notch ligands and receptors [32]. They demonstrated that BRCA1 was localized to an intronic enhancer within the *JAG1* gene, an event requiring ΔNp63. This BRCA1/ΔNp63-mediated induction of *JAG1* must play important roles in the regulation of breast stem/precursor cells since knockdown of these proteins resulted in increased tumorsphere growth and increased activity of stem cell markers [32]. Thus, BRCA1/ΔNp63-mediated transactivation of Notch signaling is a key event in the normal differentiation process in breast tissue.

Regulation of p63 and p73 by Mdm2

The interactions of p53 with Mdm2 and Mdmx, mediated via the TA domain of p53 have been well-documented [33–36]. The physiological importance of the regulation of p53 by the Mdm2 and Mdmx ubiquitin ligases as well as the role of its aberrant regulation in tumors has also been reported [37–41]. Unlike p53, which protects genomic stability, the two homologous proteins of the same family, p63 and p73, regulate developmental processes as described in this chapter.

Since all three p53 family proteins have homologous TA domains, it was speculated that p63 and p73 may be regulated by Mdm in a similar manner as has been reported for p53. The ability of Mdm2 and Mdmx to bind to p73 has been well-documented [42] and Zdzalik et al. [43] provided a detailed kinetic characterization of this interaction. The interaction of Mdm with p63 has also been studied previously, but the results were controversial due to lower affinity for such an interaction [44–47, 92]. In fact, Zdzalik et al. [43] showed that both Mdm2 and Mdmx form complexes with the p63 TA domain, however the interactions were weaker than those determined for p53 or p73. The interaction of the p63 TA domain is specific and mechanistically similar to that of the p53 TA domain since the p63(Ala) mutant peptide showed no activity in the assays performed. Although the interactions of p73 with Mdm2 and Mdmx have also been studied previously, only the affinity of p73 for Mdm2 has been reported [48, 49]. The interaction between p63 and Mdm2 is one order of magnitude weaker than those of Mdm with p53 and p73. Conversely, the affinities of both Mdm2 and Mdmx for p73 are of the same order of magnitude as those for p53, which justifies the conclusion that these proteins truly interact in cells, as has previously been suggested in other studies [42, 46, 50, 92]. The weaker interactions of both Mdm2 and Mdmx with p63 explain the inconsistent results reported by different groups on the interactions of those proteins. Clearly, at sufficiently high concentrations, these proteins will form a stable complex, but whether such concentrations are ever encountered under physiological conditions in cells remains a very intriguing question for future studies [43]. It is also noteworthy that although the affinities of p53 for Mdm2 and Mdmx are similar, both p63 and p73 interact more strongly with Mdmx. Therefore, Mdmx, but not Mdm2, may have a stronger impact on the regulation of intracellular p63 and p73.

Constitutional, All Splicing Isoforms' Knockout Mice for p63, p73

Mills et al. [22] reported the phenotypes of *p63*-deficient mice (all splicing isoforms). *p63*-null mice are born alive but have striking developmental defects. Their limbs are absent or truncated, defects that are caused by a failure of the apical ectodermal ridge differentiation. The skin of *p63*-null mice did not progress through an early developmental stage lacking stratification with no differentiation markers.

Hair follicles, teeth and mammary glands were absent in *p63*-deficient mice. Thus, in contrast to p53, p63 is essential for several aspects of ectodermal differentiation during embryogenesis. Keyes et al. [51] studied spontaneous and chemically-induced tumor development using *p63*$^{+/-}$ mice since *p63*$^{-/-}$ mice had serious developmental defects, and thus were not suitable for *in vivo* tumor development studies. They found that *p63*$^{+/-}$ mice were not tumor prone and mice heterozygous for both *p63* and *p53* had fewer tumors than *p53*$^{+/-}$ mice. Furthermore, p63 expression was maintained in carcinomas. These findings demonstrate that p63 plays a markedly different biological role in cancer than p53.

Mice deficient for all *p73* splicing isoforms also exhibited profound developmental and immunological defects, including hippocampal dysgenesis, hydrocephalus, chronic infections, and inflammation, as well as abnormalities in pheromone sensory pathways [52]. It should be noted that mice lacking *p73* showed no increased susceptibility to spontaneous tumorigenesis, in contrast to *p53*-deficient mice [23, 24]. They speculated that potentially dominant-negative, p73 variants were the predominant expression products of this gene in developing and adult tissues, explaining the mechanistic basis of the hippocampal dysgenesis and the loss of pheromone responses in *p73*-null mice. Thus p73 plays unique roles in neurogenesis, sensory pathways, and homeostatic control [52].

Flores et al. [53] explored the combined role of p63 and p73 in DNA damage-induced apoptosis. The combined absence of *p63* and *p73* severely impaired the induction of p53-dependent apoptosis in response to DNA damage in E1A-expressing cells and in the developing central nervous system in mice although the *p53* locus remained intact. This was explained by the inability of p53 to bind the promoters of apoptosis-associated target genes and to upregulate their transcription in *p63*$^{-/-}$*; p73*$^{-/-}$*;* E1A(+) cells and the developing central nervous system [53, 54].

Splicing Isoform-Specific Knockout Mouse Models for p63, p73

p63

The roles of *p63* in tumor suppression have been a hot topic of debate. The most intriguing question is whether p63 is a tumor suppressor gene or an oncogene. Many studies have shown p63 overexpression in human cancers [55]; discussed later in this chapter), while others demonstrate that a loss of p63 is associated with tumor progression and metastasis [56]. This controversy has been attributed to the existence of multiple splicing isoforms with distinct functions; the full-length TA isoform of p63 has structural and functional similarity to p53 (Fig. 2.2), whereas the ΔNp63 protein acts primarily in dominant-negative fashion against all family members of p53. To study splicing isoform–specific differences of *p63* functions *in vivo*, Su et al. [57] developed a *TAp63* conditional knockout mouse and used it to delete *TAp63* in the germline (*TAp63*$^{-/-}$; using *Zp3-cre* or *Protamine-cre*) or in K14-expressing cells in the basal layer of the epidermis (*TAp63*$^{fl/fl}$*;* using *K14cre+*).

TAp63$^{-/-}$ mice aged prematurely and developed blisters, skin ulcerations, senescence of hair follicle-associated dermal and epidermal cells, and decreased hair morphogenesis, indicating that TAp63 serves to maintain adult skin stem cells by causing cellular senescence and genomic stability, thereby preventing premature tissue aging [57]. The same group followed spontaneous tumor development in *TAp63*$^{-/-}$, *TAp63*$^{+/-}$ and wild-type mice for 2.5 years and found that both *TAp63*$^{+/-}$ mice and *TAp63*$^{-/-}$ mice developed carcinomas and sarcomas with significantly shorter lifespan than the wild-type cohort. Consistent with this finding, tumors from *TAp63*$^{+/-}$ mice retained the wild-type allele of *TAp63* suggesting that *TAp63* is haplo-insufficient tumor suppression. Both *TAp63*$^{+/-}$ and *TAp63*$^{-/-}$ mice developed highly metastatic tumors, and 10 % of these metastases were found in the brain, a rare finding in endogenous mouse tumor models. Although equivalent numbers of carcinomas metastasized in the *TAp63*$^{-/-}$ and *TAp63*$^{+/-}$ mice, a greater number of sarcomas metastasized in *TAp63*$^{+/-}$ mice than in *TAp63*$^{-/-}$ mice, indicating that heterozygous loss for *TAp63* rather than homozygous loss results in a more severe phenotype.

Keyes et al. [58] observed that ΔNp63α overexpression in mouse embryonic fibroblasts (MEFs) bypassed Ras-mediated senescence and drove tumorigenesis *in vivo*. They identified chromatin-remodeling protein Leeh as a novel target for ΔN63α that is an essential mediator of senescence bypass. This bypass of senescence by ΔNp63α promoted stem cell-like proliferation and maintained the survival of keratin 15-positive cells. Thus, ΔNp63α is a novel oncogene that cooperates with Ras to promote tumor development by initiating stem cell proliferation. By contrast, overexpression of TAp63 forms in *p53*$^{-/-}$ MEFs increased senescence and reduced tumor development *in vivo*, consistent with a p53-independent effect of TAp63 [59].

The TAp63 and ΔNp63 isoforms have special effects in epidermal tissue differentiation [60]. In murine embryonic stem cells, ΔNp63, but not TAp63, is highly expressed in epidermis and is critical for the expression of the cytokeratins K5 and K14, two markers of keratinocyte differentiation, indicating that only ΔNp63 is required for the commitment of ectodermal into epidermal cells [61, 62]. In summary, p63 and its splicing variants play specific roles in epidermal commitment, cell proliferation, and senescence bypass; alterations of this intricate balance contribute to tumor development.

p73

Mice with a complete deficiency of *p73* exhibited severe neurological and immunological defects due to the absence of all *TAp73* and *ΔNp73* isoforms as described in the previous section. To study mice deficient for specific p73 protein isoforms, Tak Mak's group created *p73* isoform-specific knockout mice [63, 64]. Tomasini et al. [63] created mice in which exons encoding the *TAp73* isoforms were specifically deleted at exons 2/3 to establish a *TAp73*-deficient (*TAp73*$^{-/-}$) mice. Mice specifically lacking in *TAp73* isoforms showed a phenotype intermediate between the phenotypes of *p73*$^{-/-}$ and *p53*$^{-/-}$ mice with respect to the incidence of spontaneous and carcinogen-induced tumors, infertility, and aging, as well as hippocampal dysgenesis. In addition, cells from *TAp73*$^{-/-}$ mice showed genomic instability associated with enhanced aneuploidy, which could account for the increased incidence of spontaneous tumors in these animals. Hence, *TAp73* isoforms exert tumor-suppressive functions indicating an emerging role for *Trp73* in the maintenance of genomic stability. Wilhlem et al. [64] generated mice that were selectively deficient for the *ΔNp73* isoform by depleting *ΔN* form-specific exon 3' (*ΔNp73*$^{-/-}$). These mice were viable and fertile, but showed signs of neurodegeneration. Cells from *ΔNp73*$^{-/-}$ mice were sensitive to DNA-damaging agents and showed an increase in p53-dependent apoptosis. They found that the ΔNp73 protein localized directly to the site of DNA damage, interacted with the DNA damage sensor protein 53BP1, and inhibited ATM activation and subsequent p53 phosphorylation. This finding may explain why human tumors with high levels of ΔNp73 expression showed resistance to chemotherapy.

In summary, these studies show that TAp63 and TAp73 proteins have specific roles in preventing tumor development *in vivo*. Conversely the ΔN forms act as oncogenes by preventing senescence and maintaining progenitor cell status. When overexpressed, both TAp63 and TAp73 proteins transactivate subsets of known p53 target genes involved in cell-cycle arrest and apoptosis, such as *p21*^{Cip1} and *Bbc3* [65–68]. Of note, both TAp63 and TAp73 also regulate distinct sets of genes that are not transcriptional targets for p53 through unique p63RE as described in the previous section. In contrast, ΔNp63 and ΔNp73 proteins have been shown to function in part as dominant-negative inhibitors of the p53 family, leading to the hypothesis that these isoforms may exhibit proto-oncogenic function. ΔN isoforms inhibit the function of TA forms through (1) direct competition for DNA-binding sites and (2) formation hetero-oligomeric complexes with TAp63/TAp73, and less strongly with p53 [11, 20, 69–72]. Interestingly, expression of the ΔNp73 is strongly up-regulated by TAp73 and p53, thus creating a feedback loop that tightly regulates the function of TAp73 and more importantly, of p53 [70].

Aberrant Expression, Altered Splicing, and Mutations of p63 and p73 in Human Cancer

Alterations of p63 Isoforms in Human Cancers

Both p63 and p73 were initially hypothesized to function as tumor suppressors based on their homology to p53. However, accumulating evidence shows that mutation of either of these genes in human cancer is quite rare [73, 74], indicating that they are not classical tumor suppressor genes like *p53* or *RB* that meet the Knudson's two-hit hypothesis [75]. Although there have been numerous studies on p63 expression in human cancers, loss of heterozygosity (LOH) of the *p63* locus has not been studied extensively in human malignancies [76], possibly because the genomic locus 3q27-28 is not the site of frequent gene deletion in cancer. Conversely, decreased p63 expression is a common feature of high-grade invasive urothelial carcinomas and associates with reduced β-catenin. Both ΔNp63 and TAp63 are frequently downregulated in bladder cancer and this reduction correlates with a poor prognosis [77]. The majority of prostate cancers show loss of p63, but it is overexpressed in some poorly differentiated tumors and correlates with a poor prognosis [78]. In addition, loss of p63 results in enhanced metastasis in prostate cancer [79]. Koga et al. [80] studied the expression of p63, β-catenin, and uroplakin III by immunohistochemistry in high-grade invasive bladder carcinomas. Lower p63 expression was significantly associated with higher TNM stage, lymph-node metastasis, and reduced β-catenin expression. Importantly, lower p63 expression was significantly associated with a poor prognosis. Impaired p63 expression was associated with biological aggressiveness of high-grade invasive urothelial carcinomas. Moreover, loss of p63 expression was a pre-requisite for uroplakin III expression. Their data suggested that p63 plays critical roles in tumor progression and biochemical terminal differentiation of urothelial neoplasms [80].

Oral lichen planus (OLP) is a relatively common chronic disease of the oral mucosa for which the etiology or pathogenesis is not fully understood. Sniezek et al. [81, 82] showed decreased expression of p63 in OLP compared to normal mucosa, a decrease they suggested could explain the hyper-differentiation, or pro-differentiation, seen in this disorder OLP. Consistent with these findings, another group reported downregulation of p63 in this disorder [83].

The *p63* gene maps to chromosome 3q27-28, a region frequently amplified in squamous cell carcinomas [55, 76, 84–86]. Most squamous cell carcinomas retain p63 expression, where it is often overexpressed [55, 87, 88]. Although some controversy exists as to whether p63 is the targeted gene driving amplification of this locus, several groups have reported increased *p63* mRNA levels that correlate with an increase in *p63* gene copy number in squamous cell carcinomas of the lung, head, and neck (HNSCCs) [84, 89, 90]. In other cases, overexpression of p63 appears to be independent of genomic DNA amplification of the locus [91]. In esophageal carcinomas, amplification of the *p63* gene was reported in ~20 % of squamous cell carcinomas and 10 % of adenocarcinomas [76]. Given that the total frequency of tumors in which p63 is upregulated is much higher (>50 %), gene amplification is unlikely to be the main mechanism underlying the increased levels of p63. Rather, transcriptional or post-transcriptional changes are involved. Multiple studies have shown that p63 overexpression occurs in up to 80 % of primary HNSCCs and also in other squamous cell carcinomas, including those in the lung, nasopharynx, and cervix [55, 92–94]. By the use of isoform-specific antibodies, Nylander et al. [95] mapped expression of the different p63 isoforms within normal oral mucosa and HNSCCs, showing increased expression of p63, mainly the ΔNp63 isoforms, in tumors compared to normal mucosa. They indicated specific roles for the individual isoforms in cell differentiation and neoplasia [95]. In invasive breast cancer, the frequency of p63 expression varies, ranging from 0 to 30 % [96–98]. It is now considered that p63 is expressed in at least a subset of breast tumors that are known to have a basal epithelial phenotype [99].

TAp63 vs. ΔNp63 in Cancer

In esophageal carcinomas, p63 isoforms are upregulated not only in carcinomas, but also in squamous dysplasias [76]. Although early studies for the detection of p63 did not differentiate among different isoforms, recent studies used isoform-specific RT-PCR coupled with Western blot analysis to quantitatively demonstrate that ΔNp63α is the predominant p63 isoform expressed in squamous cell carcinomas. Using such an approach, it has been reported that tumor-suppressive TAp63 overexpression is rare in HNSCC, and that *ΔNp63* mRNA expression was at least 100-fold more abundant than *TAp63* mRNA in all cases [87, 100]. These findings are consistent with the inability of many investigators to detect TAp63 protein isoforms by Western blot analysis in either primary keratinocytes or HNSCC cells. ΔNp63 is the predominant variant that is found in HNSCCs; however, in Barrett's esophagus, a disorder in which the stratified epithelium is replaced by a simple columnar epithelium that consists of mucosecretory cells, the *p63* gene expression is not highly prominent [76].

Tumors often have simultaneous transcriptional upregulation of both *TAp63* and *ΔNp63* isoforms, with ΔNp63 being predominant at protein levels [55, 87]. This would represent the anti-apoptotic and proliferative effects of ΔNp63 as described in the previous section. Moreover, it was reported that ΔNp63*α* expression directly correlates with a poor response to cisplatin in HNSCC [101]. In pancreatic cancer, Danilov et al. [102] showed that ΔNp63α enhanced the oncogenic potential of tumor cells through trans-activation of *EGFR* and *14-3-3σ*. Leong et al. [103] reported that the p63/p73 network mediates chemosensitivity to cisplatin in a subset of primary breast cancers. Thus, p63 is involved in chemosensitivity of multiple types of tumors. In HNSCC, DNA damage by chemotherapy caused a decrease in ΔNp63-mediated transcriptional repression by blocking p63-responsive elements or sequestering TAp63 in less active hetero-tetramers, together with increased expression of p73, thus allowing TAp73-mediated cell death [100]. Together, these reports indicate that it is not only the levels of individual p53 family members, but rather the ratio between TA (transcriptionally active, having tumor-suppressor functions) and ΔN (acting as dominant-negative over the TA isoforms, showing oncogenic properties) isoforms that determines the biological outcome.

In lung cancer, amplification of chromosomal region 3q26-3qter is frequently found in tumors. Massion et al. [55] analyzed *p63* gene copy number and expression by immunohistochemistry in tissue microarrays of >200 non-small cell lung cancers (NSCLCs) and correlated them with survival. The *p63* genomic locus was amplified in 88 % of squamous cell carcinomas, but only in 11 % of adenocarcinomas and 2 % of large cell carcinomas of the lung, indicating clear association of gene amplification with squamous cell lung cancer. The major splicing variant of p63 expressed was ΔNp63α. Furthermore, *p63* genomic amplification and protein staining was associated with better survival. They found a significant increase in *p63* copy number in pre-invasive lesions graded severe dysplasia or higher. Thus, there is early and frequent genomic amplification of *p63* in the development of squamous carcinoma of the lung and patients with NSCLC showing amplification and overexpression of p63 had prolonged survival [55]. However, two other groups have failed to demonstrate the favorable prognostic value of p63 in lung cancer. Iwata et al. [104] reported a lack of prognostic significance regarding ΔNp63 immunoreactivity in lung cancer. Uramoto et al. [105] showed that the expressions of ΔNp63 in lung cancer did not significantly affect survival while patients with a positive ΔNp73 expression had a poorer prognosis in comparison to the negative group. The differential prognostic values of p63 in these, Massion's and two other studies, can be attributed to the fact that the former study focused on gene copy number of *p63* and immunohistochemical staining of p63 (all splicing isoforms) in squamous cell lung cancer while the latter two groups studied the expression of the ΔNp63 protein and survival of non-small cell lung cancer (in Uramoto's study; squamous cell carcinoma only in Iwata's study).

ΔNp63α can act as a transcriptional repressor, but the link between the transcriptional functions of p63 and its biological role is still unclear. Barbieri et al. [106] depleted endogenous p63 by shRNA to investigate the transcriptional programs controlled by p63. Disruption of p63 in squamous cell carcinoma cell lines resulted in down-regulation of transcripts specifically expressed in squamous tissues and a significant alteration of keratinocyte differentiation. They found that depletion of p63 led to up-regulation of markers of non-epithelial tissues (mesenchyme and neural tissue) in squamous cell carcinomas, which were associated with increased capacity for invasion and metastasis in tumors. Furthermore, loss of p63 expression was accompanied by a shift toward mesenchymal morphology and an increase in motility in primary keratinocytes and squamous cell lines [106]. Thus, loss of endogenous p63 results in up-regulation of genes associated with invasion and metastasis, and predisposes to a loss of epithelial markers and acquisition of mesenchymal characteristics. Although the squamous cell carcinoma cell lines they analyzed expressed predominantly ΔNp63, the interpretation of their experimental results is controversial since their shRNA depleted both *TAp63* and *ΔNp63* at the same time. *p63* isoform-specific shRNA should be used to define the roles of each isoform in cell growth, differentiation, invasion and metastasis.

Regulation of Gene Expression by ΔNp63

Although ΔNp63 lacks the amino-terminal transactivation domain consisting of 39 amino acids that is present in TAp63, ΔNp63 still activates a group of genes that includes, but is not restricted to genes regulated by p53 [66]. Helton et al. showed that all NH2-terminally deleted p63 isoforms still retain a potential in transactivation and growth suppression [107]. Interestingly, they showed that ΔNp63β possessed a remarkable ability to suppress cell proliferation and transactivate target genes, which is consistently higher than that seen with ΔNp63α. They showed that an intact DNA-binding domain is required for ΔNp63 function. In addition, they found that the novel transactivation domain for the ΔNp63 variant was composed of the 14 unique ΔN residues along with the adjacent region, including a PXXP motif [107]. They also showed that a PPXY motif shared by ΔNp63α and ΔNp63β was required for optimal transactivation of target gene promoters [107]. Very recently, Ceraldo et al. [108] identified a novel p63 transcriptional target, caspase-1. Caspase-1 is pro-inflammatory caspase, which functions in tumor suppression. They showed that both p63 isoforms (TAp63, ΔNp63) increased caspase-1 expression through physical binding to its promoter. Consistently they also identified a direct correlation between p63 and caspase-1 expression in human cancer data sets. Functional interaction between p63 and caspase-1 represented a predictor of longer survival in human cancers. Together, in addition to dominant-negative effects of ΔNp63 on TA isoforms of p53 family proteins, regulation of gene expression by ΔNp63 variants should be re-evaluated from the viewpoint of tumor suppression.

Alterations of p73 Isoforms in Human Cancers

The *p73* gene has been speculated to be classical tumor suppressor genes like *p53* when the cDNAs were cloned [10]. In gastrointestinal tumors, LOH for *p73* has been reported in 10–40 % of the cases [76] although LOH for *p63* has not been reported in cancers. Despite these expectations, subsequent studies have demonstrated that the *TP73* locus was not the hot spot of gene deletion in cancers. Rather, studies of multiple tumor types have shown that p73 splicing variants are overexpressed, but not mutated or deleted in human malignancies [17].

To investigate the role of the *p73* gene in human carcinogenesis, Han et al. [109] studied genetic alterations of this gene by analyzing the entire coding exons as well as their surrounding exon-intron boundaries by PCR-CCSP and direct sequencing with primary samples from breast, colorectal, gastric cancers, neuroblastomas, and also with lung and pancreatic cancer cell lines since they are known to have frequent LOH in the 1p region. However, of the 185 cases, somatic missense mutation of glutamine from arginine at codon 269 was found in only one breast cancer. Monoallelic expression of p73 was observed in pancreatic cancer cell lines. Nomoto et al. [110] analyzed 61 primary lung cancer samples of the *p73* locus at 1p36.33 by PCR-SSCP and Southern blotting. Although allelic loss at the 1p36.33 locus was observed in 42 % of cases, somatic mutations of the *p73* gene were not observed in their samples, suggesting the presence of an as yet to be determined tumor suppressor gene at the locus. In summary, inactivation of the *p73* gene is very rare even in cancers involving chromosome 1p [109].

TAp73 vs. ΔTAp73 in Human Cancers

Overexpression of p73 mRNA and/or protein relative to neighbor normal tissues has been reported in a variety of tumors, such as neuroblastoma, glioma, ependymoma, breast, lung, colon, stomach, liver, ovarian, bladder, cholangiocellular carcinomas, and myelogenous leukemias [17]. Concin et al. [111] studied the expression profile of all N-terminal isoforms, distinguishing between TAp73 and ΔTAp73 (ΔNp73, ΔN'p73, Δex2p73, and Δex2/3p73) (Fig. 2.2). Ovarian cancers almost universally overexpressed ΔN'p73 compared with normal tissues (95 % of cancers). About one-third of tumors also exhibited concomitant up-regulation of TAp73, whereas only a small subgroup of tumors overexpressed ΔNp73 [111]. Thus, deregulation of the E2F-responsive P1 promoter, rather than the P2 promoter, is mainly responsible for the production of ΔTAp73 in ovarian cancer. A trend was found for better overall survival in patients with low expression of ΔN'p73/ΔNp73, compared with those with high expression. Cancers with wild-type p53 showed significantly higher deregulation of ΔNp73, ΔN'p73, and Δex2/3p73 (transdominant p73) than p53 mutant cancers. Thus, overexpression of transdominant p73 isoforms can function as epigenetic inhibitors of p53 *in vivo*, thereby alleviating selection pressure for *p53* mutations in ovarian cancer [111].

Dominguez et al. [112] analyzed 113 colon and 60 breast cancer patients' primary samples and reported the association of ΔTAp73 variants and advanced pathologic stage, lymph node metastasis, vascular invasion, presence of polyps, and tumor localization. Overexpression of TP73 variants in tumor tissues indicates that they may be involved in carcinogenesis. The association between upregulation of ΔTAp73 isoforms and poor prognosis suggests that they may be of practical clinical prognostic value. Faridoni-Laurens et al. [113] analyzed the expression of TAp73 and ΔTAp73 in HNSCC and compared them to the p53 status. They found that all of the p73 isoforms were upregulated in comparison to those in normal adjacent tissue. Although p73 belongs to the gene family of p53, *p53* mutations and *p73* transcript alterations were not mutually exclusive. All of the HNSCC specimens studied had at least one p53 mutation and/or one *ΔTAp73* transcript alteration. Although both the ΔNp73 and the TAp73 transcripts were upregulated in HNSCC, the predominant protein in the cancers expressed was ΔNp73. Furthermore, a trend was found for better overall survival in patients with a low expression of ΔNp73. Thus deregulation of both the p53 and the p73 pathways plays an important role in inducing HNSCC [113].

By using specific polyclonal ΔNp73 antiserum against the exon 3′-specific peptide for p73, ΔNp73 and ΔN'p73 expressions were studied in paraffin-embedded tumor samples from 132 lung cancer samples [105, 114]. The ΔN/ΔN'p73 protein was detected mainly in the cytoplasm of tumor cells in 77 of 132 patients (58.3 %) with lung cancer. Importantly, lung cancer patients with positive ΔN/ΔN'p73 expression had a poorer clinical outcomes than those with negative expression. In addition, multivariate analysis of the clinicopathological characteristics of lung cancer indicated that positive expression of ΔN/ΔN'p73 was a significant independent factor for predicting poor prognosis ($P<0.0001$, risk ratio = 3.39). Thus, expression of ΔN/ΔN'p73 will be a useful marker for predicting poor prognosis of patients who undergo resection of lung cancer. Consistent with these findings, overexpression of the N-terminal splice variants (Δex2p73, Δex2-3p73), but not TAp73, was shown to be associated with a poor prognosis in low-grade gliomas [115], which should be helpful in decision-making in clinics.

The truncated oncogenic isoform Δex2p73 is expressed in hepatocellular carcinomas (HCC); however, the underlying mechanisms regulating this process are unknown. Castillo et al. [116] used human normal and diseased liver tissue samples to examine the association between activation of epidermal growth factor receptor (EGFR) by its ligand amphiregulin (AR) and the alternative splicing of *p73* pre-mRNA into the tumorigenic isoform Δex2p73, via c-Jun N-terminal-kinase-1-mediated signaling. Δex2p73 was expressed in a subset of premalignant cirrhotic livers and in otherwise healthy livers that harbored a primary tumor, as well as in HCC tissues. Δex2p73 expression was correlated with that of the EGFR ligand AR, previously shown to have a role in hepatocarcinogenesis. Autocrine activation of the EGFR by AR triggered c-Jun N-terminal kinase-1 activity and inhibited the expression of the splicing regulator Slu7, leading to the accumulation of Δex2p73 transcripts in HCC cells. Their study provided a mechanism for the generation of pro-tumorigenic Δex2p73 during liver tumorigenesis via activation of EGFR signaling by AR and c-Jun N-terminal kinase-1 activity, leading to inhibition of the splicing regulator Slu7 [116]. This is a unique report that showed the specific role of a particular splicing factor in aberrant *p73* splicing.

The molecular mechanisms underlying overexpression of ΔNp63 or ΔNp73 in cancers in comparison to normal tissues need further investigation. Methylation-mediated silencing of the P1 promoter for *TAp73* was reported in lymphoblastic leukemias and Burkitt's lymphomas [117, 118]. These findings indicate that either ΔNp63 or ΔNp73 overexpression or TAp73 promoter silencing is required to inactivate the tumor-suppressive activity of TAp73. Although TAp73 isoforms were paradoxically overexpressed (18–30 folds) in HNSCC tumor cells in comparison to non-transformed keratinocytes, ΔNp63α was also overexpressed in these tumors and was physically associated with TAp73, thereby inhibiting p73-dependent pro-apoptotic activity [87, 100]. *BRCA1*-deficient tumor cells exhibit increased sensitivity to cisplatin, and patients with BRCA1-associated ovarian carcinomas had better outcomes with platinum-based chemotherapy compared with sporadic cases. Ibrahim et al. [119] reported that *BRCA1*-deficient ovarian carcinoma cells exhibited hypermethylation within the P1 promoter for *p73*, which included the binding site for the p73 transcriptional repressor ZEB1, leading to the abrogation of ZEB1-binding and increased expression of transactivating p73 isoforms (TAp73), explaining increased cisplatin sensitivity of *BRCA1*-deficient ovarian carcinomas. Thus, TAp73 might represent a response predictor and potential therapeutic target for enhancing chemosensitivity in ovarian cancer.

Although promoter methylation is the major mechanism of *p73* inactivation in hematopoietic malignancies [120], the situation is different in epithelial tumors – carcinomas. Daskalos et al. [121] studied the DNA methylation status of both P1 and P2 promoters as a means of epigenetic transcriptional control of their corresponding isoforms in 102 primary NSCLCs and reported that the P2 hypomethylation-associated overexpression of Δ*Np73* mRNA is a frequent event, particularly among squamous cell carcinomas. P2 hypomethylation strongly correlated with long interspersed nuclear element-1 element hypomethylation, indicating that ΔNp73 overexpression may be a consequence of global DNA hypomethylation. Guan and Chen analyzed p73 in prostate cancer and found that ΔNp73 was significantly increased in 20 of 33 prostate carcinomas [122]. However, none of the specimens expressed ΔN'p73. The positive expression of ΔNp73 correlated with the Gleason score in prostate cancer. Interestingly, prostate cancer samples with wild-type p53 had significantly higher expression of ΔNp73 than p53 mutant cancers. These data suggested a potential role for ΔNp73 in prostate cancer progression.

Diaz et al. [123] conducted a translational study to evaluate whether 1,25(OH)(2) vitamin D(3) downregulates TP73 variants in colon and breast cancers [123]. They reported that ectopic survivin expression led to an increase in all of the TAp73, ΔNp73, ΔEx2p73, and ΔEx2-3p73 transcripts. In these cancers, direct correlations were observed between TP73 variants and survivin levels. Interestingly, 1,25(OH)(2) vitamin D(3) negatively regulated survivin and TP73 variants in these tumors. Thus positive regulation of TP73 isoforms by survivin may exist, which raised the possibility that the downregulation of TP73 isoforms may be possible with 1,25(OH)(2)D(3) through survivin.

In summary, although somatic point mutations are rarely found in *p73* in human cancers, aberrant splicing that result in ΔTAp73 overexpression are very frequently found. Since these proteins have transdominant activity on all p53 family proteins, it is speculated that this abnormal splicing contributes to human carcinogenesis, esp. in ovarian, breast, lung, and prostate cancers, HNSCCs, and hematological malignancies. Published results indicate that ΔTAp73 overexpression is associated poor clinical outcomes at least in lung cancer and HNSCCs. Of note, it may be possible to correct aberrant expression of p73 isoforms in cancer through the use of 1,25(OH)(2)D(3).

Conclusive Remarks

Judging from the very low frequency of mutations for *p63* and *p73* in human cancers, these are not classical tumor suppressor genes, but the possibility remains that these are haplo-insufficient tumor suppressors, just like $p27^{Kip1}$, *PTEN*, or *DMP1* [124–132]. Detailed analyses with specific primers are required to determine whether these are true tumor suppressors. Accumulating pieces of evidence suggest that TA- and ΔN- isoforms play distinct roles in cell cycle progression, apoptosis, and tumor development/prevention. Detection of each isoform by Western blotting or immunohistochemistry with specific antibodies or real-time PCR-mediated quantification of each splicing isoform will be needed to determine the prognostic value of each splicing isoform in cancer. Of note, both ΔNp73 and ΔN'p73 have unique amino acid sequences generated from the exon 3' that is absent in TAp73. This has made it possible to generate ΔN/ΔN'p73-specific antibodies that can be used in diagnostic immunohistochemistry.

Not many studies have been done to elucidate the mechanisms of overexpression of ΔN isoforms of p63 and p73 in human cancers. Identification of critical splicing factors and characterization of signaling pathways that contribute to this process will be critical to correct the errors for splicing for these genes in human cancers. Finally, specific targeting of ΔN- isoforms with antisense DNA, stabilized RNA, shRNA may have therapeutic values in treating human cancer overexpressing these splicing isoforms with oncogenic activity.

Acknowledgements K. Inoue has been supported by NIH/NCI 2R01CA106314, ACS RSG-07-207-01-MGO, and KG080179.

References

1. Toledo F, Wahl GM (2006) Regulating the p53 pathway: in vitro hypotheses, in vivo veritas. Nat Rev Cancer 6:909–923, Review
2. Menendez D, Inga A, Resnick MA (2009) The expanding universe of p53 targets. Nat Rev Cancer 9:724–737, Review
3. Vousden KH, Prives C (2009) Blinded by the light: the growing complexity of p53. Cell 137:413–431, Review
4. Green DR, Chipuk JE (2006) p53 and metabolism: inside the TIGAR. Cell 126(1):30–32, Review
5. Rufini A, Tucci P, Celardo I, Melino G (2013) Senescence and aging: the critical roles of p53. Oncogene 32(43):5129–5143, Review
6. Xu Y (2008) Induction of genetic instability by gain-of-function p53 cancer mutants. Oncogene 27(25):3501–3507
7. Lai D, Visser-Grieve S, Yang X (2012) Tumour suppressor genes in chemotherapeutic drug response. Biosci Rep 32(4):361–374
8. Frezza C, Martins CP (2012) From tumor prevention to therapy: empowering p53 to fight back. Drug Resist Updat 15(5–6):258–267
9. Senzer N, Nemunaitis J, Nemunaitis D, Bedell C, Edelman G, Barve M, Nunan R, Pirollo KF, Rait A, Chang EH (2013) Phase I study of a systemically delivered p53 nanoparticle in advanced solid tumors. Mol Ther 21(5):1096–1103
10. Kaghad M, Bonnet H, Yang A, Creancier L, Biscan JC, Valent A, Minty A, Chalon P, Lelias JM, Dumont X, Ferrara P, McKeon F, Caput D (1997) Monoallelically expressed gene related to p53 at 1p36, a region frequently deleted in neuroblastoma and other human cancers. Cell 90(4):809–819
11. Yang A, Kaghad M, Wang Y, Gillett E, Fleming MD, Dötsch V, Andrews NC, Caput D, McKeon F (1998) p63, a p53 homolog at 3q27-29, encodes multiple products with transactivating, death-inducing, and dominant-negative activities. Mol Cell 2(3):305–316
12. Murray-Zmijewski F, Lane DP, Bourdon JC (2006) p53/p63/p73 isoforms: an orchestra of isoforms to harmonise cell differentiation and response to stress. Cell Death Differ 13(6):962–972, Review
13. Collavin L, Lunardi A, Del Sal G (2010) p53-family proteins and their regulators: hubs and spokes in tumor suppression. Cell Death Differ 17(6):901–911
14. Schultz J, Ponting CP, Hofmann K, Bork P (1997) SAM as a protein interaction domain involved in developmental regulation. Protein Sci 6(1):249–253
15. Thanos CD, Bowie JU (1999) p53 Family members p63 and p73 are SAM domain-containing proteins. Protein Sci 8(8):1708–1710
16. Serber Z, Lai HC, Yang A, Ou HD, Sigal MS, Kelly AE, Darimont BD, Duijf PH, Van Bokhoven H, McKeon F, Dötsch V (2002) A C-terminal inhibitory domain controls the activity of p63 by an intramolecular mechanism. Mol Cell Biol 22(24):8601–8611
17. Moll UM, Slade N (2004) p63 and p73: roles in development and tumor formation. Mol Cancer Res 2(7):371–386, From tumor prevention to therapy: empowering p53 to fight back. Review
18. Bourdon JC, Fernandes K, Murray-Zmijewski F, Liu G, Diot A, Xirodimas DP, Saville MK, Lane DP (2005) p53 isoforms can regulate p53 transcriptional activity. Genes Dev 19(18):2122–2137, Epub 2005 Aug 30. Review
19. Courtois S, Caron de Fromentel C, Hainaut P (2004) p53 protein variants: structural and functional similarities with p63 and p73 isoforms. Oncogene 23(3):631–638, Review

20. Chi SW, Ayed A, Arrowsmith CH (1999) Solution structure of a conserved C-terminal domain of p73 with structural homology to the SAM domain. EMBO J 18(16):4438–4445
21. De Laurenzi V, Costanzo A, Barcaroli D, Terrinoni A, Falco M, Annicchiarico-Petruzzelli M, Levrero M, Melino G (1998) Two new p73 splice variants, gamma and delta, with different transcriptional activity. J Exp Med 188(9):1763–1768
22. Mills AA, Zheng B, Wang XJ, Vogel H, Roop DR, Bradley A (1999) p63 is a p53 homologue required for limb and epidermal morphogenesis. Nature 398(6729):708–713
23. Donehower LA, Harvey M, Slagle BL, McArthur MJ, Montgomery CA Jr, Butel JS, Bradley A (1992) Mice deficient for p53 are developmentally normal but susceptible to spontaneous tumours. Nature 356(6366):215–221
24. Jacks T, Remington L, Williams BO, Schmitt EM, Halachmi S, Bronson RT, Weinberg RA (1994) Tumor spectrum analysis in p53-mutant mice. Curr Biol 4(1):1–7
25. Malkin D, Li FP, Strong LC, Fraumeni JF Jr, Nelson CE, Kim DH, Kassel J, Gryka MA, Bischoff FZ, Tainsky MA et al (1990) Germ line p53 mutations in a familial syndrome of breast cancer, sarcomas, and other neoplasms. Science 250(4985):1233–1238
26. Malkin D (1993) p53 and the Li-Fraumeni syndrome. Cancer Genet Cytogenet 66(2):83–92, Review
27. Strong LC, Williams WR, Tainsky MA (1992) The Li-Fraumeni syndrome: from clinical epidemiology to molecular genetics. Am J Epidemiol 135(2):190–199, Review
28. Osada M, Park HL, Nagakawa Y, Yamashita K, Fomenkov A, Kim MS, Wu G, Nomoto S, Trink B, Sidransky D (2005) Differential recognition of response elements determines target gene specificity for p53 and p63. Mol Cell Biol 25(14):6077–6089
29. Ellisen LW, Ramsayer KD, Johannessen CM, Yang A, Beppu H, Minda K, Oliner JD, McKeon F, Haber DA (2002) REDD1, a developmentally regulated transcriptional target of p63 and p53, links p63 to regulation of reactive oxygen species. Mol Cell 10(5):995–1005
30. Sasaki Y, Ishida S, Morimoto I, Yamashita T, Kojima T, Kihara C, Tanaka T, Imai K, Nakamura Y, Tokino T (2002) The p53 family member genes are involved in the Notch signal pathway. J Biol Chem 277(1):719–724, Epub 2001 Oct 18
31. Nakagawa T, Takahashi M, Ozaki T, Watanabe Ki K, Todo S, Mizuguchi H, Hayakawa T, Nakagawara A (2002) Autoinhibitory regulation of p73 by Delta Np73 to modulate cell survival and death through a p73-specific target element within the Delta Np73 promoter. Mol Cell Biol 22(8):2575–2585
32. Buckley NE, Nic An Tsaoir CB, Blayney JK, Oram LC, Crawford NT, D'Costa ZC, Quinn JE, Kennedy RD, Harkin DP, Mullan PB (2013) BRCA1 is a key regulator of breast differentiation through activation of Notch signaling with implications for anti-endocrine treatment of breast cancers. Nucleic Acids Res 41(18):8601–8614
33. Lin J, Chen J, Elenbaas B, Levine AJ (1994) Several hydrophobic amino acids in the p53 amino-terminal domain are required for transcriptional activation, binding to mdm-2 and the adenovirus 5 E1B 55 kD protein. Genes Dev 8:1235–1246
34. Kussie PH, Gorina S, Marechal V, Elenbaas B, Moreau J, Levine AJ et al (1996) Structure of the MDM2 oncoprotein bound to the p53 tumor suppressor transactivation domain. Science 274:948–953
35. Deb SP (2003) Cell cycle regulatory functions of the human oncoprotein MDM2. Mol Cancer Res 1(14):1009–1016, Review
36. Popowicz GM, Czarna A, Holak TA (2008) Structure of the human Mdmx protein bound to the p53 tumor suppressor transactivation domain. Cell Cycle 7:2441–2443
37. Kubbutat MH, Jones SN, Vousden KH (1997) Regulation of p53 stability by Mdm2. Nature 387:299–303
38. Haupt Y, Maya R, Kazaz A, Oren M (1997) Mdm2 promotes the rapid degradation of p53. Nature 387:296–299
39. Danovi D, Meulmeester E, Pasini D, Migliorini D, Capra M, Frenk R et al (2004) Amplification of Mdmx (or Mdm4) directly contributes to tumor formation by inhibiting p53 tumor suppressor activity. Mol Cell Biol 24:5835–5843

40. Linke K, Mace PD, Smith CA, Vaux DL, Silke J, Day CL (2008) Structure of the MDM2/MDMX RING domain heterodimer reveals dimerization is required for their ubiquitylation in trans. Cell Death Differ 15:841–848
41. Okamoto K, Taya Y, Nakagama H (2009) Mdmx enhances p53 ubiquitination by altering the substrate preference of the Mdm2 ubiquitin ligase. FEBS Lett 583:2710–2714
42. Ongkeko WM, Wang XQ, Siu WY, Lau AW, Yamashita K, Harris AL, Cox LS, Poon RY (1999) MDM2 and MDMX bind and stabilize the p53-related protein p73. Curr Biol 9(15):829–832
43. Zdzalik M, Pustelny K, Kedracka-Krok S, Huben K, Pecak A, Wladyka B, Jankowski S, Dubin A, Potempa J, Dubin G (2010) Interaction of regulators Mdm2 and Mdmx with transcription factors p53, p63 and p73. Cell Cycle 9(22):4584–4591
44. Kojima T, Ikawa Y, Katoh I (2001) Analysis of molecular interactions of the p53-family p51(p63) gene products in a yeast two-hybrid system: homotypic and heterotypic interactions and association with p53-regulatory factors. Biochem Biophys Res Commun 281:1170–1175
45. Little NA, Jochemsen AG (2001) Hdmx and Mdm2 can repress transcription activation by p53 but not by p63. Oncogene 20:4576–4580
46. Wang X, Arooz T, Siu WY, Chiu CH, Lau A, Yamashita K et al (2001) MDM2 and MDMX can interact differently with ARF and members of the p53 family. FEBS Lett 490:202–208
47. Calabro V, Mansueto G, Parisi T, Vivo M, Calogero RA, La Mantia G (2002) The human MDM2 oncoprotein increases the transcriptional activity and the protein level of the p53 homolog p63. J Biol Chem 277:2674–2681
48. Schon O, Friedler A, Bycroft M, Freund SM, Fersht AR (2002) Molecular mechanism of the interaction between MDM2 and p53. J Mol Biol 323:491–501
49. Burge S, Teufel DP, Townsley FM, Freund SM, Bycroft M, Fersht AR (2009) Molecular basis of the interactions between the p73 N terminus and p300: effects on transactivation and modulation by phosphorylation. Proc Natl Acad Sci U S A 106:3142–3147
50. Mavinahalli JN, Madhumalar A, Beuerman RW, Lane DP, Verma C (2010) Differences in the transactivation domains of p53 family members: a computational study. BMC Genomics 11(Suppl 1):S5. doi:10.1186/1471-2164-11-S1-S5
51. Keyes WM, Vogel H, Koster MI, Guo X, Qi Y, Petherbridge KM, Roop DR, Bradley A, Mills AA (2006) p63 heterozygous mutant mice are not prone to spontaneous or chemically induced tumors. Proc Natl Acad Sci U S A 103(22):8435–8440
52. Yang A, Walker N, Bronson R, Kaghad M, Oosterwegel M, Bonnin J, Vagner C, Bonnet H, Dikkes P, Sharpe A, McKeon F, Caput D (2000) p73-deficient mice have neurological, pheromonal and inflammatory defects but lack spontaneous tumours. Nature 404(6773):99–103
53. Flores ER, Tsai KY, Crowley D, Sengupta S, Yang A, McKeon F, Jacks T (2002) p63 and p73 are required for p53-dependent apoptosis in response to DNA damage. Nature 416(6880):560–564
54. Taneja P, Zhu S, Maglic D, Fry EA, Kendig RD, Inoue K (2011) Transgenic and knockout mice models to reveal the functions of tumor suppressor genes. Clin Med Insights Oncol 5:235–257. doi:10.4137/CMO.S7516, Epub 2011 Jul 28. Review
55. Massion PP, Taflan PM, Jamshedur Rahman SM, Yildiz P, Shyr Y, Edgerton ME, Westfall MD, Roberts JR, Pietenpol JA, Carbone DP, Gonzalez AL (2003) Significance of p63 amplification and overexpression in lung cancer development and prognosis. Cancer Res 63(21):7113–7121
56. Flores ER (2007) The roles of p63 in cancer. Cell Cycle 6(3):300–304, Epub 2007 Feb 3. Review
57. Su X, Paris M, Gi YJ, Tsai KY, Cho MS, Lin YL, Biernaskie JA, Sinha S, Prives C, Pevny LH, Miller FD, Flores ER (2009) TAp63 prevents premature aging by promoting adult stem cell maintenance. Cell Stem Cell 5(1):64–75
58. Keyes WM, Pecoraro M, Aranda V, Vernersson-Lindahl E, Li W, Vogel H, Guo X, Garcia EL, Michurina TV, Enikolopov G, Muthuswamy SK, Mills AA (2011) ΔNp63α is an oncogene that targets chromatin remodeler Lsh to drive skin stem cell proliferation and tumorigenesis. Cell Stem Cell 8(2):164–176
59. Guo X, Keyes WM, Papazoglu C, Zuber J, Li W, Lowe SW, Vogel H, Mills AA (2009) TAp63 induces senescence and suppresses tumorigenesis in vivo. Nat Cell Biol 11(12):1451–1457

60. Aberdam D, Gambaro K, Medawar A, Aberdam E, Rostagno P, de la Forest Divonne S, Rouleau M (2007) Embryonic stem cells as a cellular model for neuroectodermal commitment and skin formation. C R Biol 330(6–7):479–484. Epub 16 Apr 2007 Apr. Review
61. Medawar A, Virolle T, Rostagno P, de la Forest-Divonne S, Gambaro K, Rouleau M, Aberdam D (2008) DeltaNp63 is essential for epidermal commitment of embryonic stem cells. PLoS One 3(10):e3441
62. Shalom-Feuerstein R, Lena AM, Zhou H, De La Forest DS, Van Bokhoven H, Candi E, Melino G, Aberdam D (2011) ΔNp63 is an ectodermal gatekeeper of epidermal morphogenesis. Cell Death Differ 18(5):887–896
63. Tomasini R, Tsuchihara K, Wilhelm M, Fujitani M, Rufini A, Cheung CC, Khan F, Itie-Youten A, Wakeham A, Tsao MS, Iovanna JL, Squire J, Jurisica I, Kaplan D, Melino G, Jurisicova A, Mak TW (2008) TAp73 knockout shows genomic instability with infertility and tumor suppressor functions. Genes Dev 22(19):2677–2691
64. Wilhelm MT, Rufini A, Wetzel MK, Tsuchihara K, Inoue S, Tomasini R, Itie-Youten A, Wakeham A, Arsenian-Henriksson M, Melino G, Kaplan DR, Miller FD, Mak TW (2010) Isoform-specific p73 knockout mice reveal a novel role for delta Np73 in the DNA damage response pathway. Genes Dev 24(6):549–560
65. Zhu J, Jiang J, Zhou W, Chen X (1998) The potential tumor suppressor p73 differentially regulates cellular p53 target genes. Cancer Res 58(22):5061–5065
66. Dohn M, Zhang S, Chen X (2001) p63alpha and DeltaNp63alpha can induce cell cycle arrest and apoptosis and differentially regulate p53 target genes. Oncogene 20(25):3193–3205
67. Melino G, De Laurenzi V, Vousden KH (2002) p73: friend or foe in tumorigenesis. Nat Rev Cancer 2(8):605–615, Review
68. Melino G (2011) p63 is a suppressor of tumorigenesis and metastasis interacting with mutant p53. Cell Death Differ 18(9):1487–1499, Review
69. Davison TS, Vagner C, Kaghad M, Ayed A, Caput D, Arrowsmith CH (1999) p73 and p63 are homotetramers capable of weak heterotypic interactions with each other but not with p53. J Biol Chem 274(26):18709–18714
70. Grob TJ, Novak U, Maisse C, Barcaroli D, Lüthi AU, Pirnia F, Hügli B, Graber HU, De Laurenzi V, Fey MF, Melino G, Tobler A (2001) Human delta Np73 regulates a dominant negative feedback loop for TAp73 and p53. Cell Death Differ 8(12):1213–1223
71. Stiewe T, Zimmermann S, Frilling A, Esche H, Pützer BM (2002) Transactivation-deficient DeltaTA-p73 acts as an oncogene. Cancer Res 62(13):3598–3602
72. Chan WM, Siu WY, Lau A, Poon RY (2004) How many mutant p53 molecules are needed to inactivate a tetramer? Mol Cell Biol 24(8):3536–3551
73. Yoshikawa H, Nagashima M, Khan MA, McMenamin MG, Hagiwara K, Harris CC (1999) Mutational analysis of p73 and p53 in human cancer cell lines. Oncogene 18(22):3415–3421
74. Ichimiya S, Nimura Y, Kageyama H, Takada N, Sunahara M, Shishikura T, Nakamura Y, Sakiyama S, Seki N, Ohira M, Kaneko Y, McKeon F, Caput D, Nakagawara A (1999) p73 at chromosome 1p36.3 is lost in advanced stage neuroblastoma but its mutation is infrequent. Oncogene 18(4):1061–1066
75. Knudson AG Jr (1971) Mutation and cancer: statistical study of retinoblastoma. Proc Natl Acad Sci U S A 68(4):820–823
76. Zaika AI, El-Rifai W (2006) The role of p53 protein family in gastrointestinal malignancies. Cell Death Differ 13(6):935–940, Review
77. Park BJ, Lee SJ, Kim JI, Lee CH, Chang SG, Park JH et al (2000) Frequent alteration of p63 expression in human primary bladder carcinomas. Cancer Res 60:3370–3374
78. Grisanzio C, Signoretti S (2008) p63 in prostate biology and pathology. J Cell Biochem 103:1354–1368
79. Tucci P, Agostini M, Grespi F, Markert EK, Terrinoni A, Vousden KH et al (2012) Loss of p63 and its microRNA-205 target results in enhanced cell migration and metastasis in prostate cancer. Proc Natl Acad Sci U S A 109:15312–15317
80. Koga F, Kawakami S, Fujii Y, Saito K, Ohtsuka Y, Iwai A, Ando N, Takizawa T, Kageyama Y, Kihara K (2003) Impaired p63 expression associates with poor prognosis and uroplakin III expression in invasive urothelial carcinoma of the bladder. Clin Cancer Res 9(15):5501–5507

81. Sniezek JC, Matheny KE, Burkey BB, Netterville JL, Pietenpol JA (2002) Expression of p63 and 14-3-3sigma in normal and hyperdifferentiated mucosa of the upper aerodigestive tract. Otolaryngol Head Neck Surg 126(6):598–601
82. Sniezek JC, Matheny KE, Westfall MD, Pietenpol JA (2004) Dominant negative p63 isoform expression in head and neck squamous cell carcinoma. Laryngoscope 114(12):2063–2072
83. Ebrahimi M, Wahlin YB, Coates PJ, Sjöström B, Nylander K (2006) Decreased expression of p63 in oral lichen planus and graft-vs.-host disease associated with oral inflammation. J Oral Pathol Med 35(1):46–50
84. Hibi K, Trink B, Patturajan M, Westra WH, Caballero OL, Hill DE, Ratovitski EA, Jen J, Sidransky D (2000) AIS is an oncogene amplified in squamous cell carcinoma. Proc Natl Acad Sci U S A 97(10):5462–5467
85. Di Como CJ, Urist MJ, Babayan I, Drobnjak M, Hedvat CV, Teruya-Feldstein J, Pohar K, Hoos A, Cordon-Cardo C (2002) p63 expression profiles in human normal and tumor tissues. Clin Cancer Res 8(2):494–501
86. Brunelli M, Bria E, Nottegar A, Cingarlini S, Simionato F, Caliò A, Eccher A, Parolini C, Iannucci A, Gilioli E, Pedron S, Massari F, Tortora G, Borze I, Knuutila S, Gobbo S, Santo A, Tondulli L, Calabrò F, Martignoni G, Chilosi M (2012) True 3q chromosomal amplification in squamous cell lung carcinoma by FISH and aCGH molecular analysis: impact on targeted drugs. PLoS One 7(12):e49689
87. DeYoung MP, Johannessen CM, Leong CO, Faquin W, Rocco JW, Ellisen LW (2006) Tumor-specific p73 up-regulation mediates p63 dependence in squamous cell carcinoma. Cancer Res 66(19):9362–9368
88. King KE, Weinberg WC (2007) p63: defining roles in morphogenesis, homeostasis, and neoplasia of the epidermis. Mol Carcinog 46(8):716–724
89. Tonon G, Wong KK, Maulik G, Brennan C, Feng B, Zhang Y, Khatry DB, Protopopov A, You MJ, Aguirre AJ, Martin ES, Yang Z, Ji H, Chin L, Depinho RA (2005) High-resolution genomic profiles of human lung cancer. Proc Natl Acad Sci U S A 102(27):9625–9630
90. Tonon G, Brennan C, Protopopov A, Maulik G, Feng B, Zhang Y, Khatry DB, You MJ, Aguirre AJ, Martin ES, Yang Z, Ji H, Chin L, Wong KK, Depinho RA (2005) Common and contrasting genomic profiles among the major human lung cancer subtypes. Cold Spring Harb Symp Quant Biol 70:11–24
91. Redon R, Muller D, Caulee K, Wanherdrick K, Abecassis J, du Manoir S (2001) A simple specific pattern of chromosomal aberrations at early stages of head and neck squamous cell carcinomas: PIK3CA but not p63 gene as a likely target of 3q26-qter gains. Cancer Res 61(10):4122–4129
92. Wang TY, Chen BF, Yang YC, Chen H, Wang Y, Cviko A, Quade BJ, Sun D, Yang A, McKeon FD, Crum CP (2001) Histologic and immunophenotypic classification of cervical carcinomas by expression of the p53 homologue p63: a study of 250 cases. Hum Pathol 32(5):479–486
93. Hu H, Xia SH, Li AD, Xu X, Cai Y, Han YL, Wei F, Chen BS, Huang XP, Han YS, Zhang JW, Zhang X, Wu M, Wang MR (2002) Elevated expression of p63 protein in human esophageal squamous cell carcinomas. Int J Cancer 102(6):580–583
94. Weber A, Bellmann U, Bootz F, Wittekind C, Tannapfel A (2002) Expression of p53 and its homologues in primary and recurrent squamous cell carcinomas of the head and neck. Int J Cancer 99(1):22–28
95. Nylander K, Vojtesek B, Nenutil R, Lindgren B, Roos G, Zhanxiang W, Sjöström B, Dahlqvist A, Coates PJ (2002) Differential expression of p63 isoforms in normal tissues and neoplastic cells. J Pathol 198(4):417–427
96. Wang X, Mori I, Tang W, Nakamura M, Nakamura Y, Sato M, Sakurai T, Kakudo K (2002) p63 expression in normal, hyperplastic and malignant breast tissues. Breast Cancer 9(3):216–219
97. Reis-Filho JS, Milanezi F, Amendoeira I, Albergaria A, Schmitt FC (2003) Distribution of p63, a novel myoepithelial marker, in fine-needle aspiration biopsies of the breast: an analysis of 82 samples. Cancer 99(3):172–179
98. Koker MM, Kleer CG (2004) p63 expression in breast cancer: a highly sensitive and specific marker of metaplastic carcinoma. Am J Surg Pathol 28(11):1506–1512

99. Livasy CA, Karaca G, Nanda R, Tretiakova MS, Olopade OI, Moore DT, Perou CM (2006) Phenotypic evaluation of the basal-like subtype of invasive breast carcinoma. Mod Pathol 19(2):264–271
100. Rocco JW, Leong CO, Kuperwasser N, DeYoung MP, Ellisen LW (2006) p63 mediates survival in squamous cell carcinoma by suppression of p73-dependent apoptosis. Cancer Cell 9(1):45–56
101. Zangen R, Ratovitski E, Sidransky D (2005) DeltaNp63alpha levels correlate with clinical tumor response to cisplatin. Cell Cycle 4(10):1313–1315, Epub 2005 Oct 1
102. Danilov AV, Neupane D, Nagaraja AS, Feofanova EV, Humphries LA, DiRenzo J, Korc M (2011) DeltaNp63alpha-mediated induction of epidermal growth factor receptor promotes pancreatic cancer cell growth and chemoresistance. PLoS One 6(10):e26815
103. Leong CO, Vidnovic N, DeYoung MP, Sgroi D, Ellisen LW (2007) The p63/p73 network mediates chemosensitivity to cisplatin in a biologically defined subset of primary breast cancers. J Clin Invest 117(5):1370–1380
104. Iwata T, Uramoto H, Sugio K, Fujino Y, Oyama T, Nakata S, Ono K, Morita M, Yasumoto K (2005) A lack of prognostic significance regarding DeltaNp63 immunoreactivity in lung cancer. Lung Cancer 50(1):67–73
105. Uramoto H, Sugio K, Oyama T, Nakata S, Ono K, Nozoe T, Yasumoto K (2006) Expression of the p53 family in lung cancer. Anticancer Res 26(3A):1785–1790
106. Barbieri CE, Tang LJ, Brown KA, Pietenpol JA (2006) Loss of p63 leads to increased cell migration and up-regulation of genes involved in invasion and metastasis. Cancer Res 66(15):7589–7597
107. Helton ES, Zhu J, Chen X (2006) The unique NH2-terminally deleted (DeltaN) residues, the PXXP motif, and the PPXY motif are required for the transcriptional activity of the DeltaN variant of p63. J Biol Chem 281(5):2533–2542
108. Celardo I, Grespi F, Antonov A, Bernassola F, Garabadgiu AV, Melino G, Amelio I (2013) Caspase-1 is a novel target of p63 in tumor suppression. Cell Death Dis 4:e645
109. Han S, Semba S, Abe T, Makino N, Furukawa T, Fukushige S, Takahashi H, Sakurada A, Sato M, Shiiba K, Matsuno S, Nimura Y, Nakagawara A, Horii A (1999) Infrequent somatic mutations of the p73 gene in various human cancers. Eur J Surg Oncol 25(2):194–198
110. Nomoto S, Haruki N, Kondo M, Konishi H, Takahashi T, Takahashi T, Takahashi T (1998) Search for mutations and examination of allelic expression imbalance of the p73 gene at 1p36.33 in human lung cancers. Cancer Res 58(7):1380–1383
111. Concin N, Becker K, Slade N, Erster S, Mueller-Holzner E, Ulmer H, Daxenbichler G, Zeimet A, Zeillinger R, Marth C, Moll UM (2004) Transdominant DeltaTAp73 isoforms are frequently up-regulated in ovarian cancer. Evidence for their role as epigenetic p53 inhibitors in vivo. Cancer Res 64(7):2449–2460
112. Domínguez G, García JM, Peña C, Silva J, García V, Martínez L, Maximiano C, Gómez ME, Rivera JA, García-Andrade C, Bonilla F (2006) DeltaTAp73 upregulation correlates with poor prognosis in human tumors: putative in vivo network involving p73 isoforms, p53, and E2F-1. J Clin Oncol 24(5):805–815, Epub 2005 Dec 27
113. Faridoni-Laurens L, Tourpin S, Alsafadi S, Barrois M, Temam S, Janot F, Koscielny S, Bosq J, Bénard J, Ahomadegbe JC (2008) Involvement of N-terminally truncated variants of p73, deltaTAp73, in head and neck squamous cell cancer: a comparison with p53 mutations. Cell Cycle 7(11):1587–1596
114. Uramoto H, Sugio K, Oyama T, Nakata S, Ono K, Morita M, Funa K, Yasumoto K (2004) Expression of deltaNp73 predicts poor prognosis in lung cancer. Clin Cancer Res 10(20):6905–6911
115. Wager M, Guilhot J, Blanc JL, Ferrand S, Milin S, Bataille B, Lapierre F, Denis S, Chantereau T, Larsen CJ, Karayan-Tapon L (2006) Prognostic value of increase in transcript levels of Tp73 DeltaEx2-3 isoforms in low-grade glioma patients. Br J Cancer 95(8):1062–1069
116. Castillo J, Goñi S, Latasa MU, Perugorría MJ, Calvo A, Muntané J, Bioulac-Sage P, Balabaud C, Prieto J, Avila MA, Berasain C (2009) Amphiregulin induces the alternative splicing of p73 into its oncogenic isoform DeltaEx2p73 in human hepatocellular tumors. Gastroenterology 137(5):1805–15.e1-4

117. Corn PG, Kuerbitz SJ, van Noesel MM, Esteller M, Compitello N, Baylin SB, Herman JG (1999) Transcriptional silencing of the p73 gene in acute lymphoblastic leukemia and Burkitt's lymphoma is associated with 5′ CpG island methylation. Cancer Res 59(14):3352–3356
118. Kawano S, Miller CW, Gombart AF, Bartram CR, Matsuo Y, Asou H, Sakashita A, Said J, Tatsumi E, Koeffler HP (1999) Loss of p73 gene expression in leukemias/lymphomas due to hypermethylation. Blood 94(3):1113–1120
119. Ibrahim N, He L, Leong CO, Xing D, Karlan BY, Swisher EM, Rueda BR, Orsulic S, Ellisen LW (2010) BRCA1-associated epigenetic regulation of p73 mediates an effector pathway for chemosensitivity in ovarian carcinoma. Cancer Res 70(18):7155–7165
120. Alexandrova EM, Moll UM (2012) Role of p53 family members' p73 and p63 in human hematological malignancies. Leuk Lymphoma 53(11):2116–2129
121. Daskalos A, Logotheti S, Markopoulou S, Xinarianos G, Gosney JR, Kastania AN, Zoumpourlis V, Field JK, Liloglou T (2011) Global DNA hypomethylation-induced ΔNp73 transcriptional activation in non-small cell lung cancer. Cancer Lett 300(1):79–86
122. Guan M, Chen Y (2005) Aberrant expression of DeltaNp73 in benign and malignant tumours of the prostate: correlation with Gleason score. J Clin Pathol 58(11):1175–1179
123. Díaz R, González-Sancho JM, Soldevilla B, Silva J, García JM, García V, Peña C, Herrera M, Gómez I, Bonilla F, Domínguez G (2010) Differential regulation of TP73 isoforms by 1α,25-dihydroxyvitamin D3 and survivin in human colon and breast carcinomas. Genes Chromosomes Cancer 49(12):1135–1142
124. Quon KC, Berns A (2001) Haplo-insufficiency? Let me count the ways. Genes Dev 15(22):2917–2921, Review
125. Berger AH, Pandolfi PP (2011) Haplo-insufficiency: a driving force in cancer. J Pathol 223(2):137–146. doi:10.1002/path.2800, Epub 2010 Oct 29. Review
126. Inoue K, Zindy F, Randle DH, Rehg JE, Sherr CJ (2001) Dmp1 is haplo-insufficient for tumor suppression and modifies the frequencies of Arf and p53 mutations in Myc-induced lymphomas. Genes Dev 15(22):2934–2939
127. Inoue K, Mallakin A, Frazier DP (2007) Dmp1 and tumor suppression. Oncogene 26(30):4329–4335, Epub 2007 Jan 22. Review
128. Mallakin A, Sugiyama T, Taneja P, Matise LA, Frazier DP, Choudhary M, Hawkins GA, D'Agostino RB Jr, Willingham MC, Inoue K (2007) Mutually exclusive inactivation of DMP1 and ARF/p53 in lung cancer. Cancer Cell 12(4):381–394
129. Sugiyama T, Frazier DP, Taneja P, Kendig RD, Morgan RL, Matise LA, Lagedrost SJ, Inoue K (2008) Signal transduction involving the dmp1 transcription factor and its alteration in human cancer. Clin Med Oncol 2:209–219, Epub 2008 Apr 1
130. Taneja P, Maglic D, Kai F, Sugiyama T, Kendig RD, Frazier DP, Willingham MC, Inoue K (2010) Critical roles of DMP1 in human epidermal growth factor receptor 2/neu-Arf-p53 signaling and breast cancer development. Cancer Res 70(22):9084–9094
131. Zhu S, Mott RT, Fry EA, Taneja P, Kulik G, Sui G, Inoue K (2013) Cooperation between Dmp1 loss and cyclin D1 overexpression in breast cancer. Am J Pathol 183(4):1339–1350
132. Fry EA, Taneja P, Maglic D, Zhu S, Sui G, Inoue K (2013) Dmp1α inhibits HER2/neu-induced mammary tumorigenesis. PLoS One 8(10):e77870. doi:10.1371/journal.pone.0077870

Chapter 3
Cooperation of p53 Mutations with Other Oncogenic Alterations in Cancer

Javier E. Girardini, Dawid Walerych, and Giannino Del Sal

Abstract Following the initial findings suggesting a pro-oncogenic role for p53 point mutants, more than 30 years of research have unveiled the critical role exerted by these mutants in human cancer. A growing body of evidence, including mouse models and clinical data, has clearly demonstrated a connection between mutant p53 and the development of aggressive and metastatic tumors. Even if the molecular mechanisms underlying mutant p53 activities are still the object of intense scrutiny, it seems evident that full activation of its oncogenic role requires the functional interaction with other oncogenic alterations. p53 point mutants, with their pleiotropic effects, simultaneously activating several mechanisms of aggressiveness, are engaged in multiple cross-talk with a variety of other cancer-related processes, thus depicting a complex molecular landscape for the mutant p53 network. In this chapter revealing evidence illustrating different ways through which this cooperation may be achieved will be discussed. Considering the proposed role for mutant p53 as a driver of cancer aggressiveness, disarming mutant p53 function by uncoupling the cooperation with other oncogenic alterations, stands out as an exciting possibility for the development of novel anti-cancer therapies.

Keywords Mutant p53 • Gain of function • Post-translational modifications • Cancer-associated mutations • Oncogenic signaling • Metastasis

In retrospective, the way in which p53 was discovered was already telling us much about what has only begun to emerge in the last decade on the role of p53 point

---

J.E. Girardini
Molecular Oncology Group, Institute of Molecular and Cell Biology of Rosario, IBR-CONICET, Rosario, Argentina

D. Walerych • G. Del Sal (✉)
Laboratorio Nazionale CIB (LNCIB), Area Science Park, Trieste, Italy

Dipartimento di Scienze della Vita, Università degli Studi di Trieste, 34127 Trieste, Italy
e-mail: delsal@lncib.it

S.P. Deb and S. Deb (eds.), *Mutant p53 and MDM2 in Cancer*,
Subcellular Biochemistry 85, DOI 10.1007/978-94-017-9211-0_3,

mutants in human cancer. p53 was originally identified as an interactor of SV40 T-antigen in transformed cells, or as a frequently abundant protein in cancer cells, however, most of those original evidences was obtained from samples that contained missense mutations in the p53 gene [1]. Therefore, the p53 field started with p53 point mutants, most of which, according to the current hypothesis, may have been endowed with pro-oncogenic functions able to actively cooperate in the acquisition of aggressive tumor phenotypes. Of course, in the early days of p53 research, dominated by the enthusiasm to understand the tumor suppressor function of wt p53, the relevance of p53 mutants on tumor progression was not evident. It took further 20 years of research to start figuring out the picture.

Yet, the initial discovery of p53 in virus transformed and tumor cells was somehow anticipating another key aspect of mutant p53 oncogenic function: p53 mutants best reveal its full armory when combined with other cancer-related alterations including genetic lesions but also aberrantly regulated intracellular signaling circuits or even abnormal clues from a transforming microenvironment. The presence of point mutations radically alters p53 function causing much more than simple loss of wt function [2]. On one hand, mutant proteins may exert a dominant negative effect by binding and inhibiting wt p53, while on the other, point mutants may acquire novel activities that actively cooperate with tumor progression, collectively known as Gain of Function (GOF) activities. However, strict GOF refers to novel biochemical properties that are independent from wt p53 function. Of course, both dominant negative and GOF activities collaborate with tumor progression albeit in fairly different ways. Even if the dominant negative effect is obviously expected to be beneficial for tumor progression, it is worth noting that the ability of mutant p53 to promote tumor aggressiveness was demonstrated in elegant in vivo models for bonafide GOF lacking wt p53 [3]. In this respect, mutant p53 may be considered as a pleiotropic factor that eliminates wt p53 tumor suppressor function and simultaneously activates several mechanisms of aggressiveness. Besides, in human tumors, loss of heterozygosity is a frequent event that renders tumor cells devoid of wt p53 by eliminating the remaining wt allele.

Our current understanding of mutant p53 GOF describes mutant p53 as a protein that affects different aspects of cell behaviour by physically interacting with protein partners, thereby altering their normal function [4]. Therefore, in a defined cell context, displaying a repertoire of mutant p53 interactors under the effect of a particular combination of signaling pathways, it could be expected that mutant p53 may simultaneously activate several complementary oncogenic mechanisms that will eventually promote the reprogramming of cell behavior. According to this idea, the presence of full active mutant p53 may be regarded as a pivotal event in tumor progression that tips the balance towards the development of overt aggressive and metastatic tumor phenotypes. The diversity of mechanisms engaged by mutant p53 offers several opportunities to crosstalk with other oncogenic circuits. Oncogenic alterations may connect with the mutant p53 network at different levels. For example, altered signaling pathways may induce specific combinations of post-translational modifications on mutant p53 that enhance or inhibit its activity, stability or subcellular localization. Alternatively, mutant p53 may physically interact with deregulated protein partners that are inactive or absent in normal cells. Also, downstream effectors activated by mutant p53 function may synergize with other oncogenic mechanisms.

In this scenario, understanding how other cancer-related alterations impact on the mutant p53 network stands out as key question in cancer biology, with enormous consequences on the clinics. In this chapter we will revise the current knowledge on mutant p53 GOF trying to underline connections between oncogenic alterations and mutant p53 function.

Mutant p53 and Hyperactivated Ras Signaling

The cooperation with aberrantly activated Ras signaling is perhaps the best documented interaction of the mutant p53 network with other cancer-related alterations. The Ras family includes several small GTPases that have a complex role as upstream modulators of signaling pathways [5]. In human cancer, the genes encoding some of its members, namely *HRAS*, *KRAS*, and *NRAS*, frequently harbor missense mutations that inhibit the GTPase activity. Such mutations leave the enzyme in a constitutively activated state, which was correlated with the ability of these mutants to cooperate with cell transformation as a consequence of deregulated signaling. Genetic lesions on *RAS* genes vary according to tumor type, being *KRAS* mutations the most frequent, accounting for up to 60 % in pancreatic tumors [6]. However, mutation frequencies in *H-RAS* and *N-RAS* are remarkably elevated in experimentally induced tumors [7].

The ability of mutant Ras to promote tumorigenesis in cooperation with other genetic lesions has been widely documented in vitro and in vivo. In particular, the ability of activated Ras protein to drive transformation of cultured cells in the absence of wt p53 activity was demonstrated several years ago. Those evidences, among others, were pivotal in defining the mechanisms of multistep carcinogenesis, by showing that hyperproliferative signaling in the context of impaired mechanisms of proliferation control, such as growth arrest, apoptosis and senescence, may act as a major driving force of cell transformation. However, under the light of mutant p53 GOF, several other ways in which the crosstalk between deregulated Ras signaling and the mutant p53 network may contribute to the etiology and the progression of the disease are being unveiled.

The connection between mutant p53 and Ras signaling was initially suggested in experiments showing that p53 point mutants cooperated with constitutively activated Ras to transform normal rat embryo fibroblasts [8, 9]. Although suggestive of an acquired oncogenic function, these evidences were not conclusive since the effects observed may be ascribed to inhibition of endogenous wt p53 from rat fibroblasts by mutant p53. Later, experiments performed in embryo fibroblasts from mutant p53 knock-in mice, lacking the wt allele, confirmed the ability of mutant p53 to enhance activated Ras-driven transformation as a bonafide new acquired activity of mutant p53 [10, 11]. The study of this cooperation was further extended in vivo, in mouse models that combined tissue specific expression of activated Ras with p53 point mutants, which revealed the deleterious effect that mutant p53 may exert under particular conditions. In a revealing study, a model for pancreatic ductal carcinoma (PDA) was developed, where expression of knocked-in mouse *p53R172H* and *KRasG12D* was restricted to pancreas progenitor cells [12] . Ras activation was enough to promote tumorigenesis, however, the presence of mutant p53 dramatically accelerated tumor onset and reduced median survival. Notably, mice carrying activated Ras in combination with a mutant p53 allele developed aggressive and metastatic disease at high frequency, a feature that was not observed in mice expressing only activated Ras or mutant p53. Even if those mice harbored wt p53, the observation of consistent loss of heterozygosity that eliminated the wt allele gave strong support to the notion of mutant p53 GOF. It is interesting to note that more than 90 % of human PDAs carry point mutations in *KRAS* gene while 75 % of them harbor point mutations in *TP53* [13] . Despite this high frequency of *KRAS* mutation, mice expressing only activated Ras failed to recapitulate the aggressive characteristics of the human disease which are instead readily observed in tumors from mice carrying both activated Ras and mutant p53. Therefore, these evidences point out at *KRAS* mutation as an initiating event and suggest that later, during disease progression, mutant p53 is required to develop a full blown metastatic phenotype.

Later, a model for Non Melanoma Skin Cancer (NMSC) gave further support to the cooperation between Ras and mutant p53 but more important, provided unambiguous evidences that the observed phenotypes depended upon acquired activities of mutant p53 that are different from the ability to inhibit wt p53. More than 40 % of squamous cell carcinomas of the skin (SCCs), which account for the vast majority of NMSC, carry missense mutations in the p53 gene [14, 15]. *RAS* genes are mutated in 10–30 % of human skin SCCs [16, 17], however, in mouse models of chemically induced skin carcinogenesis, *RAS* mutations markedly increase up to 90 % of cases [7].

A model based on inducible expression of knocked-in p53 *p53R172H* and *KRasG12D* in the skin through a recombination-based strategy was used to study the cooperation between both genetic lesions [18]. Exclusive expression of mutant p53 was achieved by simultaneous deletion of the remaining wt allele by Cre-mediated recombination and those mice were compared with mice harboring deletion of both p53 alleles. Again, simultaneous expression of activated Ras and mutant p53 was followed by enhanced tumor development and increased metastasis frequency comparing with expression of activated Ras or mutant p53 alone. Moreover, close characterization revealed that tumors developed in mice bearing mutant p53 were undifferentiated spindle cell carcinomas expressing keratin markers associated with malignant progression that were not found in p53 null tumors. Noteworthy, in both the PDA and MNSC models, tumors bearing activated Ras and mutant p53 showed elevated levels of genomic instability, including pronounced centrosome amplification and enhanced aneuploidy, a feature that was absent in tumors lacking the mutant p53 allele. These observations are in line with other reports that described the ability of mutant p53 to enhance genomic instability [19–21], and suggest that Ras signaling is relevant for this GOF activity.

What the effects of hyperactivated Ras signaling on mutant p53 actually are is still unclear. The complex network of pathways activated downstream Ras makes it difficult to discern which players may be involved in the cooperation with mutant p53 in a particular cell type. One possibility is that activation of Ras signaling contributes to enhanced mutant p53 phosphorylation which may in turn enhance protein activity and/or stability. Indeed, activated Ras induces phosphorylation of ectopically expressed p53R280K on Ser33 and Ser46, while inhibition of kinases downstream of Ras reduces phosphorylation of endogenous p53R280K on S46 in MDA-MB-231 breast cancer cells [11]. Similarly, activated Ras induces phosphorylation of S/T-P sites on murine p53R172H. Inhibition of MEK1/2 also reduced phoshporylation on Ser6 and Ser9 [22]. Consistently, impaired phosphorylation on these sites leads to reduced mutant p53 oncogenic function by uncoupling the communication between the mutant p53 network and other deregulated signaling pathways as will be further discussed.

On the other hand, mutant p53 may contribute to modulate the effect of HRas signaling on gene expression. In fact distinct mutant p53 classes, structural and contact, were found to affect Ras oncogenic signaling in different ways. While contact mutants, such as R248Q or R273H, activate a transcriptional program dependent on NF-kB that is downstream of Ras, conformational p53 mutants R175H or H179R inhibit BTG2 protein that negatively regulates HRas. Therefore, even though the mechanisms may be different for specific p53 mutants and cell backgrounds, mutant p53 GOF still converges on activating Ras signaling and its phenotypic effects [23].

Mutant p53 Stabilization: How to Alleviate Inherent Instability?

As the initial evidences showing elevated mutant p53 levels in tumors suggested, protein stabilization stands out as a central aspect of mutant p53 oncogenic function. Mutant p53 is highly expressed in human tumors and cancer-derived cell lines, in fair contrast to wt p53 which is hardly detectable in unstressed cells and whose expression becomes transiently elevated only in response to distinct signals [24]. These high protein levels are thought to be responsible for large part of mutant p53 oncogenic function, since depletion of mutant p53 reduces the aggressive features associated to GOF, such as proliferation, apoptosis inhibition, migration, invasion and metastatic potential [4]. Accordingly, precocious stabilization of mutant p53 in knock-in mice is associated with reduced survival and enhanced aggressiveness [25, 26].

A prominent difference between wt and mutant p53 that became soon evident is that mutant p53 stability dramatically increases in non stimulated cell lines, approaching a half life of several hours, comparing with approximately 30 min for wt p53. These observations suggested that point mutations may alter protein structure in a way that conferred enhanced resistance to degradation or that, having lost the ability to induce transcription of *MDM2*, p53 mutants may be more stable as a consequence of reduced expression of its main E3 ubiquitin ligase. Even if these hypothesis can not be ruled out for all p53 mutants, for those cancer-related point mutants that have been studied the situation seems to be different. A growing body of evidences has now established that most p53 mutants are inherently unstable proteins in non-transformed cells, which however become extremely long-lived in tumor cells. Perhaps the most vivid manifestation of this phenomenon comes from knock-in mice, were mutant p53 protein levels are barely detectable in normal tissues but are elevated exclusively in tumors [10, 27], implying that the inability to induce *MDM2* transcription or the mutation per se were not responsible for its stabilization. Moreover, mutant p53 levels in normal tissues of knock-in mice are increased upon stimuli that stabilize wt p53, such as ionizing radiation and genotoxic insult [25, 26], suggesting that both wt and mutant p53 protein stability may be regulated by similar mechanisms. How do tumor cells then manage to stabilize mutant p53? The lessons from mouse models, instead, are consistent with the idea that other alterations during malignant transformation contribute to mutant p53 stabilization. In light of the proposed role for mutant p53 as a pro-metastatic factor, understanding how other frequent alterations in human tumors impinge on mutant p53 stabilization may unveil critical hubs in the mechanisms underlying tumor aggressiveness and may highlight points for therapeutic intervention.

Our knowledge on the mechanisms that regulate mutant p53 degradation is still fragmentary, nevertheless, accumulating evidences have shown that it may be targeted for ubiquitin-dependent 26S proteasomal degradation by MDM2 and CHIP (Fig. 3.1). As for wt p53, MDM2 seems to be a major determinant of mutant p53 levels. Indeed, compelling evidences from knock-in mice where expression of *p53R172H* was combined with *Mdm2* deletion showed that absence of Mdm2 leaded to early stabilization of mutant p53 in normal tissues, confirming that this E3 ubiquitin ligase is a physiologic regulator of mutant p53 levels [25]. Consequently, early mutant p53 stabilization in double mutant mice was associated with a strikingly elevated metastasis frequency and reduced survival. MDM2 is frequently overexpressed in human tumors and correlations between MDM2 protein levels and clinical prognosis were described in some human tumors [28, 29]. Taking into consideration the ability of MDM2 to target both wt and mutant p53 for degradation opposing effects may be expected for tumor cells with deregulated MDM2. In addition, it should be considered that other p53-independent functions of MDM2 were also reported that make even more difficult to understand its actual role in tumorigenesis [29–31]. In the case of tumors with mutant p53, even in conditions were p53 point mutants are hyperstable, they are still susceptible to degradation provided that the whole mechanism is reactivated, as suggested by experiments in cell lines, were supraphysiologic levels of ectopic MDM2 readily induced ubiquitin-dependent mutant p53 degradation [32–34]. These observations may suggest that tumor cells harboring mutant p53 should avoid high expression of MDM2 since this would reduce the levels of this potent pro-oncogenic factor. However, no such straightforward correlation has been described. Instead, some evidences suggest that aggressive tumor cells may easily combine elevated MDM2 levels with hyperstable mutant p53 by activating mechanism that selectively protect mutant

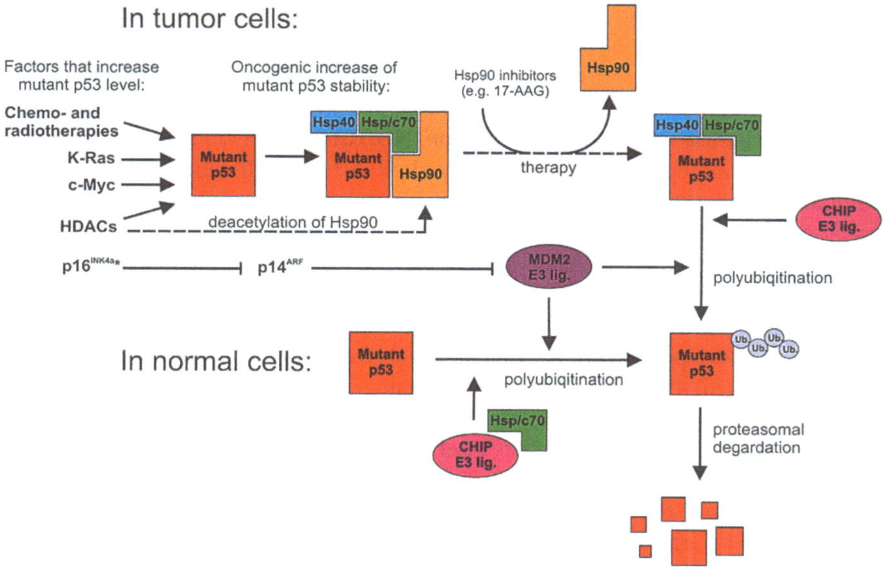

Fig. 3.1 Mechanisms of oncogenic increase of the mutant p53 level. In normal cells mutant p53 is kept at low levels due to lack of oncogenic stimuli and effective action of its main E3 ubiquitin ligases (E3 lig.) – Mdm2 and CHIP (the co-chaperone of Hsc70 or its stress-induced paralog – Hsp70). In tumor cells various signaling pathways converge on increasing the mutant p53 level. The increased level of mutant p53 (especially unfolded, structural mutants) attracts the Hsp40-Hsc/p70-Hsp90 chaperone machinery, that in attempt to reactivate the mutant p53, further stabilizes it. Hsp90 inhibitors, that dissolve the complex, allows the action of CHIP and Mdm2 E3 ubiquitin ligases, that access and target mutant p53 for degradation. (*) $p16^{INK4a}$ is often lost in tumors, what allows $p14^{ARF}$ ($p19^{ARF}$ in mice) to downregulate Mdm2 and increase the mutant p53 level with pro-oncogenic consequences

p53 proteins from degradation [35]. By doing so, tumor cells may also benefit from p53-independent tumor promoting activities of MDM2.

One such mechanism was proposed, which is based on the association of p53 point mutants with HSP90 chaperone machinery [34]. Tumor cells make use of the chaperone system in order to cope with the persistent proteotoxic stress generated by hypoxia, acidosis, high levels of reactive oxygen species and protein misfolding among other altered biochemical conditions that are present during malignant transformation [36]. Several reports have shown that mutant p53 engages in stable complexes with the HSP90 machinery, which includes HSP90, HSP70 and other co-chaperones, that protect it from degradation [37–40], and that pharmacological inhibition of HSP90 by Geldanamycin or 17AAG is able to reduce mutant p53 stability [34, 40]. Surprisingly, MDM2 and CHIP were shown to be inactivated by recruitment into the same complexes, thereby preventing degradation. Disruption of these complexes or inhibition of chaperone activity releases both E3 ubiquitin ligases from inhibition and triggers mutant p53 degradation [34, 35]. This connection between MDM2 and the chaperone system opens several possibilities to define pharmacological strategies aimed at inducing selective degradation of mutant p53 in tumor cells.

An interesting example is the ability of some histone deacetylase inhibitors (HDACi) to induce mutant p53 degradation. HDACi are being intensively studied as promising chemotherapy drugs since they elicit different anticancer responses with a remarkable specificity for tumor cells, and some of them are undergoing clinical trials. At least three HDACi have been reported to reduce mutant p53 levels in cell lines, FR901228, Trichostatin A (TSA) [41] and suberoylanilide hydroxamic acid (SAHA) [42], which however were proposed to act through different mechanisms. SAHA inhibits HDAC6, which normally activates HSP90 by promoting deacetylation of K294. Upon SAHA treatment, MDM2 and CHIP may be released from inhibition by HSP90 complex and induce mutant p53 degradation [42]. Nevertheless, an inhibitory effect of HDAC inhibitors on transcription of *TP53* was described, that may also cooperate with the observed effects [43]. In the case of FR901228 and TSA, even if they may counteract HSP90 activity at rather high concentrations, the effect on mutant p53 levels was observed at lower concentrations and was accompanied by transcriptional induction of *p21* and *MDM2* and conformational changes in mutant p53, suggestive of reactivation of wt-like functions [41].

Recent evidences suggest that establishment of oncogenic signaling circuits through oncogene activation or tumor suppressor inactivation may also provide mechanisms to stabilize mutant p53. For example, homozygous *p53R172H* knock-in mice carrying an allele for constitutively active KRas showed a selective increase of mutant p53 levels in hyperplastic lesions in the lungs before tumor onset. Conversely, this was not observed in mice with mutant p53 without Ras activation, or mice with activated Ras in a wt p53 background [26]. Notably, mutant p53 was expressed in all tumors from mice with activated Ras, but only in 75 % of spontaneous tumors harboring only p53R172H. Even though no differences in survival were observed, mutant p53 stabilization in mice with activated KRas was associated with a more severe phenotype with higher frequency of advanced and metastatic carcinomas. Similarly, *c-Myc* overexpression increased mutant p53 levels in *p53R172H* heterozygous knock-in mice carrying *c-Myc* under the control of the immunoglobulin heavy chain Eμ enhancer [26]. The *Eμ-Myc* transgene induces B-cell lymphomas and overexpression of mutant p53 was observed in all lymphomas carrying *Eμ-Myc* in mutant p53 heterozygous mice, which consistently showed loss of heterozygosity at the p53 *locus*. This effect was not appreciated in homozygous mutant p53 knock-in mice because Eu-Myc is embryonic lethal [44]. Specific upregulation of mutant p53 seems to be selected in that model, since *Eμ-myc* $p53^{+/-}$ mice failed to express any p53 due to loss of the remaining allele. These effects of oncogene activation may be rationalized considering that hyperproliferative signals that block MDM2 to induce wt p53 stabilization in normal cells may have a similar effect in tumor cells were p53 is mutated.

Loss of tumor suppressor $p16^{INK4a}$ also showed a positive effect on mutant p53, as mutant p53 levels were increased in normal tissues and tumors of homozygous *p53R172H* knock-in mice carrying a deletion in both $p16^{INK4a\ -/-}$ alleles [25], and protein stabilization was correlated with increased metastatic potential comparing with *p16*$^{INK4a -/-}$ p53$^{-/-}$ mice [26]. In this case, the observed effect may be a consequence of the released inhibition of Cyclin-D1/Cdk4 from p16^{INK4a}, that ultimately leads to Rb phosphorylation, E2F activation and transcriptional induction of the Mdm2 inhibitor p19^{ARF}.

In line with the idea that similar mechanisms may induce wt and mutant p53 stabilization it has been shown that doxorubicin and γ-radiation treatment results in increased mutant p53 levels in normal tissues of homozygous *p53R172H* knock-in mice [25, 26]. For mutant p53 mice treated with γ-radiation a significant decrease in survival was also observed comparing with treated *p53*$^{-/-}$ mice or untreated mutant p53 mice [26]. These observations claim that depending on the presence of mutations on *TP53* different clinical outcomes may be expected in patients treated with therapies based on p53 activation, with deleterious consequences in cases expressing p53 point mutants. Notably, treatment with a ROS scavenger was able to significantly reduce mutant p53 stabilization upon γ-radiation treatment in *p53R172H* knock-in mice [26], implying that oxidative stress generated by radiation may act a stimulus for mutant p53 stabilization. Oxidative stress conditions are common in tumors as a result of deregulated metabolic processes, therefore, alterations that predispose tumor cells to high levels of ROS should be consider as potential factors cooperating with mutant p53 stabilization. In agreement with the evidences from knock-in mice, mutant p53 levels where increased upon exposure to ionizing radiation in zebrafish embryos. However, protein levels were higher in embryos carrying M214K mutation comparing with wt ones, and persisted for remarkably longer periods [45]. Notably, Mdm2 downregulation was enough to induce mutant p53 accumulation but instead showed no effect on wt embryos.

Activation of Mutant p53 Function by Chronic Alteration of Intracellular Signaling Circuits

The ability to reprogram cell behavior is essential for tumorigenesis and endows tumor cells with the extraordinary phenotypic plasticity that characterizes metastatic tumors. This ability is to a large extent supported by the establishment of aberrant signaling circuits through subversion of physiologic pathways. Cancer cells often redirect the functionality of signaling pathways to sustain elevated proliferation rates and avoid cell death, but they must also remain highly responsive in order to cope with a changing microenvironment.

A major role in signal transduction is played by post-translational modifications since they allow rapid and reversible changes in protein function in response to a dynamic array of extra and intracellular signals. Wt p53 function is regulated by complex combinations of post-translational modifications including phosphorylation, acetylation, ubiquitination, methylation, sumoylation and neddylation [24]. On the contrary, we know very little about the post-translational modifications present in mutant p53 and the enzymes responsible for them. Some post-translational

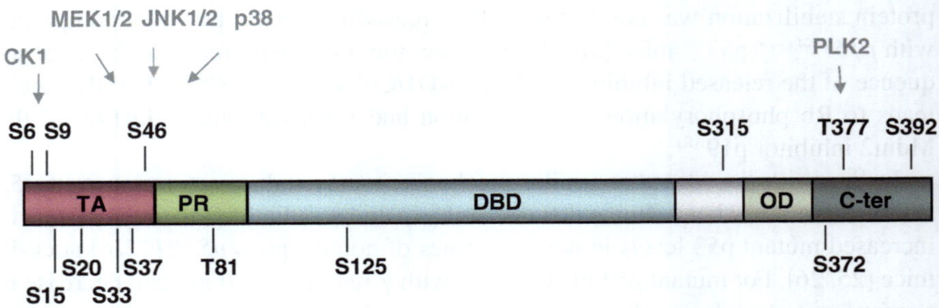

Fig. 3.2 Phosphorylation of p53 point mutants. Schematic representation of the p53 protein showing the individual domains: *TA* transactivation, *PR* proline-rich, *DBD* DNA binding domain, *OD* oligomerization domain, *C-ter* carboxi terminus. Residues that were reported to be phosphorylated in cell lines or tumor samples are shown in *black* (see text for references). On top of the protein are indicated those residues for which evidences showing an effect on mutant p53 activity were reported. For the other residues a regulatory role was not explored yet. In *grey* are shown some kinases that were proposed to phosphorylate the indicated residues basing on experimental evidences

modifications on mutant p53 were actually described both in cultured cells and in human tumors, particularly phosphorylation [46, 47] ubiquitination [33, 34] and acetylation [46, 48], however, we still lack a complete knowledge on how post-translational modification codes affect protein function.

Several residues on mutant p53 were shown to be phosphorylated (Fig. 3.2) and for some of them the modification affects protein function. For example, impairment of phosphorylation on S6, S9, S46 and S315 was associated with reduced pro-migratory activity [11, 22]. Also, PLK2 was shown to phosphorylate mutant p53 on several sites including T377 [49]. Interestingly, phosphorylation at this particular site was not described before neither in wild-type nor mutant p53. Through these modifications PLK2 enhances the transcriptional activity of the mutant p53/NF-Y complex, leading to aberrant expression of target genes increasing cell proliferation and drug resistance. In addition, PLK2 itself was identified as a mutant p53 target gene induced upon treatment with chemotherapeutic drugs, thereby establishing a positive feedback loop that contributes to amplify mutant p53 functions. This connection not only uncovers the complexity underlying the regulation of mutant p53 but also points out at PLK2 as an interesting player in the oncogenic circuits established in tumor cells. Nevertheless, the role of PLK2 in cancer biology deserves more investigation. Even if some reports support a PLK2 pro-oncogenic function [50, 51]. PLK2 promoter is silenced by hypermethylation in B-cell malignancies and ovarian cancer [52, 53], thereby suggesting that the role of PLK2 in tumorigenesis may be different according to the molecular context.

Phosphorylation on S392, instead, may play a negative regulatory role on mutant p53 function, since it reduced transformation potential and chemoresistance in vitro [54]. Conversely, mutant p53 was shown to be frequently hyperphosphorylated on

S392 in tumor-derived cell lines [46] as well as in squamous cell carcinoma (SCC) [55], actinic keratosis, Bowen's disease, basal cell carcinoma [56] and transitional cell carcinoma (TCC) [57]. Moreover, a correlation between S392 phosphorylation and poor clinical outcome was observed for SCC [55] and in the case of TCC an association between S392 phosphorylation and histological tumor grade and tumor stage was found [57], arguing for a complex role for mutant p53 phosphorylation in different cell contexts or stages of tumor development. Interestingly, phosphorylation of some residues is readily detected even in the absence of stimuli that are normally required for wt p53 phosphorylation. These modifications are attenuated upon downregulation or pharmacological inhibition of kinases [11, 22], suggesting that in tumor cells some signaling pathways are chronically activated, resulting in persistent protein modification. Therefore, it seems logical to speculate that alterations in the enzymes that catalyze these modifications will affect mutant p53 oncogenic function. Nonetheless, little is known about potential connections between altered function of kinases or phosphatases acting on mutant p53 and its oncogenic function.

In addition to enzymes directly responsible for protein modification, alteration of signaling pathways at other levels may also favor mutant p53 function. As discussed above, oncogenic Ras activates several different pathways that may ultimately lead to enhanced mutant p53 phosphorylation. Yet, several other mechanisms may be envisioned, including altered function of membrane or intracellular receptors as well as downstream transducers. A remarkable example is represented by Pin1, a peptidyl-prolyl isomerase belonging to the parvulin family, which cooperates with mutant p53 to foster tumor aggressiveness [11]. Pin1 catalizes the isomerization of the intervening peptide bond in phosphorylated S-P or T-P motifs (S/T-P motifs) that is otherwise restrained by the cyclic structure of P residue [58, 59]. Upon isomerization local conformational changes are induced that may have different consequences depending on the substrate, including altered function, stability or subcellular localization [60]. A unique feature of Pin1 is that it is the only prolil isomerase able to bind its substrates upon phosphorylation of S or T, which renders S/T-P motifs unavailable for other prolil isomerases and further blocks spontaneous isomerization. Moreover, the action of some kinases like CDKs [61] or MAPKs [62] and phosphatases as PP2A [63] depends on the isomerization state, giving the opportunity to Pin1 to regulate also the phosphorylation status of some substrates. The ability to act simultaneously on different proteins, transducing complex combinations of phosphorylation signals into functional changes, allows Pin1 to act as a global modulator of cell behavior. Despite being involved in several aspects of normal cell physiology, including cell cycle regulation, mitosis and RNA polymerase function in the last decade Pin1 has emerged as a critical factor in tumorigenesis [64]. Pin1 is frequently overexpressed in human tumors [65] and mounting evidences have demonstrated its ability to amplify oncogenic mechanisms [64, 66].

The simultaneous presence of mutant p53 and abnormally elevated levels of Pin1 endows tumor cells with the unique opportunity to establish a molecular axis that exploits deregulated phosphorylation signaling to activate downstream mechanisms of aggressiveness [11]. Upon phosphorylation on S/T-P motifs, Pin1 binds mutant p53 and amplifies several aspects of its oncogenic function in vitro and in vivo, including, cooperation with oncogenic Ras in cell transformation in mouse embryo fibroblasts and enhancement of migration and invasion in breast cancer cell lines. Noteworthy, Pin1 also binds wt p53 at the same S/T-P motifs but promoting in that case protein stabilization and apoptotic response, further underlying that wt and mutant p53 may share similar regulatory mechanisms [67–69].

The cooperation between Pin1 and mutant p53 amplifies al least two independent but complementary mechanisms. On one hand Pin1 enhances complex formation between p63 and mutant p53 [11], blocking transcriptional activation of the anti-metastatic p63 targets *CCNG2* and *SHARP1* [22]. On the other, Pin1 promotes the ability of mutant p53 to activate a transcriptional program that promotes aggressiveness. This program includes 10 genes (Pin1/mutant p53 signature) whose expression in primary human breast tumors is correlated with reduced survival and metastasis development and some of them are directly involved in the pro-migratory function of mutant p53 [11]. Alterations leading to elevated Pin1 levels and p53 missense mutations may therefore be regarded as pivotal events in tumor progression that prime tumor cells to be definitively engaged in the development of aggressive phenotypes. A prominent role in this process is played by the acquisition of *RAS* activating mutations since Ras signaling collaborates with the Pin1/mutant p53 interaction by enhancing phosphorylation of S/T-P sites. Arguing for a role of the Pin1/ mutant p53 axis as a driver of aggressiveness in human tumors, analysis of a breast cancer patient cohort showed that, while Pin1 overexpression had no prognostic value, the combination of elevated Pin1 levels with the presence of p53 missense mutations correlated with poor clinical outcome and behaved as an independent prognostic factor [11]. Further supporting the notion that mutant p53 depends on Pin1 to fully unleash its oncogenic potential in vivo, in *p53R172H* knock-in mice, lack of Pin1 was correlated with an increase in survival and a marked alteration in tumor spectrum characterized by absence of carcinomas.

Intriguingly, Pin1 overexpression did show a correlation with poor clinical outcome in other cancer types independently of p53 mutations, such us lung [70], prostate [71] and oral squamous cell carcinoma [72], however, p53 mutation was not considered in those studies. These differences in the clinical significance of Pin1 overexpression may be rationalized considering the provocative hypothesis that in breast cancer the mutant p53 network may be the major determinant of aggressiveness among the oncogenic mechanisms amplified by Pin1, while in other tumor types, other mutant p53-independent mechanisms may take over. Also, when considering the clinical relevance of Pin1 deregulation it should be kept in mind that Pin1 may also collaborate with mechanisms of tumor suppression as for example p53- or p73- induced apoptosis [67, 73] as well as degradation of CMYC [74] and CCNE [75]. Therefore it is conceivable that depending on the balance between pro-oncogenic and anti-oncogenic pathways targeted by Pin1 that are active in a particular cell context, Pin1 overexpression may also reduce cell proliferation or tumorigenic potential.

Capturing Signals to Drive Aggressiveness

In addition to the activation of cell-autonomous programs tumor cells receive signals from their microenvironment that are decisive for their ability to survive, proliferate and become invasive [76]. Recent evidences have revealed that mutant p53 can also act at this level by redefining the meaning of particular signals or by rendering tumor cells hyper responsive to positive stimuli (Fig. 3.3).

An unexpected link between TGF-beta signaling and mutant p53 was described which endows tumor cells with enhanced metastatic capability. The TGF-beta family has multiple roles in normal physiology and contributes with tumor suppression in epithelial cells by inducing cell cycle arrest or apoptosis [77]. During tumor progression however, TGF-beta signaling may switch to a pro-metastatic role through mechanisms that are not fully understood. In part, this switch depends on the activation of genes involved in migration and invasion. Upon binding to a specific membrane receptor, TGF-beta ligands trigger phosphorylation of SMAD proteins that translocate into the nucleus and regulate gene expression in combination with other transcription factors and regulators. According to cell type and context, complex regulatory networks are established that turn on different transcriptional programs.

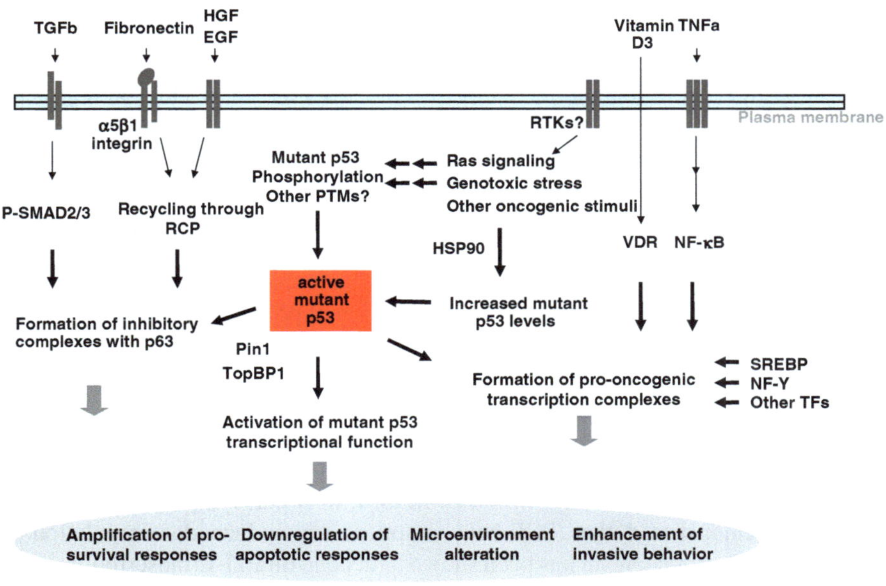

Fig. 3.3 Cooperation between mutant p53 and other alterations in cancer. Summary of the proposed connections between the mutant p53 network and other frequently altered pathways in human cancer, as discussed in the text. Membrane receptors for the indicated signaling molecules are depicted in *dark grey*, the plasma membrane is represented by a *double line. RTKs* Receptor Tyrosine Kinases, *PTMs* post-translational modifications, *P-SMAD2/3* phosphorylated SMAD2 o SMAD3

Remarkably, downregulation of endogenous mutant p53 severely impaired TGF-beta1 induced migration and invasion in vitro and reduced the metastatic potential of human tumor-derived cell lines in mice without affecting the expression of the TGF-beta pro-metastatic transcriptional program [22]. These findings suggest that mutant p53 may be necessary to allow TGF-beta1 pro-metastatic action. This functional connection relies on the ability of mutant p53 to interact with SMAD2 and p63, forming a ternary complex that blocks the role of p63 as an anti-metastatic factor. As a consequence, transcriptional induction of p63 target genes *SHARP1* and *CCGN2* is impaired which in turn promotes cell migration. Outstandingly, low expression of these two genes was found to correlate with higher risk of recurrence in breast cancer patients and was proposed as a prognostic tool. The mechanism described also collaborates to understand the cooperation between mutant Ras and TGF-beta1 in promoting metastasis [78], since complex formation requires phosphorylation of mutant p53 on S6 and S9, which is enhanced by Ras signaling. Intriguingly, SMAD proteins seem to interact with p63 in the absence of mutant p53 [22], suggesting that these complexes may be involved in other functions. In the presence of p53 mutants, however, complex formation serves to inactivate p63-dependent transcription. Therefore, mutant p53 changes the biological significance of SMAD/p63 interaction, releasing tumor cells from the anti-metastasic action of p63 and paving the way for TGF-beta -dependent pro-metastatic responses.

Nevertheless, the relationship between mutant p53 and TGF-beta signaling seems to be far more complex, as suggested by experiments where ectopic expression of p53R175H in p53 null H1299 cells opposed TGF-beta1 induced migration in vitro [79]. In this case, mutant p53 attenuated TGF-beta1 signaling by reducing SMAD2/3 phosphorylation and SMAD4 nuclear translocation, leading to decreased expression of TGF-beta targets like p21, PAI-1, SMAD7 as well as genes proposed to favor invasion as MMP2 and MMP9. Moreover, in prostate and breast cancer cells mutant p53 was also found to inhibit TGF-beta-induced cell migration. In this case even though SMAD-dependent signaling was enhanced, cell migration was inhibited by mutant p53-dependent downregulation of p52ShcA-ERK axis [80]. An hypothesis to conceal these apparently conflicting evidences may be proposed considering that mutant p53 may block TGF-beta induced cell cycle arrest, senescence or apoptosis, which require full activation of transcriptionally active SMAD complexes able to increase the expression of p21 and PAI-1, while simultaneously providing a novel mechanism of aggressiveness based on sequestration of p63 in a ternary complex with phosphor-SMADs. In turn, in B-cell lymphoma cell lines it has been demonstrated that TGF-beta treatment can induce downregulation of mutant p53. The mechanism has been shown to depend on E2F1 transcription factor and $p14^{ARF}$, the regulator of p53 stability [81]. Therefore, the anti-tumor activity of TGF-beta may impact back on the mutant p53 GOF, adding another layer to the complexity of TGF-beta – mutant p53 relationship.

Another interesting example of an external signal whose biological meaning may be subverted by p53 mutants is represented by Vitamin D3. Upon activation of the receptor VDR by binding to Vitamin D3 it regulates gene expression either by transcriptional induction or repression of selected targets. Vitamin D3 has attracted the attention of researchers as a leading compound for chemotherapy because it exerts pro-apoptotic and anti-cancer effects in a number of in vitro [82, 83] and in vivo [84–86] experimental models. Nevertheless, Vitamin D3/ VDR signaling may show anti-apoptotic effects as well [87, 88]. Moreover, VDR is upregulated in several human cancers [89, 90] and elevated VDR expression was shown to correlate with tumor stage [91, 92]. The recent finding that mutant p53 binds VDR and alters downstream signaling shed some light to understand this opposing biological outcomes [93]. Upon binding to VDR mutant p53 enhances nuclear import and alters gene expression by favoring Vitamin D3-mediated transcriptional activation of pro-survival genes and repression of pro-apoptotic ones. Accordingly, Vitamin D3 treatment reduced cell death and protected against doxorubicin or cisplatin cytotoxicity exclusively in cells harboring either endogenous or ectopically expressed mutant p53.

Taking into account the multiplicity of signals from the microenvironment that may be sensed by tumor cells it is conceivable that p53 mutants may use similar mechanisms in order to reinterpret the biological effect of other signals. Noteworthy, this ability calls extreme caution when considering potential therapeutic strategies, since the presence of mutant p53 may turn useless tumor suppressing treatments or even worse, transform them into tumor promoting processes.

The invasive behavior of cells may be favored by other external signals like growth factors and components of the extracellular matrix. Mutant p53 allows tumor cells to further exploit this trait by exacerbating the response to EGF, HGF and fibronectin. A central role in this process is played by integrin heterodimers which regulate the coordinated transduction of external signals through their ability to crosstalk with receptor tyrosine kinase (RTK)–activated signaling. Recent evidences have revealed that trafficking of integrin heterodimers through the endosomal pathway governs their signaling activity [94, 95]. Upon internalization of receptors, endosomal membranes act as platforms where different signaling complexes that regulate downstream signaling are conformed. For example, activated $\alpha 5\beta 1$ integrin heterodimers may form complexes with EGF receptor (EGFR) which are internalized and then recycled to the plasma membrane in a process that depends on the interaction of these complexes with Rab-coupling Protein (RCP), thereby activating AKT signaling [96]. Mutant p53 increases the rate of integrin and EGFR recycling to the plasma membrane, without affecting internalization, and as a consequence promotes random migration, loss of polarity and invasion [97]. Further contributing to an invasive phenotype, mutant p53 makes use of a similar mechanism to foster HGF-induced cell scattering. In this case, recycling of the HGF receptor MET is enhanced through an $\alpha 5\beta 1$ and RCP dependent process that results in activation of MAPK instead of AKT signaling [98]. Other RTKs such as IGF-1 and PDGF-β receptors are unable to bind RCP and they do not affected invasiveness in mutant-p53 expressing cells, despite being able to activate AKT, suggesting that integrin/RCP dependent recycling is specific for some RTKs.

While binding of receptors to RCP is required in both cases, other molecular events underlying this process remain to be elucidated. In the case of EGFR, mutant p53 selectively enhances the interaction between α5β1 and RCP but leaves unaffected the interaction between EGFR and RCP. Of note, an indirect mechanism was suggested for the effect of mutant p53, since no interaction was observed between mutant p53 and RCP. Interestingly, for both receptors the effect of mutant p53 may be ascribed to p63 inhibition, since p63 downregulation in p53 null H1299 cells phenocopies the effect of mutant p53 ectopic expression on EGF or HGF -driven migration and invasion while p63 overexpression opposes it [97, 98]. Moreover, p63 downregulation enhanced α5β1/EGRF recycling and RCP-dependent cell scattering. On the contrary, p73 downregulation did not affect random migration, cell scattering or invasion of H1299 towards EFG or HGF.

Another remarkable connection is the crosstalk between mutant p53 and the NF-κB pathway, which may also help tumor cells to become more sensitive to external signals. The family of NF-κB transcription factors is the central hub of a complex network able to promote apoptosis inhibition, proliferation as well as migratory and invasive behavior in response to a broad range of signals including growth factors an cytokines [99]. Persistent activation of NF-κB pathway is the hallmark of chronic inflammation and is also frequently found in human cancers [100]. Mutant p53 was reported to enhance transcription of *NF-κB2* by promoting binding of STAT2 and CBP as well as histone acetylation to its promoter [101, 102]. Accordingly, p52/p100 (NF-κB2) protein levels were found elevated in cells expressing mutant p53. More important, downregulation of p52/p100 severely impaired mutant p53 activities such as enhanced proliferation, migration and chemoresistance [102]. Therefore, by stimulating the ability to sense pro-oncogenic signals while at the same time taking advantage of NF-κB pro-inflammatory responses mutant p53 may endow tumor cells with an effective strategy to better exploit their microenvironment. Another interesting example of how this may occur was provided by evidences showing that mutant p53 enhances NF-κB signaling, albeit through a different mechanism, that involves p50/p65 nuclear translocation in response to TNFα [103]. Mutant p53 also contributed to inhibit TNFα induced apoptosis possibly due to enhanced pro-survival NF-κB signaling that counteracts activation of the extrinsic apoptotic pathway.

Recent findings also suggest how mutant p53 may cooperate with NF-κB signaling to transform tumor microenvironment. Mutant p53 enhanced transcription of chemokines like CXCL5, CXCL8 and CXCL12 [104], which are regulated by the NF-κB pathway and may cooperate with several aspects of tumor progression [105]. In particular for CXCL5, a role in sustaining mutant p53-induced migration was demonstrated. In addition, these results suggest the intriguing possibility that a positive feedback loop may be favored, since CXCL8 may activate NF-κB [106]. It is worth noting that other NF-κB-independent mechanisms were described through which mutant p53 promotes expression of CXCL1 and CXCL8 [107], suggesting a more profound alteration of the cytokine profile in cells harboring p53 mutants.

The evidences discussed herein are suggestive of a key role for p53 mutants in interpreting external signals and redefining the tumor microenvironment. However, the diversity of signaling pathways and mechanisms involved is remarkable and this scenario may be further complicated by novel findings. Considering the heterogeneity found in tumor cells and the multiplicity of extracellular situations that may be encountered, the ability of mutant p53 to act pleiotropically may be regarded as a redundant mechanism to ensure the phenotypic plasticity required to take advantage of a wide range of external signals

Do p53 Mutations Cooperate with Other Alterations to Reprogram Cancer Cell Transcriptome?

Mounting evidences have demonstrated that several p53 mutants have the ability to significantly alter gene expression and it has been proposed that these mutant p53-induced transcriptional profiles underlie their pro-oncogenic role. However, it proved difficult to define a mutant p53 specific transcription profile since there is little overlap between reported results, even for studies on the same p53 mutant. The multiplicity of mechanisms that mutant p53 seems to engage and the complex regulatory network that we are starting to figure out are nonetheless in agreement with a highly context dependent outcome, in terms of the actual genes that may be affected in a particular cell type. In other words, mutant p53 might always behave as a global regulator of gene expression, but other factors could have a decisive role in selecting which genes will be targeted. This hypothesis provides several ways through which other pathways frequently altered in human cancer may impinge on the mutant p53 transcriptome (Fig. 3.3).

Topoisomerase IIβ binding protein (TopBP1) was proposed to serve as a scaffold for several processes regarding DNA function and maintenance. TopBP1 interacts with mutant p53 and promotes its ability to complex with NF-Y [108]. The functional consequence of this interaction is an enhanced recruitment of mutant p53 and p300 on target promoters and increased expression of mutant p53/ NF-Y targets upon genotoxic insult. Intriguingly, TopBP1 also promotes the interaction between mutant p53 and p63 o p73 and contributes to the inhibition of the transcriptional activity of both p53 family members. Accordingly, downregulation of TopBP1 counteracted several aspects of mutant p53 function including resistance to apoptosis upon DNA damage and enhanced proliferation. Moreover, TopBP1 was found to be overexpressed in breast cancer and this alteration was correlated with poor clinical outcome [108, 109], suggesting that concomitant TopBP1 overexpression and p53 mutation may foster tumor progression through activation of a defined transcription profile.

Another example is provided by Pin1, which is frequently overexpressed in human cancers and cooperates with mutant p53 to enhance transcription of target genes by enhancing the recruitment of mutant p53 to promoters [11]. Of note, since Pin1 binding depends on phosphorylation of mutant p53 on S/T-P sites, it may represent a key connection between altered signaling pathways and mutant p53 transcriptional function. As the list of mutant p53 partners/regulators acting on transcriptional activity increases, more relevant oncogenic connections may be unveiled.

It is interesting to note that TopBP1 and Pin1 converge on their ability to enhance formation of p63/mutant p53 complexes, implying that both proteins may contribute to further modify the transcription profile through alteration of p63 activity. The ability of mutant p53 to form complexes with p63 and p73 was largely proposed as a GOF mechanism. Even if downregulation of both proteins may phenocopy mutant p53 effects under some circumstances [10, 22, 97] understanding of the precise mechanisms involved have been complicated by the presence of multiple isoforms for both p53 family members. Studies that specifically addressed the role of TA isoforms have showed that loss of TAp63 or TAp73 enhanced tumorigenesis in vivo [110, 111] arguing for their role as tumor suppressors. Moreover, mutant p53 was shown to repress the expression of several TAp63 target genes important for tumor suppression like *SHARP1*, *CCGN2* [22] and *DICER* [11], supporting the idea that inhibition of TAp63 transcriptional activity may be a pivotal GOF mechanism. However, the crosstalk between mutant p53 and p63 isoforms may still uncover more possibilities. Surprisingly, studies performed in H1299 cells, which express TAp63, showed that mutant p53 may also enhance the transcriptional activity of p63 on some target genes [112]. Furthermore, p63 and mutant p53 were co-recruited to both p63-induced and –repressed genes and p63 downregulation severely impaired mutant p53 recruitment. These evidences are suggestive of a more widespread role for the TAp63/mutant p53 complex than simple repression of TAp63 transcriptional activity.

Concerning DNp63 isoforms the situation is less clear. Even though DNp63 isoforms showed pro-survival roles [113, 114], the study of their relevance on in vivo tumorigenesis was complicated by the embryonic lethality of knock-out mice [113]. The analysis of promoter occupancy by ChIP-sequencing in HaCat cells, which predominantly express DNp63 revealed novel insights on the mutant p53/p63 connection. Those findings showed that endogenous mutant p53 and DNp63 are simultaneously recruited to several chromatin locations including gene promoters, and suggested that mutant p53 may alter normal p63 distribution on DNA [115]. The role of individual p63 and p73 isoforms on human cancer is an expanding field that will help us to learn more on mutant p53 activities in the near future. It should also be taken into consideration that N-terminal and C-terminal p53 isoforms exist as well [116], which were reported to be differentially expressed in normal and tumor cells [117]. In normal breast tissue, the C-terminally truncated and modified β and γ isoforms were both present, whereas p53β was only detected in 33 % of tested tumors, and p53γ in none. In contrast, N-terminally truncated Δ133p53 was not detectable in healthy controls, but it was found in 80 % of all tested breast tumors. Several questions arise regarding the biological consequences of the expression of these isoforms. For example, what is the effect of a missense mutation on each of them, and which are the consequences of simultaneous expression of different p53 proteins? Some evidences demonstrated that expression of the p53γ isoform significantly improves the outcome of breast cancer patients bearing mutant p53 [118]. The mutations were present in both full-length p53 and p53γ, leading to the conclusion that oncogenic properties induced by a mutation in TP53 are not simply transferred to a shorter p53 isoform. Likewise, how do changes on the relative levels of particular wt or mutated p53 isoforms impact on tumor progression? More information on the prevalence of p53 isoforms in specific tumor types and on the functional interactions among p53 family isoforms will be helpful to understand their role in cancer biology.

Clinical Interactions Between p53 Mutation and Other Cancer-Related Alterations

After the explosive last ten years on mutant p53 research a wealth of evidences have been gathered, that definitively confirmed the GOF hypothesis. Moreover, these studies elegantly showed that mutant p53 act as a pro-aggressiveness factor in a great variety of experimental systems including in vivo models. Those evidences predict that the presence of p53 point mutants exerting pro-oncogenic properties in patients should be associated with reduced overall survival, resistance to chemotherapy and higher probability to develop metastasis. However, when it comes to human cancer, the actual consequences of p53 mutation are still difficult to understand.

Whether if p53 mutation behaves as an independent prognostic and/or predictive factor is a long standing question that several studies attempted to address during the last two decades. The interpretation of these studies was initially obscured by the use methodological approaches that considered p53 overexpression, determined by immunohistochemistry, as a surrogate for p53 mutation. Despite the existence of a correlation between the occurrence of p53 missense mutations and high protein levels, this methodology proved to be inaccurate, since several mutations do not lead to protein accumulation and, on the other hand, accumulation of wt p53 may also occur due to alterations in regulatory circuits. More recent studies that circumvented this problem assessing p53 mutations by direct sequencing showed a general trend that confirmed the association between p53 mutation and poor clinical outcome. Nonetheless, it should be noted that some studies did not supported such correlation while others even reported a negative association. We do not intend to further discuss this topic (for a more detailed discussion see reference [4]) however some interesting considerations can be made. One of the largest studies on the clinical value of somatic p53 mutations was performed in a cohort of 1794 breast cancer patients [119]. This study confirmed that p53 mutation has an independent prognostic value in breast cancer. Curiously, non-missense mutations showed even a stronger association with poor clinical outcome than missense mutations. On the other hand, another study on non small cell lung cancer (NSCLC) including 250 cases failed to reveal any prognostic value for p53 mutations [120], despite the higher frequency of this alteration (~48 %) comparing with breast cancer (~22 %). This apparently confusing scenario may be a reflection of the complex and multifaceted effects that alterations of the p53 pathway may exert on the etiology and progression of the disease. Understanding when and where p53 point mutants exert pro-oncogenic functions is a challenge with several layers of complexity.

Indeed, different situations could be envisioned that may radically change the consequences of p53 mutation. For example, when does mutation occur? In some cases p53 mutation was proposed as an early event, possibly involved in tumor initiation, while other evidences suggest that it may be a late event associated to the progression to aggressive tumor phenotypes. Second, what happens with the remaining wt allele? Does it become lost at any time? When does LOH actually happen in those cases? These different possibilities may affect the disease in different ways, since it is not clear whether dominant negative effects are fully active in vivo. Evidences from *p53R172H* knock-in mice showed that mutant p53 inhibits apoptosis in adult heterozygous mice, but on the contrary, it was not able to rescue wtp53 lethality during embryogenesis in a mdm2 $^{-/-}$ background [10], suggesting that, at least under some circumstances, mutant p53 may not be able to completely block wt p53 function. Third, do different point mutants have different activities? Even if some differences were reported on the biochemical or biological properties of some p53 mutants [27, 121], different point mutants were shown to display similar pro-invasive and pro-metastatic functions in vitro and in vivo [11, 22, 27, 97]. Provocatively, even the possibility that different p53 mutations may coexist in a single patient, at least transiently, could be considered. Such situation could hypothetically arise during malignant transformation as a consequence of individual mutational events in different clonal outgrowths arising simultaneously under conditions of high genomic instability, before a particular clone of transformed cells becomes selected. Finally, if the activation of a particular mutant p53-dependent oncogenic mechanism depends on the presence of a defined set of protein interactors, it seems logical to expect different mutant p53 effects on different cell types or even in similar tumor types harboring different landscapes of genetic and epigenetic alterations. On this scenario, a relevant issue in clinical oncology should be how to stratify patients into clinically significant groups, even among patients harboring p53 mutations, basing on the oncogenic circuits that are activated in each case.

In the light of the evidences discussed in this chapter it seems that a fruitful approach would be to consider the clinical effects of p53 missense mutations in association with other oncogenic alterations that may foster or even oppose mutant p53 GOF. An outstanding example of the multiple clinical interactions that mutant p53 may display in a defined cancer type is represented by the distribution of p53 mutations in breast cancer and its connection with other clinicopathological features. As discussed before, despite the rather low frequency of p53 mutations found in breast cancer p53 mutation has a prognostic value considering the overall population. A breakthrough in the understanding of the molecular nature of breast tumors was the demonstration that cases can be classified basing on their expression profile [122] and that this stratification has a prognostic value, being able to distinguish groups with different clinical outcomes [123]. Strikingly, p53 mutation frequency dramatically increases in some of those subclasses, approaching values between 70 and 80 % of cases in ERBB2+ and Basal-like groups, which are correlated with reduced survival [123, 124]. Basal-like subclass is particularly relevant since it consists mostly of estrogen receptor-, progesterone receptor-, and ERBB2/Her2-negative (triple-negative) immunohistochemical phenotype, that have a higher risk of recurrence and whose management still represents a clinical challenge.

These studies also revealed a negative correlation between estrogen receptor (ER) expression and p53 mutation as ERBB2+ and Basal-like subclasses contain mostly tumors that lack ER expression [123, 124]. It is interesting to note that while ER status did not show prognostic value in the large-scale study by Olivier and colleagues, PR absence was found to be associated with an increased risk in cases bearing wtp53. However, this correlation was lost in cases with mutant p53, which showed a higher risk than wtp53 cases independently of PR status, unveiling a clinical interaction between both variables [119]. Although not much explored, the hypothesis that p53 mutants may act as epigenetic regulators was proposed basing on data from chIP-cloning studies that found mutant p53 preferentially interacting with chromatin on non-coding regions placed far away from gene promoters [125]. Further evidences for a link between mutant p53 and epigenetic regulation came from a study comprising 900 breast cancer cases showing that 85 genes are differentially methylated according to p53 status [126]. The biological meaning of these observations still remains elusive but they nevertheless are suggestive of potentially relevant links between oncogenic pathways.

Other clinical evidences may be reinterpreted under the light of the recent findings on the mechanisms that govern mutant p53 stability. Clinical studies in breast cancer have shown that a polymorphism in the MDM2 promoter, known as SNP309, affects the significance of p53 mutation. In patients harboring SNP309 a T is replaced by a G in position 309 of intron 1, leading to enhanced gene expression [127]. Patients carrying SNP309 in both alleles showed reduced p53 expression in tumors. Moreover, among patients harboring the common genotype, p53 mutation or high p53 expression in tumors was associated with decreased survival, while this correlation was lost in patients with SNP309 [128]. These evidences are in agreement with the current hypothesis on the role of MDM2 as a negative regulator of both wt and mutant p53, predicting that enhanced MDM2 function would oppose tumor progression by attenuating mutant p53 activities.

Similarly, studies on NSCLC highlighted correlations between p53 mutations and the Rb pathway. In agreement with other reports [120], p53 mutation was not associated with survival. Intriguingly, although Cyclin D1 overexpression alone showed no correlation with clinical outcome, it was associated with reduced survival among cases harboring p53 mutations [129]. Alterations on the Rb pathway as a consequence of $p16^{INK4a}$ deletion in knock-in mice was associated with precocious stabilization of mutant p53 [25]. Therefore, in tumors with mutant p53, high CyclinD1 levels may also cooperate with protein stabilization and GOF activities, by promoting Rb hyperphosphorylation and E2F1-dependent p19ARF upregulation.

The question still remains, what are the manifestations of mutant p53 GOF in the clinics? The molecular effects related to some mutant p53 activities were actually observed in several clinical settings, with particular emphasis on breast cancer, providing support to the notion that these activities put forth disease progression. For example, the expression levels of mutant p53/NF-Y targets cyclin A and cdk1 were found to be elevated in rectal tumor samples with high levels of p53 [48]. In line with those observations, high expression levels of the mutant p53 interactor TopBP1 was correlated with elevated expression levels of mutant p53/NF-Y targets in a case study of 58 primary breast tumors harboring mutant p53 and associated with increased risk of relapse and reduced survival in independent public databases of breast cancer [108]. In light of these evidences it would be important to analyze if assessing TopBP1 overexpression may contribute to distinguish groups of patients harboring mutant p53 that may potentially display different clinical outcomes.

Similarly, the presence of mutant p53 was correlated with high expression of 11 of the 17 genes from the mevalonate pathway identified as transcriptional targets of mutant p53 in several breast cancer datasets. Moreover, p53 mutations were significantly more frequent in patients displaying high expression of those 17 genes and this group was associated with worse clinical outcome [130]. Mevalonate pathway genes are induced upon mutant p53 recruitment to Sterol Regulatory Element (SRE) on target promoters, in a process that depends on SRE binding proteins −1 and −2 (SREBP). Therefore, alterations on SREBP activity may ultimately affect mutant p53 function. Little is known about mutations or altered expression of SREBPs in human cancer, however, SREBP1 and SREBP12 may be activated by AKT/mTOR pathway, providing a potential link between this pathway and mutant p53 function [131–133].

The detection of p53 missense mutations could also be relevant to identify cases were TGF-beta acts as a pro-metastatic factor. Even if the molecular events that regulate the switch of TGF-beta signaling from tumor suppressing to tumor-promoting are not completely clear, mutant p53 was suggested to play a prominent role on this process. According to this hypothesis the TGF-beta pro-metastatic function may remain inhibited by p63 even if pro-invasive TGF-beta targets are induced, and mutant p53 releases this inhibition by blocking p63 transcriptional function [22]. Therefore, patients harboring p53 missense mutations may constitute a group with higher risk to suffer a TGF-beta pro-metastatic switch. In line with these evidences, low expression of anti-metastatic p63 targets *SHARP*-1 and *CCNG2*, which are co-repressed by TGF-beta and mutant p53 also showed a correlation with poor clinical outcome in breast cancer datasets.

More indirect connections may be suggested basing on evidences reporting the clinical interactions between p53 mutation and alterations affecting growth factor receptors. For example, the analysis of 194 cases of Glioblastoma Multiforme (GBM) showed that p53 expression was significantly associated with reduced survival in tumors bearing *EGFR* amplification or overexpression. Likewise, *EGFR* alterations were only correlated with reduced survival in cases showing detectable p53 staining [134]. Another report on the clinical significance of p53 mutation and ERBB2 amplification in more than 500 breast cancer patients showed that cases bearing both alterations showed dramatically reduced survival and increased risk in multivariate analysis [135]. Similarly, a study comprising 140 bladder cancer cases detected an interaction between p53 mutation and the presence of the polymorphism G388R in the transmembrane domain of FGFR4. The authors reported higher risk of death for cases carrying the homozygous G388R FGFR4 phenotype and p53 mutation [136]. Taking into account that phosphorylation affects mutant p53 activities [11, 22, 49, 54], it is tempting to speculate that alterations on these receptors may contribute to mutant p53 function by fueling oncogenic signals, for example through Ras activation. Further research on the connection between mutant p53 oncogenic function and the alterations on growth factor receptors would be valuable, considering the well documented cooperation between oncogenic Ras signaling and mutant p53.

Further support for the relevance of p53 GOF in human cancer is provided by the aforementioned evidences showing that high expression of Pin1/ mutant p53 target genes is associated with reduced overall survival and shorter time to distant metastasis in breast cancer databases. Remarkably, Pin1 expression was shown to modulate the prognostic value of p53 mutation in a group of 212 breast cancer cases. When cases were stratified according to p53 status and Pin1 expression levels, overall survival was found to be significantly decreased in patients with tumors expressing high levels of Pin1 and p53 missense mutations, compared to cases with low Pin1 expression and p53 missense mutations, or cases bearing wild-type p53 [11]. In addition, p53 mutation correlates with shorter OS only in cases with high Pin1 expression levels. Moreover, the combination of Pin1 overexpression and p53 mutation behaved as an independent predictor of poor clinical outcome and response to chemotherapy.

These evidences suggest that combining oncogenic alterations that impinge on mutant p53 GOF with p53 mutations may provide a more accurate prediction of clinical outcome. They also provide some hints to realize under what specific conditions mutant p53 may actually become a driver of tumor aggressiveness and may be considered as a proof of principle to search novel strategies for decision making in the clinics. Although much remains to be done, the progress made in understanding the role of mutant p53 GOF in breast cancer is encouraging and underlines the urgent need to extend this analysis to other tumor types.

Concluding Remarks

The acknowledgement of the complex effects that p53 mutants may exert on tumor cells has extended our understanding of the central role of the p53 pathway in cancer biology. Moreover, the ability to transform an efficient tumor suppressor pathway into a network that promotes tumor aggressiveness by only introducing a missense mutation is one of the finest examples of the efficient use of resources that takes place in tumor cells. Still, one of the major challenges in the mutant p53 field is to rationalize the amazing variety of mechanisms that seems to be engaged by p53 mutants.

The emerging picture shows mutant p53 at the center of a highly interconnected network that links oncogenic signaling with several apparently redundant mechanisms that invariably foster tumor cell phenotypes. As we are starting to understand the overwhelming complexity or tumor progression it does not seem illogical that tumor cells could make use of complex strategies involving multifunctional proteins able to wire spare circuits in tight cooperation with other mechanisms of malignancy. Potentially, such a strategy would maximize the capacity of tumor cells to proliferate, survive and eventually spread to distant locations, under a wide range of external conditions. The examples discussed in this chapter underline some ways in which other alterations in human cancer may affect the final consequences of mutant p53 activities.

The involvement of the mutant p53 network in human cancer and in particular its possible role as a driver of aggressiveness, open up a new dimension for the development of clinical strategies to treat the disease. Cancer represents a major unresolved problem in human health because of the lack of efficient therapies. A main drawback concerning the design of cancer therapies is the molecular heterogeneity found in human tumors. This situation makes a strategy targeting a particular pathway ineffective for patients harboring oncogenic circuits that rely on alteration of other pathways. In addition, the phenotypic plasticity and the high levels of genomic instability found in tumor cells often conspire to develop resistance to therapies that may appear initially effective. Nevertheless, the enormous effort on cancer research in the last decades have also showed that some recurrent alterations, like the presence of p53 point mutants, may be at the base of the oncogenic potential of a large number of tumor types. If mutant p53 is actually acting as a driver of aggressiveness in all tumors expressing a p53 point mutant it follows that therapies targeting mutant p53 may be highly effective. Moreover, such strategies would be exquisitely specific since they would affect exclusively cells harboring p53 mutations sparing normal cells form undesired effects. It is necessary to understand how other alterations affect mutant p53 function in order to detect what are the cases where mutant p53 GOF is actually relevant and what would be the particular biological outcome expected. In addition, the crosstalk between the mutant p53 network and other alterations may provide novel ways to disarm the oncogenic mechanisms, for example blocking a mutant p53 interactor, an activating signal or a downstream effector, instead of targeting mutant p53 directly. As the concept of tailored medicine is closer to be considered in clinical practice the assessment of the p53 status in cancer patients, with particular emphasis on the conditions that may unleash mutant p53 oncogenic power may provide valuable information to improve the clinical management of the disease.

References

1. Levine A, Oren M (2009) The first 30 years of p53: growing ever more complex. Nat Rev Cancer 9:749–758
2. Oren M, Rotter V (2010) Mutant p53 gain-of-function in cancer. Cold Spring Harb Perspect Biol 2:a001107. doi:10.1101/cshperspect.a001107
3. Lozano G (2010) Mouse models of p53 functions. Cold Spring Harb Perspect Biol 2: a001115. doi: 10.1101/cshperspect.a001115
4. Brosh R, Rotter V (2009) When mutants gain new powers: news from the mutant p53 field. Nat Rev Cancer 9:701–713

5. Karnoub AE, Weinberg RA (2008) Ras oncogenes: split personalities. Nat Rev Mol Cell Biol 9:517–531
6. Catalogue of Somatic Mutations in Cancer (COSMIC) (2013) Wellcome Trust Sanger Institute. Hinxton, UK. http://cancer.sanger.ac.uk/cancergenome/projects/cosmic/
7. Balmain A, Brown K (1988) Oncogene activation in chemical carcinogenesis. Adv Cancer Res 51:147–182
8. Hinds P, Finlay C, Levine A (1989) Mutation is required to activate the p53 gene for cooperation with the ras oncogene and transformation. J Virol 63:739
9. Hinds PW et al (1990) Mutant p53 DNA clones from human colon carcinomas cooperate with ras in transforming primary rat cells: a comparison of the "hot spot" mutant phenotypes. Cell Growth Differ 1:571–580
10. Lang GA et al (2004) Gain of function of a p53 hot spot mutation in a mouse model of Li-Fraumeni syndrome. Cell 119:861–872
11. Girardini JE et al (2011) A Pin1/mutant p53 axis promotes aggressiveness in breast cancer. Cancer Cell 20:79–91
12. Hingorani S et al (2005) *Trp53R172H* and *KrasG12D* cooperate to promote chromosomal instability and widely metastatic pancreatic ductal adenocarcinoma in mice. Cancer Cell 7:469–483
13. Hruban RH, Iacobuzio-Donahue C, Wilentz RE, Goggins M, Kern SE (2001) Molecular pathology of pancreatic cancer. Cancer J 7:251–258
14. Bolshakov S et al (2003) p53 mutations in human aggressive and nonaggressive basal and squamous cell carcinomas. Clin Cancer Res 9:228–234
15. Brash DA et al (1991) A role for sunlight in skin cancer: UV-induced p53 mutations in squamous cell carcinoma. Proc Natl Acad Sci U S A 88:10124–10128
16. Pierceall WE et al (1991) Ras gene mutation and amplification in human nonmelanoma skin cancers. Mol Carcinog 4:196–202
17. Spencer JM et al (1995) Activated ras genes occur in human actinic keratoses, premalignant precursors to squamous cell carcinomas. Arch Dermatol 131:786–800
18. Caulin C et al (2007) An inducible mouse model for skin cancer reveals distinct roles for gain- and loss-of-function *p53* mutations. J Clin Invest 117:1893–1901
19. Gualberto A, Aldape K, Kozakiewicz K, Tlsty TD (1998) An oncogenic form of p53 confers a dominant, gain-of-function phenotype that disrupts spindle checkpoint control. Proc Natl Acad Sci U S A 95:5166–5177
20. El-Hizawi S, Lagowski JP, Kulesz-Martin M, Albor A (2002) Induction of gene amplification as a gain-of-function phenotype of mutant p53 proteins. Cancer Res 62:3264–3270
21. Song H, Hollstein M, Xu Y (2007) p53 gain-of-function cancer mutants induce genetic instability by inactivating ATM. Nat Cell Biol 9:573–580
22. Adorno M et al (2009) A Mutant-p53/Smad complex opposes p63 to empower TGFb-Induced metastasis. Cell 137:87–98
23. Solomon H et al (2012) Various p53 mutant proteins differently regulate the Ras circuit to induce a cancer-related gene signature. J Cell Sci 125:3144–3152
24. Murray-Zmijewski F, Slee EA, Lu X (2008) A complex barcode underlies the heterogeneous response of p53 to stress. Nat Rev Mol Cell Biol 9:702–712
25. Terzian T et al (2008) The inherent instability of mutant p53 is alleviated by Mdm2 or p16INK4a loss. Genes Dev 22:1337–1344
26. Suh YA et al (2011) Multiple stress signals activate mutant p53 in vivo. Cancer Res 71:7168–7175
27. Olive KP et al (2004) Mutant *p53 g*ain of function in two mouse models of Li-Fraumeni syndrome. Cell 119:847–860
28. Onel K, Cordon-Cardo C (2004) MDM2 and prognosis. Mol Cancer Res 2:1–8
29. Deb SP (2003) Cell cycle regulatory functions of the human oncoprotein MDM2. Mol Cancer Res 1:1009–1016
30. Iwakuma T, Lozano G (2003) MDM2, an introduction. Mol Cancer Res 1:993–1000
31. Bouska A, Eischen CM (2009) Mdm2 affects genome stability independent of p53. Cancer Res 69:1697–1701

32. Midgley CA, Lane DP (1997) p53 protein stability in tumour cells is not determined by mutation but is dependent on Mdm2 binding. Oncogene 15:1179–1189
33. Lukashchuk N, Vousden KH (2007) Ubiquitination and degradation of mutant p53. Mol Cell Biol 27:8284–8295
34. Li D et al (2011) Functional inactivation of endogenous MDM2 and CHIP by HSP90 causes aberrant stabilization of mutant p53 in human cancer cells. Mol Cancer Res 9:577–588
35. Peng Y et al (2001) Inhibition of MDM2 by hsp90 contributes to mutant p53 stabilization. J Biol Chem 276:40583–40590
36. Whitesell L, Lindquist SL (2005) HSP90 and the chaperoning of cancer. Nat Rev Cancer 5:761–772
37. Finlay CA et al (1988) Activating mutations for transformation by p53 produce a gene product that forms an hsc70-p53 complex with an altered half-life. Mol Cell Biol 8:531–539
38. Blagosklonny MV, Toretsky J, Bohen S, Neckers L (1996) Mutant conformation of p53 translated in vitro or in vivo requires functional HSP90. Proc Natl Acad Sci U S A 93:8379–8383
39. King FW, Wawrzynow A, Höhfeld J, Zylicz M (2001) Co-chaperones Bag-1, Hop and Hsp40 regulate Hsc70 and Hsp90 interactions with wild-type or mutant p53. EMBO J 20:6297–6305
40. Whitesell L et al (1988) The physical association of multiple molecular chaperone proteins with mutant p53 is altered by geldanamycin, an hsp90-binding agent. Mol Cell Biol 18:1517–1524
41. Blagosklonny MV et al (2005) Depletion of mutant p53 and cytotoxicity of histone deacetylase inhibitors. Cancer Res 65:7386–7392
42. Li D, Marchenko ND, Moll UM (2011) SAHA shows preferential cytotoxicity in mutant p53 cancer cells by destabilizing mutant p53 through inhibition of the HDAC6-Hsp90 chaperone axis. Cell Death Differ 18:1904–1913
43. Yan W et al (2013) Histone deacetylase inhibitors suppress mutant p53 transcription via histone deacetylase 8. Oncogene 32:599–609
44. Post SM et al (2010) p53-dependent senescence delays Emu-myc-induced B-cell lymphomagenesis. Oncogene 29:1260–1269
45. Guo L et al (2012) Ionizing radiation induces a dramatic persistence of p53 protein accumulation and DNA damage signaling in mutant p53 zebrafish. Oncogene. doi:10.1038/onc.2012.409
46. Minamoto T et al (2001) Distinct pattern of p53 phosphorylation in human tumors. Oncogene 20:3341–3347
47. Matsumoto M, Furihata M, Ohtsuki Y (2006) Posttranslational phosphorylation of mutant p53 protein in tumor development. Med Mol Morphol 39:79–87
48. Di Agostino S et al (2006) Gain of function of mutant p53: the mutant p53/NF-Y protein complex reveals an aberrant transcriptional mechanism of cell cycle regulation. Cancer Cell 10:191–202
49. Valenti F et al (2011) Mutant p53 oncogenic functions are sustained by Plk2 kinase through an autoregulatory feedback loop. Cell Cycle 10:4330–4340
50. Tan LB et al (2010) Identification of urine PLK2 as a marker of bladder tumors by proteomic analysis. World J Urol 28:117–122
51. Matsumoto T et al (2009) Polo-like kinases mediate cell survival in mitochondrial dysfunction. Proc Natl Acad Sci U S A 106:14542–14546
52. Strebhardt K (2010) Multifaceted polo-like kinases: drug targets and antitargets for cancer therapy. Nat Rev Drug Discov 9:643–660
53. Syed N et al (2011) Polo-like kinase Plk2 is an epigenetic determinant of chemosensitivity and clinical outcomes in ovarian cancer. Cancer Res 71:3317–3327
54. Yap D et al (2004) Ser392 phosphorylation regulates the oncogenic function of mutant p53. Cancer Res 64:4749–4754
55. Matsumoto M, Furihata M, Kurabayashi A, Ohtsuki Y (2004) Phosphorylation state of tumor-suppressor gene p53 product overexpressed in skin tumors. Oncol Rep 12:1039–1043

56. Matsumoto M et al (2004) Prognostic significance of serine 392 phosphorylation in overexpressed p53 protein in human esophageal squamous cell carcinoma. Oncology 67:143–150
57. Furihata M et al (2002) Frequent phosphorylation at serine 392 in overexpressed p53 protein due to missense mutation in carcinoma of the urinary tract. J Pathol 197:82–88
58. Ranganathan R, Lu KP, Hunter T, Noel JP (1997) Structural and functional analysis of the mitotic rotamase Pin1 suggests substrate recognition is phosphorylation dependent. Cell 89:875–886
59. Yaffe MB et al (1997) Sequence-specific and phosphorylation-dependent proline isomerization: a potential mitotic regulatory mechanism. Science 278:1957–1960
60. Liou YC, Zhou XZ, Lu KP (2011) Prolyl isomerase Pin1 as a molecular switch to determine the fate of phosphoproteins. Trends Biochem Sci 36:501–514
61. Brown NR et al (1999) Effects of phosphorylation of threonine 160 on cyclin-dependent kinase 2 structure and activity. J Biol Chem 274:8746–8756
62. Weiwad M, Kullertz G, Schutkowski M, Fischer G (2000) Evidence that the substrate backbone conformation is critical to phosphorylation by p42 MAP kinase. FEBS Lett 478:39–42
63. Zhou XZ et al (2000) Pin1-dependent prolyl isomerization regulates dephosphorylation of Cdc25C and tau proteins. Mol Cell 6:873–883
64. Yeh E, Means AR (2007) PIN1, the cell cycle and cancer. Nat Rev Cancer 7:381–388
65. Bao L et al (2004) Prevalent overexpression of prolyl isomerase Pin1 in human cancers. Am J Pathol 164:1727–1737
66. Wulf G et al (2004) Modeling breast cancer in vivo and ex vivo reveals an essential role of Pin1 in tumorigenesis. EMBO J 23:3397–3407
67. Zacchi P et al (2002) The prolyl isomerase Pin1 reveals a mechanism to control p53 functions after genotoxic insults. Nature 419:853–857
68. Zheng H et al (2002) The Prolyl isomerase Pin1 is a novel regulator of p53 in genotoxic response. Nature 419:849–853
69. Mantovani F et al (2007) The prolyl isomerase Pin1 orchestrates p53 acetylation and dissociation from the apoptosis inhibitor iASPP. Nat Struct Mol Biol 14:912–920
70. Tan X et al (2010) Pin1 expression contributes to lung cancer: prognosis and carcinogenesis. Cancer Biol Ther 9:111–119
71. Ayala G et al (2003) The Prolyl Isomerase Pin1 is a novel prognostic marker in human prostate cancer. Cancer Res 63:6244–6251
72. Leung KW et al (2009) Pin1 overexpression is associated with poor differentiation and survival in oral squamous cell carcinoma. Oncol Rep 21:1097–1104
73. Mantovani F et al (2004) Pin1 links the activities of c-Abl and p300 in regulating p73 function. Mol Cell 14:625–636
74. Yeh E et al (2004) A signalling pathway controlling c-Myc degradation that impacts oncogenic transformation of human cells. Nat Cell Biol 6:308–318
75. Yeh E, Lew BO, Means AR (2006) The loss of PIN1 deregulates cyclin E and sensitizes mouse embryo fibroblasts to genomic instability. J Biol Chem 281:241–251
76. Hanahan D, Weinberg RA (2011) Hallmarks of cancer: the next generation. Cell 144(5):646–674
77. Bierie B, Moses HL (2006) Tumour microenvironment: TGFbeta: the molecular Jekyll and Hyde of cancer. Nat Rev Cancer 6:506–520
78. Grünert S, Jechlinger M, Beug H (2003) Diverse cellular and molecular mechanisms contribute to epithelial plasticity and metastasis. Nat Rev Mol Cell Biol 4:657–665
79. Kalo E et al (2007) Mutant p53 attenuates the SMAD-dependent transforming growth factor beta1 (TGF-beta1) signaling pathway by repressing the expression of TGF-beta receptor type II. Mol Cell Biol 27:8228–8242
80. Lin S et al (2011) Mutant p53 disrupts role of ShcA protein in balancing Smad protein-dependent and -independent signaling activity of transforming growth factor-ß (TGF-ß). J Biol Chem 286:44023–44034
81. Chen G et al (2012) Transforming growth factor ß1 (TGF-ß1) suppresses growth of B-cell lymphoma cells by p14(ARF)-dependent regulation of mutant p53. J Biol Chem 287:23184–23195

82. Colston KW, Chander SK, Mackay AG, Coombes RC (1992) Effects of synthetic vitamin D analogues on breast cancer cell proliferation in vivo and in vitro. Biochem Pharmacol 44:693–702
83. Nakagawa K et al (2005) 22-Oxa-1alpha,25-dihydroxyvitamin D3 inhibits metastasis and angiogenesis in lung cancer. Carcinogenesis 26:1044–1054
84. Kerner SA, Scott RA, Pike JW (1989) Sequence elements in the human osteocalcin gene confer basal activation and inducible response to hormonal vitamin D3. Proc Natl Acad Sci U S A 86:4455–4459
85. Zinser G, Packman K, Welsh J (2002) Vitamin D(3) receptor ablation alters mammary gland morphogenesis. Development 129:3067–3076
86. Zinser G, Sundberg JP, Welsh J (2002) Vitamin D(3) receptor ablation sensitizes skin to chemically induced tumorigenesis. Carcinogenesis 23(12):2103–2109
87. Wang SH et al (1999) 1Alpha,25-dihydroxyvitamin D3 up-regulates Bcl-2 expression and protects normal human thyrocytes from programmed cell death. Endocrinology 140:1649–1656
88. Duque G et al (2004) Vitamin D inhibits Fas ligand-induced apoptosis in human osteoblasts by regulating components of both the mitochondrial and Fas-related pathways. Bone 35:57–64
89. Friedrich M et al (1998) Expression of 1,25-dihydroxy vitamin D3 receptor in breast carcinoma. J Histochem Cytochem 46:1335–1337
90. Friedrich M et al (2002) Analysis of vitamin D-receptor (VDR) and retinoid X-receptor alpha in breast cancer. Histochem J 34:35–40
91. Sahin MO et al (2005) 1,25 Dihydroxyvitamin D(3) receptor expression in superficial transitional cell carcinoma of the bladder: a possible prognostic factor? Eur Urol 47:52–57
92. Menezes RJ et al (2008) Vitamin D receptor expression in normal, premalignant, and malignant human lung tissue. Cancer Epidemiol Biomarkers Prev 17:1104–1110
93. Stambolsky P et al (2010) Modulation of the vitamin D3 response by cancer-associated mutant p53. Cancer Cell 17:273–285
94. White DP, Caswell PT, Norman JC (2007) Alpha v beta3 and alpha5beta1 integrin recycling pathways dictate downstream Rho kinase signaling to regulate persistent cell migration. J Cell Biol 177:515–525
95. Caswell PT, Norman JC (2008) Endocytic transport of integrins during cell migration and invasion. Trends Cell Biol 18:257–263
96. Caswell PT et al (2008) Rab-coupling protein coordinates recycling of alpha5beta1 integrin and EGFR1 to promote cell migration in 3D microenvironments. J Cell Biol 183:143–155
97. Muller P et al (2009) Mutant p53 drives invasion by promoting integrin recycling. Cell 139:1327–1341
98. Muller P et al (2012) Mutant p53 enhances MET trafficking and signalling to drive cell scattering and invasion. Oncogene 32:1252–1265
99. Chaturvedi MM et al (2011) NF-κB addiction and its role in cancer: 'one size does not fit all'. Oncogene 30:1615–1630
100. DiDonato JA, Mercurio F, Karim M (2012) NF-κB and the link between inflammation and cancer. Immunol Rev 246:379–400
101. Scian M et al (2005) Tumor-derived p53 mutants induce NF-kappaB2 gene expression. Mol Cell Biol 25:10097–10110
102. Vaughan CA et al (2012) p53 mutants induce transcription of NF-?B2 in H1299 cells through CBP and STAT binding on the NF-?B2 promoter and gain of function activity. Arch Biochem Biophys 518:79–88
103. Weisz L et al (2007) Mutant p53 enhances nuclear factor kappaB activation by tumor necrosis factor alpha in cancer cells. Cancer Res 67:2396–2401
104. Yeudall WA et al (2012) Gain-of-function mutant p53 upregulates CXC chemokines and enhances cell migration. Carcinogenesis 33:442–451
105. Richmond A (2002) Nf-kappa B, chemokine gene transcription and tumour growth. Nat Rev Immunol 2:664–674

106. Martin D, Galisteo R, Gutkind JS (2009) CXCL8/IL8 stimulates vascular endothelial growth factor (VEGF) expression and the autocrine activation of VEGFR2 in endothelial cells by activating NFkappaB through the CBM (Carma3/Bcl10/Malt1) complex. J Biol Chem 284:6038–6042
107. Fontemaggi G et al (2009) The execution of the transcriptional axis mutant p53, E2F1 and ID4 promotes tumor neo-angiogenesis. Nat Struct Mol Biol 16:1086–1093
108. Liu K, Ling S, Lin WC (2011) TopBP1 mediates mutant p53 gain of function through NF-Y and p63/p73. Mol Cell Biol 31:4464–4481
109. Liu K et al (2009) Regulation of p53 by TopBP1: a potential mechanism for p53 inactivation in cancer. Mol Cell Biol 29:2673–2693
110. Su X et al (2010) TAp63 suppresses metastasis through coordinate regulation of Dicer and miRNAs. Nature 467:986–990
111. Tomasini R et al (2008) TAp73 knockout shows genomic instability with infertility and tumor suppressor functions. Genes Dev 22:2677–2691
112. Neilsen PM et al (2011) Mutant p53 uses p63 as a molecular chaperone to alter gene expression and induce a pro-invasive secretome. Oncotarget 2:1203–1217
113. Romano RA et al (2012) DNp63 knockout mice reveal its indispensable role as a master regulator of epithelial development and differentiation. Development 139:772–782
114. Lee HO et al (2006) A dominant negative form of p63 inhibits apoptosis in a p53-independent manner. Biochem Biophys Res Commun 344:166–172
115. Martynova E et al (2012) Gain-of-function p53 mutants have widespread genomic locations partially overlapping with p63. Oncotarget 3(2):132–143
116. Khoury MP, Bourdon JC (2011) p53 isoforms: an intracellular microprocessor? Genes Cancer 2:453–465
117. Bourdon JC et al (2005) p53 isoforms can regulate p53 transcriptional activity. Genes Dev 19:2122–2137
118. Bourdon JC et al (2011) p53 mutant breast cancer patients expressing p53? have as good a prognosis as wild-type p53 breast cancer patients. Breast Cancer Res 13:R7. doi:10.1186/bcr2811
119. Olivier M et al (2006) The clinical calue of somatic TP53 GeneMutations in 1,794 patients with breast cancer. Clin Cancer Res 12:1157–1167
120. Scoccianti C et al (2012) Prognostic value of TP53, KRAS and EGFR mutations in nonsmall cell lung cancer: the EUELC cohort. Eur Respir J 40:177–184
121. Muller P, Vousden KH (2013) p53 mutations in cancer. Nat Cell Biol 15:2–8
122. Perou CM et al (2000) Molecular portraits of human breast tumours. Nature 406:747–752
123. Sørlie T et al (2001) Gene expression patterns of breast carcinomas distinguish tumor subclasses with clinical implications. Proc Natl Acad Sci U S A 98:10869–10874
124. Langerød A et al (2007) *TP53* mutation status and gene expression profiles are powerful prognostic markers of breast cancer. Breast Cancer Res 9:R30
125. Brázdová M et al (2009) Modulation of gene expression in U251 glioblastoma cells by binding of mutant p53 R273H to intronic and intergenic sequences. Nucleic Acids Res 37:1486–1500
126. Rønnenberg JA et al (2011) Methylation profiling with a panel of cancer related genes: association with estrogen receptor, TP53 mutation status and expression subtypes in sporadic breast cancer. Mol Oncol 5:61–76
127. Bond GL et al (2004) A single nucleotide polymorphism in the MDM2 promoter attenuates the p53 tumor suppressor pathway and accelerates tumor formation in humans. Cell 119(5):591–602
128. Boersma BJ et al (2006) Association of breast cancer outcome with status of p53 and MDM2 SNP309. J Natl Cancer Inst 98:911–919
129. Burke L et al (2005) Prognostic implications of molecular and immunohistochemical profiles of the Rb and p53 cell cycle regulatory pathways in primary non-small cell lung carcinoma. Clin Cancer Res 11:232–2431
130. Freed-Pastor WA et al (2012) Mutant p53 disrupts mammary tissue architecture via the mevalonate pathway. Cell 148:244–258

131. Yamauchi Y, Furukawa K, Hamamura K, Furukawa K (2011) Positive feedback loop between PI3K-Akt-mTORC1 signaling and the lipogenic pathway boosts Akt signaling: induction of the lipogenic pathway by a melanoma antigen. Cancer Res 71:4989–4997
132. Luu W, Sharpe LJ, Stevenson J, Brown AJ (2012) Akt acutely activates the cholesterogenic transcription factor SREBP-2. Biochim Biophys Acta 1823:458–464
133. Bakan I, Laplante M (2012) Connecting mTORC1 signaling to SREBP-1 activation. Curr Opin Lipidol 23:226–234
134. Ruano R et al (2009) Worse outcome in primary glioblastoma multiforme with concurrent epidermal growth factor receptor and p53 alteration. Am J Clin Pathol 131:257–263
135. Bull SV et al (2004) The combination of p53 mutation and neu/erbB-2 amplification is associated with poor survival in node-negative breast cancer. J Clin Oncol 22:86–96
136. Yang YC et al (2006) Joint association of polymorphism of the FGFR4 gene and mutation TP53 gene with bladder cancer prognosis. Br J Cancer 95:1455–1458

Chapter 4
p53: Its Mutations and Their Impact on Transcription

Catherine Vaughan, Isabella Pearsall, Andrew Yeudall, Swati Palit Deb, and Sumitra Deb

Abstract p53 is a tumor suppressor protein whose key function is to maintain the integrity of the cell. Mutations in p53 have been found in up to 50 % of all human cancers and cause an increase in oncogenic phenotypes such as proliferation and tumorigenicity. Both wild-type and mutant p53 have been shown to transactivate their target genes, either through directly binding to DNA, or indirectly through protein-protein interactions. This review discusses possible mechanisms behind both wild-type and mutant p53-mediated transactivation and touches on the concept of addiction to mutant p53 of cancer cells and how that may be used for future therapies.

C. Vaughan
Massey Cancer Center, Virginia Commonwealth University,
401 College Street, Richmond, VA 23298, USA

Integrated Life Sciences Program, Virginia Commonwealth University,
1000 West Cary Street, Richmond, VA 23284, USA

I. Pearsall
Massey Cancer Center, Virginia Commonwealth University,
401 College Street, Richmond, VA 23298, USA

A. Yeudall
Department of Oral and Craniofacial Molecular Biology,
Virginia Commonwealth University, 1101 East Leigh Street,
Richmond, VA 23298, USA

Massey Cancer Center, Virginia Commonwealth University,
401 College Street, Richmond, VA 23298, USA

S.P. Deb • S. Deb (✉)
Department of Biochemistry and Molecular Biology, and the Massey Cancer Center,
Virginia Commonwealth University, Richmond, VA 23298, USA
e-mail: sdeb@vcu.edu

S.P. Deb and S. Deb (eds.), *Mutant p53 and MDM2 in Cancer*,
Subcellular Biochemistry 85, DOI 10.1007/978-94-017-9211-0_4,

Keywords p53 • Tumor suppressor • Transcription • Transactivation • Mutant p53 • Gain-of-function • Oncogenesis

Discovery of p53

p53 is a tumor suppressor protein that was first detected in association with the simian virus 40 large T antigen in virus-transformed cells [21, 29, 72, 75, 90, 102, 148]. High levels of p53 were subsequently observed in cell lines transformed by a variety of agents including DNA and RNA viruses, irradiation, and chemical carcinogens [29, 38, 58, 66, 89, 133]. p53 was thought to be a nuclear oncogene when genomic and cDNA clones of p53 were found to immortalize primary cells and to cooperate with the ras oncogene in transformation of primary cells [37, 66, 124]. However, these original p53 clones were found to contain a mutation [58]. When wild-type (WT) p53 was expressed, the proliferation of transformed cells was inhibited, oncogene-mediated cell transformation was suppressed, and the tumorigenic potential of tumor-derived cell lines was eliminated [6, 7, 31, 37, 41, 105]. Both somatic and germ line (Li-Fraumeni syndrome) mutations of the p53 gene have been detected in a variety of human tumors. Databases listing all known p53 mutations published have been established (http://p53.free.fr/; http://p53.iarc.fr/). Currently, p53 mutations are the most frequently reported genetic defects in human cancers, occurring in approximately 50 % of all human cancers.

There are excellent reviews in which different aspects of the p53 tumor suppressor are discussed, and which the reader should consult to become acquainted with recent developments regarding WT p53 and its properties [4, 82, 163]. In this chapter we will outline some of p53's properties, both WT and mutant, and then focus on some new findings that underpin the significant role played by mutant p53 in oncogenesis. For a wider perspective of p53 mutations in cancer, there are many excellent reviews that should be consulted [18, 45, 51, 110, 122, 164].

WT p53

p53 is mostly localized in the nucleus. The level of WT p53 is very low in normal cells; however, levels of p53 rise rapidly when cells are exposed to stress [80, 126]. This is mostly because of post-translational modifications such as phosphorylation, in which WT p53 becomes activated and initiates a cascade of events that lead to cell growth arrest and/or cell death [42]. In most of the p53 mutants identified in human cancer, these processes are defective (loss-of-function). WT p53 is a sequence-specific DNA-binding protein and a transcription factor that activates a large group of genes whose regulatory sequences have p53 consensus binding sites

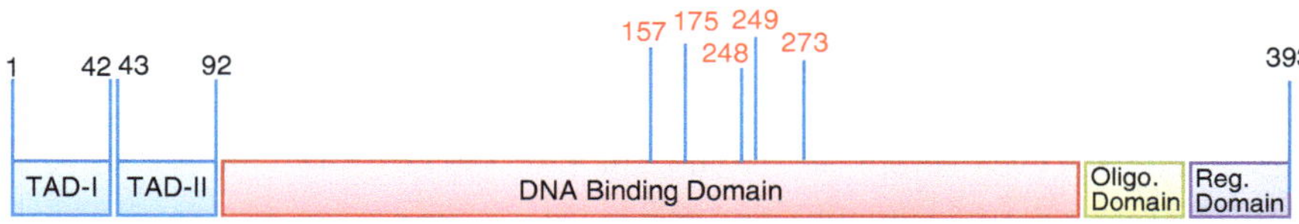

Fig. 4.1 Diagram representing p53 with some common structural and functional domains and hot spot mutations (shown in *red*) found in different cancers

[46, 180]. Although it is accepted that WT p53 has biological activities that are transcription independent, the vast array of its functions are dependent on its transcriptional ability [126]. WT p53 has been shown to be involved in many biological activities including aging [135], DNA damage response [130], autophagy [134], maintenance of genetic stability [4], metabolism [99], miRNA transcription [57], and stem cell fate determination [162], amongst others.

Human p53 is a 393 amino acid protein (Fig. 4.1) with a well-defined and worked-out structure-function relationship. The DNA bound partial structure along with full length WT p53 has been solved for some time now [24, 120]. Predictions of structural alterations by mutations in certain cancers have also been done. Different structural and functional domains indicated in Fig. 4.1 are mostly derived by mutational analysis along with functional studies.

Transcriptional Effects of WT p53

Several groups initially demonstrated that p53 has a transactivation domain [40, 117, 129]. The sequence specific transcriptional activity of WT p53 has been demonstrated with synthetic and biologically relevant promoters with regulatory sequences containing p53 consensus binding sites [35, 46]. It is now known that transcription of a series of genes involved in a variety of biological activities is modulated (activated or inhibited) by p53 [85, 132]. Most of these genes have some type of a p53 consensus site defined by 5′-PuPuPuC(A/T)(T/A)GPyPyPy-3′ [35, 46]. These genes have been identified by both gene expression analysis and in vivo WT p53 binding, although there is a lack of a one-to-one correlation between in vivo p53 binding and gene activation by WT p53 [103, 169]. In most cases, the p53 binding site is roughly within 400 base pairs (bp) of the transcription start site (TSS) in the promoter region; however, binding sites have been identified at different distances from the TSS and also within exons [154].

Several groups have investigated the mechanism of WT p53-mediated transactivation. There is a crucial step before the initiation of transcription can occur which seems to involve the opening of chromatin via modification of histones (e.g., by acetylation). p53 has been shown to be an integral part of this process by recruiting histone acetyl transferases (HATs) such as p300/CBP, Tip60, and PCAF [9, 48, 53, 81, 94, 97]. p53 has been found to be associated with chromatin in unstimulated or unstressed cells even before activation starts [10, 39, 116, 146], but not all genes associated with WT p53 are transactivated [116].

There may be differences in the molecular mechanisms by which WT p53 activates transcription for individual genes that are subtle or even indirect, as there are genes identified as being p53 targets but which lack apparent p53 binding sites in their vicinity [116]. A simplified explanation of WT p53-mediated transactivation is one in which p53 binds to its site on the chromosome, assembles HATs (and/or histone methyl transferases) and contacts the RNA polymerase II complex at the site of transcription via the "mediator" complex and TBP associated factors, depending on the particular target gene [10, 22, 95, 100, 101, 145]. Variations in the binding of p53 to different factors results in differential transactivation of genes leading to distinct phenotypic expression of p53 activation, e.g. apoptosis or G_1/S arrest [10]. A similar mechanism of action can be envisioned for the induction of transcriptional elongation steps. A significant possibility is the influence of DNA bending induced by p53 binding to its site on the DNA in transcription [10, 98, 114].

Since the early 1990s, p53 has been recognized as a transcriptional repressor as it was shown to inhibit transcription from a number of cellular and viral promoters by several groups [49, 138, 153]. Other cellular genes have been reported to be repressed by WT p53 [71, 112, 113, 141], yet the mechanism of WT p53-mediated transcriptional repression is less well understood and understudied. It would appear that there are two ways of inhibiting transcription of specific genes by WT p53: (a) indirect and (b) direct.

(a) In the case of indirect repression, a repressor such as p21 is transactivated by WT p53 which may, in turn, transcriptionally inhibit expression of a series of genes [127]. Another possibility is that as a result of processes such as WT p53-induced apoptosis, cell cycle arrest, and senescence, there will be a repression of the expression of a group of genes accompanying WT p53 induction.
(b) In direct transcriptional repression by WT p53, at least three mechanistic scenarios have been demonstrated. (i) In the first, WT p53 binds to its consensus sequence on the regulatory region of the target gene. It is unclear what the exact mechanism of repression is, although it is perhaps by competition with other essential transcriptional factors needed to interact at the site [60]. There is also some evidence to indicate that the orientation of p53 binding sites may determine whether binding results in transactivation or repression [50, 67, 84]. (ii) WT p53 physically and functionally interacts with the transcription initiation complex including TFIID, possibly through its ability to interact with TBP [10, 22, 95, 100, 101, 145] and inhibits transcription. This type of inhibition is further complicated by p53's interaction with other transcription factors such as NF-YA for cyclin B2 [62]. (iii) WT p53 also represses transcription by binding to its site on regulatory sequences and recruits co-repressors such as mSin3A which inhibits HATs, thereby repressing transcription. Examples of this model

are the repression of MAP4 and survivin by WT p53 [60, 113]. There are many details about the various interactions necessary for the transcriptional repression functions that remain unsolved.

Mutations in p53

There are currently two well-respected online resources that track reported p53 mutations in human cancers (http://p53.free.fr and http://p53.iarc.fr). All p53 mutations found in human cancer can be divided empirically into three functionally non-exclusive categories: (1) Loss of function mutations. These mutations are responsible for the loss of the tumor suppressor function of p53. Almost all of the p53 mutants that have been identified in human cancer fall into this category. In general, they are defective in sequence-specific transactivation of genes containing WT p53 consensus binding sites. (2) Dominant negative mutations. It has been shown that some p53 mutants can form hetero-oligomers with WT p53 or other p53 mutants. Sometimes this hetero-oligomerization can cause changes in the properties of WT p53, with the mutant becoming dominant over the WT protein's properties, therefore abrogating its tumor suppressor functions. An example of this is the immortalization and transformation of rodent embryo fibroblasts by mutant p53 [66, 70]. This may happen in human cancers when one allele of the p53 gene is mutated (point mutated), generating a dominant negative mutant protein that may inactivate the co-expressed WT p53 resulting in weaker tumor suppression. This is thought to underpin the increased cancer susceptibility of patients with Li-Fraumeni syndrome [181]. (3) Gain of function (GOF) mutations. This third kind of mutation, found mostly in the tumor suppressor p53, is widely under study now. In this case, mutant p53 performs a dominant oncogenic role that does not depend on complex formation with WT p53. An example of this would be expression of mutant p53 in cells where WT p53 is absent and which showed enhanced oncogenic properties compared to p53-null cells. The list of GOF properties is growing. However, the molecular mechanisms responsible for the acquisition of GOF properties are not yet clear and work is still in progress. The issue of GOF will be discussed in more depth below.

In cancers that have a GOF p53 mutation, the individual p53 mutation may have profound overall implications for the oncogenic state of the cell. These can be divided into three categories: (a) loss of WT p53 transactivation function, eliminating the ability to activate processes involved in growth suppression under stress situations. This is directly related to the loss of transactivation ability of WT p53 which in normal cells will induce genes involved in growth arrest, apoptosis, and other genes involved in a growth suppressive response. (b) Loss of transcriptional repressor function of WT p53, thereby losing the ability to regulate some of the genes that are involved in growth promotion. (c) Acquisition of new (oncogenic) properties by mutations of the p53 gene, a byproduct of that is the activation of a series of genes involved in oncogenic initiation and progression (detailed later).

p53 Gain of Function Mutation

From the outset of its discovery, it has been observed that, in the majority of cases of p53 mutations, there is a single nucleotide alteration which results in a single amino acid-substituted protein that is expressed at a relatively high level [45, 83]. For other tumor suppressor genes, this phenomenon is relatively uncommon. In a significantly large number of human cancers, there is an overexpression of point-mutated p53 proteins, suggesting a selection pressure behind the accumulated mutant protein. These observations led to the gain of function hypothesis, which predicts that mutations in p53 not only destroy the tumor suppressor function, but that the mutant proteins gain new oncogenic properties. The hypothesis also predicts that tumors that express mutant p53 may be more aggressive than p53-null tumors. There are specific p53 mutations that have been associated with poor prognosis of a number of cancers, including those of breast [5, 16, 73, 125], colon [55, 69, 136], and lung [2, 20, 149], thus supporting the gain of function hypothesis.

Gain of Function Activities and Implications

The biological response to GOF p53 mutants was observed first in mouse tumor systems [174] in which expression of a mutant p53 protein resulted in tumor formation. The main issue in demonstrating GOF activity is to distinguish between mutant p53's dominant negative (DN) and GOF activities if the cell system has a WT counterpart. To do this, biological GOF activity was confirmed using the 10(3) murine fibroblast cell line that is endogenously null in p53 expression, and forced expression of mutant p53 led to tumor formation in nude mice while 10(3) alone or 10(3) transfected with a control plasmid did not form tumors [33]. Later, more laboratories demonstrated biological functions related to GOF p53 expression and tumorigenicity [61, 77].

Two groups first demonstrated the biochemical function of GOF p53 when mutant p53 was shown to up-regulate certain cellular and viral promoters including PCNA and MDR1 [23, 28]. Promoter activation seems to be a criterion of GOF activity, as many more promoters have been shown to be activated by mutant p53. Examples of this include: epidermal growth factor receptor (EGFR), multiple drug resistance (MDR1/ABCB1), vascular endothelial growth factor (VEGF), human interleukin-6 (IL-6), basic fibroblast growth factor (BFGF), human HSP70, c-*myc*, insulin like growth factor I receptor (IGF-IR), insulin like growth factor II receptor (IGF-IIR), the anti-apoptotic factor BAG-I, collagenase-3, c-*fos*, 15-lipoxygenase [27–34], etc.

Using cell and animal model systems, expression of mutant p53 has been shown to result in oncogenic and proliferative processes [122] such as (i) increased tumorigenicity [33, 77, 88], (ii) increased growth in soft agar [147], (iii) decreased sensitivity to chemotherapeutic drugs [11, 17, 74, 142], (iv) increased resistance to γ-irradiation [96], (v) accelerated chemical carcinogenesis [168], (vi) increased homologous recombination induced by ionizing radiation [65, 79], (vii) disruption of the spindle check point [65, 131, 167], (viii) activated topoisomerase I activity [3], (ix) increased DNA synthesis and increased growth rate [15, 27, 30, 34, 56, 111], (x) induction of gene amplification [36, 64, 112], (xi) cooperation with the TGF-β pathway [1, 68], (xii) induction of cellular motility, invasive capability and metastasis [1, 109, 150, 155, 166, 177], somatic cell reprogramming [107, 139], increased tumor angiogenesis [44], and promotion of chronic inflammation and associated cancer [25].

Mouse Models for GOF p53

Several mouse models have been generated to test if the expression of p53 mutants in mice will induce GOF activity. Using a knock-in mouse line, Liu et al. [91, 92] demonstrated an increased metastatic potential in mice inheriting p53-R172H (homologous to human p53-R175H). Using two mouse models of the Li-Fraumeni Syndrome, clear and strong evidence of mutant p53 gain of function has been demonstrated, where higher metastatic spread and/or different tumor spectra have been observed in the presence of mutant p53 [76, 121]. Wang et al. [167, 168] showed some evidence of gain of function associated with genomic instability using a murine p53-R172H mutant transgenic model system. Murphy et al. [111] used the same p53 mutant in a mammary epithelial murine cell model to reach similar conclusions. Hixon et al. [59] proposed that cells carrying mutant p53 over-express Cks1, a protein that mediates activating phosphorylation of the anaphase promoting complex (APC) by cdc2, leading to chromosomal instability as cells are unstable to sustain APC inactivation. Oncogenic mutant p53 has been demonstrated to confer a dominant, gain of function phenotype that disrupts spindle checkpoint control that is not dependent on transactivation [167]. Ohiro et al. [119] demonstrated that mutant p53 inhibits stress-inducible kinase pathways, and showed its anti-apoptotic activity. This activity is also independent of the transactivation function of mutant p53. Possibly the most unequivocal demonstration of gain of function activity in mouse systems was shown when mutant p53 was stabilized by knockout of MDM2 or p16 [156].

Transcriptional Effect of GOF p53

There have been several groups that have identified sets of cellular genes that are either up- or down-regulated by GOF p53 mutants as identified mostly by microarray analyses after expressing p53 mutants in p53 null cells such as H1299 [45, 122, 142–144, 171–173]. A series of genes involved in cell growth and oncogenesis were found to be up-regulated by the expression of p53 mutants R175H, R273H, and D281G [142–144]. Similar to WT p53, mutant p53 has also been demonstrated to repress expression of a number of genes including inhibitor of differentiation (Id2)

[161, 176] p21, gadd45, PERP and PTEN [161], among others. Some of these genes are involved in pro-apoptotic activities, including CD95 (Fas/ApoI) [54, 178], caspase-3 [175] and others [140, 179].

Mechanisms to Explain p53 Gain of Function

The mechanisms underlying the different phenotypes of p53 GOF have not been fully elucidated. In this section we will attempt to give an up-to-date version of the proposed mechanisms and why they are preferred. Mutant p53 may drive GOF phenotypes by acquiring new functions that result in uncontrolled proliferation, survival and motility [19, 174]. There are two possible mechanisms to explain GOF mutant p53. (Fig. 4.2): (a) one that involves its direct binding to DNA and regulation of gene expression; (b) one in which mutant p53 does not directly come near regulatory sequences on the chromosome. This second category can be subdivided further: (1) protein-protein interactions between mutant p53 and other cellular protein(s), such as the p53 family members, p63 and p73, DNA machinery proteins, and/or proteins of the apoptotic pathway [106, 108]; (2) modulation of target genes by mutant p53 (Fig. 4.2), such as activating growth promoting genes, disrupting

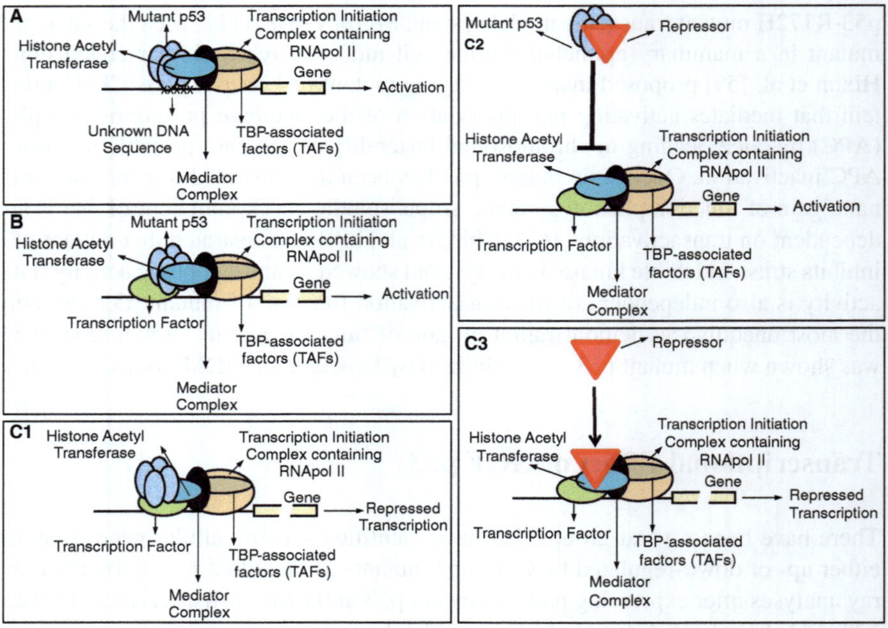

Fig. 4.2 Model showing transactivation of promoters by GOF p53

DNA repair or apoptotic activities or inhibiting growth suppressive genes [12, 13, 151, 172, 178]. These two mechanisms are not mutually exclusive.

(a) *Regulation of transcription by binding to DNA*. There is some evidence to show that GOF p53 does in fact bind to DNA on the chromosome, particularly in G/C rich areas around transcription start sites of some genes that are characterized by active chromatin marks [128]. It seems that this interaction is responsible for the control of expression of genes such as GAS1 and HTR2A. The DNA binding activity ascribed to GOF p53 seems to stem from its ability to interact with stem and loop structured DNA and non-B DNA structures [52, 165]. Further work on this model is needed in order to clarify how regulation of transactivation occurs.
(b) *Regulation of transcription by protein-protein interactions.* As mentioned above, GOF p53 can modulate the transcription of genes through its interactions with other protein factors.

    (i) *Interaction of mutant p53 with p63 and p73 to induce GOF activity.* The fact that p53 family members p63 and p73 retain the capability of interacting with tumor-derived p53 mutants led to the possibility that such interactions may inhibit p63/p73 function [26, 47, 86, 93, 110, 151]. In this model, it is thought that mutant p53 may be inhibiting p63/p73's DNA binding ability by interacting with the DNA binding domain of p63/p73, as model systems outlined in Figure 4.2 C2 predict [47]. As explained in this model, some of the GOF activities of mutant p53 assume that mutant p53 would block the inhibitory effects of p63/p73 (mostly p63). However, transactivation observed by GOF p53 cannot easily be explained by this model alone, as there are genes that are in fact transactivated by p63 as well as GOF p53 [123, 160]. There is another model that is evolving that involves mostly p63 (and perhaps p73), in which it is assumed that mutant p53 can anchor to p63 and become nucleated on promoters with p63/p73 sites. There is evidence for this possibility also [115].
    (ii) *Interaction of mutant p53 with transcription factors resulting in its nucleation on the regulatory sequences controlling gene transcription.* It was proposed about 20 years ago [14, 28, 152] that mutant p53 may transactivate genes it up-regulates indirectly through interactions with transcription factors that normally bind to the regulatory regions of those target genes. Various groups have shown interactions of p53 (both WT and mutant) with several transcription factors including Sp1, E2F1, Ets1, CREB, p300/CREBP, NFY-A [14, 30, 43, 48, 78, 87, 93, 137] and others. Although in some cases evidence has been presented that indicates nucleation of mutant p53 on DNA by anchoring onto another transcription factor, in most cases there is a lack of clear-cut evidence. In the model outlined in Figure 4.2, overall it is assumed that mutant p53 becomes situated on the promoter-regulatory sequence of a target gene by its interaction with a transcription factor, and then through its intact transactivation domain it is able to interact with the transcription initiation complex pre-

sumably near the transcription start site. It is most likely that the transactivation domain interacts with the component of the "mediator" complex [10, 63, 104]. Also, it is generally assumed that mutant p53-CREBP/p300 interaction would facilitate nucleation of these histone acetylases on the chromatin. This results in an opening up of the chromatin by acetylation of histones [8, 118].

Cancer Cells Expressing GOF p53 Are Addicted to Mutant p53

Recent evidence indicates that a number of cancer cell lines are "addicted" to the presence of the GOF p53 protein, and in some cancers such as lung carcinoma, mutations in the p53 gene act as a "driver" for lung oncogenesis [32]. This oncogene addiction is recognized to be a phenomenon whereby cancer cells are continuously dependent on the presence of "driver" genes for their sustenance [157]. By definition, activation of such a gene is not required to be involved in the initiating events of oncogenesis, but it may arise later in the process and becomes essential for cancer cells to survive and grow. The removal or inhibition of that gene/gene product could lead to the selective death of cancer cells, their cell cycle arrest, senescence, or differentiation. In this case, inhibition of mutant p53 would prevent transcriptional activation of target genes involved in cell growth and survival and would remove the mutant p53-mediated inhibition of p63/p73 leading to regulation of the cell cycle and possibly apoptosis. In animal model systems, tumor regression has been observed and this has opened up important avenues for anticancer therapy [170]. Several lung cancer cell lines with endogenous expression of mutant p53 have been tested for their addiction to mutant p53 expression. Surprisingly, the addiction seems to be allele specific in some GOF activity, though all the cell lines characteristically lose enhanced growth upon reduction of GOF p53 levels [158, 159]. The mechanism of addiction to GOF p53 for some cancer cells for their tumorigenicity is unknown [158, 159].

Summary

p53 plays a significant role in oncogenesis, whether it is present in the cell as the wildtype or mutant form of the protein. WT p53 has long been shown to be key to maintaining the integrity of the cell and therefore preventing oncogenesis; meanwhile mutant p53 causes the acquisition of new oncogenic functions. Both WT and mutant p53 have biological activities that are dependent on its transcriptional ability, and both proteins have been demonstrated to repress expression of a number of genes. Although the mechanism for how p53 activates transcription of its target genes has not been fully elucidated, there are several models that are being explored such as direct binding to DNA, recruitment of histone acetyltransferases, or protein-protein binding. Not only does mutant p53 cause cancer cells to become more oncogenic, but recently it has been shown that cancer cells can become addicted to the mutant protein. Reducing mutant p53 expression in various cell lines has been shown to be an effective way to diminish cell growth, migration, and tumorigenicity. Since the goal of cancer research aims to discover what causes a cell to become oncogenic in the hope of developing a way to cure the disease, understanding the mechanism of gain of function is essential to be able to effectively target mutant p53 for cancer therapy.

References

1. Adorno M, Cordenonsi M, Montagner M, Dupont S, Wong C, Hann B, Solari A, Bobisse S, Rondina MB, Guzzardo V, Parenti AR, Rosato A, Bicciato S, Balmain A, Piccolo S (2009) A mutant-p53/Smad complex opposes p63 to empower TGFbeta-induced metastasis. Cell 137:87–98
2. Ahrendt SA, Hu Y, Buta M, McDermott MP, Benoit N, Yang SC, Wu L, Sidransky D (2003) p53 mutations and survival in stage I non-small-cell lung cancer: results of a prospective study. J Natl Cancer Inst 95:961–970
3. Albor A, Kaku S, Kulesz-Martin M (1998) Wild-type and mutant forms of p53 activate human topoisomerase I: a possible mechanism for gain of function in mutants. Cancer Res 58:2091–2094
4. Aylon Y, Oren M (2011) p53: guardian of ploidy. Mol Oncol 5:315–323
5. Baker L, Quinlan PR, Patten N, Ashfield A, Birse-Stewart-Bell LJ, McCowan C, Bourdon JC, Purdie CA, Jordan LB, Dewar JA, Wu L, Thompson AM (2010) p53 mutation, deprivation and poor prognosis in primary breast cancer. Br J Cancer 102:719–726
6. Baker SJ, Fearon ER, Nigro JM, Hamilton SR, Preisinger AC, Jessup JM, vanTuinen P, Ledbetter DH, Barker DF, Nakamura Y, White R, Vogelstein B (1989) Chromosome 17 deletions and p53 gene mutations in colorectal carcinomas. Science 244:217–221
7. Baker SJ, Markowitz S, Fearon ER, Willson JK, Vogelstein B (1990) Suppression of human colorectal carcinoma cell growth by wild-type p53. Science 249:912–915
8. Bannister AJ, Kouzarides T (1996) The CBP co-activator is a histone acetyltransferase. Nature 384:641–643
9. Barlev NA, Liu L, Chehab NH, Mansfield K, Harris KG, Halazonetis TD, Berger SL (2001) Acetylation of p53 activates transcription through recruitment of coactivators/histone acetyltransferases. Mol Cell 8:1243–1254
10. Beckerman R, Prives C (2010) Transcriptional regulation by p53. Cold Spring Harb Perspect Biol 2:a000935
11. Blandino G, Levine AJ, Oren M (1999) Mutant p53 gain of function: differential effects of different p53 mutants on resistance of cultured cells to chemotherapy. Oncogene 18:477–485
12. Boehden GS, Akyuz N, Roemer K, Wiesmuller L (2003) p53 mutated in the transactivation domain retains regulatory functions in homology-directed double-strand break repair. Oncogene 22:4111–4117
13. Boehden GS, Restle A, Marschalek R, Stocking C, Wiesmuller L (2004) Recombination at chromosomal sequences involved in leukaemogenic rearrangements is differentially regulated by p53. Carcinogenesis 25:1305–1313
14. Borellini F, Glazer RI (1993) Induction of Sp1-p53 DNA-binding heterocomplexes during granulocyte/macrophage colony-stimulating factor-dependent proliferation in human erythroleukemia cell line TF-1. J Biol Chem 268:7923–7928

15. Bossi G, Lapi E, Strano S, Rinaldo C, Blandino G, Sacchi A (2006) Mutant p53 gain of function: reduction of tumor malignancy of human cancer cell lines through abrogation of mutant p53 expression. Oncogene 25:304–309
16. Bourdon JC, Khoury MP, Diot A, Baker L, Fernandes K, Aoubala M, Quinlan P, Purdie CA, Jordan LB, Prats AC, Lane DP, Thompson AM (2011) p53 mutant breast cancer patients expressing p53gamma have as good a prognosis as wild-type p53 breast cancer patients. Breast Cancer Res 13:R7
17. Bristow RG, Peacock J, Jang A, Kim J, Hill RP, Benchimol S (2003) Resistance to DNA-damaging agents is discordant from experimental metastatic capacity in MEF ras-transformants-expressing gain of function MTp53. Oncogene 22:2960–2966
18. Brosh R, Rotter V (2009) When mutants gain new powers: news from the mutant p53 field. Nat Rev Cancer 9:701–713
19. Cadwell C, Zambetti GP (2001) The effects of wild-type p53 tumor suppressor activity and mutant p53 gain-of-function on cell growth. Gene 277:15–30
20. Campling BG, el-Deiry WS (2003) Clinical implications of p53 mutations in lung cancer. Methods Mol Med 75:53–77
21. Chang C, Simmons DT, Martin MA, Mora PT (1979) Identification and partial characterization of new antigens from simian virus 40-transformed mouse cells. J Virol 31:463–471
22. Chen X, Farmer G, Zhu H, Prywes R, Prives C (1993) Cooperative DNA binding of p53 with TFIID (TBP): a possible mechanism for transcriptional activation. Genes Dev 7:1837–1849
23. Chin KV, Ueda K, Pastan I, Gottesman MM (1992) Modulation of activity of the promoter of the human MDR1 gene by Ras and p53. Science 255:459–462, New York, N.Y
24. Cho Y, Gorina S, Jeffrey PD, Pavletich NP (1994) Crystal structure of a p53 tumor suppressor-DNA complex: understanding tumorigenic mutations. Science 265:346–355
25. Cooks T, Pateras IS, Tarcic O, Solomon H, Schetter AJ, Wilder S, Lozano G, Pikarsky E, Forshew T, Rozenfeld N, Harpaz N, Itzkowitz S, Harris CC, Rotter V, Gorgoulis VG, Oren M (2013) Mutant p53 prolongs NF-kappaB activation and promotes chronic inflammation and inflammation-associated colorectal cancer. Cancer Cell 23:634–646
26. Cordani N, Pozzi S, Martynova E, Fanoni D, Borrelli S, Alotto D, Castagnoli C, Berti E, Vigano MA, Mantovani R (2011) Mutant p53 subverts p63 control over KLF4 expression in keratinocytes. Oncogene 30:922–932
27. Deb D, Scian M, Roth KE, Li W, Keiger J, Chakraborti AS, Deb SP, Deb S (2002) Hetero-oligomerization does not compromise 'gain of function' of tumor-derived p53 mutants. Oncogene 21:176–189
28. Deb S, Jackson CT, Subler MA, Martin DW (1992) Modulation of cellular and viral promoters by mutant human p53 proteins found in tumor cells. J Virol 66:6164–6170
29. DeLeo AB, Jay G, Appella E, Dubois GC, Law LW, Old LJ (1979) Detection of a transformation-related antigen in chemically induced sarcomas and other transformed cells of the mouse. Proc Natl Acad Sci U S A 76:2420–2424
30. Di Agostino S, Strano S, Emiliozzi V, Zerbini V, Mottolese M, Sacchi A, Blandino G, Piaggio G (2006) Gain of function of mutant p53: the mutant p53/NF-Y protein complex reveals an aberrant transcriptional mechanism of cell cycle regulation. Cancer Cell 10:191–202
31. Diller L, Kassel J, Nelson CE, Gryka MA, Litwak G, Gebhardt M, Bressac B, Ozturk M, Baker SJ, Vogelstein B et al (1990) p53 functions as a cell cycle control protein in osteosarcomas. Mol Cell Biol 10:5772–5781
32. Ding L, Getz G, Wheeler DA, Mardis ER, McLellan MD, Cibulskis K, Sougnez C, Greulich H, Muzny DM, Morgan MB, Fulton L, Fulton RS, Zhang Q, Wendl MC, Lawrence MS, Larson DE, Chen K, Dooling DJ, Sabo A, Hawes AC, Shen H, Jhangiani SN, Lewis LR, Hall O, Zhu Y, Mathew T, Ren Y, Yao J, Scherer SE, Clerc K, Metcalf GA, Ng B, Milosavljevic A, Gonzalez-Garay ML, Osborne JR, Meyer R, Shi X, Tang Y, Koboldt DC, Lin L, Abbott R, Miner TL, Pohl C, Fewell G, Haipek C, Schmidt H, Dunford-Shore BH, Kraja A, Crosby SD, Sawyer CS, Vickery T, Sander S, Robinson J, Winckler W, Baldwin J, Chirieac LR, Dutt A, Fennell T, Hanna M, Johnson BE, Onofrio RC, Thomas RK, Tonon G, Weir BA, Zhao X, Ziaugra L, Zody MC, Giordano T, Orringer MB, Roth JA, Spitz MR, Wistuba II, Ozenberger B, Good PJ, Chang AC, Beer DG, Watson MA, Ladanyi M, Broderick S, Yoshizawa A,

Travis WD, Pao W, Province MA, Weinstock GM, Varmus HE, Gabriel SB, Lander ES, Gibbs RA, Meyerson M, Wilson RK (2008) Somatic mutations affect key pathways in lung adenocarcinoma. Nature 455:1069–1075
33. Dittmer D, Pati S, Zambetti G, Chu S, Teresky AK, Moore M, Finlay C, Levine AJ (1993) Gain of function mutations in p53. Nat Genet 4:42–46
34. Duan W, Gao L, Jin D, Otterson GA, Villalona-Calero MA (2008) Lung specific expression of a human mutant p53 affects cell proliferation in transgenic mice. Transgenic Res 17:355–366
35. el-Deiry WS, Kern SE, Pietenpol JA, Kinzler KW, Vogelstein B (1992) Definition of a consensus binding site for p53. Nat Genet 1:45–49
36. El-Hizawi S, Lagowski JP, Kulesz-Martin M, Albor A (2002) Induction of gene amplification as a gain-of-function phenotype of mutant p53 proteins. Cancer Res 62:3264–3270
37. Eliyahu D, Michalovitz D, Eliyahu S, Pinhasi-Kimhi O, Oren M (1989) Wild-type p53 can inhibit oncogene-mediated focus formation. Proc Natl Acad Sci U S A 86:8763–8767
38. Eliyahu D, Raz A, Gruss P, Givol D, Oren M (1984) Participation of p53 cellular tumour antigen in transformation of normal embryonic cells. Nature 312:646–649
39. Espinosa JM, Portal D, Lobo GS, Pereira CA, Alonso GD, Gomez EB, Lan GH, Pomar RV, Flawia MM, Torres HN (2003) Trypanosoma cruzi poly-zinc finger protein: a novel DNA/RNA-binding CCHC-zinc finger protein. Mol Biochem Parasitol 131:35–44
40. Fields S, Jang SK (1990) Presence of a potent transcription activating sequence in the p53 protein. Science 249:1046–1049
41. Finlay CA, Hinds PW, Levine AJ (1989) The p53 proto-oncogene can act as a suppressor of transformation. Cell 57:1083–1093
42. Fiscella M, Ullrich SJ, Zambrano N, Shields MT, Lin D, Lees-Miller SP, Anderson CW, Mercer WE, Appella E (1993) Mutation of the serine 15 phosphorylation site of human p53 reduces the ability of p53 to inhibit cell cycle progression. Oncogene 8:1519–1528
43. Fogal V, Hsieh JK, Royer C, Zhong S, Lu X (2005) Cell cycle-dependent nuclear retention of p53 by E2F1 requires phosphorylation of p53 at Ser315. EMBO J 24:2768–2782
44. Fontemaggi G, Dell'Orso S, Trisciuoglio D, Shay T, Melucci E, Fazi F, Terrenato I, Mottolese M, Muti P, Domany E, Del Bufalo D, Strano S, Blandino G (2009) The execution of the transcriptional axis mutant p53, E2F1 and ID4 promotes tumor neo-angiogenesis. Nat Struct Mol Biol 16:1086–1093
45. Freed-Pastor WA, Prives C (2012) Mutant p53: one name, many proteins. Genes Dev 26:1268–1286
46. Funk WD, Pak DT, Karas RH, Wright WE, Shay JW (1992) A transcriptionally active DNA-binding site for human p53 protein complexes. Mol Cell Biol 12:2866–2871
47. Gaiddon C, Lokshin M, Ahn J, Zhang T, Prives C (2001) A subset of tumor-derived mutant forms of p53 down-regulate p63 and p73 through a direct interaction with the p53 core domain. Mol Cell Biol 21:1874–1887
48. Giebler HA, Lemasson I, Nyborg JK (2000) p53 recruitment of CREB binding protein mediated through phosphorylated CREB: a novel pathway of tumor suppressor regulation. Mol Cell Biol 20:4849–4858
49. Ginsberg D, Mechta F, Yaniv M, Oren M (1991) Wild-type p53 can down-modulate the activity of various promoters. Proc Natl Acad Sci U S A 88:9979–9983
50. Godar S, Ince TA, Bell GW, Feldser D, Donaher JL, Bergh J, Liu A, Miu K, Watnick RS, Reinhardt F, McAllister SS, Jacks T, Weinberg RA (2008) Growth-inhibitory and tumor-suppressive functions of p53 depend on its repression of CD44 expression. Cell 134:62–73
51. Goh W, Lane D, Ghadessy F (2010) Development of a novel multiplex in vitro binding assay to profile p53-DNA interactions. Cell Cycle 9:3030–3038
52. Gohler T, Jager S, Warnecke G, Yasuda H, Kim E, Deppert W (2005) Mutant p53 proteins bind DNA in a DNA structure-selective mode. Nucleic Acids Res 33:1087–1100
53. Grossman SR (2001) p300/CBP/p53 interaction and regulation of the p53 response. Eur J Biochem/FEBS 268:2773–2778

54. Gurova KV, Rokhlin OW, Budanov AV, Burdelya LG, Chumakov PM, Cohen MB, Gudkov AV (2003) Cooperation of two mutant p53 alleles contributes to Fas resistance of prostate carcinoma cells. Cancer Res 63:2905–2912
55. Hamelin R, Laurent-Puig P, Olschwang S, Jego N, Asselain B, Remvikos Y, Girodet J, Salmon RJ, Thomas G (1994) Association of p53 mutations with short survival in colorectal cancer. Gastroenterology 106:42–48
56. Haupt S, di Agostino S, Mizrahi I, Alsheich-Bartok O, Voorhoeve M, Damalas A, Blandino G, Haupt Y (2009) Promyelocytic leukemia protein is required for gain of function by mutant p53. Cancer Res 69:4818–4826
57. Hermeking H (2012) MicroRNAs in the p53 network: micromanagement of tumour suppression. Nat Rev Cancer 12:613–626
58. Hinds P, Finlay C, Levine AJ (1989) Mutation is required to activate the p53 gene for cooperation with the ras oncogene and transformation. J Virol 63:739–746
59. Hixon ML, Flores A, Wagner M, Gualberto A (2000) Gain of function properties of mutant p53 proteins at the mitotic spindle cell cycle checkpoint. Histol Histopathol 15:551–556
60. Hoffman WH, Biade S, Zilfou JT, Chen J, Murphy M (2002) Transcriptional repression of the anti-apoptotic survivin gene by wild type p53. J Biol Chem 277:3247–3257
61. Hsiao M, Low J, Dorn E, Ku D, Pattengale P, Yeargin J, Haas M (1994) Gain-of-function mutations of the p53 gene induce lymphohematopoietic metastatic potential and tissue invasiveness. Am J Pathol 145:702–714
62. Imbriano C, Gurtner A, Cocchiarella F, Di Agostino S, Basile V, Gostissa M, Dobbelstein M, Del Sal G, Piaggio G, Mantovani R (2005) Direct p53 transcriptional repression: in vivo analysis of CCAAT-containing G2/M promoters. Mol Cell Biol 25:3737–3751
63. Ito M, Yuan CX, Malik S, Gu W, Fondell JD, Yamamura S, Fu ZY, Zhang X, Qin J, Roeder RG (1999) Identity between TRAP and SMCC complexes indicates novel pathways for the function of nuclear receptors and diverse mammalian activators. Mol Cell 3:361–370
64. Iwamoto KS, Mizuno T, Ito T, Tsuyama N, Kyoizumi S, Seyama T (1996) Gain-of-function p53 mutations enhance alteration of the T-cell receptor following X-irradiation, independently of the cell cycle and cell survival. Cancer Res 56:3862–3865
65. Iwanaga Y, Jeang KT (2002) Expression of mitotic spindle checkpoint protein hsMAD1 correlates with cellular proliferation and is activated by a gain-of-function p53 mutant. Cancer Res 62:2618–2624
66. Jenkins JR, Rudge K, Currie GA (1984) Cellular immortalization by a cDNA clone encoding the transformation-associated phosphoprotein p53. Nature 312:651–654
67. Johnson RA, Ince TA, Scotto KW (2001) Transcriptional repression by p53 through direct binding to a novel DNA element. J Biol Chem 276:27716–27720
68. Kalo E, Buganim Y, Shapira KE, Besserglick H, Goldfinger N, Weise L, Stambolsky P, Henis YI, Rotter V (2007) Mutant p53 attenuates the SMAD-dependent TGF-{beta}1 signaling pathway by repressing the expression of TGF-{beta} receptor type II. Mol Cell Biol 27(23):8228–8242
69. Katkoori VR, Jia X, Shanmugam C, Wan W, Meleth S, Bumpers H, Grizzle WE, Manne U (2009) Prognostic significance of p53 codon 72 polymorphism differs with race in colorectal adenocarcinoma. Clin Cancer Res 15:2406–2416
70. Kikuchi-Yanoshita R, Tanaka K, Muraoka M, Konishi M, Kawashima I, Takamoto S, Hirai H, Miyaki M (1995) Malignant transformation of rat embryo fibroblasts by cotransfection with eleven human mutant p53 cDNAs and activated H-ras gene. Oncogene 11:1339–1345
71. Krause K, Haugwitz U, Wasner M, Wiedmann M, Mossner J, Engeland K (2001) Expression of the cell cycle phosphatase cdc25C is down-regulated by the tumor suppressor protein p53 but not by p73. Biochem Biophys Res Commun 284:743–750
72. Kress M, May E, Cassingena R, May P (1979) Simian virus 40-transformed cells express new species of proteins precipitable by anti-simian virus 40 tumor serum. J Virol 31:472–483
73. Lai H, Lin L, Nadji M, Lai S, Trapido E, Meng L (2002) Mutations in the p53 tumor suppressor gene and early onset breast cancer. Cancer Biol Ther 1:31–36
74. Lai SL, Perng RP, Hwang J (2000) p53 gene status modulates the chemosensitivity of non-small cell lung cancer cells. J Biomed Sci 7:64–70

75. Lane DP, Crawford LV (1979) T antigen is bound to a host protein in SV40-transformed cells. Nature 278:261–263
76. Lang GA, Iwakuma T, Suh YA, Liu G, Rao VA, Parant JM, Valentin-Vega YA, Terzian T, Caldwell LC, Strong LC, El-Naggar AK, Lozano G (2004) Gain of function of a p53 hot spot mutation in a mouse model of Li-Fraumeni syndrome. Cell 119:861–872
77. Lanyi A, Deb D, Seymour RC, Ludes-Meyers JH, Subler MA, Deb S (1998) 'Gain of function' phenotype of tumor-derived mutant p53 requires the oligomerization/nonsequence-specific nucleic acid-binding domain. Oncogene 16:3169–3176
78. Lee CW, Ferreon JC, Ferreon AC, Arai M, Wright PE (2010) Graded enhancement of p53 binding to CREB-binding protein (CBP) by multisite phosphorylation. Proc Natl Acad Sci U S A 107:19290–19295
79. Lee JM, Bernstein A (1993) p53 mutations increase resistance to ionizing radiation. Proc Natl Acad Sci U S A 90:5742–5746
80. Lees-Miller SP, Sakaguchi K, Ullrich SJ, Appella E, Anderson CW (1992) Human DNA-activated protein kinase phosphorylates serines 15 and 37 in the amino-terminal transactivation domain of human p53. Mol Cell Biol 12:5041–5049
81. Legube G, Linares LK, Tyteca S, Caron C, Scheffner M, Chevillard-Briet M, Trouche D (2004) Role of the histone acetyl transferase Tip60 in the p53 pathway. J Biol Chem 279:44825–44833
82. Levine AJ, Oren M (2009) The first 30 years of p53: growing ever more complex. Nat Rev Cancer 9:749–758
83. Levine AJ, Wu MC, Chang A, Silver A, Attiyeh EF, Lin J, Epstein CB (1995) The spectrum of mutations at the p53 locus. Evidence for tissue-specific mutagenesis, selection of mutant alleles, and a "gain of function" phenotype. Ann N Y Acad Sci 768:111–128
84. Li B, Lee MY (2001) Transcriptional regulation of the human DNA polymerase delta catalytic subunit gene POLD1 by p53 tumor suppressor and Sp1. J Biol Chem 276:29729–29739
85. Li M, He Y, Dubois W, Wu X, Shi J, Huang J (2012) Distinct regulatory mechanisms and functions for p53-activated and p53-repressed DNA damage response genes in embryonic stem cells. Mol Cell 46:30–42
86. Li Y, Prives C (2007) Are interactions with p63 and p73 involved in mutant p53 gain of oncogenic function? Oncogene 26:2220–2225
87. Lill NL, Grossman SR, Ginsberg D, DeCaprio J, Livingston DM (1997) Binding and modulation of p53 by p300/CBP coactivators. Nature 387:823–827
88. Lin J, Teresky AK, Levine AJ (1995) Two critical hydrophobic amino acids in the N-terminal domain of the p53 protein are required for the gain of function phenotypes of human p53 mutants. Oncogene 10:2387–2390
89. Linzer DI, Levine AJ (1979) Characterization of a 54 K dalton cellular SV40 tumor antigen present in SV40-transformed cells and uninfected embryonal carcinoma cells. Cell 17:43–52
90. Linzer DI, Maltzman W, Levine AJ (1979) The SV40 A gene product is required for the production of a 54,000 MW cellular tumor antigen. Virology 98:308–318
91. Liu G, McDonnell TJ, Montes de Oca Luna R, Kapoor M, Mims B, El-Naggar AK, Lozano G (2000) High metastatic potential in mice inheriting a targeted p53 missense mutation. Proc Natl Acad Sci U S A 97:4174–4179
92. Liu G, Parant JM, Lang G, Chau P, Chavez-Reyes A, El-Naggar AK, Multani A, Chang S, Lozano G (2004) Chromosome stability, in the absence of apoptosis, is critical for suppression of tumorigenesis in Trp53 mutant mice. Nat Genet 36:63–68
93. Liu K, Ling S, Lin WC (2011) TopBP1 mediates mutant p53 gain of function through NF-Y and p63/p73. Mol Cell Biol 31:4464–4481
94. Liu L, Scolnick DM, Trievel RC, Zhang HB, Marmorstein R, Halazonetis TD, Berger SL (1999) p53 sites acetylated in vitro by PCAF and p300 are acetylated in vivo in response to DNA damage. Mol Cell Biol 19:1202–1209
95. Liu X, Miller CW, Koeffler PH, Berk AJ (1993) The p53 activation domain binds the TATA box-binding polypeptide in Holo-TFIID, and a neighboring p53 domain inhibits transcription. Mol Cell Biol 13:3291–3300

96. Lotem J, Sachs L (1995) Interferon-gamma inhibits apoptosis induced by wild-type p53, cytotoxic anti-cancer agents and viability factor deprivation in myeloid cells. Leukemia 9:685–692
97. Love IM, Sekaric P, Shi D, Grossman SR, Androphy EJ (2012) The histone acetyltransferase PCAF regulates p21 transcription through stress-induced acetylation of histone H3. Cell Cycle 11:2458–2466
98. Ludes-Meyers JH, Subler MA, Shivakumar CV, Munoz RM, Jiang P, Bigger JE, Brown DR, Deb SP, Deb S (1996) Transcriptional activation of the human epidermal growth factor receptor promoter by human p53. Mol Cell Biol 16:6009–6019
99. Maddocks OD, Vousden KH (2011) Metabolic regulation by p53. J Mol Med (Berl) 89:237–245
100. Martin DW, Munoz RM, Subler MA, Deb S (1993) p53 binds to the TATA-binding protein-TATA complex. J Biol Chem 268:13062–13067
101. Martin DW, Subler MA, Munoz RM, Brown DR, Deb SP, Deb S (1993) p53 and SV40 T antigen bind to the same region overlapping the conserved domain of the TATA-binding protein. Biochem Biophys Res Commun 195:428–434
102. Melero JA, Stitt DT, Mangel WF, Carroll RB (1979) Identification of new polypeptide species (48–55 K) immunoprecipitable by antiserum to purified large T antigen and present in SV40-infected and -transformed cells. Virology 93:466–480
103. Menendez D, Nguyen TA, Freudenberg JM, Mathew VJ, Anderson CW, Jothi R, Resnick MA (2013) Diverse stresses dramatically alter genome-wide p53 binding and transactivation landscape in human cancer cells. Nucleic Acids Res 41(15):7286–7301
104. Meyer KD, Lin SC, Bernecky C, Gao Y, Taatjes DJ (2010) p53 activates transcription by directing structural shifts in Mediator. Nat Struct Mol Biol 17:753–760
105. Michalovitz D, Halevy O, Oren M (1990) Conditional inhibition of transformation and of cell proliferation by a temperature-sensitive mutant of p53. Cell 62:671–680
106. Mihara M, Erster S, Zaika A, Petrenko O, Chittenden T, Pancoska P, Moll UM (2003) p53 has a direct apoptogenic role at the mitochondria. Mol Cell 11:577–590
107. Mizuno H, Spike BT, Wahl GM, Levine AJ (2010) Inactivation of p53 in breast cancers correlates with stem cell transcriptional signatures. Proc Natl Acad Sci U S A 107:22745–22750
108. Moll UM, Zaika A (2001) Nuclear and mitochondrial apoptotic pathways of p53. FEBS Lett 493:65–69
109. Muller PA, Caswell PT, Doyle B, Iwanicki MP, Tan EH, Karim S, Lukashchuk N, Gillespie DA, Ludwig RL, Gosselin P, Cromer A, Brugge JS, Sansom OJ, Norman JC, Vousden KH (2009) Mutant p53 drives invasion by promoting integrin recycling. Cell 139:1327–1341
110. Muller PA, Vousden KH (2013) p53 mutations in cancer. Nat Cell Biol 15:2–8
111. Murphy KL, Dennis AP, Rosen JM (2000) A gain of function p53 mutant promotes both genomic instability and cell survival in a novel p53-null mammary epithelial cell model. FASEB J 14:2291–2302
112. Murphy KL, Rosen JM (2000) Mutant p53 and genomic instability in a transgenic mouse model of breast cancer. Oncogene 19:1045–1051
113. Murphy M, Hinman A, Levine AJ (1996) Wild-type p53 negatively regulates the expression of a microtubule-associated protein. Genes Dev 10:2971–2980
114. Nagaich AK, Appella E, Harrington RE (1997) DNA bending is essential for the site-specific recognition of DNA response elements by the DNA binding domain of the tumor suppressor protein p53. J Biol Chem 272:14842–14849
115. Neilsen PM, Noll JE, Suetani RJ, Schulz RB, Al-Ejeh F, Evdokiou A, Lane DP, Callen DF (2011) Mutant p53 uses p63 as a molecular chaperone to alter gene expression and induce a pro-invasive secretome. Oncotarget 2:1203–1217
116. Nikulenkov F, Spinnler C, Li H, Tonelli C, Shi Y, Turunen M, Kivioja T, Ignatiev I, Kel A, Taipale J, Selivanova G (2012) Insights into p53 transcriptional function via genome-wide chromatin occupancy and gene expression analysis. Cell Death Differ 19:1992–2002
117. O'Rourke RW, Miller CW, Kato GJ, Simon KJ, Chen DL, Dang CV, Koeffler HP (1990) A potential transcriptional activation element in the p53 protein. Oncogene 5:1829–1832

118. Ogryzko VV, Schiltz RL, Russanova V, Howard BH, Nakatani Y (1996) The transcriptional coactivators p300 and CBP are histone acetyltransferases. Cell 87:953–959
119. Ohiro Y, Usheva A, Kobayashi S, Duffy SL, Nantz R, Gius D, Horikoshi N (2003) Inhibition of stress-inducible kinase pathways by tumorigenic mutant p53. Mol Cell Biol 23:322–334
120. Okorokov AL, Sherman MB, Plisson C, Grinkevich V, Sigmundsson K, Selivanova G, Milner J, Orlova EV (2006) The structure of p53 tumour suppressor protein reveals the basis for its functional plasticity. EMBO J 25:5191–5200
121. Olive KP, Tuveson DA, Ruhe ZC, Yin B, Willis NA, Bronson RT, Crowley D, Jacks T (2004) Mutant p53 gain of function in two mouse models of Li-Fraumeni syndrome. Cell 119:847–860
122. Oren M, Rotter V (2010) Mutant p53 gain-of-function in cancer. Cold Spring Harb Perspect Biol 2:a001107
123. Osada M, Park HL, Nagakawa Y, Yamashita K, Fomenkov A, Kim MS, Wu G, Nomoto S, Trink B, Sidransky D (2005) Differential recognition of response elements determines target gene specificity for p53 and p63. Mol Cell Biol 25:6077–6089
124. Parada LF, Land H, Weinberg RA, Wolf D, Rotter V (1984) Cooperation between gene encoding p53 tumour antigen and ras in cellular transformation. Nature 312:649–651
125. Petitjean A, Achatz MI, Borresen-Dale AL, Hainaut P, Olivier M (2007) TP53 mutations in human cancers: functional selection and impact on cancer prognosis and outcomes. Oncogene 26:2157–2165
126. Pietenpol JA, Tokino T, Thiagalingam S, el-Deiry WS, Kinzler KW, Vogelstein B (1994) Sequence-specific transcriptional activation is essential for growth suppression by p53. Proc Natl Acad Sci U S A 91:1998–2002
127. Quaas M, Muller GA, Engeland K (2012) p53 can repress transcription of cell cycle genes through a p21(WAF1/CIP1)-dependent switch from MMB to DREAM protein complex binding at CHR promoter elements. Cell Cycle 11:4661–4672
128. Quante T, Otto B, Brazdova M, Kejnovska I, Deppert W, Tolstonog GV (2012) Mutant p53 is a transcriptional co-factor that binds to G-rich regulatory regions of active genes and generates transcriptional plasticity. Cell Cycle 11:3290–3303
129. Raycroft L, Wu HY, Lozano G (1990) Transcriptional activation by wild-type but not transforming mutants of the p53 anti-oncogene. Science 249:1049–1051
130. Reinhardt HC, Schumacher B (2012) The p53 network: cellular and systemic DNA damage responses in aging and cancer. Trends Genet 28:128–136
131. Restle A, Farber M, Baumann C, Bohringer M, Scheidtmann KH, Muller-Tidow C, Wiesmuller L (2008) Dissecting the role of p53 phosphorylation in homologous recombination provides new clues for gain-of-function mutants. Nucleic Acids Res 36:5362–5375
132. Riley T, Sontag E, Chen P, Levine A (2008) Transcriptional control of human p53-regulated genes. Nat Rev 9:402–412
133. Rotter V, Boss MA, Baltimore D (1981) Increased concentration of an apparently identical cellular protein in cells transformed by either Abelson murine leukemia virus or other transforming agents. J Virol 38:336–346
134. Ryan KM (2011) p53 and autophagy in cancer: guardian of the genome meets guardian of the proteome. Eur J Cancer 47:44–50
135. Sahin E, DePinho RA (2012) Axis of ageing: telomeres, p53 and mitochondria. Nat Rev 13:397–404
136. Samowitz WS, Curtin K, Ma KN, Edwards S, Schaffer D, Leppert MF, Slattery ML (2002) Prognostic significance of p53 mutations in colon cancer at the population level. Int J Cancer 99:597–602
137. Sampath J, Sun D, Kidd VJ, Grenet J, Gandhi A, Shapiro LH, Wang Q, Zambetti GP, Schuetz JD (2001) Mutant p53 cooperates with ETS and selectively up-regulates human MDR1 not MRP1. J Biol Chem 276:39359–39367
138. Santhanam U, Ray A, Sehgal PB (1991) Repression of the interleukin 6 gene promoter by p53 and the retinoblastoma susceptibility gene product. Proc Natl Acad Sci U S A 88:7605–7609

139. Sarig R, Rivlin N, Brosh R, Bornstein C, Kamer I, Ezra O, Molchadsky A, Goldfinger N, Brenner O, Rotter V (2010) Mutant p53 facilitates somatic cell reprogramming and augments the malignant potential of reprogrammed cells. J Exp Med 207:2127–2140
140. Schilling T, Kairat A, Melino G, Krammer PH, Stremmel W, Oren M, Muller M (2010) Interference with the p53 family network contributes to the gain of oncogenic function of mutant p53 in hepatocellular carcinoma. Biochem Biophys Res Commun 394:817–823
141. Scian MJ, Carchman EH, Mohanraj L, Stagliano KE, Anderson MA, Deb D, Crane BM, Kiyono T, Windle B, Deb SP, Deb S (2008) Wild-type p53 and p73 negatively regulate expression of proliferation related genes. Oncogene 27:2583–2593
142. Scian MJ, Stagliano KE, Anderson MA, Hassan S, Bowman M, Miles MF, Deb SP, Deb S (2005) Tumor-derived p53 mutants induce NF-kappaB2 gene expression. Mol Cell Biol 25:10097–10110
143. Scian MJ, Stagliano KE, Deb D, Ellis MA, Carchman EH, Das A, Valerie K, Deb SP, Deb S (2004) Tumor-derived p53 mutants induce oncogenesis by transactivating growth-promoting genes. Oncogene 23:4430–4443
144. Scian MJ, Stagliano KE, Ellis MA, Hassan S, Bowman M, Miles MF, Deb SP, Deb S (2004) Modulation of gene expression by tumor-derived p53 mutants. Cancer Res 64:7447–7454
145. Seto E, Usheva A, Zambetti GP, Momand J, Horikoshi N, Weinmann R, Levine AJ, Shenk T (1992) Wild-type p53 binds to the TATA-binding protein and represses transcription. Proc Natl Acad Sci U S A 89:12028–12032
146. Shaked H, Shiff I, Kott-Gutkowski M, Siegfried Z, Haupt Y, Simon I (2008) Chromatin immunoprecipitation-on-chip reveals stress-dependent p53 occupancy in primary normal cells but not in established cell lines. Cancer Res 68:9671–9677
147. Shi XB, Nesslinger NJ, Deitch AD, Gumerlock PH, deVere White RW (2002) Complex functions of mutant p53 alleles from human prostate cancer. Prostate 51:59–72
148. Simmons DT, Chang C, Martin MA (1979) Multiple forms of polyoma virus tumor antigens from infected and transformed cells. J Virol 29:881–887
149. Skaug V, Ryberg D, Kure EH, Arab MO, Stangeland L, Myking AO, Haugen A (2000) p53 mutations in defined structural and functional domains are related to poor clinical outcome in non-small cell lung cancer patients. Clin Cancer Res 6:1031–1037
150. Solomon H, Madar S, Rotter V (2011) Mutant p53 gain of function is interwoven into the hallmarks of cancer. J Pathol 225:475–478
151. Strano S, Fontemaggi G, Costanzo A, Rizzo MG, Monti O, Baccarini A, Del Sal G, Levrero M, Sacchi A, Oren M, Blandino G (2002) Physical interaction with human tumor-derived p53 mutants inhibits p63 activities. J Biol Chem 277:18817–18826
152. Subler MA, Martin DW, Deb S (1994) Activation of the human immunodeficiency virus type 1 long terminal repeat by transforming mutants of human p53. J Virol 68:103–110
153. Subler MA, Martin DW, Deb S (1992) Inhibition of viral and cellular promoters by human wild-type p53. J Virol 66:4757–4762
154. Tarasov V, Jung P, Verdoodt B, Lodygin D, Epanchintsev A, Menssen A, Meister G, Hermeking H (2007) Differential regulation of microRNAs by p53 revealed by massively parallel sequencing: miR-34a is a p53 target that induces apoptosis and G1-arrest. Cell Cycle 6:1586–1593
155. Taylor WR, Egan SE, Mowat M, Greenberg AH, Wright JA (1992) Evidence for synergistic interactions between ras, myc and a mutant form of p53 in cellular transformation and tumor dissemination. Oncogene 7:1383–1390
156. Terzian T, Suh YA, Iwakuma T, Post SM, Neumann M, Lang GA, Van Pelt CS, Lozano G (2008) The inherent instability of mutant p53 is alleviated by Mdm2 or p16INK4a loss. Genes Dev 22:1337–1344
157. Torti D, Trusolino L (2011) Oncogene addiction as a foundational rationale for targeted anti-cancer therapy: promises and perils. EMBO Mol Med 3:623–636
158. Vaughan CA, Frum R, Pearsall I, Singh S, Windle B, Yeudall A, Deb SP, Deb S (2012) Allele specific gain-of-function activity of p53 mutants in lung cancer cells. Biochem Biophys Res Commun 428:6–10

159. Vaughan CA, Singh S, Windle B, Sankala HM, Graves PR, Andrew Yeudall W, Deb SP, Deb S (2011) p53 mutants induce transcription of NF-kappaB2 in H1299 cells through CBP and STAT binding on the NF-kappaB2 promoter and gain of function activity. Arch Biochem Biophys 518:79–88
160. Vaughan CA, Singh S, Windle B, Yeudall WA, Frum R, Grossman SR, Deb SP, Deb S (2012) Gain-of-Function activity of mutant p53 in lung cancer through Up-Regulation of receptor protein Tyrosine Kinase Axl. Genes Cancer 3:491–502
161. Vikhanskaya F, Lee MK, Mazzoletti M, Broggini M, Sabapathy K (2007) Cancer-derived p53 mutants suppress p53-target gene expression–potential mechanism for gain of function of mutant p53. Nucleic Acids Res 35:2093–2104
162. Vilborg A, Bersani C, Wilhelm MT, Wiman KG (2011) The p53 target Wig-1: a regulator of mRNA stability and stem cell fate? Cell Death Differ 18:1434–1440
163. Vousden KH, Prives C (2009) Blinded by the light: the growing complexity of p53. Cell 137:413–431
164. Walerych D, Napoli M, Collavin L, Del Sal G (2012) The rebel angel: mutant p53 as the driving oncogene in breast cancer. Carcinogenesis 33:2007–2017
165. Walter K, Warnecke G, Bowater R, Deppert W, Kim E (2005) tumor suppressor p53 binds with high affinity to CTG.CAG trinucleotide repeats and induces topological alterations in mismatched duplexes. J Biol Chem 280:42497–42507
166. Wang SP, Wang WL, Chang YL, Wu CT, Chao YC, Kao SH, Yuan A, Lin CW, Yang SC, Chan WK, Li KC, Hong TM, Yang PC (2009) p53 controls cancer cell invasion by inducing the MDM2-mediated degradation of Slug. Nat Cell Biol 11:694–704
167. Wang XJ, Greenhalgh DA, Jiang A, He D, Zhong L, Brinkley BR, Roop DR (1998) Analysis of centrosome abnormalities and angiogenesis in epidermal-targeted p53172H mutant and p53-knockout mice after chemical carcinogenesis: evidence for a gain of function. Mol Carcinog 23:185–192
168. Wang XJ, Greenhalgh DA, Jiang A, He D, Zhong L, Medina D, Brinkley BR, Roop DR (1998) Expression of a p53 mutant in the epidermis of transgenic mice accelerates chemical carcinogenesis. Oncogene 17:35–45
169. Wei CL, Wu Q, Vega VB, Chiu KP, Ng P, Zhang T, Shahab A, Yong HC, Fu Y, Weng Z, Liu J, Zhao XD, Chew JL, Lee YL, Kuznetsov VA, Sung WK, Miller LD, Lim B, Liu ET, Yu Q, Ng HH, Ruan Y (2006) A global map of p53 transcription-factor binding sites in the human genome. Cell 124:207–219
170. Weinstein IB (2002) Cancer. Addiction to oncogenes–the Achilles heal of cancer. Science 297:63–64
171. Weisz L, Damalas A, Liontos M, Karakaidos P, Fontemaggi G, Maor-Aloni R, Kalis M, Levrero M, Strano S, Gorgoulis VG, Rotter V, Blandino G, Oren M (2007) Mutant p53 enhances nuclear factor kappaB activation by tumor necrosis factor alpha in cancer cells. Cancer Res 67:2396–2401
172. Weisz L, Oren M, Rotter V (2007) Transcription regulation by mutant p53. Oncogene 26:2202–2211
173. Weisz L, Zalcenstein A, Stambolsky P, Cohen Y, Goldfinger N, Oren M, Rotter V (2004) Transactivation of the EGR1 gene contributes to mutant p53 gain of function. Cancer Res 64:8318–8327
174. Wolf D, Harris N, Rotter V (1984) Reconstitution of p53 expression in a nonproducer Ab-MuLV-transformed cell line by transfection of a functional p53 gene. Cell 38:119–126
175. Wong RP, Tsang WP, Chau PY, Co NN, Tsang TY, Kwok TT (2007) p53-R273H gains new function in induction of drug resistance through down-regulation of procaspase-3. Mol Cancer Ther 6:1054–1061
176. Yan W, Liu G, Scoumanne A, Chen X (2008) Suppression of inhibitor of differentiation 2, a target of mutant p53, is required for gain-of-function mutations. Cancer Res 68:6789–6796
177. Yeudall WA, Vaughan CA, Miyazaki H, Ramamoorthy M, Choi MY, Chapman CG, Wang H, Black E, Bulysheva AA, Deb SP, Windle B, Deb S (2012) Gain-of-function mutant p53 upregulates CXC chemokines and enhances cell migration. Carcinogenesis 33:442–451

178. Zalcenstein A, Stambolsky P, Weisz L, Muller M, Wallach D, Goncharov TM, Krammer PH, Rotter V, Oren M (2003) Mutant p53 gain of function: repression of CD95(Fas/APO-1) gene expression by tumor-associated p53 mutants. Oncogene 22:5667–5676
179. Zalcenstein A, Weisz L, Stambolsky P, Bar J, Rotter V, Oren M (2006) Repression of the MSP/MST-1 gene contributes to the antiapoptotic gain of function of mutant p53. Oncogene 25:359–369
180. Zauberman A, Barak Y, Ragimov N, Levy N, Oren M (1993) Sequence-specific DNA binding by p53: identification of target sites and lack of binding to p53–MDM2 complexes. EMBO J 12:2799–2808
181. Zerdoumi Y, Aury-Landas J, Bonaiti-Pellie C, Derambure C, Sesboue R, Renaux-Petel M, Frebourg T, Bougeard G, Flaman JM (2013) Drastic effect of germline TP53 missense mutations in Li-Fraumeni patients. Hum Mutat 34:453–461

Chapter 5
Transcriptional Regulation by Mutant p53 and Oncogenesis

Raffaela Santoro, Sabrina Strano, and Giovanni Blandino

Abstract More than half of all human cancers carry p53 gene mutations whose resulting proteins are mostly full-length with a single aminoacid change, abundantly present in cancer cells and unable to exert oncosuppressor activities. Frequently, mutant p53 proteins gain oncogenic functions through which they actively contribute to the establishment, the maintenance and the spreading of a given cancer cell. Intense research effort has been devoted to the deciphering of the molecular mechanisms underlying the gain of function of mutant p53 proteins. Here we mainly review the oncogenic transcriptional activity of mutant p53 proteins that mainly occurs through the aberrant cooperation with bona-fide transcription factors and leads to either aberrant up-regulation or down-regulation of selected target genes. Thus, mutant p53 proteins are critical components of oncogenic transcriptional networks that have a profound impact in human cancers.

Keywords Mutant p53 • Transcription • Cancer • Mutation • Target

Introduction

The tetrameric transcription factor p53 consists of three structural and functional domains:the amino-terminal transcriptional activation domain (TAD, residues 1–42); the DNA binding domain (DBD, aminoacidic residues 101–306); the carboxyl-terminal oligomerization domain (OD, residues 307–355) (http://p53.free.fr).

R. Santoro • S. Strano
Molecular Chemoprevention Group, Molecular Medicine, Regina Elena National Cancer Institute, Via Elio Chianesi, 53, 00144 Rome, Italy

G. Blandino, M.D. (✉)
Translational Oncogenomics Unit, Molecular Medicine, Regina Elena National Cancer Institute, Via Elio Chianesi, 53, 00144 Rome, Italy
e-mail: blandino@ifo.it

S.P. Deb and S. Deb (eds.), *Mutant p53 and MDM2 in Cancer*,
Subcellular Biochemistry 85, DOI 10.1007/978-94-017-9211-0_5,

Upon several stress signals, such as DNA damage and activation of oncogenes, p53 undergoes post-translational modifications such as phosphorylation, acetylation, sumoylation, ubiquitination [3], mediated by interaction with stress-sensing molecules like TAT Interacting Protein 60 (Tip60) [72], Ataxia Telangiectasia Mutant (ATM) and Peptidyl-prolyl cis-trans isomerase NIMA-interacting 1 (PIN1) [82, 83, 86, 88]. Once activated, p53 can induce cell cycle arrest, senescence and apoptosis through transcriptional regulation of its target genes (Fig. 5.1a), as well as orchestrating a DNA damage response [1].

About 50 % of human cancers harbour p53 mutations and the majority of them are missense mutations, mainly located within the very fragile DNA binding domain. Missense mutations generate stable proteins incapable of binding canonical p53 consensus sequences [31, 65]. Indeed, any mutation in this region causes a loss of p53 transcriptional activity [35], since it leads to either an alteration in the residues which determine a sequence-specific binding to the DNA (class I mutants, e.g. mutants at codons 248 or 273), or a change of p53 folding, abolishing completely the binding to the DNA and changing the capability to interact with other proteins (class II mutants, e.g. mutants at codon 175) (http://p53.free.fr). Intuitively, class II mutants are associated with a more severe phenotype *in vitro*. In fact, although they cannot bind canonical p53 consensus elements by direct binding, it is known that they can be recruited to regulative elements on the chromatin (so far, attempts to identify a specific mutant-p53 consensus sequence have failed [79]), through interaction with other transcription factors, such as NF-Y [10], ETS1 [54, 57], ETS2 [14], p73 and p63 [12, 24, 67–69] and contribute to the regulation of their target genes (Fig. 5.1b).

GOF of Mutant p53

For many years the arising of tumors has been ascribed to the mere loss of p53 activities through mutations or deletions. Indeed mutants of p53 loose the capability of activating p53-responsive genes and to exert antiproliferative and proapoptotic activities. However, this hypothesis does not solve the question why other tumour suppressors could not make up for the loss of p53 functions in cancer cells. Recent work has shown that mutants of p53 gain new functions and contribute to the arising and maintenance of cancer [13, 36]. The generation of knock-in mice containing p53 mutations corresponding to the most common p53 missense mutations in humans, and the comparison to mice with a p53-null allele, gave new insights on the mechanisms of GOF of mutant p53 [34, 65].

In particular, both $p53^{+/-}$ and $p53^{-/-}$ mice develop tumours. $p53^{-/-}$ mice develop tumours earlier than heterozygous, with a median tumour incidence of 4–5 months and lifespan of about 10 months [27]. Heterozygous mice develop tumours with a median incidence of 18 months, when 50 % of mice succumbs [27]. Osteosarcomas arise with the highest frequency in p53 heterozygous mice, while 70 % of p53 *null* mice harbour malignant lymphomas [27]. The difference in the tumour spectrum between homozygous and heterozygous mice remains unclear.

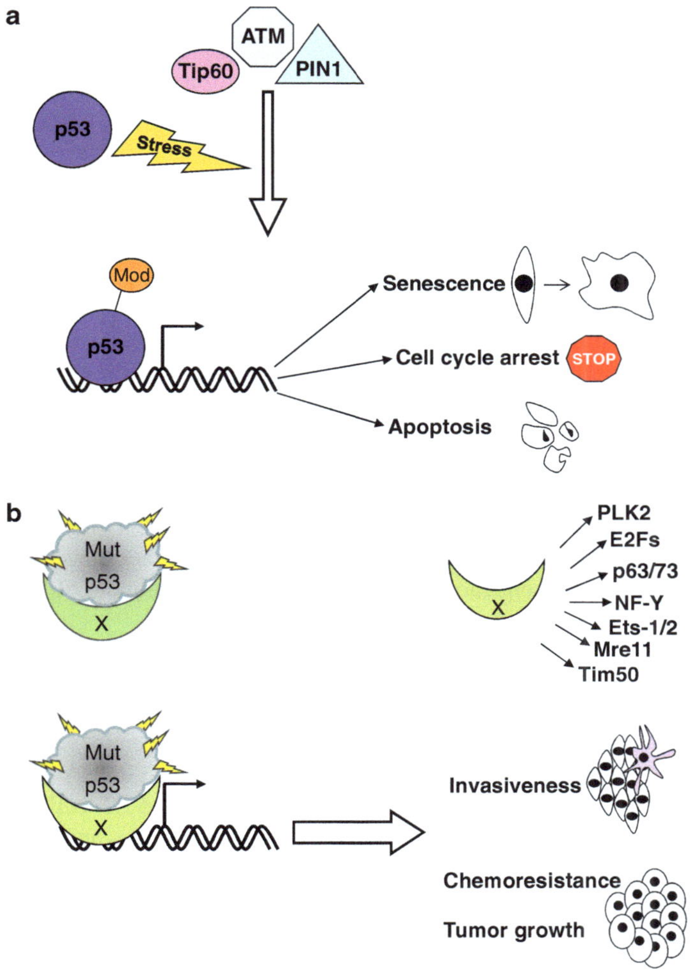

Fig. 5.1 Activation of transcription by p53 wt and mutant p53. (**a**) stress signals, such as DNA damage and activation of oncogenes, cause posttranslational modifications of p53, mediated by interaction with stress-sensing molecules like Tip60, ATM and PIN1. Once activated, p53 can induce cell cycle arrest, senescence and apoptosis through transcriptional regulation of its target genes. (**b**) mutant p53 acquires transcriptional activities by interacting with transcription factors such as NF-Y, Ets-1/2, p63/p73, E2Fs, PLK2, Mre11 and Tim50, which induce the recruitment of mutant p53 to regulatory regions on the chromatin

Knock-in mouse models have been produced for a variety of mutant p53 alleles [15]. p53R175H/+mice, corresponding to the hotspot R172H in humans, show no significant difference in lifespan as compared to $p53^{+/-}$ mice, while their tumours show a highly metastatic phenotype [40, 42]. Similar data are obtained with other gain of function transgenic mice [15].

Mutant p53 Protein Complexes

In vitro experiments have shown that, unlike wild-type (wt) p53, mutant p53 is able to interact with the p53 family members p63 and p73 through its DBD, and that this interaction inhibits p63 and p73 transcriptional functions [12, 18, 24, 40, 67]. In addition, it has recently been shown that mutant p53 can be recruited onto a subset of p63-responsive elements in the promoters of genes upregulated and downregulated by p63 [49].

Knockout mice $p53^{+/-}p63^{+/-}$ and $p53^{+/-}p73^{+/-}$ develop tumours at higher frequencies than $p53^{+/-}$ mice which are also wild-type for p63 and p73, supporting the hypothesis that inactivation of p63 and p73 by mutant p53 contributes to the metastatic phenotype of mutant p53 mice [33, 52]. Moreover, knockdown of mutant p53 renders cells more sensitive to chemotherapy [68], while knockdown or inactivation of p73 reduces their sensitivity to chemotherapeutic agents [32]. However, while p53 is ubiquitously expressed, p63 and p73 expression is tissue-specific and the ratios of TA and ΔN isoforms vary from one tissue to another, which leads to different p63- and p73-dependent phenotypes in cancers of different origin.

In addition to p63 and p73, mutant p53 interacts with other transcription factors and the functional consequence of this binding is the recruitment of mutant p53 to regulative regions on the chromatin.

The transcription factor NF-Y has been identified as an interaction partner for mutant p53 [10]. NF-Y binds to the CCAAT consensus motif present in the promoters of several cell cycle regulating genes, such as E2Fs and cyclines, and contributes to the modulation of their expression, through selective recruitment of acetylases and/or deacetylases [10]. Upon DNA damage, the complex mutant p53/NF-Y drives the transcription of the cell cycle progressing genes cyclin A, B1 and B2, cdk1 and cdc25C, resulting in an increased DNA replication [10], which constitutes an important aspect of mutant p53 GOF activities. In addition, both NF-Y and NFκB are recruited onto the MAP2K3 promoter and required for MAP2K3 transcription. Mutant p53 engaging in a complex with NF-Y and NFκB contributes to MAP2K3 upregulation [26].

The Polo-Like Kinase 2 (PLK2) protein plays a critical role in cell cycle progression in response to DNA damage, when it is activated by p53 wt and activates a G2 checkpoint, thereby inducing cell cycle arrest and inhibiting aberrant DNA replication and mitotic catastrophe [4]. Like p53 wt, mutant p53 is able to induce PLK2 expression as well [74]. In turn, PLK2 physically interacts with mutant p53, mediates its phosphorylation and promotes its acetylation, thereby inducing its recruitment onto CCAAT consensus motifs [29, 74].

Mre11 has been identified as an interactor of mutant p53 as well [63]. Together with Rad50 and NBS1, Mre11 forms the MRN complex, which is involved in sensing DNA damage and recruiting the ATM kinase to the sites of DNA damage. The binding of mutant p53 to Mre11 seems to prevent the recruitment of the MRN complex to the sites of DNA double strand brakes (DSBs), thus leading to an impaired ATM response, generating higher genetic instability [63].

It has recently been shown that mutant p53 interacts with the transcription factors ETS1 and ETS2, which bind the consensus motif GGAAR [14]. This binding results in the upregulation of the TDP2 gene, which is responsible for Etoposide resistance [14]. Fontemaggi et al. have shown that mutant p53 can engage into a complex with E2F1 and regulate the expression of the transcription factors ID2 and ID4 [19], which can in turn induce the expression of cytokines, thereby contributing to neo-angiogenesis [19]. Immunoprecipitation followed by protein identification by mass spectrometry has shown Tim50 (translocator of the inner mitochondrial membrane 50) as a novel interaction partner of mutant p53 [58]. In this manuscript, Tim50 is also identified as a mutant p53-target gene, which contributes to the growth of mutant-p53 expressing cells [58].

Since the interaction with novel partners confers mutant p53 new functions, many efforts are being made to identify additional interaction partners for mutant p53, especially in different contexts and in response to different stimuli [7]. This is even more important if we take into account that mutant p53 binds with high affinity to the nuclear matrix, suggesting that mutant p53 could perturb and rearrange the nuclear structure [37, 80].

Mutant p53 Target Genes

Several laboratories have performed chromatin immunoprecipitation (ChIP) experiments to reveal whether mutant p53 could be specifically recruited *in vivo* onto regulative regions on the chromatin [10, 78, 87]. However, so far no specific consensus sequence for mutant p53 has been identified [79] but, as already mentioned, recruitment of mutant p53 onto the promoters can be mediated by other canonical transcription factors. A first step to unravel the specific set of targets for each mutant p53 is the use of "ChIP-on-chip" and Chip-seq analyses and it could contribute to the identification of transcriptional signatures for p53 mutants [9, 25, 76, 81].

In addition, a series of microarray analyses performed in different cell lines have shown that mutant p53 can modulate the expression of several genes [51, 59, 60, 73, 78]. However, the picture appears very complex due to the great heterogeneity of p53 mutations. In fact, each mutant of p53 can regulate the expression of a pattern of genes only partially overlapping with the ones of other p53 mutants. This seems reasonable since different mutations in p53 cause interaction with different sets of partners.

As many efforts have been made in the past to identify transcriptional targets of p53 wt, so many are being made for mutant p53, in order to understand which pathways are deregulated and responsible for mutant p53-induced carcinogenesis.

The most known gain of function activity of mutant p53 is the induction of cell proliferation. Several studies have been conducted and the genes c-fos [55], c-myc [21], PCNA [8, 61], cyclin A, B1, B2, cdk1 and cdc25c [10] have been identified as mutant p53 specific targets, which can promote cell proliferation. Mutant p53 can exert its pro-invasive and pro-angiogenic activities mainly through the activation of its target genes ID4 [19] and the chemokines CXCL1 [62], CXCL5, CXCL8 and CXCL12 [85], as well as MMP-13 [70] and EGFR [43]. Of note, both CXCL1 and CXCL8 are transcriptional targets of ID4 [19].

Given the importance of microRNA deregulation in cancer, an increasing number of manuscripts has been published on the role of mutant p53 in this process. Donzelli et al. have shown that mutant p53R175H can induce the expression of miR-128-2, upregulated in many chemoresistant cancers, through binding to its host gene ARPP1, which results in chemoresistance [17].

It has been described that p53 mutants R175H, R273H and C135Y can downregulate the expression of miR-130b, which leads to upregulation of the transcriptional repressor ZEB1. This activates the ZEB1 signaling pathway which induces Epithelial-Mesenchymal-Transition (EMT) [16].

miR-155, which is upregulated in breast cancers, is activated by mutant p53 and promotes EMT through inhibition of genes such as ZNF652 [50]. In addition, mutant p53 is recruited by ZEB1 onto the promoter region of miR-223, resulting in miR-223 down-regulation. Consequently, the mir-223 target Stathmin-1 is upregulated, leading to mutant p53-mediated chemoresistance [47].

Conclusions and Perspectives

There is growing evidence that mutants of p53 acquire an oncogenic role. Being p53 mutated in more than 50 % of human cancers, restoration of p53 wt functions in cancers bearing mutant p53 represents a tantalizing strategy for cancer therapy. In the era of translational cancer research, when the aim is to design ad hoc therapy for each patient, the establishment of expression signatures, being them based on proteomics, miRNA or gene expression profiling, is becoming of main importance. For this reason, many efforts have been made to establish differential signatures between normal and cancerous tissues and among the patients, in relation to various parameters altered in cancer, among which p53. Though a molecular signature for p53 mutated cancers has not been established yet, some groups have found a correlation between mutant p53 and poor survival in breast cancer [2, 30, 45, 48, 53, 64]. However, the GOF of p53 mutants has not been completely unravelled yet, and further studies are needed in this field.

Several groups have shown that reactivation of p53 in p53-deficient cells can induce senescence or apoptosis [46, 66, 75, 77, 84]. Other groups have instead used small compounds in order to induce a correct folding of p53 mutants (reviewed in [28]): PRIMA-1 imposes a specific DNA binding and transcriptional activation of the p53 target genes mdm2, p21 and PUMA [5, 6, 41]; Cp-31398 confers wild-type conformation to some mutants by unknown mechanisms [20, 39, 44, 56, 71]; the CDB3 peptide renders wild-type and mutant p53 more thermostable by binding to

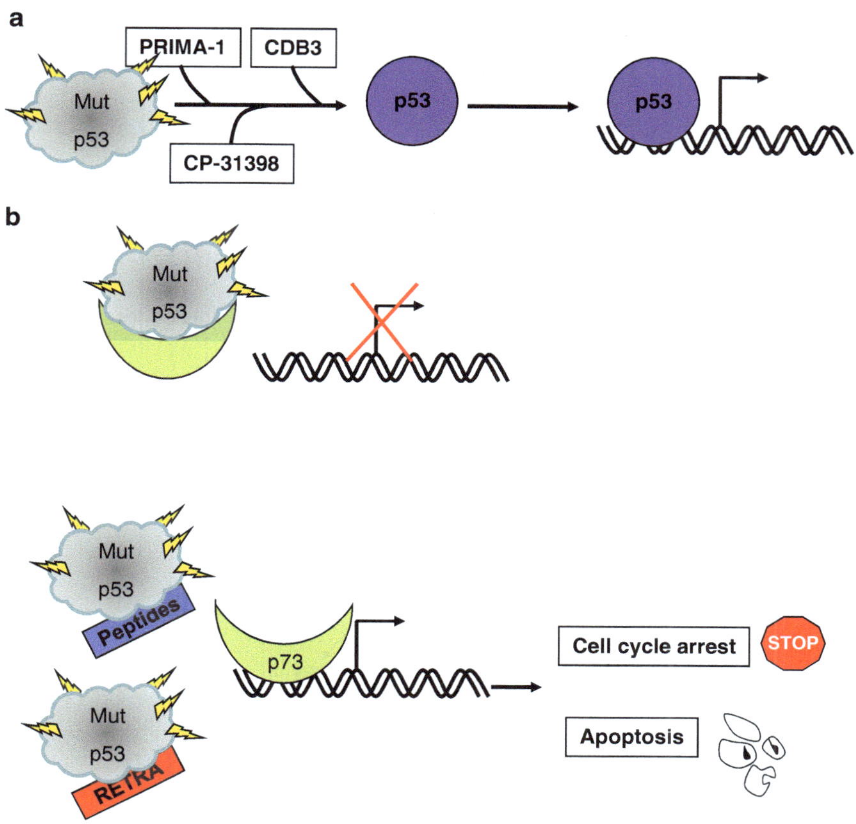

Fig. 5.2 Manipulation of mutant p53. (**a**) The small compounds PRIMA-1, CP-31398 and CDB3 confer native conformation to mutant p53; hence p53 is able to bind canonical consensus elements and activate the transcription of its target genes, thereby regaining its transcriptional activities. (**b**) Binding of mutant p53 to p73 prevents the recruitment of p73 to responsive promoters; the small compound RETRA and small synthetic peptides, which bind to the core domain of mutant p53, impair the interaction between p73 and mutant p53; p73 is then available to bind to the chromatin and drive the transcription of its target genes

their DBD and promoting the proper folding, without compromising the protein-DNA interaction [22, 23] (Fig. 5.2a).

A more recent approach is designed to inhibit the interaction of mutant p53 to its binding partner p73. To this aim two kinds of molecules have been used: (a) the small compound RETRA, which is able to displace p73 from the inactive complex with mutant p53 allowing it to activate transcription [38], and (b) synthetic peptides designed to interact with the mutant p53 core domain, which are able to destroy the interaction of some mutants of p53 with the DNA binding domain of p73, restoring the capability of p73 to activate transcription and induce cell cycle arrest and apoptosis [11] (Fig. 5.2b).

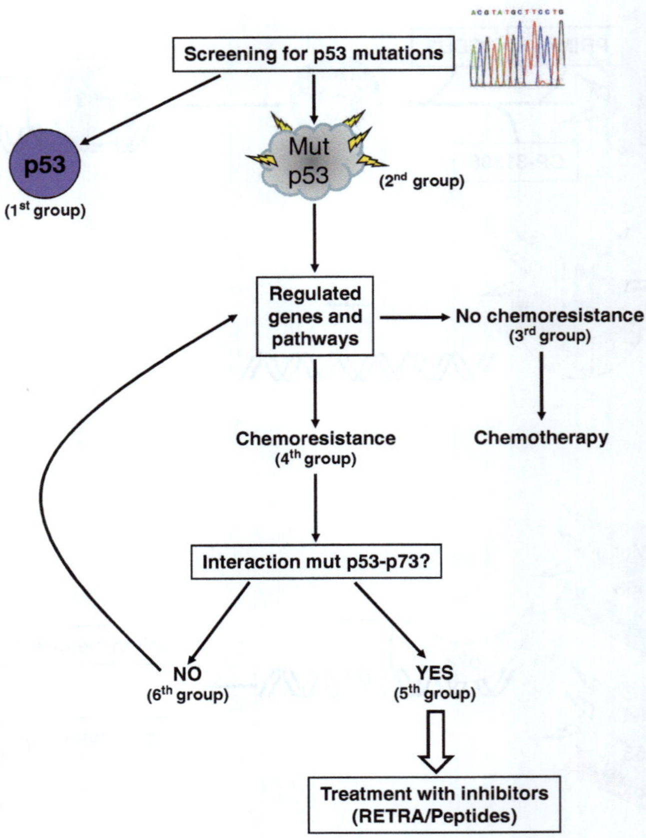

Fig. 5.3 Example of molecular stratification. Cancer patients are screened for p53 mutations and divided into two groups according to p53 status (p53 wt-1st group, p53 mutated-2nd group); patients from the 2nd group are subjected to expression profiling to determine pathways regulated by mutant p53 and predict a response to chemotherapy as an example; two more groups of patients are identified: a 3rd group responding to chemotherapy and a 4th group, which is predicted to show chemoresistance and which is therefore not subjected to chemotherapy. The hypothesis of interaction mutant p53-p73 is tested in the latter group and patients are again divided into two subgroups: in the 5th group mutant p53 and p73 interact with each other, therefore patients are subjected to treatment with inhibitors of the interaction, such as RETRA and small peptides binding to mutant p53; the 6th group is instead subjected to further analyses to find the proper treatment

Knowing the binding partners of p53 mutants which mediate its recruitment onto noncanonical p53 promoters, the transcriptionally deregulated pathways, the mechanisms by which p53 mutants confer chemoresistance [32, 68], invasiveness and metastatic potential [13, 36], together with the discovery of novel strategies to interfere with mutant p53 GOF activities, will allow us to stratify cancer patients and choose the correct treatment, ranging from radio- and chemotherapy to specific inhibitors of activated pathways and of mutant p53 itself (Fig. 5.3).

References

1. Bartek J, Bartkova J, Lukas J (2007) DNA damage signalling guards against activated oncogenes and tumour progression. Oncogene 26:7773–7779
2. Bonnefoi H, Diebold-Berger S, Therasse P, Hamilton A, van de Vijver M, MacGrogan G et al (2003) Locally advanced/inflammatory breast cancers treated with intensive epirubicin-based neoadjuvant chemotherapy: are there molecular markers in the primary tumour that predict for 5-year clinical outcome? Ann Oncol 14:406–413
3. Brooks CL, Gu W (2003) Ubiquitination, phosphorylation and acetylation: the molecular basis for p53 regulation. Curr Opin Cell Biol 15:164–171
4. Burns TF, Fei P, Scata KA, Dicker DT, El-Deiry WS (2003) Silencing of the novel p53 target gene Snk/Plk2 leads to mitotic catastrophe in paclitaxel (taxol)- exposed cells. Mol Cell Biol 23:5556–5571
5. Bykov VJ, Issaeva N, Selivanova G, Wiman KG (2002) Mutant p53-dependent growth suppression distinguishes PRIMA-1 from known anticancer drugs: a statistical analysis of information in the National Cancer Institute database. Carcinogenesis 23:2011–2018
6. Bykov VJ, Issaeva N, Shilov A, Hultcrantz M, Pugacheva E, Chumakov P et al (2002) Restoration of the tumor suppressor function to mutant p53 by a low-molecular-weight compound. Nat Med 8:282–288
7. Coffill CR, Muller PA, Oh HK, Neo SP, Hogue KA, Cheok CF et al (2012) Mutant p53 interactome identifies nardilysin as a p53R273H-specific binding partner that promotes invasion. EMBO Rep 13:638–644
8. Deb S, Jackson CT, Subler MA, Martin DW (1992) Modulation of cellular and viral promoters by mutant human p53 proteins found in tumor cells. J Virol 66:6164–6170
9. Dell'Orso S, Fontemaggi G, Stambolsky P, Goeman F, Voellenkle C, Levrero M et al (2011) ChIP-on-chip analysis of in vivo mutant p53 binding to selected gene promoters. OMICS 15:305–312
10. Di Agostino S, Strano S, Emiliozzi V, Zerbini V, Mottolese M, Sacchi A et al (2006) Gain of function of mutant p53: the mutant p53/NF-Y protein complex reveals an aberrant transcriptional mechanism of cell cycle regulation. Cancer Cell 10:191–202
11. Di Agostino S, Cortese G, Monti O, Dell'Orso S, Sacchi A, Eisenstein M, Citro G, Strano S, Blandino G (2008) The disruption of the protein complex mutant p53/p73 increases selectively the response of tumor cells to anticancer drugs. Cell Cycle 7:3440–3447
12. Di Como CJ, Gaiddon C, Prives C (1999) p73 function is inhibited by tumor-derived p53 mutants in mammalian cells. Mol Cell Biol 19:1438–1449
13. Dittmer D, Pati S, Zambetti G, Chu S, Teresky AK, Moore M et al (1993) Gain of function mutations in p53. Nat Genet 4:42–46
14. Do PM, Varanasi L, Fan S, Li C, Kubacka I, Newman V et al (2012) Mutant p53 cooperates with ETS2 to promote etoposide resistance. Genes Dev 26:830–845
15. Donehower LA, Lozano G (2009) 20 years studying p53 functions in genetically engineered mice. Nat Rev Cancer 9:831–841
16. Dong P, Karaayvaz M, Jia N, Kaneuchi M, Hamada J, Watari H et al (2012). Mutant p53 gain-of-function induces epithelial-mesenchymal transition through modulation of the miR-130b-ZEB1 axis. Oncogene 32:3286–3295
17. Donzelli S, Fontemaggi G, Fazi F, Di Agostino S, Padula F, Biagioni F et al (2012) MicroRNA-128-2 targets the transcriptional repressor E2F5 enhancing mutant p53 gain of function. Cell Death Differ 19:1038–1048
18. Flores ER, Tsai KY, Crowley D, Sengupta S, Yang A, McKeon F et al (2002) p63 and p73 are required for p53-dependent apoptosis in response to DNA damage. Nature 416:560–564
19. Fontemaggi G, Dell'Orso S, Trisciuoglio D, Shay T, Melucci E, Fazi F et al (2009) The execution of the transcriptional axis mutant p53, E2F1 and ID4 promotes tumor neo-angiogenesis. Nat Struct Mol Biol 16:1086–1093

20. Foster BA, Coffey HA, Morin MJ, Rastinejad F (1999) Pharmacological rescue of mutant p53 conformation and function. Science 286:2507–2510
21. Frazier MW, He X, Wang J, Gu Z, Cleveland JL, Zambetti GP (1998) Activation of c-myc gene expression by tumor-derived p53 mutants requires a discrete C-terminal domain. Mol Cell Biol 18:3735–3743
22. Friedler A, Hansson LO, Veprintsev DB, Freund SM, Rippin TM, Nikolova PV et al (2002) A peptide that binds and stabilizes p53 core domain: chaperone strategy for rescue of oncogenic mutants. Proc Natl Acad Sci U S A 99:937–942
23. Friedler A, Veprintsev DB, Hansson LO, Fersht AR (2003) Kinetic instability of p53 core domain mutants: implications for rescue by small molecules. J Biol Chem 278:24108–24112
24. Gaiddon C, Lokshin M, Ahn J, Zhang T, Prives C (2001) A subset of tumor- derived mutant forms of p53 down-regulate p63 and p73 through a direct interaction with the p53 core domain. Mol Cell Biol 21:1874–1887
25. Goeman F, Fontemaggi G, Blandino G (2012) ChIP-on-chip to identify mutant p53 targets. Methods Mol Biol 962:211–226
26. Gurtner A, Starace G, Norelli G, Piaggio G, Sacchi A, Bossi G (2010) Mutant p53-induced up-regulation of mitogen-activated protein kinase kinase 3 contributes to gain of function. J Biol Chem 285:14160–14169
27. Harvey M, McArthur MJ, Montgomery CA Jr, Butel JS, Bradley A, Donehower LA (1993) Spontaneous and carcinogen-induced tumorigenesis in p53-deficient mice. Nat Genet 5:225–229
28. Haupt S, Haupt Y (2004) Improving cancer therapy through p53 management. Cell Cycle 3:912–916
29. Haupt S, Haupt Y (2012) Mutant p53 subverts PLK2 function in a novel, reinforced loop of corruption. Cell Cycle 11:217–218
30. Hensel M, Schneeweiss A, Sinn HP, Egerer G, Solomayer E, Haas R et al (2002) P53 is the strongest predictor of survival in high-risk primary breast cancer patients undergoing high-dose chemotherapy with autologous blood stem cell support. Int J Cancer 100:290–296
31. Hussain SP, Harris CC (1998) Molecular epidemiology of human cancer. Toxicol Lett 102–103:219–225
32. Irwin MS, Kondo K, Marin MC, Cheng LS, Hahn WC, Kaelin WG Jr (2003) Chemosensitivity linked to p73 function. Cancer Cell 3:403–410
33. Iwakuma T, Lozano G, Flores ER (2005) Li-Fraumeni syndrome: a p53 family affair. Cell Cycle 4:865–867
34. Jacks T, Remington L, Williams BO, Schmitt EM, Halachmi S, Bronson RT et al (1994) Tumor spectrum analysis in p53-mutant mice. Curr Biol 4:1–7
35. Kato S, Han SY, Liu W, Otsuka K, Shibata H, Kanamaru R et al (2003) Understanding the function-structure and function-mutation relationships of p53 tumor suppressor protein by high-resolution missense mutation analysis. Proc Natl Acad Sci U S A 100:8424–8429
36. Kim E, Deppert W (2004) Transcriptional activities of mutant p53: when mutations are more than a loss. J Cell Biochem 93:878–886
37. Koga H, Deppert W (2000) Identification of genomic DNA sequences bound by mutant p53 protein (Gly245→Ser) in vivo. Oncogene 19:4178–4183
38. Kravchenko JE, Ilyinskaya GV, Komarov PG, Agapova LS, Kochetkov DV, Strom E et al (2008) Small-molecule RETRA suppresses mutant p53-bearing cancer cells through a p73-dependent salvage pathway. Proc Natl Acad Sci U S A 105:6302–6307
39. Lane DP, Hupp TR (2003) Drug discovery and p53. Drug Discov Today 8:347–355
40. Lang GA, Iwakuma T, Suh YA, Liu G, Rao VA, Parant JM et al (2004) Gain of function of a p53 hot spot mutation in a mouse model of Li-Fraumeni syndrome. Cell 119:861–872
41. Lehmann S, Bykov VJ, Ali D, Andren O, Cherif H, Tidefelt U et al (2012) Targetingp53 in vivo: a first-in-human study with p53-targeting compound APR- 246 in refractory hematologic malignancies and prostate cancer. J Clin Oncol 30:3633–3639
42. Liu G, McDonnell TJ, Montes de Oca Luna R, Luna R, Kapoor M, Mims B, El-Naggar AK et al (2000) High metastatic potential in mice inheriting a targeted p53 missense mutation. Proc Natl Acad Sci U S A 97:4174–4179

43. Ludes-Meyers JH, Subler MA, Shivakumar CV, Munoz RM, Jiang P, Bigger JE et al (1996) Transcriptional activation of the human epidermal growth factor receptor promoter by human p53. Mol Cell Biol 16:6009–6019
44. Luu Y, Bush J, Cheung KJ Jr, Li G (2002) The p53 stabilizing compound CP- 31398 induces apoptosis by activating the intrinsic Bax/mitochondrial/caspase-9 pathway. Exp Cell Res 276:214–222
45. Malamou-Mitsi V, Gogas H, Dafni U, Bourli A, Fillipidis T, Sotiropoulou M et al (2006) Evaluation of the prognostic and predictive value of p53 and Bcl-2 in breast cancer patients participating in a randomized study with dose-dense sequential adjuvant chemotherapy. Ann Oncol 17:1504–1511
46. Martins CP, Brown-Swigart L, Evan GI (2006) Modeling the therapeutic efficacy of p53 restoration in tumors. Cell 127:1323–1334
47. Masciarelli S, Fontemaggi G, Di Agostino S, Donzelli S, Carcarino E, Strano S et al (2013). Gain of function mutp53 down-regulates miR223 contributing to chemoresistance of cultured tumor cells. Oncogene 33:1601–1608
48. Miller LD, Smeds J, George J, Vega VB, Vergara L, Ploner A et al (2005) An expression signature for p53 status in human breast cancer predicts mutation status, transcriptional effects, and patient survival. Proc Natl Acad Sci U S A 102:13550–13555
49. Neilsen PM, Noll JE, Suetani RJ, Schulz RB, Al-Ejeh F, Evdokiou A et al (2011) Mutant p53 uses p63 as a molecular chaperone to alter gene expression and induce a pro-invasive secretome. Oncotarget 2:1203–1217
50. Neilsen PM, Noll JE, Mattiske S, Bracken CP, Gregory PA, Schulz RB et al (2012) Mutant p53 drives invasion in breast tumors through up-regulation of miR-155. Oncogene 32:2992–3000
51. O'Farrell TJ, Ghosh P, Dobashi N, Sasaki CY, Longo DL (2004) Comparison of the effect of mutant and wild-type p53 on global gene expression. Cancer Res 64:8199–8207
52. Olive KP, Tuveson DA, Ruhe ZC, Yin B, Willis NA, Bronson RT et al (2004) Mutant p53 gain of function in two mouse models of Li-Fraumeni syndrome. Cell 119:847–860
53. Olivier M, Langerod A, Carrieri P, Bergh J, Klaar S, Eyfjord J et al (2006) The clinical value of somatic TP53 gene mutations in 1,794 patients with breast cancer. Clin Cancer Res 12:1157–1167
54. Pastorcic M, Das HK (2000) Regulation of transcription of the human presenilin-1 gene by ets transcription factors and the p53 protooncogene. J Biol Chem 275:34938–34945
55. Preuss U, Kreutzfeld R, Scheidtmann KH (2000) Tumor-derived p53 mutant C174Y is a gain-of-function mutant which activates the fos promoter and enhances colony formation. Int J Cancer 88:162–171
56. Rippin TM, Bykov VJ, Freund SM, Selivanova G, Wiman KG, Fersht AR (2002) Characterization of the p53-rescue drug CP-31398 in vitro and in living cells. Oncogene 21:2119–2129
57. Sampath J, Sun D, Kidd VJ, Grenet J, Gandhi A, Shapiro LH et al (2001) Mutant p53 cooperates with ETS and selectively up-regulates human MDR1 not MRP1. J Biol Chem 276:39359–39367
58. Sankala H, Vaughan C, Wang J, Deb S, Graves PR (2011) Upregulation of the mitochondrial transport protein, Tim50, by mutant p53 contributes to cell growth and chemoresistance. Arch Biochem Biophys 512:52–60
59. Scian MJ, Stagliano KE, Deb D, Ellis MA, Carchman EH, Das A et al (2004) Tumor-derived p53 mutants induce oncogenesis by transactivating growth-promoting genes. Oncogene 23:4430–4443
60. Scian MJ, Stagliano KE, Ellis MA, Hassan S, Bowman M, Miles MF et al (2004) Modulation of gene expression by tumor-derived p53 mutants. Cancer Res 64:7447–7454
61. Shivakumar CV, Brown DR, Deb S, Deb SP (1995) Wild-type human p53 transactivates the human proliferating cell nuclear antigen promoter. Mol Cell Biol 15:6785–6793
62. Solomon H, Buganim Y, Kogan-Sakin I, Pomeraniec L, Assia Y, Madar S, Goldstein I, Brosh R, Kalo E, Beatus T, Goldfinger N, Rotter V (2012) Various p53 mutant proteins differently regulate the Ras circuit to induce a cancer-related gene signature. J Cell Sci. 125:3144–3152

63. Song H, Xu Y (2007) Gain of function of p53 cancer mutants in disrupting critical DNA damage response pathways. Cell Cycle 6:1570–1573
64. Sorlie T, Perou CM, Tibshirani R, Aas T, Geisler S, Johnsen H et al (2001) Gene expression patterns of breast carcinomas distinguish tumor subclasses with clinical implications. Proc Natl Acad Sci U S A 98:10869–10874
65. Soussi T, Lozano G (2005) p53 mutation heterogeneity in cancer. Biochem Biophys Res Commun 331:834–842
66. Soussi T, Wiman KG (2007) Shaping genetic alterations in human cancer: the p53 mutation paradigm. Cancer Cell 12:303–312
67. Strano S, Fontemaggi G, Costanzo A, Rizzo MG, Monti O, Baccarini A et al (2002) Physical interaction with human tumor-derived p53 mutants inhibits p63 activities. J Biol Chem 277:18817–18826
68. Strano S, Blandino G (2003) p73-mediated chemosensitivity: a preferential target of oncogenic mutant p53. Cell Cycle 2:348–349
69. Strano S, Dell'Orso S, Di Agostino S, Fontemaggi G, Sacchi A, Blandino G (2007) Mutant p53: an oncogenic transcription factor. Oncogene 26:2212–2219
70. Sun Y, Cheung JM, Martel-Pelletier J, Pelletier JP, Wenger L, Altman RD et al (2000) Wild type and mutant p53 differentially regulate the gene expression of human collagenase-3 (hMMP-13). J Biol Chem 275:11327–11332
71. Takimoto R, Wang W, Dicker DT, Rastinejad F, Lyssikatos J, el-Deiry WS (2002) The mutant p53-conformation modifying drug, CP-31398, can induce apoptosis of human cancer cells and can stabilize wild-type p53 protein. Cancer Biol Ther 1:47–55
72. Tang Y, Luo J, Zhang W, Gu W (2006) Tip60-dependent acetylation of p53 modulates the decision between cell-cycle arrest and apoptosis. Mol Cell 24:827–839
73. Tepper CG, Gregg JP, Shi XB, Vinall RL, Baron CA, Ryan PE et al (2005) Profiling of gene expression changes caused by p53 gain-of-function mutant alleles in prostate cancer cells. Prostate 65:375–389
74. Valenti F, Fausti F, Biagioni F, Shay T, Fontemaggi G, Domany E et al (2012) Mutant p53 oncogenic functions are sustained by Plk2 kinase through an autoregulatory feedback loop. Cell Cycle 10:4330–4340
75. Vassilev LT, Vu BT, Graves B, Carvajal D, Podlaski F, Filipovic Z et al (2004) In vivo activation of the p53 pathway by small-molecule antagonists of MDM2. Science 303:844–848
76. Vaughan C, Windle B, Deb S (2013) ChIP sequencing to identify p53 targets. Methods Mol Biol 962:227–236
77. Ventura A, Kirsch DG, McLaughlin ME, Tuveson DA, Grimm J, Lintault L et al (2007) Restoration of p53 function leads to tumour regression in vivo. Nature 445:661–665
78. Weisz L, Zalcenstein A, Stambolsky P, Cohen Y, Goldfinger N, Oren M et al (2004) Transactivation of the EGR1 gene contributes to mutant p53 gain of function. Cancer Res 64:8318–8327
79. Weisz L, Oren M, Rotter V (2007) Transcription regulation by mutant p53. Oncogene 26:2202–2211
80. Will K, Warnecke G, Wiesmuller L, Deppert W (1998) Specific interaction of mutant p53 with regions of matrix attachment region DNA elements (MARs) with a high potential for base-unpairing. Proc Natl Acad Sci U S A 95:13681–13686
81 Wu J, Smith LT, Plass C, Huang TH (2006) ChIP-chip comes of age for genome-wide functional analysis. Cancer Res 66:6899–6902
82. Wulf GM, Liou YC, Ryo A, Lee SW, Lu KP (2002) Role of Pin1 in the regulation of p53 stability and p21 transactivation, and cell cycle checkpoints in response to DNA damage. J Biol Chem 277:47976–47979
83. Xu Y (2003) Regulation of p53 responses by post-translational modifications. Cell Death Differ 10:400–403
84. Xue W, Zender L, Miething C, Dickins RA, Hernando E, Krizhanovsky V et al (2007) Senescence and tumour clearance is triggered by p53 restoration in murine liver carcinomas. Nature 445:656–660

85. Yeudall WA, Vaughan CA, Miyazaki H, Ramamoorthy M, Choi MY, Chapman CG et al (2012) Gain-of-function mutant p53 upregulates CXC chemokines and enhances cell migration. Carcinogenesis 33:442–451
86. Zacchi P, Gostissa M, Uchida T, Salvagno C, Avolio F, Volinia S et al (2002) The prolyl isomerase Pin1 reveals a mechanism to control p53 functions after genotoxic insults. Nature 419:853–857
87. Zalcenstein A, Weisz L, Stambolsky P, Bar J, Rotter V, Oren M (2006) Repression of the MSP/MST-1 gene contributes to the antiapoptotic gain of function of mutant p53. Oncogene 25:359–369
88. Zheng H, You H, Zhou XZ, Murray SA, Uchida T, Wulf G et al (2002) The prolyl isomerase Pin1 is a regulator of p53 in genotoxic response. Nature 419:849–853

Chapter 6
p53 Mutation in the Genesis of Metastasis

W.A. Yeudall

Abstract Development of metastatic cancer is a complex series of events that includes genesis of tumor-related vascular and lymphatic systems, enhanced cellular motility, and the capacity to invade and survive at distant sites, as well as evasion of host defences. The wild-type p53 protein plays key roles in controlling these facets of tumor progression, and loss of normal p53 function can be sufficient to predispose tumor cells to gain metastatic properties. In contrast, dominant p53 mutants that have gained oncogenic functions can actively drive metastasis through a variety of mechanisms. This chapter aims to highlight these processes.

Keywords Extracellular matrix • Motility • Epithelial-mesenchymal transition • G-protein • Chemokine • Transforming growth factor beta • microRNA

Introduction

Invasion of the surrounding or underlying tissues is a crucial step in the progression to a malignant phenotype, and likely requires altered cellular interactions with the extracellular matrix (ECM) and enhanced motility. Development of metastatic disease is a late and often fatal process through which the tumor cells become established at a site distant to that of the primary lesion. This requires multiple biological steps, including development of capillary networks and/or lymphatic vessels adjacent to the tumor (angiogenesis or lymphangiogenesis), intravasation

W.A. Yeudall (✉)
Department of Oral and Craniofacial Molecular Biology, Virginia Commonwealth University, P.O. Box 980566, Richmond, VA 23298, USA

Massey Cancer Center, Virginia Commonwealth University, P.O. Box 980566, Richmond, VA 23298, USA
e-mail: wayeudall@vcu.edu

S.P. Deb and S. Deb (eds.), *Mutant p53 and MDM2 in Cancer*, Subcellular Biochemistry 85, DOI 10.1007/978-94-017-9211-0_6,

into these vessels, transport through the blood or lymphatic system, extravasation from the circulation, and survival as a secondary tumor deposit. In the case of epithelial neoplasms, this may be preceded by a phenotypic change in the tumor cells by means of epithelial-to-mesenchymal transition (EMT), although this remains somewhat controversial [1]. More recent work has highlighted the formation of a pre-metastatic niche in potential target organs as a result of mobilization and accumulation of bone-marrow derived cells which prepare the secondary site to accommodate tumor cells.

More than 100 years ago [2], in an effort to explain why some tumors metastasize preferentially to specific target organs, Paget proposed the "seed-and-soil" hypothesis, in which tumor cells with metastatic potential (the "seed") would only metastasize to, and survive in, sites with which they had compatibility (the "soil"). In an excellent review article [3], the modern concepts of seed and soil are highlighted, which include tumor heterogeneity, selective metastasis of cells with pre-existing metastatic potential, and the consequences of the interactions between the microenvironment (metastatic niche) and the tumor cells. Current models suggest that subpopulations of cells exist within the heterogeneous primary tumor which have gained mutations that confer the propensity to metastasize and survive in particular organs, and that these mutations may be acquired early during tumorigenesis [4]. Two elegant studies highlight this using sub-lines of MDA-MB-231 breast cancer cells that metastasize predominantly either to lung or bone [5, 6]. Microarray-based comparison of gene expression in primary tumors and lung metastases revealed that the products of some genes were implicated in growth at the secondary site only, whereas others also played a role in primary tumor growth [6]. In the case of cells that were metastatic to bone, differentially-expressed genes encoded products involved in osteolysis and angiogenesis, amongst others [5]. However, the authors reported that the "bone metastasis gene signature" was additional to a previously-identified "poor prognosis gene signature" [7] that was present in the primary tumor, suggesting that additional biological functions over and above those required for primary tumorigenesis are needed to facilitate metastatic spread. Potentially, a subpopulation of cells within the primary tumor might harbor these metastasis-specific mutations, or they may be acquired as a result of further genetic insult of the primary lesion.

p53 Mutation and Development of Metastasis

It is well accepted that wild-type p53 is a key regulator of cellular homeostasis, and that its loss through deletion or mutation underpins the development of many human malignancies by abrogating cell cycle checkpoints, cell death pathways and various other functions, some of which are pertinent to metastatic spread. One such mechanism involves regulation of neovascularization, a critical early step in metastasis, as outlined above. Some years ago, Van Meir and coworkers reported that expression of wild-type p53 in glioblastoma cells resulted in release of an angiogenesis inhibitor [8], while separate studies identified TSP-1, a potent inhibitor of angiogenesis, as a direct target of wild-type p53 [9], which impacts on survival [10]. Therefore, its reduction in tumors with p53 loss-of-function would enhance the formation of a tumor blood supply. Wild-type p53 is now known to regulate angiogenesis through a number of mechanisms – by upregulating expression of angiogenesis inhibitors that include EphA2 [11–13] and BAI1 [14]. Conversely, wild-type p53 is reported to repress expression of proangiogenic molecules such as matrix metalloprotease (MMP)-1, hypoxia-inducible factor (HIF)-1α [15], the HIF-1α target vascular endothelial cell growth factor (VEGF)-A [16, 17], and cyclo-oxygenase (COX)-2 [18]. Other studies also indicate that some pro-angiogenic chemokines [19], including CXCL12, CXCL8 and CXCL5, are repressed by wild-type p53 [20, 21]. Indeed, in a cellular progression model of sarcomagenesis in which wild-type p53 was inactivated, Milyavsky et al. reported elevated expression of CXCL1 and CXCL8, amongst other genes, in the latter stages of tumor progression [22].

Wild-type p53 is also important in attenuating cellular migration and invasion. Of course, there is some overlap between genes involved in regulation of angiogenesis and cell motility. MMP-1 (collagenase IV) is a key enzyme required for degradation of extracellular matrix (ECM) as a component of the metastatic process, in addition to its role in angiogenesis. Other examples include the chemokines CXCL5, CXCL8 and CXCL12, which are repressed by wild-type p53 [20, 21] and which are key players in cell migration and metastasis [23–25]. Considerable insight into the role of p53 was provided by expression profiling studies conducted by Zhao and colleagues [26], who examined wild-type p53-dependent gene expression in a physiological setting. Amongst a cohort of targets that were either activated or repressed following induction of wild-type p53, they found activation of genes encoding α1 collagens type II and type VI, as well as structural proteins including actin and several keratins. Increased expression of plasminogen activator inhibitor-1 (PAI-1, SERPINE1), an inhibitor of the protease urokinase plasminogen activator (PLAU) was also reported. However, this protein may be involved in potentiation of the metastatic process as well as its inhibition, and its actual function may depend on the concentration in the microenvironment and the context in which it is expressed. For example, at physiological levels it acts to promote angiogenesis and cellular invasion, whereas elevated concentrations are inhibitory [27], and may act to aid cellular detachment [28] in an integrin-dependent manner [29]. Zhao et al. also found p53-dependent induction of endothelin-2 (EDN2), a modulator of vasoconstriction. Yet, EDN2 was demonstrated to function as a chemoattractant for macrophages [30] and may modulate the inflammatory infiltrate of tumors as well as enhance invasion [31]. Why this would be induced by a tumor suppressor protein such as p53 is, thus far, unclear, but may be unrelated to its role in tumor biology.

Studies in a mouse model system of hepatocellular carcinoma also provide further understanding of the role of wild-type p53 in suppression of metastasis. Delivery of a polyomavirus middle T antigen using the RCAS system into the livers of transgenic mice expressing the viral receptor (TVA) was found to result in formation of hepatic adenomas. However, when this was performed on a p53-null background, invasive and metastatic tumors developed, with differential expression of 105 genes between benign and malignant tumors [32]. These included insulin-like growth factor (IGF)-2, cathepsin E, and the chemokines CCL8 (MCP-2, SCYA8) and CCL5 (RANTES, SCYA5), all of which have recognized roles in metastatic spread of tumors.

Wild-type p53 also suppresses prometastatic activity through inhibition of small GTPase activation. Using mouse embryo fibroblasts from p53-null and $p19^{ARF}$-null animals, Guo and colleagues reported changes in actin reorganization, accompanied by activation of phosphoinositide 3-kinase (PI3K) and Rac1 [33]. Further work by this group [34] indicated that p53 loss led to increased focal adhesion formation, and that ROCK activation only partially mimicked the RhoA phenotype. Promotion of an invasive phenotype by expression of active RhoA, Rac1, or Cdc42, but not an activated Ras mutant, was also observed in p53-null cells. However, subsequent studies by Xia and Land revealed that loss of p53 function combined with activated Ras resulted in enhanced cell motility [35]. Co-expression of mutant $p53^{R175H}$ and $H\text{-}Ras^{G12V}$ led to more profound migration in both wound-closure and transwell assays, with increased activation of RhoA, but not Rac1 or Cdc42. Similar GTP loading of RhoA was observed when endogenous wild-type p53 was repressed by shRNA, as well as in p53-null HCT116 cells, indicating the importance of loss of function. Moreover, wild-type p53 was shown to enhance activity of p190 RhoGAP, thus minimizing accumulation of the GTP-bound (active) form of RhoA.

p53 Gain-of-Function Mutants

Over and above the undoubted importance of loss-of-function mutations in the development of metastatic disease, many common mutations in the p53 gene in human cancer result in expression of proteins with dominant transforming properties that actively drive tumor progression. These gain-of-function (GOF) mutants may endow the cells with many properties that impart growth or survival advantages, and may include functions pertinent to metastatic spread. Indeed, mouse models with knock-in GOF p53 alleles exhibit increased metastasis compared to p53-null animals [36–38]. GOF mutant p53 may also be associated with EMT, as mice expressing a mutant KRAS gene together with p53R172H were found to develop spindle cell carcinomas and frequent (>60 %) spread to secondary organs, including lungs and lymph nodes [38]. Nude mouse xenografts of fibroblasts expressing GOF p53 (H179L) were also shown to undergo metastatic spread, with deposits in lung and mediastinum from subcutaneous primary tumors [39].

Gain-of-Function Mechanisms

Although there is debate over the mechanism through which GOF p53 mutants act, several models have been proposed. Some years ago, transcriptomic profiling provided clues to the aberrant functions of this class of proteins. Studies in lung cancer cells expressing specific p53 mutants revealed key differences (as well as some similarities) compared to the same cells expressing wild-type p53 [40, 41], and at least in some cases this appeared to be dependent upon the transactivation properties of p53 as L22Q/W23S mutants (*i.e.*, in one of the transactivation domains) showed greatly reduced potency. Amongst the genes identified by this method, NF-κB2 (p100/p52) was upregulated by GOF p53 proteins [42], raising the possibility that transcription factors such as this may be activated by mutant p53, thereby leading to a secondary level of gene induction or repression that mediates the biological responses.

An important study by Weisz and colleagues investigated the effects of GOF p53 on the response of cancer cells to tumor necrosis factor (TNF)-α [43]. Whereas TNF-α is potentially cytotoxic, as it is able to induce an apoptotic response, it is also well-recognized that this cytokine can promote tumorigenesis through the activation of NF-κB on a background of inflammation [44, 45]. GOF p53 was shown to promote NF-κB activity in response to TNF-α, with increased nuclear translocation of p65, and to inhibit TNF-α-induced apoptosis, whereas cells lacking p53 or treated with siRNA showed increased cell death in the presence of ligand. Moreover, co-expression of mutant p53 and NF-κB was found in human tumors, further suggesting a functional relationship. Thus, GOF p53 may act as molecular switch that toggles the response to an otherwise cytotoxic factor through activation of NF-κB.

Another proposed mechanism to explain p53 gain-of-function mutation is through interaction with other members of the p53 family, such as p63 and p73. Whereas p73 may be of more importance in apoptosis, p63 isoforms have been linked to tumor progression and metastasis, although there is considerable controversy in the existing literature as to whether p63 proteins function as tumor suppressors or tumor promoters, which isoforms are responsible, and in which tissue types [46]. The presence of two promoters leads to expression of full-length TAp63 and ΔNp63 which lacks the amino terminus. Each of these undergoes alternative splicing at the 3′ end to generate α, β and γ forms and, potentially, δ and ε. ΔNp63 is required for normal epithelial differentiation, and its absence is lethal due to deficient formation of the epidermis and associated structures.

Despite some apparent contradictory functions and activities reported for p63, emerging evidence seems to support the hypothesis that p63 acts as a suppressor of metastasis, and that the balance of expression between TA and ΔN isoforms is important, with ΔNp63 (or GOF mutant p53) being capable of oligotetramerizing with TAp63 and suppressing its anti-tumorigenic and anti-metastatic properties. Mice in which TAp63 is inactivated lose the Ras-dependent senescence response and show increased sarcomagenesis in the absence of p53 [47]. Consistent with this suppressive effect, gene knockdown of p63 in squamous carcinoma cell lines led to increased expression of a cohort of genes involved in invasion and metastasis that included α4 integrin, N-cadherin, tenascin C and two Wnt proteins, Wnt-4 and -5a, with a concomitant increase in cellular migration rate [48]. Moreover, mutant p53 has been shown to increase Rab-dependent recycling of the EGFR and α5β1 integrin, inhibit the function of TAp63, and promote random migration, invasion and metastasis [49].

Maspin, also known as SERPINB5, is well-known for its function as an inhibitor of cell migration, invasion, metastasis and angiogenesis [50]. Although multiple factors contribute to altered maspin expression in human cancer, Kim et al.

identified TAp63 as an important activator of maspin expression in lung cells through binding to a p53 binding site in the maspin promoter [51]. These authors found a striking correlation between expression of maspin and p63 in lung cancers, with loss of both in adenomacarcinomas and reduced expression of maspin in lung cancer metastases to lymph node. TAp63 was found to activate maspin expression in reporter gene assays, and maspin levels correlated inversely with invasion, implying that loss (or inactivation) of TAp63 during tumor progression results in enhanced capacity for metastatic spread. Consistent with this notion of p63-dependent suppression of invasion by maspin, studies in endometrial cancer cells showed that expression of the GOF p53 mutant R273H, but not a loss of function p53 mutant, inhibited maspin expression and promoted migration and invasion [52].

Studies by Adomo and colleagues showed co-operativity between transforming growth factor (TGF)-β and GOF p53, in which GOF p53 and Smad2 enter into a ternary complex with p63 and repress its anti-metastatic activity, in part through modulation of five target genes: Cyclin G2, Sharp-1, Follistatin, ADAMTS9 and GPR87 [53]. They found that raising the level of p63 altered the balance of GOF p53 to p63, and suppressed lung colonization by tumor cells introduced *via* the mouse tail vein, further establishing p63 as a suppressor of metastasis. Moreover, TGF-β treatment of cells expressing GOF p53 enhanced cell migration. The model proposed involves interaction of the α-domain of p63 with the MH2 domain of Smad2, and binding of the transactivation domain of GOF p53 by the Smad2 MH1 domain. Thus, this may be specific for p63α isoforms. Interestingly, tumor suppressive functions of TGF-β through the repression of maspin expression and consequent inhibition of cell migration are dependent upon the presence of wild-type p53 [54].

A third mechanism to explain mutant p53 gain of function is direct recruitment, together with other transcription factors, onto the promoters of specific target genes. For example, it has been reported that GOF p53 interacts with NF-Y and p300, activating NF-Y targets [55]. A pertinent example related to invasion and metastasis is given by the elegant work of Fontemaggi et al. [56], in which they examined regulation of ID4 by GOF p53 – E2F1 complexes. They found that expression of R175H and R273H isoforms in H1299 cells led to elevated expression of ID4, whereas repression of mutant p53, but not wild-type p53, by siRNA resulted in lower ID4 levels. Subsequent analyses revealed that E2F1 was required for GOF p53-mediated activation of ID4 expression. This was shown to facilitate stabilization of mRNAs encoding the pro-angiogenic (and pro-metastatic) chemokines IL-8 (CXCL8) and Gro-α (CXCL1). Furthermore, co-expression of p53 and ID4 in breast tumors was found to correlate with increased microvessel density, a measure of angiogenesis. Enhanced recruitment of CBP and STAT onto the NF-κB2 promoter by GOF p53 mutants [57] also explains some of the earlier observations that this transcription factor is upregulated by mutant p53 [42]. Direct nucleation of mutant p53 onto the promoter of the receptor tyrosine kinase AXL has also been shown recently [58], but this appears to be independent of transactivation ability of the mutant p53, as L22Q/W23S mutants in one of the transactivation domains were still able to enhance AXL expression. Stimulation of AXL expression by GOF p53 was correlated with enhanced motility of lung cancer cells, supporting a role for this axis in aspects of tumor metastasis.

p53 and the Chemokine Network

It is now well accepted that NF-κB signaling is central to the chronic inflammatory response that promotes tumor progression and metastasis [59, 60]. The chemokine network is a complex system of ligands and receptors whose primary roles are in immune cell activation and their recruitment to inflammatory foci through production of chemokines at the inflammatory site, with establishment of a chemokine gradient towards which inflammatory cells expressing the cognate receptor migrate [61]. However, chemokines and their receptors are now firmly established as major players in tumorigenesis, angiogenesis and metastasis [23–25, 62–65]. For example, Muller et al. reported upregulation of CXCR4 and CCR7 receptors on the surface of breast cancer cells and demonstrated that this enabled homing of tumor cells to target organs expressing the ligands for these receptors [66]. Multiple studies have also reported elevated expression of pro-angiogenic chemokines such as CXCL1, CXCL2, CXCL3, CXCL5 and CXCL8 in a wide range of tumor types [67–72]. Thus, the roles of different chemokine-receptor interactions in tumor progression are varied and widespread.

As mentioned earlier, an interesting study by Moskovits et al. [20] reported that wild-type p53 represses CXCL12 expression, thereby reducing cell migration. Moreover, subsequent studies from our own laboratory [21], which focused on the contribution of GOF p53 to cell motility, highlighted a positive influence of these aberrant proteins on chemokine expression. Expression of several chemokines is regulated, at least in part, by NF-κB family transcription factors [73–83], consistent with a promoting role for inflammation in carcinogenesis, and it is clear that GOF p53 proteins activate a transcriptional response that includes NF-κB2 [41, 42]. Thus, it is perhaps no surprise that chemokines are upregulated by GOF p53. This is a clear gain of oncogenic function, as p53-null cells show higher levels of chemokine expression compared to cells expressing wild-type p53, yet substantial increases in chemokine expression occur when GOF p53 proteins are present [21]. Moreover, levels of CXCL5, CXCL8 and CXCL12 are increased differentially, depending upon the amino acid substitution present in p53, suggesting a degree of allele-specificity, and appear to depend upon elevated transcription. Further, these enhanced chemokine levels correlated with increased cellular motility, consistent with a role in invasion and metastasis. However, it is likely that deregulation of NF-κB activity is not the only mechanism responsible for enhancing chemokine expression. At least in the case of CXCL1 (Gro-α) and CXCL8 (IL-8), enhanced mRNA stabilization through GOF p53 – E2F1 activation of ID4, as noted above, is crucial [56]. In addition, inactivation of p63 may also be a key mechanism, either directly or indirectly [84]. Thus, multiple biochemical mechanisms triggered by GOF p53 may cooperate to deregulate the chemokine network in cancer cells and enhance their progression to a metastatic phenotype.

MicroRNA-Mediated Control of Metastasis

Another key mechanism that regulates processes involved in tumor cell metastasis is the action of microRNAs (miRNAs). These are small, non-coding RNAs present throughout the genome, which bind to target sequences in messenger RNAs (mRNAs), effecting their degradation or repressing their translation. They are generated in a step-wise process, which involves, first, expression of a pri-miRNA that contains a characteristic stem-loop structure. This entity is processed by an enzyme – Drosha – into a pre-miRNA, generally between 70 base pairs (bp) and 100 bp in length. Following export into the cytoplasm, the pre-miRNA is cleaved by Dicer to generate the mature miRNA consisting of two strands of 20–25 bp in length, one of which becomes incorporated into the RNA-induced silencing complex (RISC).

A number of miRNAs have been implicated in suppression of metastasis, including miR-31, miR126, miR-206, miR-335, miR-130, and the miR-200 family [85, 86]. Processing of these requires Dicer, which is transactivated by TAp63 [87]. Therefore, transdominant inhibition of p63 by GOF p53 represents yet another mode of action through which mutant p53 may promote tumor metastasis, as well as other wide-ranging effects on cellular biology [88, 89]. In addition, miR-200 species target the EMT-associated transcription factors ZEB1 and ZEB2 for degradation [90, 91]; therefore inhibition of miR-200 processing through loss of Dicer activity may explain how GOF p53 can promote EMT [92].

A recent study has provided further evidence for GOF p53 regulation of microRNA that leads to enhanced invasion in breast cancer [93]. Expression of either miR-155 or GOF p53 in p53-null cells led to increased invasion and EMT. Subsequent experiments indicated a direct role for p63 in transcriptional repression of miR-155, which was relieved by the presence of GOF p53. Interestingly, these authors identified the transcriptional repressor, ZNF652, as a target of miR-155. ZNF652 blocks expression of multiple genes involved in EMT, including vimentin, EGFR, TGF-β, and TGFβR2. Thus, inactivation of p63 by GOF p53 enables miR-155 to inactivate ZNF652, thereby promoting emergence of the mesenchymal phenotype that is characteristic of many invasive epithelial malignancies. Unsurprisingly, low levels of ZNF652 were found to correlate with breast tumor invasion in clinical samples.

Conclusions

p53 mutation impacts metastatic processes on multiple levels. Loss of wild-type p53 function relieves repression of angiogenesis and enhances motility. However, dominant oncogenic p53 proteins actively drive metastasis by promoting angiogenesis through upregulating the expression of chemokines and other angiogenic factors in tumor cells. GOF p53 mutants enhance metastasis by switching on transcriptional programs that promote a more aggressive biological phenotype, and by interfering with the metastasis-suppressive functions of p63. Our emerging understanding of microRNAs in control of angiogenesis and metastasis provides a further layer of complexity to p53 function. However, the central role of p53 in development of aggressive cancers may provide opportunities for targeted therapeutic approaches, either of p53 directly or one or more of its mediators, to improve clinical outcome.

References

1. Tarin D, Thompson EW, Newgreen DF (2005) The fallacy of epithelial mesenchymal transition in neoplasia. Cancer Res 65:5996–6001
2. Paget S (1889) The distribution of secondary growths in cancer of the breast. Lancet 133:571–573
3. Fidler IJ (2002) Critical determinants of metastasis. Semin Cancer Biol 12:89–96
4. Bernards R, Weinberg RA (2002) A progression puzzle. Nature 418:823
5. Kang Y, Siegel PM, Shu W, Drobnjak M, Kakonen SM, Cordon-Cardo C, Guise TA, Massague J (2003) A multigenic program mediating breast cancer metastasis to bone. Cancer Cell 3:537–549
6. Minn AJ, Gupta GP, Siegel PM, Bos PD, Shu W, Giri DD, Viale A, Olshen AB, Gerald WL, Massague J (2005) Genes that mediate breast cancer metastasis to lung. Nature 436:518–524
7. van't Veer LJ, Dai H, van de Vijver MJ, He YD, Hart AA, Mao M, Peterse HL, van der Kooy K, Marton MJ, Witteveen AT, Schreiber GJ, Kerkhoven RM, Roberts C, Linsley PS, Bernards R, Friend SH (2002) Gene expression profiling predicts clinical outcome of breast cancer. Nature 415:530–536
8. Van Meir EG, Polverini PJ, Chazin VR, Su Huang H-J, de Tribolet N, Cavenee WK (1994) Release of an inhibitor of angiogenesis upon induction of wild type p53 expression in glioblastoma cells. Nat Genet 8:171–176
9. Dameron K, Volpert O, Tainsky M, Bouck N (1994) Control of angiogenesis in fibroblasts by p53 regulation of thrombospondin-1. Science 265:1582–1584
10. Lawler J, Miao WM, Duquette M, Bouck N, Bronson RT, Hynes RO (2001) Thrombospondin-1 gene expression affects survival and tumor spectrum of p53-deficient mice. Am J Pathol 159:1949–1956
11. Brantley DM, Cheng N, Thompson EJ, Lin Q, Brekken RA, Thorpe PE, Muraoka RS, Cerretti DP, Pozzi A, Jackson D, Lin C, Chen J (2002) Soluble Eph A receptors inhibit tumor angiogenesis and progression in vivo. Oncogene 21:7011–7026
12. Dohn M, Jiang J, Chen X (2001) Receptor tyrosine kinase EphA2 is regulated by p53-family proteins and induces apoptosis. Oncogene 20:6503–6515
13. Cheng N, Brantley DM, Liu H, Lin Q, Enriquez M, Gale N, Yancopoulos G, Cerretti DP, Daniel TO, Chen J (2002) Blockade of EphA receptor tyrosine kinase activation inhibits vascular endothelial cell growth factor-induced angiogenesis. Mol Cancer Res 1:2–11
14. Nishimori H, Shiratsuchi T, Urano T, Kimura Y, Kiyono K, Tatsumi K, Yoshida S, Ono M, Kuwano M, Nakamura Y, Tokino T (1997) A novel brain-specific p53-target gene, BAI1, containing thrombospondin type 1 repeats inhibits experimental angiogenesis. Oncogene 15:2145–2150
15. Ravi R, Mookerjee B, Bhujwalla ZM, Sutter CH, Artemov D, Zeng Q, Dillehay LE, Madan A, Semenza GL, Bedi A (2000) Regulation of tumor angiogenesis by p53-induced degradation of hypoxia-inducible factor 1Î±. Genes Dev 14:34–44
16. Mukhopadhyay D, Tsiokas L, Sukhatme VP (1995) Wild-type p53 and v-Src exert opposing influences on human vascular endothelial growth factor gene expression. Cancer Res 55:6161–6165
17. Pal S, Datta K, Mukhopadhyay D (2001) Central role of p53 on regulation of Vascular Permeability Factor/Vascular Endothelial Growth Factor (VPF/VEGF) expression in mammary carcinoma. Cancer Res 61:6952–6957

18. Subbaramaiah K, Altorki N, Chung WJ, Mestre JR, Sampat A, Dannenberg AJ (1999) Inhibition of cyclooxygenase-2 gene expression by p53. J Biol Chem 274:10911–10915
19. Strieter RM, Burdick MD, Gomperts BN, Belperio JA, Keane MP (2005) CXC chemokines in angiogenesis. Cytokine Growth Factor Rev 16:593–609
20. Moskovits N, Kalinkovich A, Bar J, Lapidot T, Oren M (2006) p53 attenuates cancer cell migration and invasion through repression of SDF-1/CXCL12 expression in stromal fibroblasts. Cancer Res 66:10671–10676
21. Yeudall WA, Vaughan CA, Miyazaki H, Ramamoorthy M, Choi MY, Chapman CG, Wang H, Black E, Bulysheva AA, Deb SP, Windle B, Deb S (2012) Gain-of-function mutant p53 upregulates CXC chemokines and enhances cell migration. Carcinogenesis 33:442–451
22. Milyavsky M, Tabach Y, Shats I, Erez N, Cohen Y, Tang X, Kalis M, Kogan I, Buganim Y, Goldfinger N, Ginsberg D, Harris CC, Domany E, Rotter V (2005) Transcriptional programs following genetic alterations in p53, INK4A, and H-Ras genes along defined stages of malignant transformation. Cancer Res 65:4530–4543
23. Yeudall WA, Miyazaki H (2007) Chemokines and squamous cancer of the head and neck: targets for therapeutic intervention? Expert Rev Anticancer Ther 7:351–360
24. Zlotnik A (2004) Chemokines in neoplastic progression. Semin Cancer Biol 14:181–185
25. Miyazaki H, Takabe K, Yeudall WA (2013) Chemokines, chemokine receptors and the gastrointestinal system. World J Gastroenterol 19:2847–2863
26. Zhao R, Gish K, Murphy M, Yin Y, Notterman D, Hoffman WH, Tom E, Mack DH, Levine AJ (2000) Analysis of p53-regulated gene expression patterns using oligonucleotide arrays. Genes Dev 14:981–993
27. Bajou K, Maillard C, Jost M, Lijnen RH, Gils A, Declerck P, Carmeliet P, Foidart JM, Noel A (2004) Host-derived plasminogen activator inhibitor-1 (PAI-1) concentration is critical for in vivo tumoral angiogenesis and growth. Oncogene 23:6986–6990
28. Czekay RP, Aertgeerts K, Curriden SA, Loskutoff DJ (2003) Plasminogen activator inhibitor-1 detaches cells from extracellular matrices by inactivating integrins. J Cell Biol 160:781–791
29. Czekay RP, Loskutoff DJ (2009) Plasminogen activator inhibitors regulate cell adhesion through a uPAR-dependent mechanism. J Cell Physiol 220:655–663
30. Grimshaw MJ, Wilson JL, Balkwill FR (2002) Endothelin-2 is a macrophage chemoattractant: implications for macrophage distribution in tumors. Eur J Immunol 32:2393–2400
31. Grimshaw MJ, Hagemann T, Ayhan A, Gillett CE, Binder C, Balkwill FR (2004) A role for endothelin-2 and its receptors in breast tumor cell invasion. Cancer Res 64:2461–2468
32. Lewis BC, Klimstra DS, Socci ND, Xu S, Koutcher JA, Varmus HE (2005) The absence of p53 promotes metastasis in a novel somatic mouse model for hepatocellular carcinoma. Mol Cell Biol 25:1228–1237
33. Guo F, Gao Y, Wang L, Zheng Y (2003) p19Arf-p53 tumor suppressor pathway regulates cell motility by suppression of phosphoinositide 3-Kinase and Rac1 GTPase Activities. J Biol Chem 278:14414–14419
34. Guo F, Zheng Y (2004) Rho family GTPases cooperate with p53 deletion to promote primary mouse embryonic fibroblast cell invasion. Oncogene 23:5577–5585
35. Xia M, Land H (2007) Tumor suppressor p53 restricts Ras stimulation of RhoA and cancer cell motility. Nat Struct Mol Biol 14:215–223
36. Lang GA, Iwakuma T, Suh YA, Liu G, Rao VA, Parant JM, Valentin-Vega YA, Terzian T, Caldwell LC, Strong LC, El-Naggar AK, Lozano G (2004) Gain of function of a p53 hot spot mutation in a mouse model of Li-Fraumeni syndrome. Cell 119:861–872
37. Olive KP, Tuveson DA, Ruhe ZC, Yin B, Willis NA, Bronson RT, Crowley D, Jacks T (2004) Mutant p53 gain of function in two mouse models of Li-Fraumeni syndrome. Cell 119:847–860
38. Caulin C, Nguyen T, Lang GA, Goepfert TM, Brinkley BR, Cai WW, Lozano G, Roop DR (2007) An inducible mouse model for skin cancer reveals distinct roles for gain- and loss-of-function p53 mutations. J Clin Invest 117:1893–1901
39. Cardinali M, Kratochvil FJ, Ensley JF, Robbins KC, Yeudall WA (1997) Functional characterization in vivo of mutant p53 molecules derived from squamous cell carcinomas of the head and neck. Mol Carcinog 18:78–88

40. Scian MJ, Stagliano KE, Deb D, Ellis MA, Carchman EH, Das A, Valerie K, Deb SP, Deb S (2004) Tumor-derived p53 mutants induce oncogenesis by transactivating growth-promoting genes. Oncogene 23:4430–4443
41. Scian MJ, Stagliano KE, Ellis MA, Hassan S, Bowman M, Miles MF, Deb SP, Deb S (2004) Modulation of gene expression by tumor-derived p53 mutants. Cancer Res 64:7447–7454
42. Scian MJ, Stagliano KE, Anderson MA, Hassan S, Bowman M, Miles MF, Deb SP, Deb S (2005) Tumor-derived p53 mutants induce NF-kappaB2 gene expression. Mol Cell Biol 25:10097–10110
43. Weisz L, Damalas A, Liontos M, Karakaidos P, Fontemaggi G, Maor-Aloni R, Kalis M, Levrero M, Strano S, Gorgoulis VG, Rotter V, Blandino G, Oren M (2007) Mutant p53 enhances nuclear factor kappaB activation by tumor necrosis factor alpha in cancer cells. Cancer Res 67:2396–2401
44. Luo JL, Maeda S, Hsu LC, Yagita H, Karin M (2004) Inhibition of NF-kappaB in cancer cells converts inflammation-induced tumor growth mediated by TNFalpha to TRAIL-mediated tumor regression. Cancer Cell 6:297–305
45. Pikarsky E, Porat RM, Stein I, Abramovitch R, Amit S, Kasem S, Gutkovich-Pyest E, Urieli-Shoval S, Galun E, Ben-Neriah Y (2004) NF-kappaB functions as a tumour promoter in inflammation-associated cancer. Nature 431:461–466
46. Melino G (2011) p63 is a suppressor of tumorigenesis and metastasis interacting with mutant p53. Cell Death Differ 18:1487–1499
47. Guo X, Keyes WM, Papazoglu C, Zuber J, Li W, Lowe SW, Vogel H, Mills AA (2009) TAp63 induces senescence and suppresses tumorigenesis in vivo. Nat Cell Biol 11:1451–1457
48. Barbieri CE, Tang LJ, Brown KA, Pietenpol JA (2006) Loss of p63 leads to increased cell migration and up-regulation of genes involved in invasion and metastasis. Cancer Res 66:7589–7597
49. Muller PA, Caswell PT, Doyle B, Iwanicki MP, Tan EH, Karim S, Lukashchuk N, Gillespie DA, Ludwig RL, Gosselin P, Cromer A, Brugge JS, Sansom OJ, Norman JC, Vousden KH (2009) Mutant p53 drives invasion by promoting integrin recycling. Cell 139:1327–1341
50. Zhang M (2004) Multiple functions of maspin in tumor progression and mouse development. Front Biosci 9:2218–2226
51. Kim S, Han J, Kim J, Park C (2004) Maspin expression is transactivated by p63 and is critical for the modulation of lung cancer progression. Cancer Res 64:6900–6905
52. Dong P, Tada M, Hamada J-I, Nakamura A, Moriuchi T, Sakuragi N (2007) p53 dominant-negative mutant R273H promotes invasion and migration of human endometrial cancer HHUA cells. Clin Exp Metastasis 24:471–483
53. Adorno M, Cordenonsi M, Montagner M, Dupont S, Wong C, Hann B, Solari A, Bobisse S, Rondina MB, Guzzardo V, Parenti AR, Rosato A, Bicciato S, Balmain A, Piccolo S (2009) A mutant-p53/Smad complex opposes p63 to empower TGF[beta]-induced metastasis. Cell 137:87–98
54. Wang SE, Narasanna A, Whitell CW, Wu FY, Friedman DB, Arteaga CL (2007) Convergence of p53 and Transforming Growth Factor beta (TGFbeta) signaling on activating expression of the tumor suppressor gene maspin in mammary epithelial cells. J Biol Chem 282:5661–5669
55. Di Agostino S, Strano S, Emiliozzi V, Zerbini V, Mottolese M, Sacchi A, Blandino G, Piaggio G (2006) Gain of function of mutant p53: the mutant p53/NF-Y protein complex reveals an aberrant transcriptional mechanism of cell cycle regulation. Cancer Cell 10:191–202
56. Fontemaggi G, Dell'Orso S, Trisciuoglio D, Shay T, Melucci E, Fazi F, Terrenato I, Mottolese M, Muti P, Domany E, Del Bufalo D, Strano S, Blandino G (2009) The execution of the transcriptional axis mutant p53, E2F1 and ID4 promotes tumor neo-angiogenesis. Nat Struct Mol Biol 16:1086–1093
57. Vaughan CA, Singh S, Windle B, Sankala HM, Graves PR, Andrew Yeudall W, Deb SP, Deb S (2012) p53 mutants induce transcription of NF-kappaB2 in H1299 cells through CBP and STAT binding on the NF-kappaB2 promoter and gain of function activity. Arch Biochem Biophys 518:79–88

58. Vaughan CA, Singh S, Windle B, Yeudall WA, Frum R, Grossman SR, Deb SP, Deb S (2012) Gain-of-function activity of mutant p53 in lung cancer through up-regulation of receptor protein tyrosine kinase Axl. Genes Cancer 3:491–502
59. Karin M, Greten FR (2005) NF-kappaB: linking inflammation and immunity to cancer development and progression. Nat Rev Immunol 5:749–759
60. Karin M, Cao Y, Greten FR, Li ZW (2002) NF-kappaB in cancer: from innocent bystander to major culprit. Nat Rev Cancer 2:301–310
61. Locati MDM, Murphy MDPM (1999) Chemokines and chemokine receptors: biology and clinical relevance in inflammation and AIDS1. Annu Rev Med 50:425–440
62. Balkwill F (2004) Cancer and the chemokine network. Nat Rev Cancer 4:540–550
63. Balkwill F, Charles KA, Mantovani A (2005) Smoldering and polarized inflammation in the initiation and promotion of malignant disease. Cancer Cell 7:211–217
64. Strieter RM, Belperio JA, Phillips RJ, Keane MP (2004) CXC chemokines in angiogenesis of cancer. Semin Cancer Biol 14:195–200
65. Strieter RM, Burdick MD, Mestas J, Gomperts B, Keane MP, Belperio JA (2006) Cancer CXC chemokine networks and tumour angiogenesis. Eur J Cancer 42:768–778
66. Muller A, Homey B, Soto H, Ge N, Catron D, Buchanan ME, McClanahan T, Murphy E, Yuan W, Wagner SN, Barrera JL, Mohar A, Verastegui E, Zlotnik A (2001) Involvement of chemokine receptors in breast cancer metastasis. Nature 410:50–56
67. Balentien E, Mufson BE, Shattuck RL, Derynck R, Richmond A (1991) Effects of MGSA/GRO alpha on melanocyte transformation. Oncogene 6:1115–1124
68. Dhawan P, Richmond A (2002) Role of CXCL1 in tumorigenesis of melanoma. J Leukoc Biol 72:9–18
69. Keane MP, Belperio JA, Xue YY, Burdick MD, Strieter RM (2004) Depletion of CXCR2 inhibits tumor growth and angiogenesis in a murine model of lung cancer. J Immunol 172:2853–2860
70. Kryczek I, Lange A, Mottram P, Alvarez X, Cheng P, Hogan M, Moons L, Wei S, Zou L, Machelon V, Emilie D, Terrassa M, Lackner A, Curiel TJ, Carmeliet P, Zou W (2005) CXCL12 and vascular endothelial growth factor synergistically induce neoangiogenesis in human ovarian cancers. Cancer Res 65:465–472
71. Strieter RM, Belperio JA, Burdick MD, Sharma S, Dubinett SM, Keane MP (2004) CXC chemokines: angiogenesis, immunoangiostasis, and metastases in lung cancer. Ann N Y Acad Sci 1028:351–360
72. Miyazaki H, Patel V, Wang H, Edmunds RK, Gutkind JS, Yeudall WA (2006) Downregulation of CXCL5 inhibits squamous carcinogenesis. Cancer Res 66:4279–4284
73. Kunsch C, Rosen CA (1993) NF-kappa B subunit-specific regulation of the interleukin-8 promoter. Mol Cell Biol 13:6137–6146
74. Kunsch C, Lang RK, Rosen CA, Shannon MF (1994) Synergistic transcriptional activation of the IL-8 gene by NF-kappa B p65 (RelA) and NF-IL-6. J Immunol 153:153–164
75. Yang J, Richmond A (2001) Constitutive IkappaB kinase activity correlates with nuclear factor-kappaB activation in human melanoma cells. Cancer Res 61:4901–4909
76. Wang D, Richmond A (2001) Nuclear factor-kappa B activation by the CXC chemokine melanoma growth-stimulatory activity/growth-regulated protein involves the MEKK1/p38 mitogen-activated protein kinase pathway. J Biol Chem 276:3650–3659
77. Chandrasekar B, Mummidi S, Perla RP, Bysani S, Dulin NO, Liu F, Melby PC (2003) Fractalkine (CX3CL1) stimulated by nuclear factor kappaB (NF-kappaB)-dependent inflammatory signals induces aortic smooth muscle cell proliferation through an autocrine pathway. Biochem J 373:547–558
78. Dhawan P, Richmond A (2002) A novel NF-kappa B-inducing kinase-MAPK signaling pathway up-regulates NF-kappa B activity in melanoma cells. J Biol Chem 277:7920–7928
79. Richmond A (2002) NF-[kappa]B, chemokine gene transcription and tumour growth. Nat Rev Immunol 2:664–674
80. Rehman AO, Wang C-Y (2008) SDF-1{alpha} promotes invasion of head and neck squamous cell carcinoma by activating NF-{kappa}B. J Biol Chem 283:19888–19894

81. Furuya M, Suyama T, Usui H, Kasuya Y, Nishiyama M, Tanaka N, Ishiwata I, Nagai Y, Shozu M, Kimura S (2007) Up-regulation of CXC chemokines and their receptors: implications for proinflammatory microenvironments of ovarian carcinomas and endometriosis. Hum Pathol 38:1676–1687
82. Maroni P, Bendinelli P, Matteucci E, Desiderio MA (2007) HGF induces CXCR4 and CXCL12-mediated tumor invasion through Ets1 and NF-kappaB. Carcinogenesis 28:267–279
83. Sun Q, Matta H, Lu G, Chaudhary PM (2006) Induction of IL-8 expression by human herpesvirus 8 encoded vFLIP K13 *via* NF-kappaB activation. Oncogene 25:2717–2726
84. Yang X, Lu H, Yan B, Romano R-A, Bian Y, Friedman J, Duggal P, Allen C, Chuang R, Ehsanian R, Si H, Sinha S, Van Waes C, Chen Z (2011) DeltaNp63 versatilely regulates a broad NF-kappaB gene program and promotes squamous epithelial proliferation, migration, and inflammation. Cancer Res 71:3688–3700
85. Valastyan S, Reinhardt F, Benaich N, Calogrias D, Szasz AM, Wang ZC, Brock JE, Richardson AL, Weinberg RA (2009) A pleiotropically acting microRNA, miR-31, inhibits breast cancer metastasis. Cell 137:1032–1046
86. Valastyan S, Weinberg RA (2010) Metastasis suppression: a role of the Dice(r). Genome Biol 11:141
87. Su X, Chakravarti D, Cho MS, Liu L, Gi YJ, Lin YL, Leung ML, El-Naggar A, Creighton CJ, Suraokar MB, Wistuba I, Flores ER (2010) TAp63 suppresses metastasis through coordinate regulation of Dicer and miRNAs. Nature 467:986–990
88. Tucci P, Agostini M, Grespi F, Markert EK, Terrinoni A, Vousden KH, Muller PAJ, Dötsch V, Kehrloesser S, Sayan BS, Giaccone G, Lowe SW, Takahashi N, Vandenabeele P, Knight RA, Levine AJ, Melino G (2012) Loss of p63 and its microRNA-205 target results in enhanced cell migration and metastasis in prostate cancer. Proc Natl Acad Sci U S A 109:15312–15317
89. Tran MN, Choi W, Wszolek MF, Navai N, Lee ILC, Nitti G, Wen S, Flores ER, Siefker-Radtke A, Czerniak B, Dinney C, Barton M, McConkey DJ (2013) The p63 protein isoform ΔNp63α inhibits epithelial-mesenchymal transition in human bladder cancer cells: role of MIR-205. J Biol Chem 288:3275–3288
90. Park S-M, Gaur AB, Lengyel E, Peter ME (2008) The miR-200 family determines the epithelial phenotype of cancer cells by targeting the E-cadherin repressors ZEB1 and ZEB2. Genes Dev 22:894–907
91. Korpal M, Lee ES, Hu G, Kang Y (2008) The miR-200 family inhibits epithelial-mesenchymal transition and cancer cell migration by direct targeting of E-cadherin transcriptional repressors ZEB1 and ZEB2. J Biol Chem 283:14910–14914
92. Ohashi S, Natsuizaka M, Wong GS, Michaylira CZ, Grugan KD, Stairs DB, Kalabis J, Vega ME, Kalman RA, Nakagawa M, Klein-Szanto AJ, Herlyn M, Diehl JA, Rustgi AK, Nakagawa H (2010) Epidermal growth factor receptor and mutant p53 expand an esophageal cellular subpopulation capable of epithelial-to-mesenchymal transition through ZEB transcription factors. Cancer Res 70:4174–4184
93. Neilsen PM, Noll JE, Mattiske S, Bracken CP, Gregory PA, Schulz RB, Lim SP, Kumar R, Suetani RJ, Goodall GJ, Callen DF (2013) Mutant p53 drives invasion in breast tumors through up-regulation of miR-155. Oncogene 32:2992–3000

Chapter 7
Structural Studies on Mechanisms to Activate Mutant p53

Hector Viadiu, Gilberto Fronza, and Alberto Inga

Abstract The design of a broad-spectrum cancer drug would provide enormous clinical benefits to treat cancer patients. Most of cancerous cells have a mutation in the p53 gene that results in an inactive mutant p53 protein. For this reason, p53 is a prime target for the development of a broad-spectrum cancer drug. To provide the atomic information to rationally design a drug to recover p53 activity is the main goal of the structural studies on mutant p53. We review three mechanisms that influence p53 activity and provide information about how reactivation of mutant p53 can be achieved: stabilization of the active conformation of the DNA-binding domain of the protein, suppression of missense mutations in the DNA-binding domain by a second-site mutation, and increased transactivation.

Keywords Mutant p53 • p53 structure • Crystallography • p53 activation • Suppression p53 missense mutations • p53 transactivation

Every year more than seven million people die of cancer in the world [27]. Enormous clinical benefits would result from the discovery of a drug able to treat any kind of

H. Viadiu (✉)
Instituto de Química, Universidad Nacional Autónoma de México (UNAM),
Mexico City, D.F., Mexico
e-mail: viadiu@unam.mx

G. Fronza
Mutagenesis Unit, IRCSS Azienda Ospedaliera Universitaria San Martino-IST-Istituto Nazionale per la Ricerca sul Cancro, 16132 Genoa, Italy
e-mail: gilberto.fronza@hsanmartino.it

A. Inga
Laboratory of Transcriptional Networks, Centre for Integrative Biology (CIBIO),
University of Trento, 38060 Trento, Italy
e-mail: inga@science.unitn.it

S.P. Deb and S. Deb (eds.), *Mutant p53 and MDM2 in Cancer*,
Subcellular Biochemistry 85, DOI 10.1007/978-94-017-9211-0_7,

cancer. As the most common genetic changes found in cancerous cells are mutations in the transcription factor p53, efforts to design a broad-spectrum cancer drug have focused in reactivating the p53 pathway [13, 52]. To pharmacologically intervene the p53 pathway is essential to understand p53 activation in molecular terms [10]. The protein codified by the p53 gene is a transcription factor that regulates how cells respond to stress, in particular, to DNA damage [61].

In normal cells, p53 promotes the transcription of genes that arrest the cell cycle to repair damaged DNA, or of genes that kill stressed cells [62]. Nonetheless, p53 is the most commonly mutated gene in cancer. Until now, nearly 2,000 different single amino acid changes in p53 have been reported in tumors; and, the occurrence frequency of each mutation varies dramatically, from extremely rare to very frequent hotspots (codons 175, 245, 248, 249, 273, or 282) [37]. About 30 % of the mutations retain, at least, some transcriptional modulation function, including an altered transactivation specificity [24]. A few of those mutations can also exhibit gain-of function, oncogenic-like properties, as revealed by various data, including transgenic mice experiments [48]. However, most of the mutations in the p53 gene result in a p53 protein unable to promote the transcription of the more than 100 genes that wild-type p53 regulates [42].

The main goal of the structural studies of the transcription factor p53 is to provide atomic information to design a drug that will recover the transcriptional activity of mutant p53 [28]. Alternative non-structural approaches are also possible. For example, cell-based assays with phenotypic read-outs that could identify p53 modulators [6, 9, 56]. In cell-based approaches, the potential for false positives is high and the effort to understand the mode of action of the selected molecules has proven to be a lengthy and difficult process [23, 55]. Another approach that has more recently been explored is to focus on the p53/cofactor interactions that a gain-of-function mutant p53 protein disrupts; blocking such mutant p53 interactions might regain p53 activity, [6, 16]. Nonetheless, even to properly design the cell-based or biochemical screenings, the structural understanding of wild type and mutant full-length p53 is critical to empower the search for high selectivity molecules. For example, to provide drugs that would increase the thermodynamic stability of the p53 DNA-binding domain (DBD), or to describe the cooperative interactions between p53 monomers and/or dimers. In all these cases, the atomic understanding of p53 tetrameric structure, its protein/DNA and protein/protein interactions would foster the development of effective rational therapies. Ideally, a structural and functional understanding of commonalities in the impact of different p53 mutations could identify lead compounds to act on entire groups of p53 mutations.

In this chapter, after providing an introduction to the structure of the transcription factor p53, we will focus on how mutant p53 activity can be recovered. We will describe three mechanisms that affect the function of transcription activation in p53: stabilization of the active conformation of the DBD of the protein, suppression of missense mutations in the DBD by second-site mutations, and increased transactivation. Specifically, we will detail, in structural terms, how a small molecule stabilizes the structure of a p53 mutant; how a second mutation cancels the effect of a formerly deleterious mutation; and, how a mutation leads to an increase in the transcription levels of target genes.

Structural Studies on the Transcription Factor p53

To rationally design a drug that could reactivate mutant p53, structural knowledge on the effect of mutations in p53 is required [28]. The human p53 protein has 393 amino acids divided in three domains: the N-terminus transactivation domain, the central DNA-binding domain (DBD), and the C-terminus with the nuclear localization sequence (NLS), the oligomerization domain (OD) and regulatory domain (RD) (Fig. 7.1a) [60]. Although there have been electron microscopy and small-angle X-ray scattering studies to define the full-length structure of p53 in the absence of DNA [51], and in the presence of DNA [2, 40, 59], the inherent flexibility of the N- and C-termini of the full-length protein has limited the resolution of these studies. Alternatively, X-ray crystallography has succeeded in providing atomic information on individual domains that could be useful to design a p53 reactivating drug (reviewed in [29, 45, 60]). In this review, we will focus on the structural work done in mutants of the DBD of p53 and its homolog protein, p73. There are two reasons to focus on the crystallographic studies of the DBD. First, ninety percent of the mutations found in tumors affect codons in the DBD [46], and, second, except for NMR structures of mutant oligomerization domain [17, 18, 43], all the existing structural knowledge on mutant p53 is on the DBD.

The pioneer work on the structure of the p53 DBD determined the fold of the DBD monomer and its contacts to DNA [14]. The p53 DBD has an immunoglobulin β-sandwich fold and recognizes the DNA with two long loops and a loop-sheet-helix motif at one edge of the β-sandwich (Fig. 7.1b). The first structure of the p53 DBD in complex with DNA classified the most-commonly found cancer mutations in two groups, as either affecting direct DNA recognition (so called "contact mutants", for example: Arg248 or Arg 273) or destabilizing the two loops and the loop-sheet-helix motif involved in DNA binding (so called "structural mutants", for example: Arg175, Gly245, Arg 249, or Arg282) [14]. Moreover, although the crystal packing did not show the expected p53 tetramer structure, the authors postulated a model of the p53 DBD tetramer bound to DNA that was confirmed to be correct by subsequent studies. The experimental evidence of the dimer and tetramer structures of the p53 DBD bound to DNA has been determined more recently [11, 12, 19, 25, 34, 35, 39, 50]. These structures define a dimer of dimers with the four DBD monomers bound to the DNA major groove in the same DNA face (Fig. 7.1c) [35].

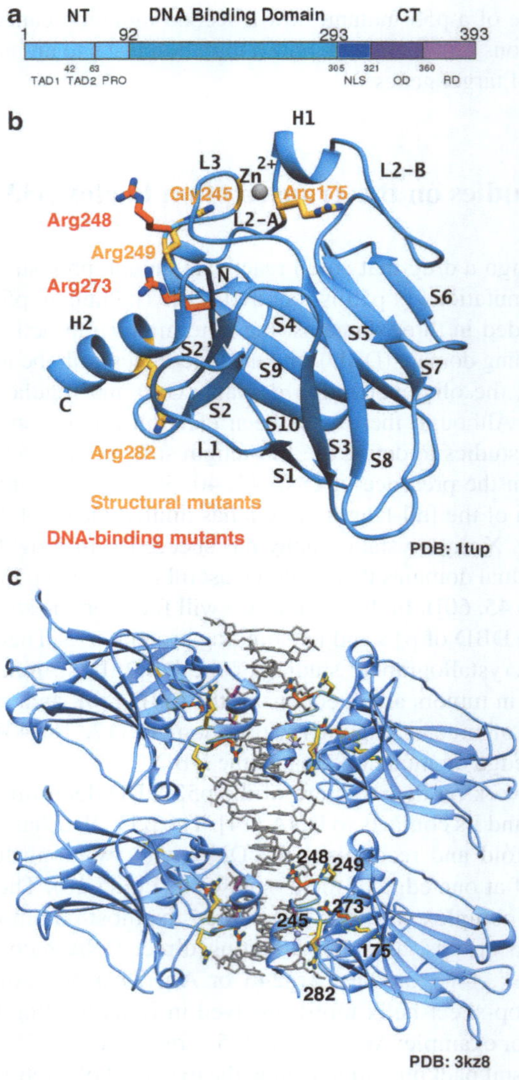

Fig. 7.1 Protein domains of human p53 and location in the structure of the p53 DNA-binding domain of the most-commonly found cancer mutations. (**a**) Scheme of the protein domains in human p53. The N-terminus is divided in three subdomains: the transactivation domains 1 and 2 (TAD1 and TAD2) and the proline-rich domain (PRD); the central domain is the DNA-binding domain (DBD); and, the C-terminus is formed by the nuclear localization sequence (NLS), the oligomerization domain (OD) and the regulatory domain (RD) [60]. (**b**) Human p53 DBD structure (PDB: 1tup) [14]. Ribbon representation of the human p53 DBD with the six most-commonly mutated amino acids shown in stick representation (arginines in positions 175, 248, 249, 273 and 282 and gly in 245). Arg248 and Arg273 are considered DNA-binding mutants (*orange*), while Arg175, Gly245, Arg249, and Arg282 are considered structural mutants that affect the fold of the DBD (*yellow*). (**c**) Tetramer of human p53 DBD bound to DNA with most-commonly found mutations shown as stick representation (PDB: 3kz8) [35]. As in (**b**), DNA-binding mutants are in *orange*, while mutations affecting the protein fold are in *yellow*

As mentioned before, the pharmacological interest of the structural studies on mutant p53 lies on understanding at the molecular level how to revert mutant p53 to display its normal transactivation profile, or at least re-enable it to activate some target genes to induce cell cycle arrest or cell death. The field of structural studies on mutant p53 is still vastly unexplored. In one hand, one could study the mutations that affect DNA binding, like Arg248 or Arg273, but the DNA-binding activity of these mutations is difficult to revert because both residues are exposed to the solvent and the mutant proteins lack a drug-binding cavity [32]. In the other hand, one could determine the structure of mutations like Arg175 that destabilize the structure, with the aim to recover its native conformation; unfortunately, this goal is experimentally very difficult to achieve due to the inherent structural heterogeneity of the unstable mutants.

In the following sections, we will focus on describing three structural studies on mutant p53 that suggest molecular strategies to how mutant p53 could regain activity. First, we will describe how a small molecule could bind to a p53 mutant and increase its thermal stability, presumably recovering DNA-binding [5]. Second, we will summarize how a second mutation can revert the deleterious effect of an initial p53 cancer-associated hotspot mutation [57]. Finally, we will describe how a mutation in the p53-homolog p73 results in a mutant with increased transactivation [15].

Stabilization of Mutant p53 by Small Organic Molecules

The idea that a small organic molecule could bind to mutant p53 DBD and revert the inactivating effect of a mutation has driven a large number of drug screenings. These screenings have resulted in numerous potential cancer drugs [56]. PRIMA-1 is the molecule that has progressed farther in clinical trials, and hydrolytic products of PRIMA-1 might covalently react to cysteine residues in the mutated DBD to activate p53 [36]. Nonetheless, the majority of these screening efforts have not been based on structural knowledge that could accelerate the search for an effective drug to reactivate mutant p53. In here, we review an example of how structure-based drug design can identify small molecules to stabilize oncogenic p53 mutants [5]. The Tyr220Cys mutation is present in an estimated ~75,000 sporadic new cancer cases per year and is a germline mutation in at least eight Li-Fraumeni families [47]. This mutation results in a protein with a largely reduced thermostability [7, 38]. The comparison of the structures of wild-type p53 DBD with Tyr220Cys p53 DBD shows that the Tyr220Cys mutation creates a cavity in the protein surface (Fig. 7.2a, b) [7, 38].

The binding of small molecules to this cavity has been postulated as a paradigm to demonstrate that a drug could stabilize mutant p53 structure. Drug screening protocols can be *in silico* or experimental. While structure-based *in silico* screenings might allow to explore a large chemical space, they result in more false positives [7]. A fast experimental protocol to screen for chemical groups with the potential to stabilize p53 mutant structure can be carried out with a relatively small library of fragments of organic molecules. These fragments comprehend a diverse

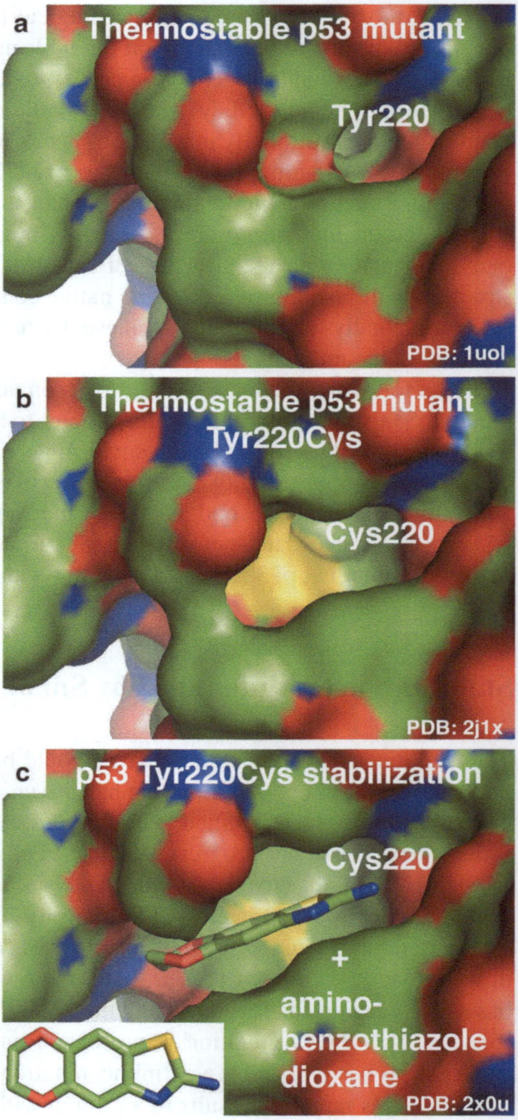

Fig. 7.2 Stabilization of the p53 DNA-binding domain by a small molecule. (**a**) Surface representation of the region formed by the end of sheet S3 (residue 147) and the beginning of sheets S1 (residue 110) and S8 (residue 230) in a thermostable quadruple mutant (Met133Leu, Val203Ala, Asn239Tyr and Asn268Asp) that closely resembles the structure of wild-type p53 DBD (PDB: 1uol) [30]. (**b**) Surface representation of the same region of the thermostable quadruple p53 DBD mutant with the additional Tyr220Cys mutation. The tyrosine to cysteine mutation results in a cavity in the surface of the protein (PDB: 2j1x) [31]. (**c**) Surface representation of the same region, but with the stabilizing 2-amino-1,3-benzothiazole-dioxane molecule bound to the expanded cavity (PDB: 2x0u) [5]

group of chemical moieties with many distinct chemical groups. Once a series of presumably weakly bound molecules are identified, those binding sites become targets for drug-design. Then, a larger molecule that incorporates as many complementary moieties to the identified protein binding surfaces can be synthesized with the expectation that binding to the protein is significantly stronger [53].

By screening such fragment library for binding to the Tyr220Cys p53 DBD mutant, the cavity that results from the replacement of a large tyrosine side chain for a short thiol group of cysteine was screened for binding-sites [5]. When 1,895 fragments were screened for binding to the Tyr220Cys p53 DBD mutant, several small organic molecules were found to interact with the mutant. Two methods to measure drug binding to the p53 DBD were used. One method was Water-Ligand-Observed via Gradient SpectroscopY (WaterLOGSY), where protein-ligand complex formation is identified by the magnetization passed through the substrate-bound water to the ligand-free water in solution. The second method was thermal denaturation scanning fluorimetry, where the binding to a small molecule is detected because it increases the protein's thermal stability as measured by the tendency of a fluorescent dye to bind the exposed hydrophobic core of the unfolded protein. Using this fragment library of 1,895 compounds, these methods showed that about 4 and 1 % of the fragments were able to bind to the Tyr220Cys p53 mutant, respectively. These binding hits were further validated in NMR Heteronuclear Single Quantum Coherence (HSQC) experiments with ^{15}N-labeled protein. To identify the binding sites, three co-crystal structures were solved. The resulting crystal structures defined the possible interactions between atoms in the protein and atoms in the small molecules. For example, 2-amino-1,3-benzothiazole-dioxane binds to the cavity expanding it by 2 Å, and making numerous hydrophobic and specific hydrogen-bond contacts between the protein and the drug (Fig. 7.2c). In the bottom of the cavity, the thiol group from the Cys220 mutant side chain contacts the sulfur atom in the thiazole ring of the drug; and, in one side of the cavity, a carbonyl-oxygen from the protein establishes a hydrogen-bond with the amino group of the drug. The discovery of small molecules that bind to the Tyr220Cys cavity demonstrates that it is possible to compensate the destabilizing effect of a mutation that creates a cavity by inserting a small molecule with complementary electrostatics to fill the cavity.

Suppression by a Second-Site Mutation

Another mechanism that helps us to explore the pharmacological options to recover the activity of mutant p53 is the fact that, after an initial mutation has inactivated p53, a second mutation could recover, at least partially, the activity of p53. This phenomenon is called intragenic suppression by a second-site mutation. Pharmacologically, this suppression mechanism is important because, in the context of a given first mutation, it points to a larger number of protein sites that could be modified to recover p53 activity.

The second-site suppression mechanism has been demonstrated in the context of one of the most-commonly found mutations, the Arg249Ser mutation [57]. This mutation is the result of the transversion G to T in the Arg249 codon (AGG) to a serine codon (AGT) that occurs in hepatocytes due to food contaminated with the mycotoxin aflatoxin B1 [1]. The resulting protein is non-functional and unable to bind DNA [8, 44]. Yeast-based studies have discovered that transactivation of the Arg249Ser mutant can be recovered by two mutations in His168Arg and Thr123Ala [8].

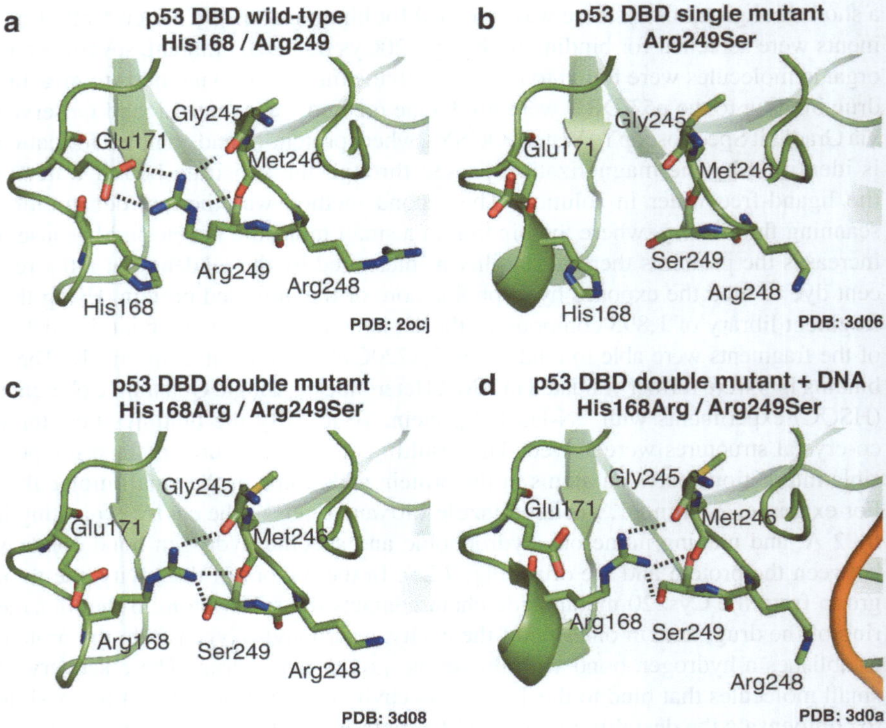

Fig. 7.3 Suppression by a second-site mutation. (**a**) Hydrogen-bond network around Arg249 in the wild-type p53 DBD that maintains the structure between loops L2A and L3 (PDB: 2ocj) [63]. (**b**) Single Arg249Ser p53 DBD mutant with the hydrogen-bond network between loops L2A and L3 broken (PDB: 3d06) [57]. (**c**) Double His168Arg/Arg249Ser p53 DBD mutant shows a reestablished hydrogen-bond network with Arg168 having a similar stabilizing role between loops L2A and L3 as Arg249 in the wild-type protein (PDB: 3d08) [57]. (**d**) Double His168Arg/Arg249Ser p53 DBD mutant in complex with DNA shows that reestablishing the hydrogen-bond network allows Arg248 to make DNA contact as in the wild-type protein (PDB: 3d0a) [57]

In vitro studies showed that the His168Arg mutation is sufficient to recover DNA binding [44].

Structural studies have described a mechanism for the second-site suppression of the Arg249Ser mutant [32, 57]. In wild-type p53, Arg249 stabilizes two flexible loops, L2 and L3. Arg248 in loop L3 establishes a network of contacts that include a bidentated salt-bridge from Glu171 in loop L2 and hydrogen-bonds between the carbonyl oxygens of Gly245 and Met246 and loop L3 (Fig. 7.3a) [63]. Instead, the mutation Arg249Ser destroys the network of contacts between loops L2 and L3 that stabilizes the protein, explaining the drastically reduced DNA binding and the inhibition of transactivation (Fig. 7.3b) [57]. The double mutant Arg249Ser/His168Arg recovers DNA binding through, the arginine in position 168 of loop L2 that forms a network of hydrogen-bonds equivalent to the one observed in wild-type p53 (Fig. 7.3c). In the new network of contacts of the double mutant, the carbonyl oxygens of Gly245 and Met246 continue forming hydrogen bonds with an arginine, although this time with Arg168 from loop L2. Moreover, Glu171 does not make any contact, but Arg168 makes another new contact with the hydroxyl group of Ser249 that maintains the stability of loop L2 and L3. These conclusions, drawn from studies of mutations in a wild-type context, were initially also shown to be valid in the context of a quadruple thermostable mutant [32]. Moreover, once the conformation of loops L2 and L3 is reestablished in the double mutant Arg249Ser/His168Arg, the DNA-binding residue Arg248 acquires a similar conformation than in the wild-type protein and contacts the DNA (Fig. 7.3d). While a detailed structural understanding has not been obtained yet, results from screening assays for second-site suppressor mutations suggest that conditions can be found to reactivate at least in part the activity of multiple cancer-associated hotspot p53 mutations [4, 8, 49].

In summary, to study the structural consequences of secondary mutations that suppress an inactive p53 mutant further defines the essential contacts required to maintain p53 activity. The potential to achieve restoration using small molecules building upon the structural knowledge gained from genetic screenings for second-site suppressor mutations has not yet been explored.

Supertransactivation

Another pharmacological possibility, inspired by yeast studies, is the idea to reactivate p53 mutants that retain DNA binding with a drug that bypasses the effect of mutations by inducing a higher transactivation level. Genetic screenings in yeast have been invaluable to study the functional characteristics of p53 mutations. In particular, genetic assays have shown that some p53 mutations result in phenotypes with higher transactivation levels than the wild-type p53 protein [26, 33, 41, 54]. Interestingly, some of those mutations can also act as intragenic second-site suppressors [8]. The mechanism that connects the initial recognition of the RE by p53 and the subsequent recruitment of transcription activating proteins can be studied by mutations that increase the transactivation level or change the specificity of transactivation.

A structural explanation that links DNA recognition with activation of transcription remains largely unexplored. In recent structural data, a picture is emerging where the conformation of loop L1 is important for the differential recognition of response elements (REs) [21]. In a not-yet-determined mechanism, loop L1 conformation appears to link DNA recognition to an increase in gene expression. In the case of p73, one of the three transcription factors of the p53 protein family, Lys138 (equivalent of Lys120 in p53 and Lys149 in p63) helps to distinguish between different REs [21]. Acetylation of the same lysine in p53 has been linked to induction of cell death [3, 58]. Target genes such as p21 and GADD45 that control DNA-repair and cell cycle progression, and are generally expressed with faster kinetics

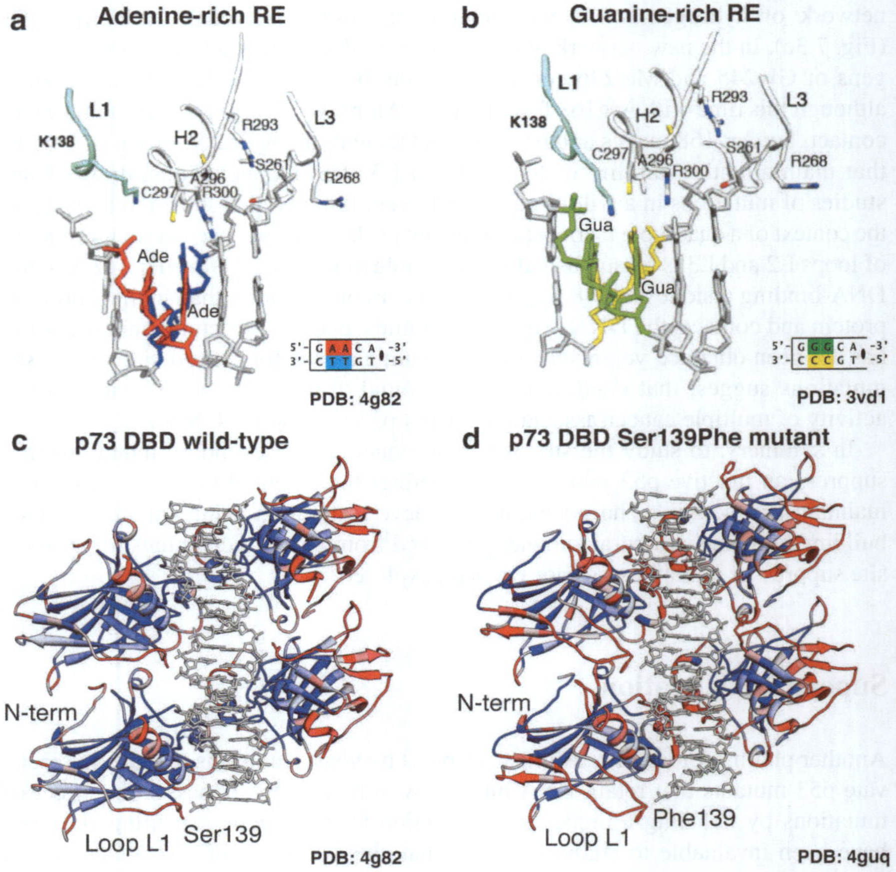

Fig. 7.4 Conformation of loop L1 and increased transactivation. (**a**) Conformation of loop L1 when adenines are present in positions 2 and 3 of the quarter-site RE, as it more frequently happens in DNA repair and cell cycle arrest REs. In this case, Lys138 is not in contact with the DNA bases and loop L1 is more flexible (PDB: 4g82) [21]. (**b**) Conformation of loop L1 when guanines are present in positions 2 and 3 of the quarter-site RE, as it more frequently happens in cell death REs. In this case, Lys138 is in hydrogen-bond contact with the DNA bases and loop L1 is less flexible (PDB: 3vd1) [22]. (**c**) Tetramer of wild type p73 DBD bound to DNA with backbone ribbon colored according to the B-factor value (*low-blue* to *high-red*) (PDB: 4g82) [21]. (**d**) Tetramer of Ser139Phe p73 DBD bound to DNA with the backbone ribbon colored according to the B-factor value (*low-blue* to *high-red*). The mutation Ser139Phe destabilizes the loop L1 resulting in an increased flexibility of loop L1 and sheet S1 that correlates with an increase in transactivation ability for a specific group of REs (PDB: 4guq) [15]

compared to genes that control cell death [64], have a higher content of adenines in positions 2 and 3 of the quarter-site RE (Fig. 7.4a) [21]. While target genes that control programmed cell death are more likely to have guanines in positions 2 and 3 (Fig. 7.4b) [21], although overall these gene have lower affinity REs because of mismatches from the optimal p53 RE consensus [20]. The sequence in positions 2 and 3 of the RE determines the conformation of loop L1. When guanines are present, Lys138 is bound to them and the conformation of loop L1 is more rigid (Fig. 7.4b); instead, when adenines are in positions 2 and 3, the conformation of loop L1 is more flexible (Fig. 7.4a) [22, 21].

We have recently characterized the Ser139Phe p73 mutant, which has an enhanced transactivation activity compared to the activity of the wild-type p73 for REs present in the promoters of several programmed cell-death, like: KILLER, AIP1, FAS and BAX. The crystal structure of the Ser139Phe p73 DBD mutant in complex with a 20 bp full-site RE shows a flexible loop L1 that has significantly higher B-factors than the same loop L1 in the structure of the wild-type p73 DBD in complex with DNA [15]. As the mutant has an enhanced transactivation towards cell-death REs, we hypothesize that the introduction of the Ser139Phe mutation triggers the activation of the normally dormant loop that is sequestered by hydrogen bonds to the guanines found in the positions 2 and 3 of cell-death REs, which can result in a significant increase in DNA binding cooperativity. The activation that results from an increasing loop L1 flexibility in the Ser139Phe p73 mutant appears to mimic the acetylation of the conserved lysine in p53 [3, 58]. As an acetylated lysine would not be able to form hydrogen bonds to the guanines in the positions 2 and 3 of the RE, we postulate that the reported need to acetylate Lys120 in p53 to activate transactivation is due to the need in the proteins of the p53 family to have a flexible loop L1 for optimal transactivation; as occurs, in the wild-type p73 protein when bound to adenine-rich REs, or in the Ser139Phe p73 mutant [15, 22, 21].

Conclusions

We have described three structural mechanisms that show how it is possible to increase the level of transactivation in mutants of the p53 protein family. Clearly, from the structural point of view, one would expect that each unique p53 mutation would result in a protein with potentially very different structural characteristics that will depend on the nature of the newly incorporated amino acid. Such structural diversity needs to be explored because is likely to help to explain the phenotypic diversity seen in tumors. Although the structural information in p53 mutants is still scarce, in here we have shown that, if one wishes to rationally intervene when mutations occur in p53, a molecular understanding of the mutant p53 structures is essential. Future structural work on p53 mutants should continue to guide the development of a cancer drug of general use.

References

1. Aguilar F, Hussain SP, Cerutti P (1993) Aflatoxin B1 induces the transversion of G→T in codon 249 of the p53 tumor suppressor gene in human hepatocytes. Proc Natl Acad Sci U S A 90:8586–8590

2. Aramayo R, Sherman MB, Brownless K, Lurz R, Okorokov AL, Orlova EV (2011) Quaternary structure of the specific p53-DNA complex reveals the mechanism of p53 mutant dominance. Nucleic Acids Res 39:8960–8971
3. Arbely E, Natan E, Brandt T, Allen MD, Veprintsev DB, Robinson CV, Chin JW, Joerger AC, Fersht AR (2011) Acetylation of lysine 120 of p53 endows DNA-binding specificity at effective physiological salt concentration. Proc Natl Acad Sci U S A 108:8251–8256
4. Baroni TE, Wang T, Qian H, Dearth LR, Truong LN, Zeng J, Denes AE, Chen SW, Brachmann RK (2004) A global suppressor motif for p53 cancer mutants. Proc Natl Acad Sci USA 101:4930–4935
5. Basse N, Kaar JL, Settanni G, Joerger AC, Rutherford TJ, Fersht AR (2010) Toward the rational design of p53-stabilizing drugs: probing the surface of the oncogenic Y220C mutant. Chem Biol 17:46–56
6. Bassett EA, Wang W, Rastinejad F, El-Deiry WS (2008) Structural and functional basis for therapeutic modulation of p53 signaling. Clin Cancer Res 14:6376–6386
7. Boeckler FM, Joerger AC, Jaggi G, Rutherford TJ, Veprintsev DB, Fersht AR (2008) Targeted rescue of a destabilized mutant of p53 by an in silico screened drug. Proc Natl Acad Sci U S A 105:10360–10365
8. Brachmann RK, Yu K, Eby Y, Pavletich NP, Boeke JD (1998) Genetic selection of intragenic suppressor mutations that reverse the effect of common p53 cancer mutations. EMBO J 17:1847–1859
9. Brown CJ, Cheok CF, Verma CS, Lane DP (2011) Reactivation of p53: from peptides to small molecules. Trends Pharmacol Sci 32:53–62
10. Brown CJ, Lain S, Verma CS, Fersht AR, Lane DP (2009) Awakening guardian angels: drugging the p53 pathway. Nat Rev Cancer 9:862–873
11. Chen Y, Dey R, Chen L (2010) Crystal structure of the p53 core domain bound to a full consensus site as a self-assembled tetramer. Structure 18:246–256
12. Chen Y, Zhang X, Dantas Machado AC, Ding Y, Chen Z, Qin PZ, Rohs R, Chen L (2013) Structure of p53 binding to the BAX response element reveals DNA unwinding and compression to accommodate base-pair insertion. Nucleic Acids Res 41:8368–8376
13. Cheok CF, Verma CS, Baselga J, Lane DP (2011) Translating p53 into the clinic. Nat Rev Clin Oncol 8:25–37
14. Cho Y, Gorina S, Jeffrey PD, Pavletich NP (1994) Crystal structure of a p53 tumor suppressor-DNA complex: understanding tumorigenic mutations. Science 265:346–355
15. Ciribilli Y, Monti P, Bisio A, Nguyen HT, Ramos A, Ethayathulla AS, Foggetti G, Menichini P, Menendez D, Resnick MA, Viadiu H, Fronza G, Inga A (2013) Transactivation specificity is conserved among p53 family proteins and depends on a response element sequence code. Nucleic Acids Res 41:8637–8653
16. Coffill CR, Muller PAJ, Oh HK, Neo SP, Hogue KA, Cheok CF, Vousden KH, Lane DP, Blackstock WP, Gunaratne J (2012) Mutant p53 interactome identifies nardilysin as a p53R273H-specific binding partner that promotes invasion. EMBO Rep 13:638–644
17. Davison TS, Nie X, Ma W, Lin Y, Kay C, Benchimol S, Arrowsmith CH (2001) Structure and functionality of a designed p53 dimer. J Mol Biol 307:605–617
18. Davison TS, Yin P, Nie E, Kay C, Arrowsmith CH (1998) Characterization of the oligomerization defects of two p53 mutants found in families with Li-Fraumeni and Li-Fraumeni-like syndrome. Oncogene 17:651–656
19. Emamzadah S, Tropia L, Halazonetis TD (2011) Crystal structure of a multidomain human p53 tetramer bound to the natural CDKN1A (p21) p53-response element. Mol Cancer Res 9:1493–1499
20. Espinosa JM (2008) Mechanisms of regulatory diversity within the p53 transcriptional network. Oncogene 27:4013–4023
21. Ethayathulla AS, Nguyen HT, Viadiu H (2013) Crystal structures of the DNA-binding domain tetramer of the p53 tumor suppressor family member p73 bound to different full-site response elements. J Biol Chem 288:4744–4754

22. Ethayathulla AS, Tse P-W, Monti P, Nguyen S, Inga A, Fronza G, Viadiu H (2012) Structure of p73 DNA-binding domain tetramer modulates p73 transactivation. Proc Natl Acad Sci U S A 109:6066–6071
23. Foster BA, Coffey HA, Morin MJ, Rastinejad F (1999) Pharmacological rescue of mutant p53 conformation and function. Science 286:2507–2510
24. Freed-Pastor WA, Prives C (2012) Mutant p53: one name, many proteins. Genes Dev 26:1268–1286
25. Ho WC, Fitzgerald MX, Marmorstein R (2006) Structure of the p53 core domain dimer bound to DNA. J Biol Chem 281:20494–20502
26. Inga A, Monti P, Fronza G, Darden T, Resnick MA (2001) p53 mutants exhibiting enhanced transcriptional activation and altered promoter selectivity are revealed using a sensitive, yeast-based functional assay. Oncogene 20:501–513
27. Jemal A, Bray F, Center MM, Ferlay J, Ward E, Forman D (2011) Global cancer statistics. CA Cancer J Clin 61:69–90
28. Joerger AC, Fersht AR (2010) The tumor suppressor p53: from structures to drug discovery. Cold Spring Harb Perspect Biol 2:a000919
29. Joerger AC, Fersht AR (2008) Structural biology of the tumor suppressor p53. Annu Rev Biochem 77:557–582
30. Joerger AC, Allen MD, Fersht AR (2004) Crystal structure of a superstable mutant of human p53 core domain. Insights into the mechanism of rescuing oncogenic mutations. J Biol Chem 279:1291–1296
31. Joerger AC, Ang HC, Fersht AR (2006) Structural basis for understanding oncogenic p53 mutations and designing rescue drugs. Proc Natl Acad Sci U S A 103:15056–15061
32. Joerger AC, Ang HC, Veprintsev DB, Blair CM, Fersht AR (2005) Structures of p53 cancer mutants and mechanism of rescue by second-site suppressor mutations. J Biol Chem 280:16030–16037
33. Kato S, Han S-Y, Liu W, Otsuka K, Shibata H, Kanamaru R, Ishioka C (2003) Understanding the function-structure and function-mutation relationships of p53 tumor suppressor protein by high-resolution missense mutation analysis. Proc Natl Acad Sci U S A 100:8424–8429
34. Kitayner M, Rozenberg H, Kessler N, Rabinovich D, Shaulov L, Haran TE, Shakked Z (2006) Structural basis of DNA recognition by p53 tetramers. Mol Cell 22:741–753
35. Kitayner M, Rozenberg H, Rohs R, Suad O, Rabinovich D, Honig B, Shakked Z (2010) Diversity in DNA recognition by p53 revealed by crystal structures with Hoogsteen base pairs. Nat Struct Mol Biol 17:423–429
36. Lambert JMR, Gorzov P, Veprintsev DB, Söderqvist M, Segerbäck D, Bergman J, Fersht AR, Hainaut P, Wiman KG, Bykov VJN (2009) PRIMA-1 reactivates mutant p53 by covalent binding to the core domain. Cancer Cell 15:376–388
37. Leroy B, Fournier JL, Ishioka C, Monti P, Inga A, Fronza G, Soussi T (2013) The TP53 website: an integrative resource centre for the TP53 mutation database and TP53 mutant analysis. Nucleic Acids Res 41:D962–D969
38. Liu X, Wilcken R, Joerger AC, Chuckowree IS, Amin J, Spencer J, Fersht AR (2013) Small molecule induced reactivation of mutant p53 in cancer cells. Nucleic Acids Res 41:6034–6044
39. Malecka KA, Ho WC, Marmorstein R (2009) Crystal structure of a p53 core tetramer bound to DNA. Oncogene 28:325–333
40. Melero R, Rajagopalan S, Lázaro M, Joerger AC, Brandt T, Veprintsev DB, Lasso G, Gil D, Scheres SHW, Carazo JM et al (2011) Electron microscopy studies on the quaternary structure of p53 reveal different binding modes for p53 tetramers in complex with DNA. Proc Natl Acad Sci U S A 108:557–562
41. Menendez D, Inga A, Resnick MA (2006) The biological impact of the human master regulator p53 can be altered by mutations that change the spectrum and expression of its target genes. Mol Cell Biol 26:2297–2308

42. Menendez D, Inga A, Resnick MA (2009) The expanding universe of p53 targets. Nat Rev Cancer 9:724–737
43. Mora P, Carbajo RJ, Pineda-Lucena A, Sánchez del Pino MM, Pérez-Payá E (2008) Solvent-exposed residues located in the beta-sheet modulate the stability of the tetramerization domain of p53 – a structural and combinatorial approach. Proteins 71:1670–1685
44. Nikolova PV, Wong KB, DeDecker B, Henckel J, Fersht AR (2000) Mechanism of rescue of common p53 cancer mutations by second-site suppressor mutations. EMBO J 19:370–378
45. Okorokov AL, Orlova EV (2009) Structural biology of the p53 tumour suppressor. Curr Opin Struct Biol 19:197–202
46. Olivier M, Eeles R, Hollstein M, Khan MA, Harris CC, Hainaut P (2002) The IARC TP53 database: new online mutation analysis and recommendations to users. Hum Mutat 19:607–614
47. Olivier M, Goldgar DE, Sodha N, Ohgaki H, Kleihues P, Hainaut P, Eeles RA (2003) Li-Fraumeni and related syndromes. Cancer Res 63:6643–6650
48. Oren M, Rotter V (2010) Mutant p53 gain-of-function in cancer. Cold Spring Harb Perspect Biol 2:a001107
49. Otsuka K, Kato S, Kakudo Y, Mashiko S, Shibata H, Ishioka C (2007) The screening of the second-site suppressor mutations of the common p53 mutants. Int J Cancer 121:559–566
50. Petty TJ, Emamzadah S, Costantino L, Petkova I, Stavridi ES, Saven JG, Vauthey E, Halazonetis TD (2011) An induced fit mechanism regulates p53 DNA binding kinetics to confer sequence specificity. EMBO J 30:2167–2176
51. Pham N, Lucumi A, Cheung N, Viadiu H (2012) The tetramer of p53 in the absence of DNA forms a relaxed quaternary state. Biochemistry 51:8053–8055
52. Razzak M (2013) Targeted therapies: one step closer to drugging p53. Nat Rev Clin Oncol 10:246
53. Rees DC, Congreve M, Murray CW, Carr R (2004) Fragment-based lead discovery. Nat Rev Drug Discov 3:660–672
54. Resnick MA, Inga A (2003) Functional mutants of the sequence-specific transcription factor p53 and implications for master genes of diversity. Proc Natl Acad Sci U S A 100:9934–9939
55. Rippin TM, Bykov VJN, Freund SMV, Selivanova G, Wiman KG, Fersht AR (2002) Characterization of the p53-rescue drug CP-31398 in vitro and in living cells. Oncogene 21:2119–2129
56. Selivanova G (2010) Therapeutic targeting of p53 by small molecules. Semin Cancer Biol 20:46–56
57. Suad O, Rozenberg H, Brosh R, Diskin-Posner Y, Kessler N, Shimon LJW, Frolow F, Liran A, Rotter V, Shakked Z (2009) Structural basis of restoring sequence-specific DNA binding and transactivation to mutant p53 by suppressor mutations. J Mol Biol 385:249–265
58. Sykes SM, Mellert HS, Holbert MA, Li K, Marmorstein R, Lane WS, McMahon SB (2006) Acetylation of the p53 DNA-binding domain regulates apoptosis induction. Mol Cell 24:841–851
59. Tidow H, Melero R, Mylonas E, Freund SMV, Grossmann JG, Carazo JM, Svergun DI, Valle M, Fersht AR (2007) Quaternary structures of tumor suppressor p53 and a specific p53 DNA complex. Proc Natl Acad Sci U S A 104:12324–12329
60. Viadiu H (2008) Molecular architecture of tumor suppressor p53. Curr Top Med Chem 8:1327–1334
61. Vousden KH, Lane DP (2007) p53 in health and disease. Nat Rev Mol Cell Biol 8:275–283
62. Vousden KH, Prives C (2009) Blinded by the light: the growing complexity of p53. Cell 137:413–431
63. Wang Y, Rosengarth A, Luecke H (2007) Structure of the human p53 core domain in the absence of DNA. Acta Crystallogr D Biol Crystallogr 63:276–281
64. Zhao R, Gish K, Murphy M, Yin Y, Notterman D, Hoffman WH, Tom E, Mack DH, Levine AJ (2000) Analysis of p53-regulated gene expression patterns using oligonucleotide arrays. Genes Dev 14:981–993

Chapter 8
Mutant p53 and the Response to Chemotherapy and Radiation

Leila Tchelebi, Hani Ashamalla, and Paul R. Graves

Abstract In addition to playing roles in the genesis and progression of cancer, mutant p53 also appears to play a significant role in the response to cancer therapy. In response to chemotherapy and radiation, two mainstays of cancer treatment, most cancer cells harboring p53 mutations show a reduced sensitivity compared to cells lacking p53 or those with wild type p53. However, there are also many instances where mutant p53 has shown no effect or enhances cellular sensitivity to chemotherapy and radiation. Similar to the *in vitro* cellular studies, the majority of clinical studies show a correlation between the presence of mutant p53 in patient tumors and adverse outcomes following treatment with chemotherapy agents or radiation in comparison to tumors with wild-type p53. However, it still remains unclear whether the presence of mutant p53 in tumors can serve as a reliable prognostic factor and aid in treatment planning. Thus, as genomic analysis of patient tumors becomes more cost effective, the role of mutant p53 in tumor responses from cancer therapy ultimately needs to be addressed. This chapter will discuss current mechanisms of how p53 mutations affect cellular responses to chemotherapy and radiation and discuss patient outcomes based on p53 status.

Keywords p53 • Chemotherapy • Radiation • DNA damage • Genomics • Radioresistance • Chemoresistance • Prognostic factor

L. Tchelebi • H. Ashamalla • P.R. Graves (✉)
Department of Radiation Oncology, New York Methodist Hospital, 506 6th Street, Brooklyn, NY 11215, USA

Weill Cornell Medical College, 445 E 69th St, New York, NY 10021, USA
e-mail: pgraves35@gmail.com

S.P. Deb and S. Deb (eds.), *Mutant p53 and MDM2 in Cancer*,
Subcellular Biochemistry 85, DOI 10.1007/978-94-017-9211-0_8,

Part 1: The Role of p53 Status in Response to Chemotherapy and γ-Irradiation: *In Vitro* Studies and Mechanisms

Introduction

Some mutations in p53 are documented to impart "gain of function" properties to the cells that harbor them including enhanced oncogenesis, tumorigenesis, transformation, increased cell growth rates, and metastasis. Another gain of function property that has been well studied over the years is the response of cells to chemotherapy and radiotherapy. A large number of studies have examined the role of p53 mutants in the response to many commonly used anti-cancer drugs and to ionizing radiation *in vitro*. Using cell lines that express different forms of mutant p53, it has been possible to begin to characterize the role of mutant p53 in the response to chemotherapeutic agents or γ-irradiation. Moreover, these *in vitro* studies have also been extremely useful for identifying mechanisms of action of mutant p53 and have allowed for the identification of signaling pathways and genes that contribute directly or indirectly to the p53 gain of function phenotype.

In addition to the *in vitro* studies, a large number of clinical studies have been performed to determine if p53 status can be used to predict patient outcome from treatment with chemo or radiotherapy. For the patient that has developed cancer and for doctors treating these patients, this topic is of great importance as the presence of mutant p53 in tumors may influence outcome from cancer therapies. Therefore, this chapter is organized into two main parts. Part one focuses on the *in vitro* studies describing the role of mutant p53 in response to chemotherapeutic agents and γ-irradiation and underlining mechanisms that may contribute to the cellular responses. The second part of the chapter summarizes the role of p53 status in prediction of patient outcome following treatment with chemo or radiotherapy. These studies, organized by tumor site and treatment modality, aim to determine whether p53 status can be used as a prognostic factor in cancer treatment.

The Role of WT p53

Although this chapter does not focus on the role of WTp53 per se, its role obviously must be considered when discussing the effects of mutant p53. The WTp53 protein performs at least three major functions in response to a variety of forms of genotoxic stress: induction of apoptosis, cell-cycle arrest and growth control, and induction of DNA repair processes. Thus, lack of these cellular functions, through mutation of p53, will have profound affects on cellular responses to chemo and radiotherapy. As an example, it is well documented that treatment with chemotherapy agents or γ-irradiation results in the induction of $p21^{Waf1}$, a mediator of the G1 cell cycle checkpoint in cells that contain WTp35 but not in cells with mutant p53 [1].

Thus, the absent functions of WTp53 when mutant p53 is present must be accounted for. Are the cellular effects observed in cells that contain mutant p53 due to the presence of mutant p53 or to the lack of WT p53? Indeed, many targets have been identified for WT p53 that are known to sensitize cells to killing by anticancer agents or radiation [2, 3]. Thus, these targets may not be activated if p53 is mutated. It is also possible that mutant p53 can heterodimerize with WTp53 proteins and act in a dominant negative fashion. So are the effects of mutant p53 due to the mutant proteins themselves or due to a dominant negative effect? These questions are not easily answered, especially *in vivo*, where it may not be possible to manipulate p53 genetics. The studies *in vitro* using isogenic p53 null cell lines have been more successful in this regard as mutant p53 can be directly compared to the absence of p53, or to WTp53.

The Nature of p53 Mutations

To understand how p53 mutations impact the response to chemo and radiotherapy, it is important to be aware of the different types of p53 mutations and the consequences of these mutations. Although a great many p53 mutations have been described, the majority of mutations in cancer are missense mutations located in the DNA-binding domain of p53. These mutations most often result in the production of full-length mutant protein that appears to exhibit effects over and above its loss of WT activity. Some of these effects include a dominant negative activity by interfering with WTp53, interference with the function of p63 and p73 [4, 5] and a 'gain of function' activity, in which p53 mutant proteins display oncogenic properties in their own right [6]. A large proportion of the missense mutations are associated with gain of function activities and arise in the 'hotspot' residues, of p53: R175, G245, R248, R273, R249 and R282. However, as might be expected, different p53 mutations do not have the same biological effects and it is likely due to the differential effects on protein function and conformation. For example, mutations at residues R248 and R273 interfere with DNA binding, whereas G245 and R249 mutations produce local distortion of the protein, and mutations at R175 and R282 produce global distortion of the protein structure [6]. Therefore, it is not surprising to see different results in the literature on the topic of the effects of mutant forms of p53. A comparison of the effects of two different p53 mutants may not be expected to yield similar results.

The Role of Mutant p53 in the Response to Chemotherapy

The p53 gene is one of the most frequently mutated genes in cancer. As such it is important to determine how it affects the response of tumors to chemotherapy. To this end, efforts have been directed to understand how p53 status impacts cell survival at the molecular level in response to a number of anticancer agents. It is important to note that the literature is not in agreement on the results of these studies and this perhaps is not surprising given all of the variables involved in these studies. One must consider cell type, chemotherapy agent and dose, p53 mutation, and genetic background of the cells in these experiments. It also important to note that p53 is only one gene and analysis of one gene may not be enough to predict cellular outcomes. The majority of reports in the literature show that mutant p53 confers decreased sensitivity to chemotherapy to the cells in which it is expressed [1, 7–15]. On the other hand, there are also reports that claim the opposite result; that is, that mutant p53 confers increased sensitivity to chemotherapy [16–18]. Still other reports suggest that WTp53 may confer increased resistance to chemotherapy agents [19]. A summary of the most significant reports is provided here.

Several studies provide evidence that cells that express mutant forms of p53 exhibit increased resistance (or decreased sensitivity) to chemotherapy agents. This phenomenon was described as early as 1997 by O'Connor and colleagues [1] and similar findings have been reported by several other research groups in different cell types [7, 9, 20–23]. O'Connor and colleagues conducted an extensive study of the role of p53 in a large spectrum of cancer cell lines against the growth inhibitory action of 123 anticancer agents [1]. A total of 39 p53 mutant and 18 wild type cell lines from a number of tissue types were examined. The overall conclusion was that cells with mutant p53 tended to exhibit less growth inhibition compared to cells with WTp53 when exposed to a number of clinically relevant anticancer agents including DNA cross-linking agents, antimetabolites, and topoisomerase inhibitors. Specifically, it was shown that mutant p53 conferred a median resistance to cisplatin, 5-FU, and bleomycin of 3–10 times to that of wild-type p53 cell lines [1]. Interestingly though, there was no difference observed between mutant p53 and WTp53 containing cells in response to treatment with the antimitotic drugs such as the Taxol family of agents [1].

Although a comparison of different cancer cell lines is a critically important piece of information as shown by the O'Connor study [1], differences in the genetic backgrounds of different cancer cell lines can obscure the overall role of p53. To address the question in isogenic cell lines, Blandino and colleagues introduced different p53 mutants into the p53-null H1299 lung adenocarcinoma cell line and measured their sensitivity to the chemotherapy agent, etoposide [9]. It was found that cells that expressed the R175H p53 mutant but not the R273H mutant exhibited a decreased sensitivity to etoposide compared to cells that expressed WTp53 or cells that expressed vector alone (null-p53). In contrast, both p53 mutants, R175H and R273H exerted similar effects with regard to cisplatin treatment, that is, both mutants conferred increased resistance to the drug [9]. Thus, not only are the effects of p53 mutants specific to the type of p53 mutant but also to the form of chemotherapeutic agent applied as well. Similar results were reported by Deb and colleagues who showed that introduction of p53 mutants into the H1299 cell line conferred decreased sensitivity to the chemotherapeutic agent etoposide [21]. However, in contrast to the study by Blandino [9], the R273H p53 mutant was also found to confer reduced sensitivity to etoposide treatment as measured by clonogenic assay [21].

Several additional studies have used different cell types and agents to address the question regarding the role of mutant p53 in the response to chemotherapy. Wong and colleagues found that expression of the p53 mutant, R273H, conferred doxorubicin resistance in an A431 human squamous carcinoma cell line [23]. It was also shown that expression of the R273H mutant into the p53-null human osteosarcoma cell line, Saos-2 reduced sensitivity to doxorubicin and methotrexate [23]. In another study it was shown that p53-null murine leukemic cells that expressed a temperature sensitive p53 mutant, V135A, exhibited decreased sensitivity to doxorubicin or cisplatin compared to cells expressing no p53 or WT p53 [7]. As p53 overexpression in cell lines could represent an artificial level of mutant p53 overexpression, the opposite technique, of reducing mutant p53 has also been explored. To validate this concept, the level of mutant p53 was reduced in human cancer cell lines by siRNA and it was found that depletion of mutant p53 reduced the resistance to anticancer drugs [24].

The Role of Mutant p53 and the Response to γ-Irradiation

Radiation is one of the main treatment modalities for cancer. However, a great number of tumor types exhibit resistance to radiation. As p53 is one of the most commonly mutated genes in cancer, it is important to determine the role of p53 status in radioresistance. The role of p53 GOF mutations and their response to radiation has been studied in a variety of cancer cell lines derived from different tissue types. A study of the role of p53 status and radiation in 60 different cancer cell lines was conducted by O'Connor and colleagues [1]. In this report it was shown that, perhaps as expected, the majority of cell lines expressing WTp53 showed a functional induction of mRNA for $p21^{Waf1}$, GADD45 and Mdm2 in response to γ-irradiation whereas the majority of cell lines expressing mutant p53 did not [1]. This indicates that mutation of p53 abrogates, for the most part, the radiation induced G1 checkpoint in cell lines harboring mutant p53. But how does p53 status affect survival from radiation treatment? In this area, it is clear that the results of the studies are mixed. In some cells, evidence shows that the presence of mutant p53 reduces sensitivity to radiation. In other cells, no effect is reported or the presence of mutant p53 has been shown to increase radiosensitivity. There are number of factors that could contribute to the variability observed in these studies including difference in genetic background of the cell lines, the type of p53 mutation present, the cellular environment at the time of irradiation and the radiation dose. It is also clear that although p53 is involved in the cellular response to radiation, it is certainly not the only factor involved. These studies are summarized in several excellent reviews and the reader is directed to these for more information on radiation and p53 [25–27].

One of the earliest reports to examine the role of p53 mutants and radiosensitivity was by Lee and colleagues [28]. In this study, transgenic mice were generated that overexpressed the p53 mutants R193P or A135V. Hematopoietic cell lines derived from the transgenic mice were then compared to WT littermate cell lines for radiosensitivity. It was shown that overexpression of either R193P or A135V increased the radiation resistance of mouse hematopoietic cell lineages by 45–57 % [28]. Although the hematopoietic cells harboring the p53 mutants showed increased radioresistance, there was no difference compared to WT cells when treated with EMS, an alkylating agent [28]. In another early report, Li and colleagues utilized temperature-sensitive myeloid cell lines that allowed permissive expression of no p53, WT p53 or the p53 mutant, A135V [7]. It was shown that induction of WT p53 expression greatly enhanced γ-irradiation induced apoptosis relative to non-p53 producing cells. In contrast, induction of the p53 mutant A135V increased cell viability following irradiation 3-fold relative to non-p53 producing cells [7]. Interestingly, treatment with Actinomycin D, a potent inhibitor of transcription, abrogated the reduced apoptosis from γ-irradiation in the cells expressing mutant p53 [7]. This result was one of the first hints that mutant p53 may act at the transcriptional level to mediate its gain of function properties.

Bristow and colleagues employed rat embryo fibroblast (REF) clones to examine how mutant p53 contributes to the radiation response [29]. It was shown that the REF clones expressing p53 mutants (H273, N190, V135, P193, D236, A143) showed increased clonogenic resistance in response to γ-irradiation relative to the non-mutant p53 expression REF clones which expressed low levels of p53 [29]. In another report, Bristow and colleagues showed that cells that expressed the R193P p53 mutant were observed to have a significantly higher survival fraction after 2 gray (SF2) (0.86) than the parental p53-null cell line (0.65) [20]. However, with regard to mechanism, no differences were observed in apoptosis rates between the mutant p53 and control cell lines following doses of both 2 and 10 Gy [20]. Similarly, and unlike the report by Li and colleagues [7], the relative radio-resistance of the REF clones expressing mutant p53 compared to REF clones that did not express mutant p53 was not explained by decreased apoptosis based on a number of morphologic and biochemical end points [29]. The authors explain this finding by citing evidence, that in general, REF clones do not undergo apoptosis in response to treatment with γ-irradiation [29].

A few recent studies have addressed the role of mutant p53 and radiosensitivity on a much larger, more systematic scale. One of the first studies to compare the effect of different p53 GOF mutants on radiosensitivity was by Okaichi and colleagues. In this study, isogenic, stable cell lines were generated by transferring different p53 mutants into the Saos-2 cell line, an osteosarcoma, which is null for p53 [30]. A total of 16 different p53 mutants were analyzed (T123A, L130V, Q143A, V157F, H168R, R175H, I195T, C238Y, C242F, G244C, G245S, R273H, C277F, R280T, R282W, E286K) and compared to WTp53 and vector only (no p53). Cell lines were then treated with γ-irradiation and clonogenic assays performed to measure radiosensitivity. The parental Saos-2 cell line and vector only transformant (p53 null) were more radioresistant than Saos-2 transfected with WTp53. The p53 mutants exhibited a range of radiosensitivities [30]. The 175H, 244C, 245S, 273H and 282W transformants were similar in radiosensitivity to the parental and control vector transformants but much more radioresistant than WTp53 transformants. In contrast, the C242F transformants were similar to WTp53 in their radiosensitivity and the p53 mutants T123A, I195T and C238Y were actually more radiosensitive than WTp53. Thus, although most of the p53 mutants were more radioresistant than WTp53 (12/16), they were not more radioresistant than cells lacking p53 [30]. Therefore in the context of this particular cell line, the p53 mutants do not seem to exhibit a dramatic "gain of function" phenotype at least with regard to radioresistance. Although in this study only one cell type was examined, the use of isogenic cell lines was a major strength as analysis of cell lines with different genetic backgrounds can make identification of the role of mutant p53 difficult.

A comprehensive study examined the role of p53 in radiosensitivity using a wide range of tumor cell lines that varied in histological type [31]. In this study a total of 39 human tumor cell lines from 9 histological types were analyzed for p53 status, radiosensitivity by clonogenic assay, and level of p21 expression. On the basis of survival fraction after 2 GY (SF2), on average, cells that express WT53 exhibited more cell killing than cell lines that express mutant p53. However, when similar comparisons were made for cell killing at higher radiation doses, there was no significant difference between cells that expressed WTp53 versus mutant p53. When compared within each histological cell type, the cell lines expressing mutant p53 exhibited less cell killing (as measuring by SF2) than those cell lines that expressed WTp53 [31]. However, again when cell killing is measured at higher doses of irradiation, the differences between WT and mutant p53 expressing cell lines, even within the same histological type are not significant. In summary, the authors conclude that the cell lines fall into four radiosensitivity groups: very sensitive (VS), sensitive (S), resistant (R) and very resistant (VR). Using this classification, 16/17 cell lines with WTp53 were sensitive (S) and only 1/16 was very resistant (VR). In contrast, only 2/15 of the cell lines expressing mutant p53 were sensitive (S), while 13/17 were resistant (R) and 2/17 were very resistant (VR). The cell lines lacking p53 were also more radiosensitive than cell lines expressing mutant p53 suggesting some "gain of function" activity of mutant p53 in this setting [31].

In head and neck cancers, p53 is one of the most commonly altered genes. Thus, it has been examined with regard to p53 status. In a very recent report on head and neck cancer, Skinner and colleagues completed both a clinical study of patient outcome and p53 status as well as an *in vitro* study of HNSCC cell lines [32]. The clinical arm of this study found that patients with disruptive p53 mutations fared worse than those with WT p53 or non-disruptive p53 mutations with regard to local regional recurrence (LRR) and overall survival. The clinical aspects of this study are covered in more detail in part 2 of this chapter. In the second part of the study, a total of 38 HNSCC cell lines of known p53 status were analyzed by clonogenic assay for response to γ-irradiation. It was shown that cell lines harboring disruptive p53 mutations were more radioresistant than those with WTp53 or non-disruptive p53 mutations. Confirmation that the radioresistance was due to mutant p53 was confirmed by silencing mutant p53 in those cell lines expressing mutant p53 and showing that the radioresistance was reduced [32]. These data argue that a 'gain of function' activity for mutant p53, which has been observed for other cellular processes, also exists for the response to radiation. An interesting aspect to this study was that the *in vitro* radiosensitivity did not correlate with apoptosis or mitotic cell death but rather to radiation-induced senescence. Overall, the presence of disruptive p53 mutants strongly inhibited radiation induced senescence [32]. Thus, it may be important to analyze the correct response to radiation of a particular cell type when studying the role of p53 mutations.

In another recent study of head and neck cancer, the role of mutant p53 was examined using an isogenic head and neck squamous carcinoma cell line pair generated to express WT p53 (HN30) or mutant p53 (HN31) respectively. It was shown that the HN31 cells demonstrated increased radioresistance compared with their wild type p53 (HN30) counterparts [33]. Interestingly, HNSCC cells expressing mutant p53 displayed decreased mitochondrial respiratory capacity and increased sensitivity to 2-DG inhibition of glycolysis [33]. This finding suggests that mutant p53 may impact mitochondrial function and that head and neck tumors expressing mutant p53 may be more susceptible to anti-metabolic strategies such as treatment with 2-DG. As there are no clinically available treatment strategies designed to specifically address mutant p53 containing head and neck tumors currently available, and the associated radioresistance, this finding may allow for the development of novel therapeutic approaches [33]. Although the above reports do support a role for mutant p53 in modulating the response to radiation, others do not [12, 17]. There is also some evidence that WTp53 can contribute to resistance from radiation [19].

Mechanisms of Mutant p53 in the Response to Chemotherapy or Radiation

An investigation into the mechanism of how mutant p53 affects cellular functions may allow for a better understanding of mutant p53 in cancer and promote the identification of new cancer targets. Thus, the underlying mechanism of mutant p53's gain of function properties has been extensively investigated. However, in this section, we restrict the discussion of mutant p53's mechanisms to those associated with alteration of the response to chemo or radiotherapy acknowledging that these mechanisms may not necessarily be dissociated from other gain of function properties of mutant p53.

The mechanism of how the cellular response to chemo or radiotherapy is affected by p53 status has been widely investigated and hypothesized to be due to a number of factors. In theory, signaling pathways involved in mediating cell survival, growth, apoptosis, drug resistance or DNA repair could be involved. Indeed many of these pathways and the genes within these pathways have emerged as candidates for the action of mutant p53. For example, mutant p53 has been shown to be involved in regulating apoptosis [6, 7, 9, 23], genomic instability [34], DNA repair [11, 25, 29], senescence [32], autophagy [35], gene transcription [6, 21, 36–40], mitochondrial function [41], drug resistance [39, 42], protein kinase signaling [43, 44] and the microRNA pathway [45]. Some of the possible mechanisms for how mutant p53 could impart resistance to chemo or radiotherapy are summarized in Fig. 8.1.

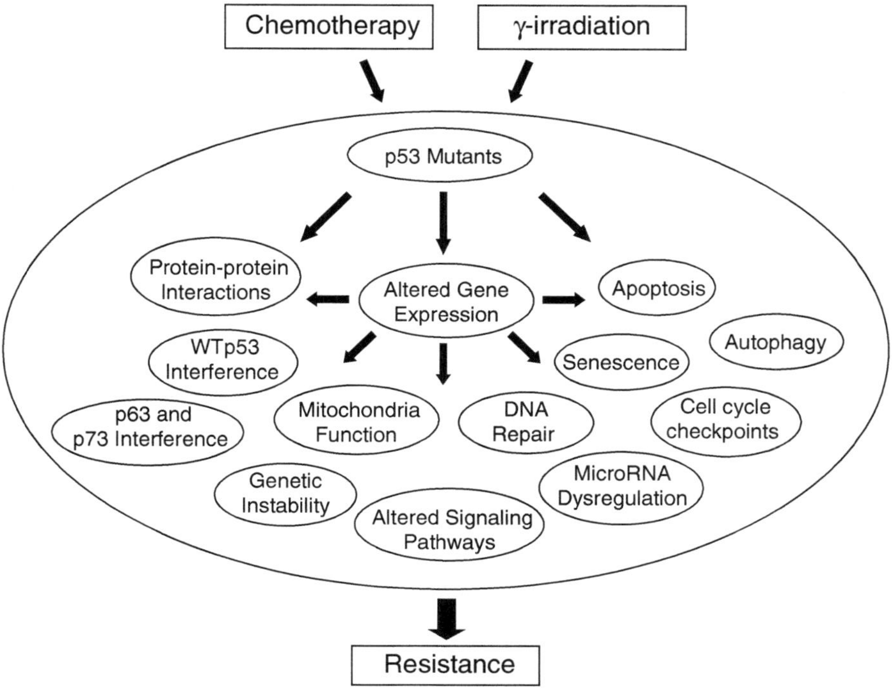

Fig. 8.1 A schematic of possible mechanisms of mutant p53 involved in the cellular response to chemotherapy agents or γ-irradiation is shown

As WTp53 is a critical regulator of apoptosis, perhaps it is not surprising that several reports have shown that expression of mutant p53 correlates with altered apoptotic pathways. As one example, it was shown that expression of the p53 mutant, R273H, in A431 cells correlated with doxorubicin resistance and lower rates of apoptosis [23]. Moreover, the drug resistance could be reduced by siRNA directed against p53, and this correlated with an increase in the expression of procaspase-3 and apoptosis [23]. Similar results were observed after introduction of p53-R273H into the p53-null human osteosarcoma cell line, Saos-2. Induction of expression of p53-273H in Saos-2 cells reduced sensitivity to doxorubicin and methotrexate, reduced procaspase-3 expression, and reduced DNA fragmentation, a marker of apoptosis [23]. Another link to apoptosis was established by Li and colleagues in myeloid cell lines [7]. It was shown that p53-null murine leukemic cells that expressed a temperature sensitive p53 mutant, V135A, exhibited decreased apoptosis rates in response to doxorubicin or cisplatin compared to cells expressing no p53 or WT p53 [7]. In a study by Blandino and colleagues, it was shown that expression of the p53 mutants 175H or 179H greatly reduced the rate of etoposide-induced apoptosis in H1299 cells compared to vector transfected controls [9]. Other p53 mutants, such as 273H and 248W had a much milder protective effect.

Considering the evidence for dysregulation of apoptosis in mutant p53 expressing cells, it is not surprising that several genes related to apoptosis have also been found to be altered in their expression levels in these cells. Notable examples include upregulation of BAG-1 [46] and NF-κB2 [21] and downregulation of FAS [47] and MST-1 [48]. For a more comprehensive review of genes dysregulated by mutant p53, the reader is directed to the review by Brosh and Rotter [6].

The biological assay used to measure p53 function is critically important in determining if a 'gain of function' effect of mutant p53 is present. For example, Skinner and colleagues found that two modes of cell death commonly associated with irradiation, apoptosis and mitotic cell death, were unaffected by p53 status [32]. However, if radiation-induced senescence was assayed, then p53 status played a significant role [32]. It was found that the presence of p53 mutations correlating with decreased radiation-induced senescence, p21 expression, and release of ROS [32]. In addition to apoptosis, a role for mutant p53 in modulating autophagy was identified as well [35]. In this study, it was shown that although γ-irradiation increased the level of autophagy in the p53-null lung cell line, H1299, expression of the R175H p53 mutant in H1299 cells greatly attenuated the level of γ-irradiation induced autophagy. Consistent with this result, the expression of Beclin-1, a marker of autophagy, also increased in H1299 cells in response to γ-irradiation but not in H1299 p53-R175H expressing cells [35]. These results suggest a gain of function role for p53 mutants through inhibition of autophagy in response to γ-irradiation [35].

Another possibility that has been explored is that mutant p53 alters gene expression of a wide array of genes involved in cancer pathways. Early evidence that transcription dysregulation may be one mechanism of mutant p53 included the finding that Actinomycin D, a potent inhibitor of transcription, blocked the reduction in apoptosis rate mediated by mutant p53 in response to treatment with doxorubicin or cisplatin [7]. One of the first studies to address the role of gene transcription in the mechanism of action of p53 GOF mutants was reported by Deb and colleagues in 2005 [21]. To avoid difficulties inherent in comparing different cancer cell lines, Deb and colleagues generated isogenic stable cell lines of a non-small cell cancer cell line, H1299, which is devoid of p53, that expressed either vector alone, or the GOF p53 mutants R175H, R273H or D281G. Relative to vector-transfected cells, H1299 cells expressing mutant forms of p53 showed a survival advantage when treated with etoposide as measured by clonogenic assay [21]. Interestingly, however, cells expressing the transactivation-deficient triple mutant p53-D281G (L22Q/W23S) had significantly lower resistance to etoposide. As the L22Q/W23S mutant is shown to be deficient in transactivation, this result suggested that the p53 mutants were acting at the level of gene transcription. To explore this further, RNA was extracted from H1299 or 21PT stable cell lines that expressed the p53 mutants R175H, R273H, D281G, or vector alone and analyzed for gene expression using an Affymetrix gene array chip. Analysis of the gene array data indicated that all three p53 mutants upregulated a common set of genes involved in a diverse array of processes including in cell cycle control, oncogenesis, invasion, metastasis, DNA replication, cell survival, and transcription. One of the genes found to be upregulated by p53 GOF proteins was NF-κB2 (p100/p52), a member of a family of sequence specific DNA binding transcription factors. This result raised the possibility that transcription factors themselves may be altered by mutant p53 leading to a secondary level of gene induction or repression. To explore the role of NF-κB2 in chemosensitivity, it was overexpressed in H1299 cells. It was also shown that H1299 cells overexpressing NF-κB2 were less sensitive to etoposide and siRNAs directed against NF-κB2 increased etoposide sensitivity [21]. As all three p53 mutants activated the NF-κB2 pathway, one possible pathway through which p53 mutants induce loss of drug sensitivity is via upregulation of the NF-κB2 pathway. Although evidence suggests that mutant p53 may act to induce the expression of genes apart from WTp53, the mechanism of transactivation by mutant p53 is not yet clear. One hypothesis is that mutant p53 may interact with other transcription factors and activate different promoters than when WTp53 is present. Evidence in support of this theory was shown by using ChIP assays [38]. In this report, the NF-κB2 promoter showed increased interaction with CBP and STAT2 in the presence of mutant p53 [38]. Thus, in H1299 cells, mutant p53 may induce gain of function activities by enhancing recruitment of CBP and STAT2 on the promoters of target genes [38].

In another approach to uncover mechanisms underlying p53 GOF mutant activities, a proteomic analysis was used to identify mutant p53 interacting proteins [41, 49]. From this analysis, MCM7, a protein involved in DNA replication, was shown to specifically interact with mutant p53 but not WTp53 [49]. Another protein, Tim50, which forms part of the mitochondrial protein import machinery, although not shown to physically interact with mutant p53, was found to be highly overexpressed in cancer cell lines that also express p53 mutants. Analysis of the Tim50 promoter revealed that mutant p53, but not WTp53 was able to upregulate Tim50 transcription [41]. Interestingly, reduction of Tim50 expression by siRNA reduced the resistance of cells harboring the p53 mutant, R175H, to paclitaxel but had no effect upon cells lacking p53. These findings identify the Tim50 gene as a transcriptional target of mutant p53 and suggest a novel mechanism by which p53 mutants enhance chemoresistance [41].

Chemoresistance may also be attributed, at least in part, by the action of mutant p53 in the dysregulation of the microRNA pathway. It was shown that expression of the p53 mutant, R275H, in the p53-null lung cell line H1299, resulted in the down regulation of the expression of miR-223. Moreover, in a colon and breast cancer cell line that expressed the p53 mutant, R273H, down-regulation of mutant p53 by shRNAi increased miR-223 expression [45]. Chromatin immunoprecipitation (ChIP) assays showed that mutant p53 was capable of binding to the miR-223 promoter [45]. Consistent with these results, overexpression of miR-223 sensitized cells to cisplatin or 5-fluorouracil. Moreover, down-regulation of mutant p53 also upregulated the levels of the Stathmin protein, a known target of miR-223. Thus from this data a model was proposed whereby mutant p53, through the down-regulation of miR-223, upregulated the Stathmin protein which contributed to chemoresistance [45].

Mutant p53 may also act to alter protein phosphorylation signaling pathways. One of the genes found to be upregulated in H1299 cells following expression of the p53 mutants R175H, R273H, and D281G was the protein tyrosine kinase Axl [44].

Consistent with this result, knockdown of endogenous mutant p53 in two different human lung cancer cell lines, H1048 (p53-R273C) and H1437 (p53-R267P) reduced Axl expression [44]. It was suggested that mutant p53 may act directly at the Axl promoter as ChIP assays demonstrated the presence of mutant p53 at the Axl promoter [44]. As it is known that Axl expression affects some of the same pathways as mutant p53 such as apoptosis, cell adhesion, and motility, the role of mutant p53 was investigated in these processes. Interestingly, knockdown of Axl by RNAi resulted in a reduction of mutant p53 gain of function activities in lung cancer cells expressing endogenous mutant p53, including growth rate and cellular motility. Taken together, these results suggest that mutant p53 may act to upregulate the Axl protein tyrosine kinase which then executes, at least in part, some of the p53 gain of function activities.

Conclusions and Possible Reasons for Discrepancies Among the Studies

The results of the studies described here, which notably represents only a small fraction of the total studies in the literature on mutant p53, gives a somewhat mixed verdict with regard to how mutant p53 affects the cellular response to chemotherapeutic agents or γ-irradiation. Although the majority of mutant p53 studies suggest that the presence of mutant p53 reduces sensitivity to chemotherapy or radiation, there are many studies that show no effect or show the opposite effect; that is, that mutant p53 enhances sensitivity to these agents. Thus, it is important to examine some of the possible reasons for these discrepancies. First, it is important to note that unless isogenic cancer cell lines containing mutant p53 are compared to each other, the same result after treatment with chemo or radiotherapy should not be expected. Cancer cell lines that are derived from different tissues are known to respond differently to different chemotherapeutic agents regardless of p53 status. This is because the cancer cell lines, much like different patient tumors, show a great amount of genetic variability. Another important point, as mentioned earlier, is that p53 gain of function mutants act differently depending upon the location and type of mutation in p53. It is well documented that the degree of chemoresistance depends critically on the type of mutation present in p53 [1]. Another important variable to consider is the type of biological assay performed to measure “gain of function” of p53 in response to treatment with chemo or radiotherapy. For some p53 mutants and cell types, a clonogenic assay may be appropriate. Chemo or radiotherapy responses are typically measured using this assay which measures the reproductive integrity of cells regardless of the specific mode of cell death involved [50]. For this reason, the clonogenic assay has long been the ‘gold standard’ for measuring responses to chemo or radiotherapy. Unfortunately, not all cell lines form colonies *in vitro*. For example, some HNSCC cell lines do not form colonies and therefore other viability assays, such as the MTT growth assay, are employed. Although most of the studies discussed here employ the clonogenic assay, in the cases where it is not used, it is important to recognize that this could be one source of variability in the results observed. Finally, there may also simply be variability due to how different investigators conduct their experiments. The larger question that remains is what is the significance of these *in vitro* studies and the mechanisms uncovered for the action of mutant p53 in a clinical setting? Inevitably, as these new targets are discovered, it will become important to evaluate the significance of these findings in animal models or in clinical settings.

Part 2: The Role of p53 Status in Response to Chemotherapy and γ-Irradiation: Clinical Studies and Patient Outcome

Introduction

The p53 gene and its protein product have been extensively studied *in vitro* and in animal models. But how can what is known about p53 at a molecular level translate to a clinical setting? Numerous studies have been conducted to analyze the relevance of a tumor's p53 status in regards to patient outcome. Special attention has been, and continues to be, paid towards understanding the implications of having WT versus mutant p53 in response to cancer therapy. Ultimately, if we are able to establish a relationship between p53 mutational status and response to therapy, we will be able to sequence each patient's tumor and ascertain which patients will respond to which therapies. In so doing, the morbidity associated with ineffective treatment modalities, and the delay in achieving response with effective ones, can be avoided.

This chapter aims to review what has been shown with respect to p53 mutation status and the response to treatment of various tumor types. The review is limited specifically to p53 mutation, not p53 over-expression, deletion, or loss of function. It is also limited to chemotherapy and radiation therapy as the treatment modalities studied. Unlike surgery, which acts to physically remove cancer cells, chemo and radiation rely on intrinsic and extrinsic cellular pathways, involving genes such as p53, to cause cancer cell death.

With very few exceptions, this chapter is concerned exclusively with studies that analyze p53 status via gene sequencing methods, as opposed to immunohistochemical staining (IHC). Studies involving IHC to assess p53 status were avoided because IHC lacks standardization and is subject to a number of biases including observer bias and variation in scoring methods among institutions. IHC is also more likely to result in false positive or false negative results and is therefore not as sensitive for detecting p53 mutations as is direct gene sequencing [51]. Another limitation of IHC is that it relies on a small number of pre-defined protein markers per tissue section to discover the presence or absence of the mutated protein. In order to develop a clinically relevant molecular indicator of response to therapy, a standardized, unbiased, and sensitive mechanism for detecting the status of the molecular marker, in the present case p53, is needed.

Gynecological Malignancies

Changes in p53 are the most frequent genetic event described in advanced ovarian cancer [52]. Understanding the role of mutant p53 in response to therapy, therefore, may be of particular importance in the treatment of this cancer. In advanced disease, where chemotherapy is the mainstay of treatment, knowledge of the relationship between p53 status and response to therapy could help to determine which chemotherapeutic agent would be most effective and which agents should be avoided in a particular patient.

Taxane- and platinum-based chemotherapeutic agents are routinely used in the treatment of ovarian cancer. There is evidence, however, that the presence of mutant p53 within tumor cells can impact a patient's response to these drugs. Existing data seems to show, however, that the effect of p53 mutational status on response to each of these therapies is not concordant. With respect to platinum-based chemotherapy, several studies have shown that tumors harboring a p53 mutation are more likely to be resistant to treatment [53–57]. Indeed, not only the presence of mutant p53, but also the specific type of mutation, has been shown to play a role in response to platinum-based chemotherapy [53]. However, despite the large number of studies that have found a correlation between p53 status and response to platinum-based chemotherapy, there is some data suggesting that no such association exists [58]. In the case of taxol-based chemotherapy, on the other hand, it has been shown that harboring a p53 mutation in tumor cells predicts a favorable response to treatment [59]. These results imply that certain patients, namely those without a mutant p53, will respond better to platinum-based chemotherapy, while those who harbor a mutation are more likely to respond to taxol-based agents. More research in this area will enable clinicians to tailor each patient's therapy in order to avoid subjecting cancer patients to taxing chemotherapeutic regimens which confer no clinical benefit.

The role of p53 status in other gynecological malignancies has also been studied, however, results are less conclusive than in the case of ovarian cancer. In cervical cancer, for instance, there remains a controversial correlation between p53 status and response to therapy [60]. Moreover, the data that is available assesses p53 status via IHC and gene expression as opposed to gene sequencing [60] which makes it difficult to draw meaningful conclusions given operator bias and lack of standardization, as previously discussed. There is some evidence, however, that cervical cancers with mutant p53 detected by gene sequencing are more likely to be radio-resistant [61]. Radiation therapy may, therefore, not be the appropriate modality of treatment for patients with these tumors. However, further investigation is needed before meaningful conclusions can be drawn.

Breast Cancer

Breast cancer is the most common non-cutaneous cancer among women and the second most common cause of cancer death among women. As in the case of many other malignancies, p53 is the most commonly mutated gene in breast cancer [62]. There are a number of treatment approaches that can be used in the management of breast cancer, some of which are interchangeable. For instance, lumpectomy with radiation is equivalent to mastectomy in many cases of early stage cancer. In the case of larger tumors, neoadjuvant chemotherapy can be used to decrease tumor volume in an attempt to avoid what may otherwise be a disfiguring surgery. Given this arsenal of treatment approaches and the morbidity associated with each one – including toxic radiation and chemotherapy – the ability to predict response to therapy would be invaluable in the management of patients with breast cancer. Determining the relationship between p53 mutational status and response to these various therapies could save many patients from undergoing unsuccessful treatments, and their associated complications, from the outset.

Neo-adjuvant chemotherapy is often employed in the management of locally-advanced invasive breast cancer; however, response to chemo varies and is often unpredictable [63]. Furthermore, there exist a number of agents to choose from, some of which may be more successful than others. Patients who fail to respond to a particular neo-adjuvant regimen are not only subjected to the toxic effects of an ineffective therapy, but, their tumors are potentially given time to grow and spread until an effective regimen is initiated. Predicting a response to chemotherapy in the neoadjuvant setting via assessment of tumor markers, such as p53, would be beneficial to a great number of patients. The studies published to date which utilized gene sequencing to assess p53 status and response to chemotherapy in the neoadjuvant setting, however, have unfortunately shown conflicting results [64–66].

Hormonal therapy is another treatment modality often used in breast cancer therapy. Tamoxifen, which is commonly used, works by binding to estrogen receptors, decreasing DNA synthesis, and inhibiting estrogen effects. Because it is involved in regulating the cell cycle, it is highly plausible that a mutation in p53 may play a role in response to the drug. The current consensus in the literature seems to be that mutated p53 causes resistance to tamoxifen [62, 67, 68]. Currently, whether or not tamoxifen is prescribed to patients is largely predicated on a tumor's hormone receptor status. However, even in patients whose tumors are highly estrogen or progesterone receptor positive, tamoxifen therapy has been known to fail, thus exposing women to a drug that can cause osteoporosis and increase the risk of uterine cancer, while conferring no benefit in the management of their disease. It would, therefore, be valuable to find an additional means of predicting response to this drug. One of these may be via assessment of p53 status.

There is limited data in the literature that studies the role of p53 mutation in response to radiation therapy in breast cancer, another mainstay in the treatment of this disease. Moreover, the available studies seem to have conflicting results. In the case of combined treatment, for instance, involving neoadjuvant chemo, surgery, and radiation therapy, it was found that a mutation in p53 predicted a poor response to therapy [69], while in the case of radiation therapy alone, a mutant p53 was shown to sensitize tumors to therapy [70]. More studies are needed before conclusions linking p53 mutation to treatment response can be drawn.

Head and Neck Cancer

Genetic changes in p53 have been reported to occur in approximately 45 % of head and neck cancers [71, 72]. Treatment for head and neck cancer is selected on the basis of site and stage of disease. Patients' response to therapy can vary widely for any given site at any given stage, however. Due to the prevalence of p53 mutations in head and neck cancers and the heterogeneity in treatment response, there has been significant interest in attempting to find molecular markers that can predict response to therapy. In the case of neoadjuvant chemotherapy and the response of patients with various stages of head and neck cancer, it has been found that p53 mutations predict failure to respond to treatment [14]. In patients treated with radiation therapy, either in the adjuvant setting or as primary therapy, it has also been found on numerous occasions that a mutation in p53 makes tumor cells less likely to respond to treatment [15, 32, 73, 74]. Finally, in one study in which both treatment modalities were assessed, p53 mutational status again predicted an unfavorable response to therapy [75].

While, in head and neck cancers overall, p53 mutation appears to portend poor response to therapy, some studies focusing on specific sites have found that no correlation exists between treatment and p53 status. For instance, in the case of nasopharyngeal cancers, one study found that p53 mutations were infrequent and were not associated with failure of radiation therapy [76]. Another study that focused only on laryngeal cancers found that, while p53 mutation was a common occurrence in these cancers, the presence of a mutation had no bearing on response to radiation [77]. In the case of oropharyngeal cancers, on the other hand, p53 mutation was associated with radiation resistance [78]. These findings may imply that the effect of p53 mutation on response to various therapies varies by site. Further investigation is needed to determine for which tumor sites the knowledge of p53 mutational status would be applicable. More evidence is also needed to assess whether mutation in p53 predicts a favorable or unfavorable response to therapy in those sites where it plays a role.

Prostate Cancer

Like breast cancer in women, prostate cancer is the most common non-cutaneous cancer in men and the second most common cause of cancer-related death among men. P53 mutations, however, have only been reported in approximately twenty percent of prostate cancers [79]. This may, at least in part, explain why there is so little data regarding the role of p53 mutations in response to therapy for this disease. With respect to prostate cancer therapy, radiation or surgery are the first-line therapies used to treat the disease. However, there is a subset of patients who fail radiation therapy and subsequently need to undergo salvage prostatectomy. Conversely, patients who fail surgery undergo salvage radiation which can result in good tumor control. It would, therefore, be beneficial to determine whether or not a particular patient's tumor is radiosensitive prior to initiating therapy.

There is little available data regarding prostate cancer, p53 status by sequencing techniques, and treatment outcomes. Most of the available studies have analyzed p53 status by IHC, to look for the presence of mutated p53 [79, 80]. One study that used both genetic sequencing and IHC to determine p53 status found that p53 immunoreactivity by IHC predicted failure to respond to radiation therapy, while the presence of a true gene mutation had no significant association with response to treatment [81]. This result has been replicated, seeming to indicate that in the case of prostate cancer, p53 mutational status does not predict response to radiation therapy [82]. More research is needed, however, before it can be concluded that no association between p53 and response to therapy exists.

Hematologic Malignancies

Hematologic malignancies encompass those tumors that affect the blood, bone marrow, and the lymph nodes. They affect both pediatric and adult populations. Chemotherapy is the mainstay of treatment for these malignancies but response to therapy is unpredictable. While some patients achieve disease-free survival using certain chemotherapeutic regimens, other patients with the same disease fail to adequately respond to the same treatment regimen.

The frequency of p53 mutations in hematologic malignancies has been reported as anywhere from 5 to 50 % [83]. The disparity in frequency of mutations among researchers might explain why results regarding the effect of p53 mutation status on response to treatment are also varied. In one study that looked at patients with acute myelogenous leukemia (AML), myelodysplastic syndrome (MDS), and chronic lymphoblastic leukemia (CLL), p53 mutation predicted a favorable response to chemotherapy in all three diseases [84]. Another study involving exclusively patients with CLL found that p53 mutations were significantly correlated with poor response to treatment [85]. In the case of lymphomas, studies have failed to show a significant relationship between p53 status and response to treatment [86, 87]. There are a number of explanations as to why these results are so varied. One explanation is that p53 status and response to therapy in hematologic malignancies is dependent upon the specific disease entity - ALL versus CLL versus CML, and so on. Another explanation may be that it depends on the specific chemotherapeutic agents used. Ultimately, the answer is most likely to be a combination of both.

Pulmonary Malignancies

Non-small cell lung cancer is the most common non-cutaneous cancer worldwide and it is the number one cause of cancer death worldwide. In certain cases, the response to therapy can be as low as fifty percent. This is especially true in advanced disease, which portends the worst prognosis. Not only are responses to therapy among patients varied, but there are a number of different chemotherapeutic regimens that can be employed in treating the disease. Therefore, identifying tumor markers that predict response to therapy in the case of pulmonary malignancies, such as non-small cell lung cancer, would be of great benefit to patients and clinicians, alike.

In locally advanced disease, chemotherapy with or without the addition of radiation is increasingly being used in the neo-adjuvant setting but response to treatment is unpredictable at the present time. Furthermore, there are different options that can be used for induction chemotherapy, among them are cisplatinum and paclitaxel. Results in the literature have shown that while cisplatinum-based induction chemotherapy either alone [88] or in combination with radiation therapy [89] is ineffective in the case of p53-mutated cancers, response to paclitaxel either alone [90] or in combination with radiation therapy [91] was unrelated to p53 status. These results indicate that, for patients whose tumors harbor a p53 mutation, cisplatinum-based regimens should be avoided. Meanwhile, the fact that p53 status is unrelated to response to paclitaxel suggests that the drug can be used regardless of whether or not a mutation is present. It also signals a need to identify more tumor markers that can help us guide treatment.

Tumors of the Central Nervous System

In the case of central nervous system (CNS) tumors, which encompasses a number of different tumor histologies, data analyzing p53 mutational status and response to therapy are limited. This may be explained by the fact that p53 mutation is a highly infrequent event in certain CNS tumors [92, 93]. Another explanation may be that that mutation status is difficult to assess. CNS tumors are frequently unresected, as in the case of primary lymphoma, or are unresectable, as in the case of brainstem gliomas, so there is limited tissue available for study.

It is important to realize that p53 may not play a role in the tumorigenesis of every tumor type. Therefore, determining the mutational status of these tumors would not be beneficial in predicting response to therapy. For example, Yeung et al. attempt to summarize what is known about the mechanism of radioresistance in vestibular Schwannomas (VS) because radiation therapy has emerged as an alternative treatment modality to surgery for these tumors [94]. The authors hoped that in determining the mechanisms of radioresistance in VS, tumor markers could eventually be identified to guide treatment planning. In identifying markers that predict tumor response to treatment, patients with radioresistant tumors could eventually be identified prior to initiating radiation therapy where treatment failure is likely to ensue. These patients could be selected for microsurgical resection instead. However, the authors found that in these tumors, p53 mutations do not contribute to tumor pathogenesis and, therefore, mutational status is unlikely to be a helpful marker in predicting treatment response.

In other tumor types, p53 mutation may be so infrequent that knowing a tumor's p53 status may be of low yield in terms of predicting response to therapy. In a study looking at glioblastomas and response to either radiation therapy or temazolomide, p53 mutations were only observed in 15 % of tumors out of a total of 301 tumors analyzed [95]. The authors found a trend towards increased response to temazolomide in the presence of p53 mutations, but their finding was not statistically significant, likely because of the small number of p53 mutated tumors.

Knowing a tumor's p53 status may be helpful, not only in predicting who will respond to which therapies, but also in identifying those patients who no longer require additional therapy. Choroid plexus tumors are pediatric tumors with poor survival rates which can be treated with chemotherapy, radiation therapy, or both. In a study by Tabori et al. [96], the authors found that patients without p53 mutations who were treated only with chemotherapy had excellent survival rates. The authors concluded that adding radiation therapy to the treatment regimen was unnecessary. If other studies were conducted also supporting this finding, that tumors with mutant p53 have an excellent response to chemotherapy, children diagnosed with choroid plexus tumors could be spared an additional, unnecessary treatment modality that increases morbidity with no survival benefit.

Tumors of the Gastrointestinal Tract

P53 status has been widely studied in tumors of the gastrointestinal tract. These focus primarily on colon cancer, which is the third most frequently diagnosed cancer in the United States in both men and women. Unfortunately, however, a great number of these studies analyze p53 status by IHC rather than by gene sequencing, as with many of the other tumors sites we have seen. Again, because IHC is a less sensitive means of detecting p53 mutation it is not very helpful in determining a clinically useful means of assessing the relationship between p53 and response to therapy. Thus, these individual studies will not be discussed here.

A number of review articles and meta-analyses have been published attempting to uncover the potential role of p53 status as predictive of GI tumors' response to therapy. These review articles highlight the fact that studies using gene sequencing to determine p53 status are few and far between. In their review of the available literature which assesses predictors of histological response to neo-adjuvant radiation and chemo-radiation, for instance, Smith et al. note that there are three times as many studies using IHC to detect p53 mutational status as there are studies using either gene sequencing or single strand conformational polymorphism analysis. Another review article by Peterson et al. [97] similarly found that, when either chemotherapy or radiation was the treatment modality used, analysis by IHC was much more frequently employed than gene sequencing. Not only is detection by IHC less reliable, but the use of this method in place of gene sequencing can result in different, even conflicting, results. In a review by Munro et al. [51], for instance, the authors found that when studies used IHC to detect p53 abnormalities, they found no relationship between p53 and treatment response to chemo, while they found that p53 mutation was associated with a poor response to radiation therapy. The same review article found that when studies used gene sequencing to detect alterations in the gene, p53 mutation was associated with an unfavorable response to *both* chemotherapy and to radiation.

While many of these review articles focus on the relationship between p53 mutation and treatment outcome, another approach is to draw conclusions regarding response to therapy in wild-type tumors. In a review of the available literature studying p53 status as a predictive biomarker in response to neo-adjuvant chemoradiation, Chen et al. [98] found that wild-type tumors were associated with good response to neo-adjuvant treatment. Unfortunately, it is not possible to extrapolate the data and conclude that having a mutated p53 would therefore predict a poor response to therapy. This is not possible because in certain cases, as we have seen with tumors in other organ systems, there may be no predictable relationship between p53 mutation and treatment response and other tumor markers may be needed.

More research is undoubtedly needed before any definitive conclusions can be drawn. However, in looking at the actual studies included in the reviews and the meta-analyses mentioned above using gene sequencing, the current evidence shows that p53 mutation is associated with poor response to therapy in the case of gastrointestinal tumors. In the case of neo-adjuvant radiation, the presence of a p53 mutation detected by gene sequencing has been found to predict a poor response to treatment [99, 100]. Adjuvant chemotherapy, too, in the presence of a p53 mutation has been shown to be less effective [101]. Finally, in the case of chemotherapy given for advanced, Stage IV colorectal cancer, studies have shown a poor response to treatment when p53 is mutated in tumor cells [102, 103].

P53 and response to treatment has also been studied in other gastrointestinal tumors, such as esophageal carcinoma. Like with many other cancers, the treatment of esophageal carcinoma employs a number of treatment modalities, none of which have been overwhelmingly effective in controlling this disease. There is, therefore, an interest in trying to identify tumor characteristics which can predict treatment response in order to avoid unnecessary, costly, and taxing therapies. In one study looking at esophageal carcinoma alone, p53 mutation was associated with poor response to chemo-radiation [104]. More studies are needed to support this finding.

Conclusions

Tumor markers are increasingly being studied as a means to determine tumor response to therapy. If tumor markers could be correctly identified which predict therapeutic response, unnecessary treatments, and the cost and morbidity associated with them, could be avoided. Effective therapies could also be identified early on in

Table 8.1 Studies that analyzed the role of mutant p53 and patient outcome to treatment with chemotherapy or radiation were reviewed

Tumor type	Treatment modality	p53 mutation → good patient response to treatment	p53 mutation → poor patient response to treatment	No correlation	Conclusion
Ovarian	Chemo: Platinum-based	0	5	1	Poor response
	Chemo: Taxol-based	1	0	0	Good response
Cervical	RT	0	1	0	Poor response
Breast	Chemo	2	0	2	Inconclusive
	Tamoxifen	0	3	0	Poor response
	RT	1	0	0	Good response
	Combined	0	1	0	Poor response
Head and neck	Chemo	0	1	0	Poor response
	RT	0	5	2	Poor response
	Both	0	1	0	Poor response
Prostate	RT	0	0	2	No correlation
Leukemia	Chemo	1	1	0	Inconclusive
Lymphoma	Chemo	0	0	2	No correlation
NSCLC	Chemo: Platinum-based	0	2	0	Poor response
	Chemo: Taxol-based	0	0	2	No correlation
CNS	Both	0	0	2	No correlation
GIT	RT	0	2	0	Poor response
	Chemo	0	2	0	Poor response

The table shows the number of studies identified, organized by tumor site and treatment modality, that were performed using p53 gene sequencing as the principal method to identify mutant p53. The conclusion was poor response if the majority of studies indicated a correlation with mutant p53 and poor patient outcome, good response if the majority of studies indicated a good patient response, and inconclusive if no majority was found. *NSCLC* non-small cell lung carcinoma, *CNS* central nervous system, *GIT* gastrointestinal

the management of aggressive tumors, allowing for more effective tumor control. Knowledge of tumor markers and response to therapy can, therefore, be used to guide treatment selection, resulting in better patient outcomes. In the case of p53, this review has shown that in the majority of tumor sites including ovarian, breast, head and neck, lung and gastrointestinal tumors, research has shown that p53 mutation predicts an unfavorable response to both chemotherapy and radiation (Please refer to Table 8.1 for a summary of the data.). Despite all that has been done thus far, much more research is needed before p53 mutational status can be adapted to the clinical setting. However, with further research, sequencing of the p53 gene may allow us to identify those patients harboring p53 mutations and spare them from unnecessary, costly, taxing, and ultimately ineffective treatments.

References

1. O'Connor PM, Jackman J, Bae I, Myers TG, Fan S, Mutoh M, Scudiero DA, Monks A, Sausville EA et al (1997) Characterization of the p53 tumor suppressor pathway in cell lines of the National Cancer Institute anticancer drug screen and correlations with the growth-inhibitory potency of 123 anticancer agents. Cancer Res 57:4285–4300
2. MacLachlan TK, El-Deiry WS (2002) Apoptotic threshold is lowered by p53 transactivation of caspase-6. Proc Natl Acad Sci U S A 99:9492–9497
3. Sax JK, Fei P, Murphy ME, Bernhard E, Korsmeyer SJ, El-Deiry WS (2002) BID regulation by p53 contributes to chemosensitivity. Nat Cell Biol 4:842–849
4. Strano S, Fontemaggi G, Costanzo A, Rizzo MG, Monti O, Baccarini A, Del Sal G, Levrero M, Sacchi A et al (2002) Physical interaction with human tumor-derived p53 mutants inhibits p63 activities. J Biol Chem 277:18817–18826
5. Strano S, Munarriz E, Rossi M, Cristofanelli B, Shaul Y, Castagnoli L, Levine AJ, Sacchi A, Cesareni G et al (2000) Physical and functional interaction between p53 mutants and different isoforms of p73. J Biol Chem 275:29503–29512
6. Brosh R, Rotter V (2009) When mutants gain new powers: news from the mutant p53 field. Nat Rev Cancer 9:701–713
7. Li R, Sutphin PD, Schwartz D, Matas D, Almog N, Wolkowicz R, Goldfinger N, Pei H, Prokocimer M, Rotter V (1998) Mutant p53 protein expression interferes with p53-independent apoptotic pathways. Oncogene 16:3269–3277
8. Fan S, el-Deiry WS, Bae I, Freeman J, Jondle D, Bhatia K, Fornace AJ Jr, Magrath I, Kohn KW, O'Connor PM (1994) p53 gene mutations are associated with decreased sensitivity of human lymphoma cells to DNA damaging agents. Cancer Res 54:5824–5830
9. Blandino G, Levine AJ, Oren M (1999) Mutant p53 gain of function: differential effects of different p53 mutants on resistance of cultured cells to chemotherapy. Oncogene 18:477–485
10. Asada N, Tsuchiya H, Tomita K (1999) De novo deletions of p53 gene and wild-type p53 correlate with acquired cisplatin-resistance in human osteosarcoma OST cell line. Anticancer Res 19:5131–5137
11. Fan J, Bertino JR (1999) Modulation of cisplatinum cytotoxicity by p53: effect of p53-mediated apoptosis and DNA repair. Mol Pharmacol 56:966–972
12. Bartussek C, Naumann U, Weller M (1999) Accumulation of mutant p53(V143A) modulates the growth, clonogenicity, and radiochemosensitivity of malignant glioma cells independent of endogenous p53 status. Exp Cell Res 253:432–439
13. Chang FL, Lai MD (2001) Various forms of mutant p53 confer sensitivity to cisplatin and doxorubicin in bladder cancer cells. J Urol 166:304–310
14. Cabelguenne A, Blons H, de Waziers I, Carnot F, Houllier AM, Soussi T, Brasnu D, Beaune P, Laccourreye O, Laurent-Puig P (2000) p53 alterations predict tumor response to neoadjuvant chemotherapy in head and neck squamous cell carcinoma: a prospective series. J Clin Oncol 18:1465–1473
15. Koch WM, Brennan JA, Zahurak M, Goodman SN, Westra WH, Schwab D, Yoo GH, Lee DJ, Forastiere AA, Sidransky D (1996) p53 mutation and locoregional treatment failure in head and neck squamous cell carcinoma. J Natl Cancer Inst 88:1580–1586
16. Fan S, Smith ML, Rivet DJ 2nd, Duba D, Zhan Q, Kohn KW, Fornace AJ Jr, O'Connor PM (1995) Disruption of p53 function sensitizes breast cancer MCF-7 cells to cisplatin and pentoxifylline. Cancer Res 55:1649–1654
17. Bradford CR, Zhu S, Ogawa H, Ogawa T, Ubell M, Narayan A, Johnson G, Wolf GT, Fisher SG, Carey TE (2003) P53 mutation correlates with cisplatin sensitivity in head and neck squamous cell carcinoma lines. Head Neck 25:654–661
18. Jackson JG, Pant V, Li Q, Chang LL, Quintas-Cardama A, Garza D, Tavana O, Yang P, Manshouri T et al (2013) p53-mediated senescence impairs the apoptotic response to chemotherapy and clinical outcome in breast cancer. Cancer Cell 21:793–806

19. Martinez-Rivera M, Siddik ZH (2012) Resistance and gain-of-resistance phenotypes in cancers harboring wild-type p53. Biochem Pharmacol 83:1049–1062
20. Bristow RG, Peacock J, Jang A, Kim J, Hill RP, Benchimol S (2003) Resistance to DNA-damaging agents is discordant from experimental metastatic capacity in MEF ras-transformants-expressing gain of function MTp53. Oncogene 22:2960–2966
21. Scian MJ, Stagliano KE, Anderson MA, Hassan S, Bowman M, Miles MF, Deb SP, Deb S (2005) Tumor-derived p53 mutants induce NF-kappaB2 gene expression. Mol Cell Biol 25:10097–10110
22. Vikhanskaya F, Lee MK, Mazzoletti M, Broggini M, Sabapathy K (2007) Cancer-derived p53 mutants suppress p53-target gene expression – potential mechanism for gain of function of mutant p53. Nucleic Acids Res 35:2093–2104
23. Wong RP, Tsang WP, Chau PY, Co NN, Tsang TY, Kwok TT (2007) p53-R273H gains new function in induction of drug resistance through down-regulation of procaspase-3. Mol Cancer Ther 6:1054–1061
24. Bossi G, Lapi E, Strano S, Rinaldo C, Blandino G, Sacchi A (2006) Mutant p53 gain of function: reduction of tumor malignancy of human cancer cell lines through abrogation of mutant p53 expression. Oncogene 25:304–309
25. Bristow RG, Benchimol S, Hill RP (1996) The p53 gene as a modifier of intrinsic radiosensitivity: implications for radiotherapy. Radiother Oncol 40:197–223
26. El-Deiry WS (2003) The role of p53 in chemosensitivity and radiosensitivity. Oncogene 22:7486–7495
27. Dahm-Daphi J (2000) p53: biology and role for cellular radiosensitivity. Strahlenther Onkol 176:278–285
28. Lee JM, Bernstein A (1993) p53 mutations increase resistance to ionizing radiation. Proc Natl Acad Sci U S A 90:5742–5746
29. Bristow RG, Hu Q, Jang A, Chung S, Peacock J, Benchimol S, Hill R (1998) Radioresistant MTp53-expressing rat embryo cell transformants exhibit increased DNA-dsb rejoining during exposure to ionizing radiation. Oncogene 16:1789–1802
30. Okaichi K, Ide-Kanematsu M, Izumi N, Morita N, Okumura Y, Ihara M (2008) Variations in sensitivity to ionizing radiation in relation to p53 mutation point. Anticancer Res 28:2687–2690
31. Williams JR, Zhang Y, Zhou H, Gridley DS, Koch CJ, Russell J, Slater JS, Little JB (2008) A quantitative overview of radiosensitivity of human tumor cells across histological type and TP53 status. Int J Radiat Biol 84:253–264
32. Skinner HD, Sandulache VC, Ow TJ, Meyn RE, Yordy JS, Beadle BM, Fitzgerald AL, Giri U, Ang KK, Myers JN (2012) TP53 disruptive mutations lead to head and neck cancer treatment failure through inhibition of radiation-induced senescence. Clin Cancer Res 18:290–300
33. Sandulache VC, Skinner HD, Ow TJ, Zhang A, Xia X, Luchak JM, Wong LJ, Pickering CR, Zhou G, Myers JN (2012) Individualizing antimetabolic treatment strategies for head and neck squamous cell carcinoma based on TP53 mutational status. Cancer 118:711–721
34. Xu Y (2008) Induction of genetic instability by gain-of-function p53 cancer mutants. Oncogene 27:3501–3507
35. Cheng G, Kong D, Hou X, Liang B, He M, Liang N, Ma S, Liu X (2013) The tumor suppressor, p53, contributes to radiosensitivity of lung cancer cells by regulating autophagy and apoptosis. Cancer Biother Radiopharm 28:153–159
36. Kim E, Deppert W (2004) Transcriptional activities of mutant p53: when mutations are more than a loss. J Cell Biochem 93:878–886
37. Scian MJ, Stagliano KE, Ellis MA, Hassan S, Bowman M, Miles MF, Deb SP, Deb S (2004) Modulation of gene expression by tumor-derived p53 mutants. Cancer Res 64:7447–7454
38. Vaughan CA, Singh S, Windle B, Sankala HM, Graves PR, Andrew Yeudall W, Deb SP, Deb S (2012) p53 mutants induce transcription of NF-kappaB2 in H1299 cells through CBP and STAT binding on the NF-kappaB2 promoter and gain of function activity. Arch Biochem Biophys 518:79–88

39. Sampath J, Sun D, Kidd VJ, Grenet J, Gandhi A, Shapiro LH, Wang Q, Zambetti GP, Schuetz JD (2001) Mutant p53 cooperates with ETS and selectively up-regulates human MDR1 not MRP1. J Biol Chem 276:39359–39367
40. Strano S, Dell'Orso S, Di Agostino S, Fontemaggi G, Sacchi A, Blandino G (2007) Mutant p53: an oncogenic transcription factor. Oncogene 26:2212–2219
41. Sankala H, Vaughan C, Wang J, Deb S, Graves PR (2011) Upregulation of the mitochondrial transport protein, Tim50, by mutant p53 contributes to cell growth and chemoresistance. Arch Biochem Biophys 512:52–60
42. Strauss BE, Shivakumar C, Deb SP, Deb S, Haas M (1995) The MDR1 downstream promoter contains sequence-specific binding sites for wild-type p53. Biochem Biophys Res Commun 217:825–831
43. Valenti F, Fausti F, Biagioni F, Shay T, Fontemaggi G, Domany E, Yaffe MB, Strano S, Blandino G, Di Agostino S (2011) Mutant p53 oncogenic functions are sustained by Plk2 kinase through an autoregulatory feedback loop. Cell Cycle 10:4330–4340
44. Vaughan CA, Singh S, Windle B, Yeudall WA, Frum R, Grossman SR, Deb SP, Deb S (2012) Gain-of-function activity of mutant p53 in lung cancer through up-regulation of receptor protein tyrosine kinase Axl. Genes Cancer 3:491–502
45. Masciarelli S, Fontemaggi G, Di Agostino S, Donzelli S, Carcarino E, Strano S, Blandino G (2013) Gain-of-function mutant p53 downregulates miR-223 contributing to chemoresistance of cultured tumor cells. Oncogene 32:1–8
46. Yang X, Pater A, Tang SC (1999) Cloning and characterization of the human BAG-1 gene promoter: upregulation by tumor-derived p53 mutants. Oncogene 18:4546–4553
47. Zalcenstein A, Stambolsky P, Weisz L, Muller M, Wallach D, Goncharov TM, Krammer PH, Rotter V, Oren M (2003) Mutant p53 gain of function: repression of CD95(Fas/APO-1) gene expression by tumor-associated p53 mutants. Oncogene 22:5667–5676
48. Zalcenstein A, Weisz L, Stambolsky P, Bar J, Rotter V, Oren M (2006) Repression of the MSP/MST-1 gene contributes to the antiapoptotic gain of function of mutant p53. Oncogene 25:359–369
49. Deb S, Graves PR (2013) Identification of novel mutant p53 interacting proteins by proteomic analysis. Methods Mol Biol 962:85–94
50. Ramamoorthy M, Vaughan C, Deb S, Deb SP (2012) Measurement of chemosensitivity and growth rate in p53 expressing cells. Methods Mol Biol 962:127–133
51. Munro AJ, Lain S, Lane DP (2005) P53 abnormalities and outcomes in colorectal cancer: a systematic review. Br J Cancer 92:434–444
52. Havrilesky L, Darcy kM, Hamdan H, Priore RL, Leon J, Bell J, Berchuck A (2003) Prognostic significance of p53 mutation and p53 overexpression in advanced epithelial ovarian cancer: a Gynecologic Oncology Group Study. J Clin Oncol 21:3814–3825
53. Righetti SC, Della Torre G, Pilotti S, Menard S, Ottone F, Colnaghi MI, Pierotti MA, Lavarino C, Cornarotti M et al (1996) A comparative study of p53 gene mutations, protein accumulation, and response to cisplatin-based chemotherapy in advanced ovarian carcinoma. Cancer Res 56:689–693
54. Buttitta F, Marchetti A, Gadducci A, Pellegrini S, Morganti M, Carnicelli V, Cosio S, Gagetti O, Genazzani AR, Bevilacqua G (1997) p53 alterations are predictive of chemoresistance and aggressiveness in ovarian carcinomas: a molecular and immunohistochemical study. Br J Cancer 75:230–235
55. Reles A, Wen WH, Schmider A, Gee C, Runnebaum IB, Kilian U, Jones LA, El-Naggar A, Minguillon C et al (2001) Correlation of p53 mutations with resistance to platinum-based chemotherapy and shortened survival in ovarian cancer. Clin Cancer Res 7:2984–2997
56. Calvert AH, Ghokul S, Al-Azraqi A, Wright J, Lind M, Bailey N, Highley M, Siddiqui N, Lunec J et al (1999) Carboplatin and paclitaxel, alone and in combination: dose escalation, measurement of renal function, and role of the p53 tumor suppressor gene. Semin Oncol 26:90–94
57. Kigawa J, Sato S, Shimada M, Takahashi M, Itamochi H, Kanamori Y, Terakawa N (2001) p53 gene status and chemosensitivity in ovarian cancer. Hum Cell 14:165–171

58. Bauerschlag DO, Schem C, Weigel MT, Von Kaisenberg C, Strauss A, Bauknecht T, Maass N, Meinhold-Heerlein I (2010) The role of p53 as a surrogate marker for chemotherapeutical responsiveness in ovarian cancer. J Cancer Res Clin Oncol 136:79–88
59. Lavarino C, Pilotti S, Oggionni M, Gatti L, Perego P, Bresciani G, Pierotti MA, Scambia G, Ferrandina G et al (2000) p53 gene status and response to platinum/paclitaxel-based chemotherapy in advanced ovarian carcinoma. J Clin Oncol 18:3936–3945
60. Petera J, Sirak I, Beranek M, Vosmik M, Drastikova M, Paulikova S, Soumarova R (2012) Molecular predictive factors of outcome of radiotherapy in cervical cancer. Neoplasma 58:469–475
61. Ishikawa H, Mitsuhashi N, Sakurai H, Maebayashi K, Niibe H (2001) The effects of p53 status and human papillomavirus infection on the clinical outcome of patients with stage IIIB cervical carcinoma treated with radiation therapy alone. Cancer 91:80–89
62. Berns EM, Foekens JA, Vossen R, Look MP, Devilee P, Henzen-Logmans SC, van Staveren IL, van Putten WL, Inganas M et al (2000) Complete sequencing of TP53 predicts poor response to systemic therapy of advanced breast cancer. Cancer Res 60:2155–2162
63. Chen MB, Zhu YQ, Xu JY, Wang LQ, Liu CY, Ji ZY, Lu PH (2012) Value of TP53 status for predicting response to neoadjuvant chemotherapy in breast cancer: a meta-analysis. PLoS One 7:e39655
64. Kandioler-Eckersberger D, Ludwig C, Rudas M, Kappel S, Janschek E, Wenzel C, Schlagbauer-Wadl H, Mittlbock M, Gnant M et al (2000) TP53 mutation and p53 overexpression for prediction of response to neoadjuvant treatment in breast cancer patients. Clin Cancer Res 6:50–56
65. Oshima K, Naoi Y, Kishi K, Nakamura Y, Iwamoto T, Shimazu K, Nakayama T, Kim SJ, Baba Y et al (2011) Gene expression signature of TP53 but not its mutation status predicts response to sequential paclitaxel and 5-FU/epirubicin/cyclophosphamide in human breast cancer. Cancer Lett 307:149–157
66. Geisler S, Lonning PE, Aas T, Johnsen H, Fluge O, Haugen DF, Lillehaug JR, Akslen LA, Borresen-Dale AL (2001) Influence of TP53 gene alterations and c-erbB-2 expression on the response to treatment with doxorubicin in locally advanced breast cancer. Cancer Res 61:2505–2512
67. Bergh J, Norberg T, Sjogren S, Lindgren A, Holmberg L (1995) Complete sequencing of the p53 gene provides prognostic information in breast cancer patients, particularly in relation to adjuvant systemic therapy and radiotherapy. Nat Med 1:1029–1034
68. Aas T, Borresen AL, Geisler S, Smith-Sorensen B, Johnsen H, Varhaug JE, Akslen LA, Lonning PE (1996) Specific P53 mutations are associated with de novo resistance to doxorubicin in breast cancer patients. Nat Med 2:811–814
69. Faille A, De Cremoux P, Extra JM, Linares G, Espie M, Bourstyn E, De Rocquancourt A, Giacchetti S, Marty M, Calvo F (1994) p53 mutations and overexpression in locally advanced breast cancers. Br J Cancer 69:1145–1150
70. Jansson T, Inganas M, Sjogren S, Norberg T, Lindgren A, Holmberg L, Bergh J (1995) p53 status predicts survival in breast cancer patients treated with or without postoperative radiotherapy: a novel hypothesis based on clinical findings. J Clin Oncol 13:2745–2751
71. Boyle JO, Hakim J, Koch W, van der Riet P, Hruban RH, Roa RA, Correo R, Eby YJ, Ruppert JM, Sidransky D (1993) The incidence of p53 mutations increases with progression of head and neck cancer. Cancer Res 53:4477–4480
72. Caamano J, Zhang SY, Rosvold EA, Bauer B, Klein-Szanto AJ (1993) p53 alterations in human squamous cell carcinomas and carcinoma cell lines. Am J Pathol 142:1131–1139
73. Alsner J, Sorensen SB, Overgaard J (2001) TP53 mutation is related to poor prognosis after radiotherapy, but not surgery, in squamous cell carcinoma of the head and neck. Radiother Oncol 59:179–185
74. Gallo O, Chiarelli I, Boddi V, Bocciolini C, Bruschini L, Porfirio B (1999) Cumulative prognostic value of p53 mutations and bcl-2 protein expression in head-and-neck cancer treated by radiotherapy. Int J Cancer 84:573–579

75. Hegde PU, Brenski AC, Caldarelli DD, Hutchinson J, Panje WR, Wood NB, Leurgans S, Preisler HD, Taylor SG 4th et al (1998) Tumor angiogenesis and p53 mutations: prognosis in head and neck cancer. Arch Otolaryngol Head Neck Surg 124:80–85
76. Chang KP, Hao SP, Lin SY, Tsao KC, Kuo TT, Tsai MH, Tseng CK, Tsang NM (2002) A lack of association between p53 mutations and recurrent nasopharyngeal carcinomas refractory to radiotherapy. Laryngoscope 112:2015–2019
77. Saunders ME, MacKenzie R, Shipman R, Fransen E, Gilbert R, Jordan RC (1999) Patterns of p53 gene mutations in head and neck cancer: full-length gene sequencing and results of primary radiotherapy. Clin Cancer Res 5:2455–2463
78. Obata A, Eura M, Sasaki J, Saya H, Chikamatsu K, Tada M, Iggo RD, Yumoto E (2000) Clinical significance of p53 functional loss in squamous cell carcinoma of the oropharynx. Int J Cancer 89:187–193
79. Grignon DJ, Caplan R, Sarkar FH, Lawton CA, Hammond EH, Pilepich MV, Forman JD, Mesic J, Fu KK et al (1997) p53 status and prognosis of locally advanced prostatic adenocarcinoma: a study based on RTOG 8610. J Natl Cancer Inst 89:158–165
80. Prendergast NJ, Atkins MR, Schatte EC, Paulson DF, Walther PJ (1996) p53 immunohistochemical and genetic alterations are associated at high incidence with post-irradiated locally persistent prostate carcinoma. J Urol 155:1685–1692
81. Rakozy C, Grignon DJ, Li Y, Gheiler E, Gururajanna B, Pontes JE, Sakr W, Wood DP Jr, Sarkar FH (1999) p53 gene alterations in prostate cancer after radiation failure and their association with clinical outcome: a molecular and immunohistochemical analysis. Pathol Res Pract 195:129–135
82. Incognito LS, Cazares LH, Schellhammer PF, Kuban DA, Van Dyk EO, Moriarty RP, Wright GL Jr, Somers KD (2000) Overexpression of p53 in prostate carcinoma is associated with improved overall survival but not predictive of response to radiotherapy. Int J Oncol 17:761–769
83. Imamura J, Miyoshi I, Koeffler HP (1994) p53 in hematologic malignancies. Blood 84:2412–2421
84. Wattel E, Preudhomme C, Hecquet B, Vanrumbeke M, Quesnel B, Dervite I, Morel P, Fenaux P (1994) p53 mutations are associated with resistance to chemotherapy and short survival in hematologic malignancies. Blood 84:3148–3157
85. Gonzalez D, Martinez P, Wade R, Hockley S, Oscier D, Matutes E, Dearden CE, Richards SM, Catovsky D, Morgan GJ (2012) Mutational status of the TP53 gene as a predictor of response and survival in patients with chronic lymphocytic leukemia: results from the LRF CLL4 trial. J Clin Oncol 29:2223–2229
86. Preudhomme C, Dervite I, Wattel E, Vanrumbeke M, Flactif M, Lai JL, Hecquet B, Coppin MC, Nelken B et al (1995) Clinical significance of p53 mutations in newly diagnosed Burkitt's lymphoma and acute lymphoblastic leukemia: a report of 48 cases. J Clin Oncol 13:812–820
87. Klumb CE, Furtado DR, de Resende LM, Carrico MK, Coelho AM, de Meis E, Maia RC, Rumjanek FD (2003) DNA sequence profile of TP53 gene mutations in childhood B-cell non-Hodgkin's lymphomas: prognostic implications. Eur J Haematol 71:81–90
88. Kandioler-Eckersberger D, Kappel S, Mittlbock M, Dekan G, Ludwig C, Janschek E, Pirker R, Wolner E, Eckersberger F (1999) The TP53 genotype but not immunohistochemical result is predictive of response to cisplatin-based neoadjuvant therapy in stage III non-small cell lung cancer. J Thorac Cardiovasc Surg 117:744–750
89. Kandioler D, Stamatis G, Eberhardt W, Kappel S, Zochbauer-Muller S, Kuhrer I, Mittlbock M, Zwrtek R, Aigner C et al (2008) Growing clinical evidence for the interaction of the p53 genotype and response to induction chemotherapy in advanced non-small cell lung cancer. J Thorac Cardiovasc Surg 135:1036–1041
90. King TC, Akerley W, Fan AC, Moore T, Mangray S, Hsiu Chen M, Safran H (2000) p53 mutations do not predict response to paclitaxel in metastatic nonsmall cell lung carcinoma. Cancer 89:769–773
91. Safran H, King T, Choy H, Gollerkeri A, Kwakwa H, Lopez F, Cole B, Myers J, Tarpey J, Rosmarin A (1996) p53 mutations do not predict response to paclitaxel/radiation for nonsmall cell lung carcinoma. Cancer 78:1203–1210

92. Irving RM, Moffat DA, Hardy DG, Barton DE, Xuereb JH, Maher ER (1993) Molecular genetic analysis of the mechanism of tumorigenesis in acoustic neuroma. Arch Otolaryngol Head Neck Surg 119:1222–1228
93. Monoh K, Ishikawa K, Yasui N, Mineura K, Andoh H, Togawa K (1998) p53 tumor suppressor gene in acoustic neuromas. Acta Otolaryngol Suppl 537:11–15
94. Yeung AH, Sughrue ME, Kane AJ, Tihan T, Cheung SW, Parsa AT (2009) Radiobiology of vestibular schwannomas: mechanisms of radioresistance and potential targets for therapeutic sensitization. Neurosurg Focus 27:E2
95. Weller M, Felsberg J, Hartmann C, Berger H, Steinbach JP, Schramm J, Westphal M, Schackert G, Simon M et al (2009) Molecular predictors of progression-free and overall survival in patients with newly diagnosed glioblastoma: a prospective translational study of the German Glioma Network. J Clin Oncol 27:5743–5750
96. Tabori U, Shlien A, Baskin B, Levitt S, Ray P, Alon N, Hawkins C, Bouffet E, Pienkowska M et al (2010) TP53 alterations determine clinical subgroups and survival of patients with choroid plexus tumors. J Clin Oncol 28:1995–2001
97. Petersen S, Thames HD, Nieder C, Petersen C, Baumann M (2001) The results of colorectal cancer treatment by p53 status: treatment-specific overview. Dis Colon Rectum 44:322–333. discussion 333–334
98. Chen MB, Wu XY, Yu R, Li C, Wang LQ, Shen W, Lu PH (2012) P53 status as a predictive biomarker for patients receiving neoadjuvant radiation-based treatment: a meta-analysis in rectal cancer. PLoS One 7:e45388
99. Kandioler D, Zwrtek R, Ludwig C, Janschek E, Ploner M, Hofbauer F, Kuhrer I, Kappel S, Wrba F et al (2002) TP53 genotype but not p53 immunohistochemical result predicts response to preoperative short-term radiotherapy in rectal cancer. Ann Surg 235:493–498
100. Rebischung C, Gerard JP, Gayet J, Thomas G, Hamelin R, Laurent-Puig P (2002) Prognostic value of P53 mutations in rectal carcinoma. Int J Cancer 100:131–135
101. Westra JL, Schaapveld M, Hollema H, de Boer JP, Kraak MM, de Jong D, ter Elst A, Mulder NH, Buys CH et al (2005) Determination of TP53 mutation is more relevant than microsatellite instability status for the prediction of disease-free survival in adjuvant-treated stage III colon cancer patients. J Clin Oncol 23:5635–5643
102. Benhattar J, Cerottini JP, Saraga E, Metthez G, Givel JC (1996) p53 mutations as a possible predictor of response to chemotherapy in metastatic colorectal carcinomas. Int J Cancer 69:190–192
103. Lenz HJ, Hayashi K, Salonga D, Danenberg KD, Danenberg PV, Metzger R, Banerjee D, Bertino JR, Groshen S et al (1998) p53 point mutations and thymidylate synthase messenger RNA levels in disseminated colorectal cancer: an analysis of response and survival. Clin Cancer Res 4:1243–1250
104. Ribeiro U Jr, Finkelstein SD, Safatle-Ribeiro AV, Landreneau RJ, Clarke MR, Bakker A, Swalsky PA, Gooding WE, Posner MC (1998) p53 sequence analysis predicts treatment response and outcome of patients with esophageal carcinoma. Cancer 83:7–18

Chapter 9
The p53-Mdm2 Loop: A Critical Juncture of Stress Response

Yaara Levav-Cohen, Zehavit Goldberg, Kah Hin Tan, Osnat Alsheich-Bartok, Valentina Zuckerman, Sue Haupt, and Ygal Haupt

Abstract The presence of a functional p53 protein is a key factor for the proper suppression of cancer development. A loss of p53 activity, by mutations or inhibition, is often associated with human malignancies. The p53 protein integrates various stress signals into a growth restrictive cellular response. In this way, p53 eliminates cells with a potential to become cancerous. Being a powerful decision maker, it is imperative that p53 will be activated properly, efficiently and temporarily in response to stress. Equally important is that p53 activation will be extinguished upon recovery from stress, and that improper activation of p53 will be avoided. Failure to achieve these aims is likely to have catastrophic consequences for the organism. The machinery that governs this tight regulation is largely based on the major inhibitor of p53, Mdm2, which both blocks p53 activities and promotes its destabilization. The interplay between p53 and Mdm2 involves a complex network of positive and negative feedback loops. Relief from Mdm2 suppression is required for p53 to be stabilized and activated in response to stress. Protection from Mdm2 entails a concerted action of modifying enzymes and partner proteins. The association of p53 with the PML-nuclear bodies may provide an infrastructure in which this complex regulatory network can be orchestrated. In this chapter we use examples to illustrate the regulatory machinery that drives this network.

Y. Levav-Cohen • Z. Goldberg • O. Alsheich-Bartok • V. Zuckerman
Lautenberg Center, IMRIC, The Hebrew University-Hadassah Medical School, Jerusalem, Israel

K.H. Tan • S. Haupt • Y. Haupt (✉)
Research Division, The Peter MacCallum Cancer Centre,
St. Andrew's Place, East Melbourne 3002, VIC, Australia
e-mail: ygal.haupt@petermac.org

S.P. Deb and S. Deb (eds.), *Mutant p53 and MDM2 in Cancer*,
Subcellular Biochemistry 85, DOI 10.1007/978-94-017-9211-0_9,

Keywords p53 • Mdm2 (Hdm2) • c-Abl • Chk • Ubiquitination • Degradation • Phosphorylation • Nuclear export • PML nuclear bodies

Introduction

The tumour suppressor p53 protein is pivotal in the prevention of cancer development. P53 determines cell fate through its activities as a transcription factor, and by engagement in critical protein interactions at the mitochondria (reviewed in [1–3]). P53 is normally labile, but in response to external and internal stress signals, it is triggered to become stable and active within the nucleus. As a transcription factor it controls the expression of genes that regulate cell growth and cell death (reviewed in refs. [2–4]). Stabilized p53 induces either cell growth arrest (temporary, or permanent "senescence"), or programmed cell death (apoptosis). The growth restrictive activities of p53 prevent the proliferation of cells with damaged DNA or with a potential for neoplastic transformation; while p53-mediated permanent cell growth inhibition (apoptosis or senescence) drives tumour suppression. Given these functions of p53, it is not surprising that p53 serves as a serious obstacle to the step-by-step progression of cancer development. This barrier is very frequently removed at one of the steps, either by direct mutation of the p53 gene, or by indirect mechanisms, such as an elevation in the expression levels of p53 inhibitors, or by down-regulation of p53 co-activators, such as ARF [3].

The lability of p53 under normal cellular conditions is largely attributable to its inhibitor Mdm2 (Hdm2 in humans), which ensures that p53 has a short half-life and consequent low basal activity. Stresses that dramatically elicit a change in p53 status include: DNA damage, untimely expression of oncogenes, hypoxia, and nucleotide depletion among others [5, 6] (also reviewed in [7–9, 2]). The activation of p53 involves stabilization of the protein, which is mediated by extensive post-translational modifications, and protein-protein interactions with cooperating factors. Once stable, p53 engages in enhanced DNA binding and transcriptional activity. The summation of the incoming signals and the cellular context, dictates whether activated p53 will direct cells to growth arrest, senescence or apoptosis (reviewed in ref. [10, 11]). This chapter focuses on the regulation of p53 by Mdm2. Particular emphasis will be given to current models explaining how the p53/Mdm2 auto-regulatory loop is modulated or interrupted in response to stress. The different mechanisms involved will be illustrated by specific examples. An attempt will be made to explain how such a busy network of regulation may be coordinated within a cell in a spatial and temporal manner in response to a given stress signal.

The p53-Mdm2 Feedback Loop

Over two decades of research have passed since the identification of *mdm2* as a p53 target gene [12, 13] (also reviewed in [2, 14]). The revelation that p53 induces Mdm2 expression, which then inhibits the biochemical and biological activities of

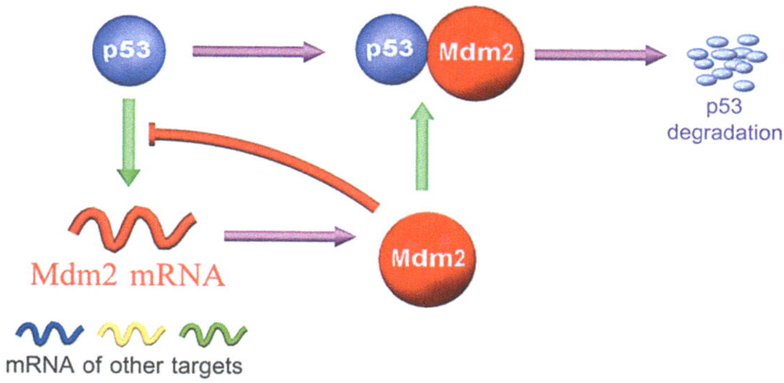

Fig. 9.1 The p53/Mdm2 autoregulatory loop. Activated p53 induces the expression of multiple target genes. One of the genes is Mdm2 which binds p53, inhibits its transcriptional activity and promotes it for proteasomal degradation

p53, defined the first and the most important auto-regulatory loop that governs p53 regulation. This loop proves to be even more powerful than initially thought, as additional multiple regulatory loops are being found to interweave with it [15–19]. Several of these loops will be described in this chapter. Mdm2 binds p53 in the transactivation domain and blocks its ability to induce or suppress transcription (reviewed in [20, 21]). The major and most efficient inhibitory effect of Mdm2 is to destroy the p53 protein via the ubiquitin-proteasome pathway [22, 23] (also reviewed in [24–26, 21]). Thus, through this negative feedback loop, Mdm2 shuts off its own expression (Fig. 9.1). The physiological significance of this auto-regulatory feedback loop was demonstrated by the clinical observation that amplification of Hdm2 in human cancers often correlates with wild type p53 status, supporting the notion that high expression of Hdm2 is sufficient for relieving a cell from p53 regulation, in the absence of p53 mutation (reviewed in [27, 28, 2]). Further, a single nucleotide polymorphism (SNP) in the *hdm2* gene that leads to increase Hdm2 expression, results in reduced levels of p53-dependent apoptosis and correlates with accelerated tumour formation [29] (for review see [30, 31]). Consistently, reduced expression of Mdm2 protein in mice carrying a hypomorphic *mdm2* allele, results in increased frequency of p53 dependent apoptosis and leads to many defects in mice [32] (or reviewed in [30, 33]). Even more convincing was the finding that mouse embryos lacking *mdm2* die early during development, but if crossed with p53 null mice this phenotype are rescued [27, 34–36] (also reviewed in [20, 33]).

Intriguingly, the presence of the Mdm2 analogue, Mdmx (also called Mdm4 or Hdmx in humans), is critical for proper mouse development. The embryonic lethality of *mdmx* null mice is rescued by the removal of p53, as in the case of *mdm2* deficient mice, supporting critical and non-overlapping roles for Mdm2 and Mdmx in the inhibition of p53, at least during early mouse development [37–39] (or reviewed in [40, 14, 33]). Moreover, analogous to Hdm2, Hdmx expression is elevated in a considerable fraction of human cancers carrying wild-type p53, implying that high

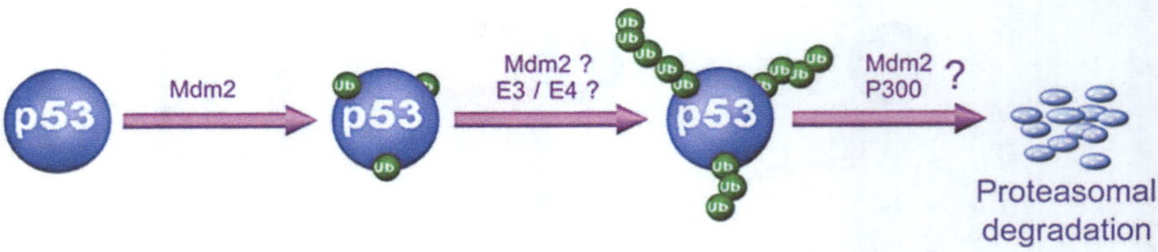

Fig. 9.2 A model for the proteasomal degradation of p53 by Mdm2. Mdm2 is important for the mono-ubiquitination of p53. The possible involvement of Mdm2 and other factors in the subsequent steps leading to p53 degradation is shown. For simplicity the sub-cellular compartmentalization of this process is not shown

level of Mdmx expression is sufficient to inactivate p53, without concomitant p53 mutations [41, 42]. Interestingly, in sharp contrast to Mdm2, the inhibition of p53 by Mdmx, does not directly involve protein degradation and there is no evidence for existence of a regulatory loop between Mdmx and p53 [43]. Mdmx binds p53 in its transactivation domain and inhibits p53 transcriptional activity [44]. Importantly, Mdmx forms heterodimers with Mdm2 and, although Mdmx lacks its own E3 ubiquitin ligase activity, it can assist Mdm2 in p53 degradation (reviewed in [45]) (Fig. 9.2).

Mdm2-Mediated p53 Ubiquitination and Degradation

The majority of studies investigating the inhibition of p53 by Mdm2 support the current model that Mdm2 promotes the ubiquitination and subsequent degradation of p53 through the proteasomal machinery (Fig. 9.1). This promotion of p53 degradation requires the E3 ligase activity of Mdm2, which is mediated by the RING-finger domain [46, 47]. Although, a number of studies identified Mdm2 as the principal endogenous E3-ligase that promotes the efficient degradation of p53 (e.g. [48, 22, 23, 46, 47]), p53 is also degraded in *mdm2* deficient cells [49], suggesting that other E3 ligase/s can promote p53 degradation in an Mdm2-independent manner, *in vivo*. Indeed, a number of other E3 ligases were shown to regulate p53 protein levels in tissue culture and in biochemical studies *in vitro*, these include: Pirh2 [50], Cop1 [51], and Arf-BP1 [52], ICP0 [53], TOPORS [54], CHIP [55], Ubc13 [56], Synoviolin [57], E4F1 [58], CARP1 [59], CARP2 [59], WWP1 [60], MSL2 [61], (also reviewed in [62, 25, 21]). However, the exact cellular contexts and level of contribution of these ligases remain to be defined.

Confirmation that the primary physiological function of Mdm2 is to promote p53 degradation via its E3 ligase activity was shown by Itahana et al. [63]. In this study, the loss of p53 rescued the early lethality of mice bearing a C462A mutation in Mdm2, a mutation that causes the abolishment of the E3 ligase activity of Mdm2, without affecting its interaction with p53. Nonetheless, accumulating evidence over the last decade indicates that Mdm2 can repress p53 activity by additional mechanisms, besides its direct interaction with p53. For example, Mdm2 can inhibit p53 acetylation mediated by p300/CBP [64, 65] as well as by inhibiting and degrading PCAF [66]. In addition, Mdm2 can recruit the histone deacetylase HDAC1 [67] and the nuclear corepressor KAP1 [68], providing alternative ways by which Mdm2 can repress the acetylation of either p53 or histones surrounding the p53 binding sites. Moreover, Mdm2 was also reported to promote NEDD8 conjugation of p53, a modification that inhibits its transcriptional activity [69]. Finally, Mdm2 induces monoubiquitination of histones in the vicinity of the p53 response elements, resulting in the transcriptional repression of p53 [70].

Mdm2 can also regulate the levels of nuclear p53 by a degradation- independent ubiquitination. Using an *in vitro* ubiquitination reconstitution assay, it has been shown that Mdm2 mediates the monomeric ubiquitination of p53 on multiple lysine residues, rather than adding ubiquitin chains onto one or few lysine residues [71]. It was found, that when expressed in low levels, Mdm2 can mediate p53-monoubiquitination, which signals for p53 nuclear export and leads to its accumulation in the cytoplasm, thereby inhibiting its role as transcription factor [72, 73]. However, proteasome-dependent degradation of nuclear p53 can also effectively prevent its transactivation function [73, 74], suggesting monoubiquitination of cytoplasmic p53 is required for other processes. Indeed, the role that p53 plays in the cytoplasm, including direct signalling at the mitochondria and the induction of transcription-independent apoptosis [75], may involve p53 modified in this manner.

P53 is subject to ubiquitination on at least 6 C-terminal lysines [48]. The six C-terminal lysines of p53 are the predominant sites for mdm2-mediated ubiquitination [76]. Although the *in vitro* data certainly demonstrate the importance of the six C-terminal lysines of p53 for Mdm2- mediated ubiquitination, knock-in studies in which the equivalent lysines have been mutated did not dramatically alter p53 protein levels [77, 78]. These studies suggests that the C-terminal lysines are not essential for efficient p53 degradation *in vivo* and that additional E3 ligases, as well as ubiquitination of additional p53 lysines, are required for effective p53 regulation. *In vitro* data suggest that lysine residues located in the DNA-binding domain and in the N-terminus may be ubiquitinated [79].

An additional mechanism by which Mdm2 may regulate p53 has been suggested. Yin et al. [80] have shown that Mdm2, by virtue of its binding to p53, induces p53 translation from an internal initiation site, generating a smaller product of 47Kd (termed p53/47). This product lacks the Mdm2 binding site and hence is more stable than wt p53 and has altered specificity towards the apoptotic target gene, *bax*. The p47 form is still subject to degradation, presumably through oligomerization with wt p53, as previously shown for other Mdm2-binding deficient mutant of p53 [81]. Surprisingly, the overall apoptotic activity of the p53/47 does not differ from that of wt p53. Hence, the physiological relevance of this mode of regulation is yet to be defined.

It is difficult to weigh the relative contribution of p53 degradation to the overall down-regulation of p53 by Mdm2 (reviewed in [24]). The inhibition of p53 transcriptional and apoptotic activities by Mdm2 without promoting p53 degradation supports dual inhibitory mechanisms by Mdm2 (e.g. refs. [82, 83]). However, genetic studies support the notion that the major physiological importance of Mdm2 relates to its ability to regulate p53 protein abundance and that Mdm2 has little effect on p53 transcriptional activity on a "per molecule" basis (reviewed in [84]). Moreover, *in vivo*, Mdm2 that binds p53 but lacks its ubiquitination ligase activity is unable to efficiently suppress p53 functions (reviewed in [85]).

Breaking the p53/Mdm2 Regulatory Loop

The p53/Mdm2 auto-regulatory loop provides explanations for: how the low basal level of p53 is maintained and how p53 returns to its basal level during the recovery from stress. At the same time, however, it raises the question of how the p53 protein temporarily escapes these intensive restrictive activities of Mdm2 (as well as of Mdmx) when cells are exposed to stress? Considerable effort has been devoted over the past decade to unravel the mechanisms by which the p53/Mdm2 loop is regulated and can be interrupted under stress conditions (e.g. [17, 18, 24, 86, 87]). Three major mechanisms for this loop interruption have been identified; first, stress induces post-translational modifications of both p53 and Mdm2; second, interacting proteins act to disrupt p53/Mdm2 binding; third, transportation of the two proteins leads to their spatial separation. A variety of proteins participate in these levels of regulation, where the choice of the particular regulator is largely dictated by the type of the incoming stress signal (e.g. [17]). It is not our intention to cover every known case, but rather to use selected examples to illustrate the principles of three major levels of regulation.

Stress Induced Phosphorylation

P53 Activation by the ATM/ATR-Chk Pathways

The p53 protein is subject to extensive post-translational modifications (reviewed in [88–90, 8, 5]), some of which affect its sensitivity to inhibition by Mdm2. Here we will focus on the role of phosphorylation in this regulation. Phosphorylation of p53, in particular within the N-terminal side, has critical impact on its functional interaction with Mdm2. Phosphorylation on threonine 18 (Thr18) by casein kinase I reduce the interaction between p53 and Mdm2 [91, 92]. This phosphorylation appears to require a preceding phosphorylation on serine 15 (Ser15) by several protein kinases including ATM and ATR (reviewed in [41, 88, 86]), which activate p53, but do not stabilize it [93–96]. Moreover, phosphorylation of threonine-proline motifs enables the binding of the prolyl isomerase PIN1 to induce cis-trans prolyl isomerizations. The phosphorylation of p53 on three sites, Ser33, Thr81 and Ser315, was shown to affect the sensitivity of p53 to Mdm2 through the involvement of Pin1 (see below). Although several PIN1 sites exist in human p53, the PIN1 site in the proline-rich

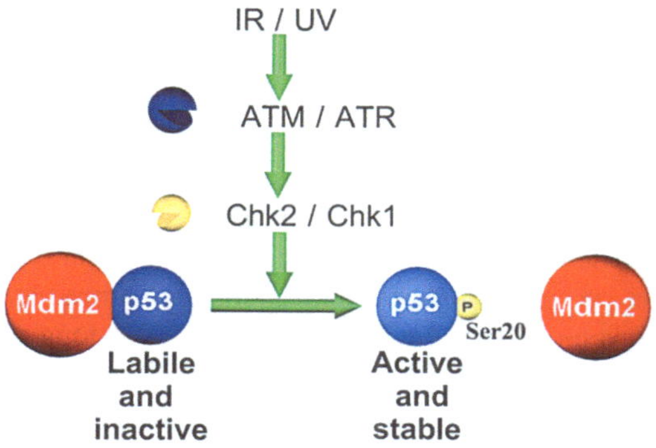

Fig. 9.3 Interruption of the p53/Mdm2 loop by DNA damage-mediated phosphorylation. A major pathway by which p53 is protected from Mdm2 in response to DNA damage. A role for Ser20 phosphorylation is shown

domain (threonine 81-proline 82) seems to be essential because proline 82 is isomerized by PIN1, enabling the recruitment of Chk2 to phosphorylate serine 20 and consequently reduce Mdm2 binding [97].

Much attention has been focused on the phosphorylation of human p53 on Ser20 (Ser23 in the mouse), as it resides within the Mdm2 binding site. This phosphorylation is mediated by the checkpoint kinase (Chk) 1, MAPKAPK2 or JNK in response to UV-light and by Chk2 in response to γ-irradiation (IR) (reviewed in [41, 88, 8]). The phosphorylation of Ser20 reduces the binding affinity between p53 and Mdm2, and consequently p53 is activated and stabilized (reviewed in [98]). Thus, one mechanism by which p53 is protected from Mdm2 in response to DNA damage involves Ser20 phosphorylation (Fig. 9.3). This temporal protection would last as long as the DNA damage signal persists. How this signal is terminated is not clear. For instance does it involve dephosphorylation of Ser20 and other sites?

Another mechanism which allows p53 induction involves stimulation of ATM and ATR in response to DNA damage, leading to phosphorylation of Mdm2 and MdmX, mediating their rapid degradation [99, 100].

Biochemical analysis and studies using cultured cells indicate that phosphorylation of these p53 sites stimulates the recruitment of key transcriptional proteins, such as p300 and CBP [101–107], leading to the acetylation of several key lysine residues in the carboxy-terminus of p53 that are normally targets for ubiquitination, this process is thought to help stabilize p53 [67, 108].

The physiological role of Ser20 in human was also tested in the transgenic mice bearing a substitution from Ser23 to alanine 23 (Ala23). Introduction of this substitution mutant p53 into ES cells or MEFs had no significant effect on the activation or accumulation of p53 in response to DNA damage in contrast to the case in the *Chk2* null mice [109, 110]. Further, Ser23 is phosphorylated in *Chk2* null mice in response to IR [111]. Can this apparent controversy be reconciled? Several possible explanations may be suggested. First, the regulatory role of Ser20 in human p53 may differ from that of Ser23 in the mouse. The lack of conservation of the Ser46 phosphorylation site in mouse p53 exemplifies this notion (see below). Second, Ser20 (and Ser23) is not the only Chk2 phosphorylation site in p53 [97, 112] (reviewed in [41, 113, 8, 9, 5]).

Third, Chk2 may also activate p53 by phosphorylation-independent mechanisms, for instance by direct interaction with p53 [114]. Whereas the precise mechanistic explanation for p53 activation by Chk2 needs further exploration, the evidence for the physiological relevance of Chk2 appears solid. The activation of p53 in response to DNA damage is severely impaired in *Chk2* null mice, and p53 mutations are not or rarely found in cancers bearing germ-line or somatic mutations in the Chk2 gene (reviewed in [115]). Despite the subtle phenotype, one study reported a partial defect in p53 accumulation and apoptosis in irradiated thymocytes [109]. Mutant mice developed B-cell lymphomas, but with a long period of latency (around 18 months, compared with 6–10 months in Trp53–/– mice).

It was also surprising that the mutation of Ser18 in mouse p53 (human Ser15) led to a modest phenotype [116, 117] that had normal p53 stability in unstressed and DNA-damaged cells, normal cell-cycle control, cell-type-specific partial defects in apoptosis and normal tumour suppression. A targeted double mutation in S18A/S23A was analysed in-vivo [118]. The double mutant knock-in mice, display reduced apoptosis in thymocytes and develop some malignancies, lending support to the physiological importance of these two key phosphorylation sites. Overall, these mice provide evidence to support the idea that DNA damage pathways can, at least partially, influence tumour suppressor function.

Mdm2 Inactivation by the ATM-c-Abl Pathway

Mdm2 is a phosphoprotein, that is subject to both phosphorylation and de-phosphorylation of specific sites in response to DNA damage [119, 120] (also reviewed in [121–123]). A search for phosphorylation sites in Mdm2 relevant to p53 regulation revealed multiple sites that are targeted by several protein kinases (reviewed in [121–123]). The interaction between Mdm2 and p53 is impaired upon phosphorylation of Mdm2 by DNA-dependent protein kinase [124]. A similar effect is observed when cyclin A-Cdk2 phosphorylates Mdm2 on threonine 216, which also augments the Mdm2/ARF binding [125].

An important antagonist of Mdm2 is the c-Abl tyrosine kinase. C-Abl is essential for the efficient accumulation of p53 in response to DNA damage [126], (reviewed in [127]). This is achieved by protecting p53 from Mdm2-induced nuclear export, ubiquitination, and degradation [83, 126, 127]. The kinase activity of c-Abl is essential for the neutralization of Mdm2. Indeed, c-Abl phosphorylates Mdm2 on tyrosine 394 [128] and on tyrosine 276 [129]. The latter modification enhances interaction of Mdm2 with ARF and leads to decreased p53 turnover [129]. Prevention of phosphorylation on tyrosine 394 enhances the ability of Mdm2 to promote p53 degradation and to inhibit p53 transcriptional and apoptotic activities [128]. Intriguingly, the adjacent amino acid Ser395 is phosphorylated by ATM in response to DNA damage [130]. This phosphorylation impairs the nuclear export and degradation of p53 [131]. Interestingly, c-Abl is activated by ATM in response to DNA damage (reviewed in [132, 133]), raising the possible scenario that ATM and c-Abl may work in concert to neutralize Mdm2 under certain stress conditions (Fig. 9.4).

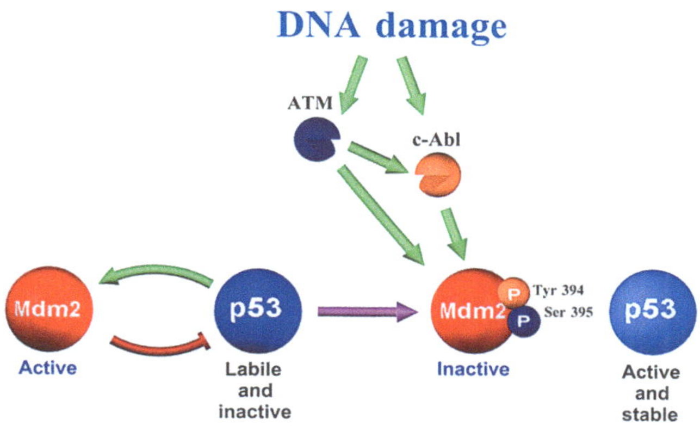

Fig. 9.4 A role for c-Abl and ATM in the protection of p53 from Mdm2 in response to DNA damage. The role for the phosphorylation of Hdm2 on Tyr394 by c-Abl and Ser395 by ATM is shown. Whether these adjacent phosphorylations have a synergistic effect is not known yet

Although Tyr394 phosphorylation occurs independently of the phosphorylation of Ser395 [128], the effect of these two kinases on the neutralization of Mdm2 maybe synergistic. Additional studies show that the complex phosphorylation of Mdm2 upon DNA damage by PI-3 kinases including ATM, ATR and DNA-PK may increase auto-ubiquitination of Mdm2 and enhance its degradation, resulting in augmented p53 abundance and transcription [100]. Additionally, it was proposed that phosphorylation of Mdm2 by ATM on multiple sites near its RING domain may disrupt Mdm2 oligomerization thus specifically suppressing p53 poly-ubiquitination [134, 135].

It is important to note that the inhibition of p53 by Mdmx is also restrained by ATM- and c-Abl-mediated phosphorylations that accelerate Mdmx degradation by Mdm2 and inhibit Mdmx binding with p53, respectively (Reviewed in [136, 28, 24, 86, 137, 138, 2]).

Stimulation of Mdm2 by the Akt/PTEN Pathway

The phosphorylation of Mdm2 can also be stimulatory as is the case with the mitogen-activated kinase, Akt. Upon growth stimulation, Mdm2 is phosphorylated by Akt on Ser166, Ser186 and Ser188, which enhance the nuclear accumulation of Mdm2 and its ability to inhibit p53 [139–141]. These phosphorylations also augment Mdm2 interaction with p300, reduce the affinity of Mdm2 for p19ARF [142] and inhibit Mdm2 self-ubiquitination [143] (Reviewed in [143–147]). Consequently, Akt stimulates the inhibition and destabilization of p53 via Mdm2. Interestingly p53 can counteract this inhibitory axis by promoting the cleavage and degradation of the Akt protein [148]. This feedback loop generates a survival signal when Akt is activated, whereas under death inducing conditions, p53 opposes survival signals by

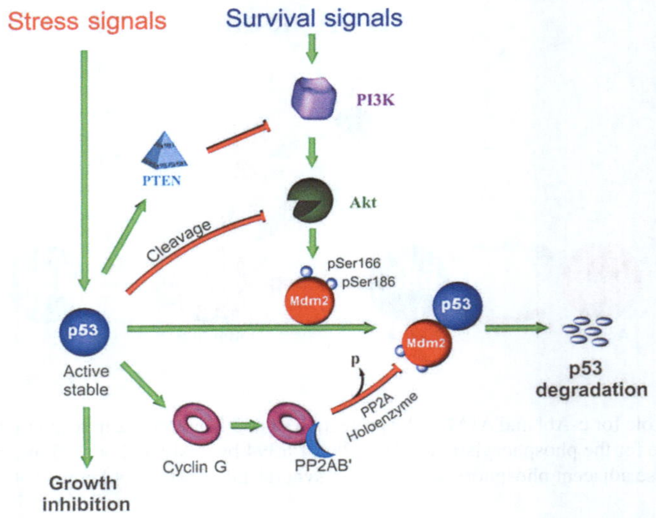

Fig. 9.5 A model for the p53/Mdm2/Akt regulatory loop. The negative regulation of p53 by Akt is induced in response to survival signals an involves the activation of Mdm2. p53 counteracts this pathways through at least 3 different loops: the cleavage of Akt, the inhibition of PI3K through PTEN, and dephosphorylation of Mdm2 on Akt sites through the induction of cyclin G and the subsequent recruitment of PP2AB'. In this pathway survival is achieved by inhibition of p53 by Akt, whereas apoptosis is achieved by counteracting Akt by p53

eliminating Akt [149] (Fig. 9.5). The regulatory loop involving p53, Mdm2 and Akt is further regulated by additional feedback loops (reviewed in [16, 150, 151]). The first involves the p53 target gene, Cyclin G, which recruits the phosphatase PP2AB' to the Mdm2/p53 complex where it de-phosphorylates Mdm2 on the Akt site Ser166 [152]. Phosphorylation of Mdm2 by cyclin A/cdk2 inhibits its activity, thus the cyclin G-PP2A phosphatase enhances Mdm2 activity and inhibits p53. Mice with the cyclin G gene knocked out are viable [153], and cyclin G null mouse embryo fibroblasts have elevated p53 protein levels in the absence of stress [152], demonstrating that this feedback loop is operational *in vivo* and acts upon the basal levels of p53 in a cell not only the higher p53 activated levels after stress.

The kinase activity of Akt is constitutively activated in human cancer as a result of dysregulation of its regulators, including loss of the tumour suppressor PTEN [154], which involves in the second loop. PTEN is a phosphatase that de-phosphorylates the Akt activating kinase, PI3-Kinase (Reviewed in [16, 155–158]). PTEN can inhibit Akt, thereby affecting the sub-cellular localization of Mdm2, PTEN can also downregulate Mdm2 and increase p53 stability [159] and finally, PTEN is a transcriptional target of p53 [160] generating another positive feedback loop. Thus, activation of p53, with the subsequent induction of effectors target genes, counteracts the survival action of Akt at multiple levels, shifting the balance towards growth inhibition (Fig. 9.5).

It was shown that the PTEN–Akt pathway participates in checkpoint control in response to DNA damage. PTEN–/– cells have high Akt activity, and are defective in checkpoint control in response to DNA damage [161] Importantly, activated Akt is able to phosphorylate Chk1 at Ser 280, thereby reduced Chk1 nuclear localization [162]. In addition, Chk1 phosphorylated at Ser 280 is located in the cytoplasm. Together, these observations indicate that loss of PTEN and subsequent activation of Akt can lead to Chk1 phosphorylation and reduced the nuclear localization of Chk1, which in turn will compromise the DNA damage response.

Overall, different stress or mitogenic signals dictates the patterns of Mdm2 phosphorylation. The summation of these phosphorylation events determines the extent by which p53 exerts its biological activities or being suppressed by Mdm2.

A Role for the Proline Rich Region of p53

The polyproline region of p53 (PPR; also referred to as the Proline-Rich-Doman, PRD; residues 62–91) consists of 5 PXXP motifs, which are partially conserved in the evolution. The regulatory role of the PPR in tumour suppression by p53 was initially noted by A. Levine and colleagues [163]. This followed a series of studies attributing various regulatory functions to the PPR. These include the induction of p53-mediated apoptosis, but notably not growth arrest [164], reflecting altered specificity of p53 for apoptotic target genes versus growth arrest promoting genes [165]. Subsequent attempts to clarify the genes that are activated both in humans and mice have led to some contradictory findings [166]. Since there was no clear functional distinction among the affected target genes, it suggests that the impact of the PPR on the transcriptional specificity of p53 does not provide the major explanation for the impaired apoptotic activity of p53ΔPro [167]. Further, since the lack of the PPR has little effect on transcription in a gene reporter assay, it is possible that the PPR may affect gene expression at the chromatin level [167]. Interestingly, the PPR is required for p53-induced apoptosis in response to chemotherapeutic treatment, and this cell death is transcriptionally-independent [168].

Searching for an additional explanation for the impaired activities of p53ΔPro, Berger et al. [169] demonstrated a link between the PPR of p53 and p53 regulation by Mdm2. p53ΔPro mutant was shown to be excessively sensitive to Mdm2-mediated ubiquitination and degradation, as well as to Mdm2-mediated inhibition of transcriptional and apoptotic activities [169]. This sensitivity of p53Δpro for inhibition by Mdm2 results from enhanced affinity of p53 to Mdm2 relative to that of wt p53 [169]. Further studies in mice confirmed the importance of the PPR of p53 for Mdm2 interactions and p53 stability [166].

This suggests that the PPR may serve as an anchor for p53 stabilizing proteins. Indeed, a region within the PPR serves as a binding site for the corepressor mSin3A. This interaction is important for transcriptional repression by p53 and for the stabilization of p53 [170]. However, this stabilization appears to be independent of Mdm2, suggesting an additional mechanism for the PPR in p53 degradation [170].

An interesting insight into this story is based on the demonstrated role for Pin1 in the regulation of p53. Pin1 is a peptidylprolyl isomerase, which converts a cis-trans configuration of a peptide bond between proline and an adjacent residue. A special case is when the preceding residue serves as a phosphorylation site, such as serine or threonine. Interestingly, one of the residues that govern the binding of Pin1 to p53 resides within the PPR, proline 82, which is affected by the phosphorylation of threonine 81 (in addition to two other residues) in response to UV light [171, 172]. Importantly, in response to Pin1 isomerization p53 is relieved from Mdm2 [171]. This may represent an additional mechanism for how the PPR of p53 influence the p53/Mdm2 loop.

Additional studies have emphasized the importance of the human p53 PPR for interactions with the ASPP family of p53 regulators, including the p53 inhibitor iASPP and its relatives ASPP, which are activators of p53. Importantly, these proteins preferentially bind to p53 codon-72 in the PPR and modulate p53 apoptotic function, through selective enhancement of pro-apoptotic gene-target activation (ie. *PIG3*). Intriguingly, the evolutionarily conserved p53 inhibitor iASPP particularly, binds more efficiently to the p53Pro72 polymorphism, which is a poorer activator of apoptosis than p53 bearing the Arg codon [173].

Modulation by Protein-Protein Interactions

Both p53 and Mdm2 form complexes with various modulators, which can be classified into two large groups. Proteins from the first group enhance specific biochemical activities of p53. For instance, Ref1 increases the DNA binding and repair activities of p53 in the presence of selenomethionine [174]. In most cases, however, the mechanisms by which members of this group activate p53 are not clear. Members of the other groups confer protection for p53 from the inhibitory effects of Mdm2. It is likely that contribution from both groups is required for maximal activation of p53. For example, the mere stabilization of p53, by proteasomal degradation, is insufficient for p53 activation. Likewise, the activation of a labile p53 protein may not provide sufficient signal for triggering growth inhibition. In accord with the topic of this chapter we shall focus here on the second group of proteins, using examples to illustrate the major mechanisms employed.

In principle, prevention of p53-Mdm2 interaction ought to be sufficient for the protection of p53 from Mdm2. The proof of principle was demonstrated by introducing to cells antibodies or peptides, directed to the interaction site in p53, and observing a reduction at the levels and activities of the p53 protein (e.g. [175]). Surprisingly, this mechanism is employed by a minority of p53 regulators or co-factors. For instance, the TAF(II)31 transcriptional co-activator of p53 competes with Mdm2 for p53 binding [176]. Other proteins impair their physical separation by imposing spatial separation (see below). Perhaps the reason why this mechanism is not widely used is because it involves a considerable risk of unscheduled activation of p53 with severe consequences. In fact, the majority of p53 co-activators that protect p53 from Mdm2 do so without interrupting their physical interaction. These include pRb, ZBP-89, ARF, Werner's syndrome protein (WRN), β-catenin, and c-Abl (reviewed in [48, 177–179]). The type of stress signals dictates, at least in part, which of the co-activators will come to action. DNA damage trigger proteins such as c-Abl to protect p53 from Mdm2, whereas deregulation of oncogenes, such as Myc and beta-catenin, trigger the ARF pathway (reviewed in [17, 180–183]). The fact that Mdm2 masks the transactivation domain of p53 makes it difficult to explain how proteins activate p53 when bound to Mdm2. It is possible that the nature of the interaction between p53 and Mdm2 is altered in a manner that alleviates the restrictive effect of Mdm2 from p53 N-terminal domain. A more trivial explanation is that upon binding of the co-activator to p53, a small pool of p53 is relieved from Mdm2 and is sufficient to perform its biological role.

The ARF Oncogenic Pathway

An important regulatory loop of p53 is that with the tumour suppressor product of the CDKN2A locus, ARF (Alternative Reading Frame) (also called $p19^{ARF}$ in mouse and $p14^{ARF}$ in human), that normally expressed at low levels in cells (reviewed in [16, 180–183]). ARF serves as the prime nodal point integrating oncogenic signals into growth inhibition through the activation of p53 [184]. This is the case when the expression of oncogenes such as c-Myc, Ras, or β-Catenin is deregulated [185, 186]. Activation of ARF leads to growth arrest, senescence or even apoptosis under certain conditions. These effects are achieved by activation of p53 through several mechanisms. ARF promotes the nucleolar localization of Mdm2, thereby spatially separating Mdm2 from nucleoplasmic p53 [184]. Also, a nucleolar-independent mechanisms has also been identified [187, 188] in which ARF directly inhibit the E3 ubiquitin ligase activity of Mdm2. Moreover, in addition to suppressing Mdm2-mediated effects on p53, ARF modulate the activity of other E3 ligases such as ARF-BP1 and ARF/ARF-BP1 interaction was found to be involved in both p53-dependent and p53-independent functions of ARF [52]. ARF also enhances p53 function by promoting the phosphorylation and inhibiting the transcriptional activity of the RelA NF-kB subunit. The NF-κB family of transcription factors display anti-apoptotic activity and antagonize the p53 pathway through induction of Mdm2 and repression of p53. Thus, by counteracting the functions of Rel A, ARF increases the effectiveness of the p53 pathway [189].

Because ARF is also activated by mitogenic signals it is imperative that unscheduled activation of p53 under growth promoting conditions will be avoided. One compensatory mechanism was demonstrated for the Ras-Raf growth promoting pathway. Activation of p53 through the Ras-Raf-ARF axis is counteracted by the parallel induction of Mdm2 [190]. Failure of the p53-ARF axis is a common event in most human cancer. Generally, mutations in both genes within the same tumour are not common events [15]. However, at least in certain tissues or cell types the p53 and ARF pathways may function independently (reviewed in [191, 180, 182]).

Mice lacking ARF are highly prone to tumour development [192], underscoring the role of ARF in tumour suppression in mice. In humans, however, mutations at the CDK2A locus (which encodes INK4A, also known as p16, and ARF in overlapping reading frames) target mainly INK4A (p16) and rarely target ARF [193, 194] suggesting that ARF may be less crucial to tumour suppression in humans. Two studies concluded that p53 did not have a tumour suppressor function in ARF-null mice [195, 196]. Using a knock-in mouse that expresses a wild-type p53–oestrogen receptor fusion protein ($p53ER^{TAM}$) which is dependent upon 4-hydroxytamoxifen (4-OHT) for activity, Christophorou and colleagues [195] showed that the restoration of p53 function 6 days before administering a single whole-body dose of ionizing radiation led to widespread p53-dependent cell death in radiosensitive tissues in a manner similar to that observed in wild-type mice. However, although there was a substantial p53 response, it provided no protection against the subsequent onset of lymphoma development. By contrast, when p53 function was absent during irradiation but was restored for a 6-day period 8 days after administering the radiation, when precancrerous cells are presumbly present, a significant level of protection from tumour formation was observed. Notably, this acquired protection was lost when the mice were crossed onto an ARF-null background. A similar conclusion has been reached by Serrano's group [196], who studied the role of ARF in tumour suppression in transgenic mice that expressed an additional copy of *Trp53* (known as $p53^{super}$ mice), which is known to provide added protection against the development of cancer [197]. They showed that wild-type and $p53^{super}$ mice, regardless of whether they are in an ARF-competent or ARF-null background, respond normally to DNA damage as measured by the number of apoptotic thymocytes detected following a high dose of ionizing radiation. However, although the $p53^{super}$ mice have extra protection against spontaneous and drug-induced tumour development, they are not protected in the absence of ARF. Moreover, when MEFs from these animals were used in a two-oncogene focus assay, focus formation was detectable only in the absence of either p53 or ARF, suggesting that ARF is required to suppress the transformed phenotype arising from oncogene expression.

The Spatial Distribution Mode of Regulation

The Nuclear Cytoplasmic Boundary

It has long been shown that p53 is a dynamic protein being shuttled between the nuclear and the cytoplasm in a cell-cycle-dependent manner (reviewed in [198, 199]). Deregulation of p53, by mutations or elevation of inhibitory proteins, such as Mdm2 and HPV-E6, can bias the shuttle towards the cytoplasm. Whereas stress conditions, on the other hand, promote the nuclear accumulation of p53.

Many studies demonstrated that the nuclear versus cytoplasmic effects of p53 are determined by multiple post-translational modifications that affect its interaction with other proteins, it's shuttling between the cytoplasm and the nucleus and its biological activities. Poly(ADP)ribosylation of p53 leads to its nuclear accumulation [200]. In contrast, monoubiquitylation by Mdm2 stimulates the nuclear export of p53, which on arrival at mitochondria is deubiquitylated by mitochondrial HAUSP, thus generating the apoptotically active non-ubiquitylated p53 [75]. Other post-translational modifications of p53 (such as phosphorylation and sumoylation of carboxy-terminal serines and lysines) can stimulate nuclear export and/or mitochondrial association. Moreover, the transcription factor Foxo3a (Foxo3) promotes p53 cytoplasmic accumulation by increasing its nuclear export [201]. This indicates that the entire context of post-transcriptional p53 modifications and protein interactions can affect the precise subcellular localization and function of p53.

Many studies have indicated that p53 play active roles in the cytoplasm, such as direct signalling at the mitochondria and the induction of apoptosis [75]. Cytoplasmic p53 can localize to the mitochondria, and induce apoptosis via interactions with antiapooptotic members of the Bcl family such as Bcl-XL and Bcl2 resulting in the permeabilization of the outer mitochondrial membrane, the release of cytochrome c and other apoptotic activators from the mitochondria [202–204]. In addition, p53 can interact with the proapoptotic factor Bak, releasing it from the negative inhibition of the anti-apoptotic Bcl2-family member Mcl1 [205]. The pro apoptotic effects of cytoplasmic p53 are not dependent on transcription, in principle. However, the control of transcription by nuclear p53 contributes to the function of cytoplasmic p53. For instance, p53 target protein, PUMA controls the sequestration of cytoplasmic p53 by the anti apoptotic Bcl-XL protein, releasing p53 to activate Bax [206]. Understanding the extra-nuclear activities of p53 will likewise furnish new opportunities to pharmacologically modulate the p53 system.

PML Nuclear Bodies as a Regulatory Junction

A dynamic redistribution of p53 within the nucleus may provide a mean by which p53 regulation is coordinated in response to a given stress signal. This transportation of p53 within the nucleus is mediated by the promyelocytic leukemia protein (PML), which forms small structures termed PML nuclear bodies (PML-NB). These nuclear structures increase in numbers and size when cells are exposed to stress, such as γIR or deregulation of oncogene expression (reviewed in [207, 208]). Several lines of evidence strongly link p53 with the PML-NBs. First, p53 is recruited to these structures in response to IR or UV light, and Ras activation [209–212]. Second, p53 interacts with PML (isoform IV), an interaction that is required for the activation of p53 by PML [210, 213]. Third, PML is critical for the activation of p53 in response to stress, and for p53-dependent apoptosis [213]. Indeed, p53 transcriptional activity is impaired in PML-null primary cells, and these mice are radioresistant, even to lethal doses of γIR [213]. Further, p53 activity is compromised in acute promyelocytic leukaemia, which explains the low frequency of p53 mutations in this form of cancer [214]. Fourth, PML itself is a p53 target gene [215], suggesting a positive feedback loop between the two proteins.

Fifth, a growing list of p53 regulators has been demonstrated to be localized to the PML-NBs. In response to oncogenic Ras activation, p53 is co-localized along with the CBP acetyltransferase into the PML-NBs, inducing the formation of a trimeric p53-PML-CBP complex and the acetylation of p53 on lys 382, leading to p53-induced senescence [212]. Interestingly, the histone deacetylase SIRT1 can also localizes into the PML-NBs and reduce the acetylation of p53 at Lys382 and thereby antagonize the induction of p53 activity [216, 217]. In addition to CBP co-localization in the PML-NBs, it was shown that UV irradiation also induces the accumulation of another acetyl transferase into the PML-NBs, TIP60 [218], which induces p53-dependent apoptosis by acetylating p53 at Lys120 [219, 220]. Apart from regulating p53 acetylation, PML was shown to promote the Ser46 phosphorylation of p53 by HIPK2 [221], a modification that increases p53-mediated apoptosis in response to UV light exposure [222, 223]. All together, these findings raise the attractive hypothesis that the PML-NBs may serve not only as a meeting junction for p53 and its regulators in response to DNA damage, but also could block the physical interaction between p53 and Mdm2. Indeed, PML can promote the phosphorylation of p53 on Thr18 and Ser20, mediated by Chk2 and CK1, respectively [224, 225]. These residues resides within the N′-terminal transactivation domain of p53 and are known to be the most important in attenuating the p53-Mdm2 interaction [98, 226].

Interestingly, CK1 can directly phosphorylate Mdm2 in its acidic domain, a modification that can further weaken the p53-Mdm2 interaction [227]. Likewise, while recruited into the PML-NBs in response to stress, Chk2 can also phosphorylate PML itself and activate its apoptotic function [228]. In addition, it was published that PML can facilitate Chk2 autophosphorylation and activation [229], revealing even more complex interaction between PML and Chk2.

Finally, PML can also regulate Mdm2. It was shown that PML can bind Mdm2 [230, 231], by that protect p53 from Mdm2 mediated degradation through additional mechanisms, involving a physical inhibition of there interaction by forming a trimeric PML-p53-Mdm2 complex [232] or by sequestrating Mdm2 to the nucleoli upon DNA damage [233]. PML can also affect Mdm2 expression indirectly, for example by inhibiting the eIF4E dependent mRNA export of Mdm2, leading to reduced levels of Mdm2 protein [234]. Interestingly, p53 overexpression results in transcriptional repression of eIF4E [235]. Consistently, Mdm2 overexpression, leads to reduced p53 and increased eIF4E levels [235]. This might suggest the existence of a feedback loop between eIF4E, Mdm2, p53 and PML. Another example involved the transcriptional repressor Daxx, which is also associates with the PML-NBs, may also play a role in the PML-NBs/p53- induced apoptosis [236]. It was shown that Daxx association with Axin, can increase the phosphorylation of p53 at Ser46, mediated by HIPK2 [237]. In addition, under unstressed conditions, Daxx can form a ternary complex with Mdm2 and the deubiquitinating enzyme HAUSP, enhancing Mdm2 stability. Upon DNA damage, this complex is dissociated, resulting in the degradation of Mdm2, which in turn stabilizes p53 [238, 239].

Thus, several competing p53 regulation mechanisms converge at PML NBs and it is becoming clear that a lot of this action involves PML directly and also is attributed to the recruitment of key p53 modulators. It is likely that many other modulators are

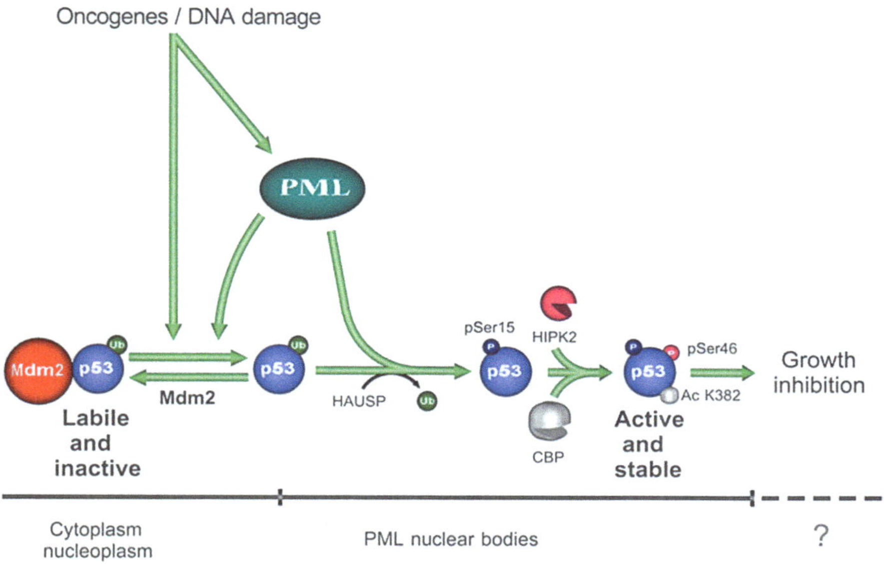

Fig. 9.6 A model for the role of PML in the integrated regulation of p53 in the PML-NBs. In response to oncogenic activation or DNA damage PML is activated by SUMOylation (s) and recruits p53 into the PML-NBs. Consequently, p53 undergoes series of modifications, including acetylation (Ac) by CBP, phosphorylation on S46 by HIPK2 and de-ubiquitination by HAUSP. In addition, PML protects p53 from Mdm2 mediated degradation. These effects of PML lead to the activation and stabilization of p53. It is not clear yet whether these various modifications occur within the PML-NBs

involved and additional PML-dependent modifications of p53. The identification of this link between p53 and the PML-NBs shed new light on how the complex network of p53 regulation may be coordinated in a timely manner within the nucleus (Fig. 9.6).

Acknowledgment Due to space limitations many original important studies have not been cited directly but rather through reviews. We are grateful to Mati Goldberg for drawing the illustrations. Work in the authors laboratory is supported by NHMRC project grants (1026990, 1026988, 1049179, 1063389) and NHMRC Fellowship to YH (9628426), by the VESKI award, and by the CASS Foundation.

References

1. Zuckerman V et al (2009) Tumour suppression by p53: the importance of apoptosis and cellular senescence. J Pathol 219(1):3–15
2. Wade M, Li YC, Wahl GM (2013) MDM2, MDMX and p53 in oncogenesis and cancer therapy. Nat Rev Cancer 13(2):83–96
3. Berkers CR et al (2013) Metabolic regulation by p53 family members. Cell Metab 18(5):617–633

4. Wang J, Yang J (2009) Interaction of tumor suppressor p53 with DNA and proteins. Curr Pharm Biotechnol 11(1):122–127
5. Gu B, Zhu WG (2012) Surf the post-translational modification network of p53 regulation. Int J Biol Sci 8(5):672–684
6. He G et al (2014) AMP-activated protein kinase induces p53 by phosphorylating MDMX and inhibiting its activity. Mol Cell Biol 34(2):148–157
7. Appella E, Anderson CW (2000) Signaling to p53: breaking the posttranslational modification code. Pathol Biol (Paris) 48(3):227–245
8. Meek DW, Anderson CW (2009) Posttranslational modification of p53: cooperative integrators of function. Cold Spring Harb Perspect Biol 1(6):a000950
9. Taira N, Yoshida K (2012) Post-translational modifications of p53 tumor suppressor: determinants of its functional targets. Histol Histopathol 27(4):437–443
10. Levine AJ, Oren M (2009) The first 30 years of p53: growing ever more complex. Nat Rev Cancer 9(10):749–758
11. Rivlin N, Koifman G, Rotter V (2014) p53 orchestrates between normal differentiation and cancer. Semin Cancer Biol. doi: 10.1016/j.semcancer.2013.12.006. [Epub ahead of print]
12. Barak Y et al (1993) mdm2 expression is induced by wild type p53 activity. EMBO J 12(2):461–468
13. Perry ME, Levine AJ (1993) Tumor-suppressor p53 and the cell cycle. Curr Opin Genet Dev 3(1):50–54
14. Li Q, Lozano G (2013) Molecular pathways: targeting Mdm2 and Mdm4 in cancer therapy. Clin Cancer Res 19(1):34–41
15. Vogelstein B, Lane D, Levine AJ (2000) Surfing the p53 network. Nature 408(6810):307–310
16. Harris SL, Levine AJ (2005) The p53 pathway: positive and negative feedback loops. Oncogene 24(17):2899–2908
17. Horn HF, Vousden KH (2007) Coping with stress: multiple ways to activate p53. Oncogene 26(9):1306–1316
18. Lavin MF, Gueven N (2006) The complexity of p53 stabilization and activation. Cell Death Differ 13(6):941–950
19. Haupt Y (2004) p53 regulation: a family affair. Cell Cycle 3(7):884–885
20. Bond GL, Hu W, Levine AJ (2005) MDM2 is a central node in the p53 pathway: 12 years and counting. Curr Cancer Drug Targets 5(1):3–8
21. Hock AK, Vousden KH (2014) The role of ubiquitin modification in the regulation of p53. Biochim Biophys Acta 1843(1):137–149
22. Haupt Y et al (1997) Mdm2 promotes the rapid degradation of p53. Nature 387(6630):296–299
23. Kubbutat MH, Jones SN, Vousden KH (1997) Regulation of p53 stability by Mdm2. Nature 387(6630):299–303
24. Marine JC, Lozano G (2010) Mdm2-mediated ubiquitylation: p53 and beyond. Cell Death Differ 17(1):93–102
25. Brooks CL, Gu W (2011) p53 regulation by ubiquitin. FEBS Lett 585(18):2803–2809
26. Satija YK, Bhardwaj A, Das S (2013) A portrayal of E3 ubiquitin ligases and deubiquitylases in cancer. Int J Cancer 133(12):2759–2768
27. Momand J, Wu HH, Dasgupta G (2000) MDM2 – master regulator of the p53 tumor suppressor protein. Gene 242(1–2):15–29
28. Marine JC (2010) Pharmacological rescue of p53 in cancer therapy: widening the sensitive tumor spectrum by targeting MDMX. Cancer Cell 18(5):399–400
29. Bond GL et al (2004) A single nucleotide polymorphism in the MDM2 promoter attenuates the p53 tumor suppressor pathway and accelerates tumor formation in humans. Cell 119(5):591–602
30. Bond GL, Hu W, Levine A (2005) A single nucleotide polymorphism in the MDM2 gene: from a molecular and cellular explanation to clinical effect. Cancer Res 65(13):5481–5484
31. Levine AJ, Hu W, Feng Z (2006) The P53 pathway: what questions remain to be explored? Cell Death Differ 13(6):1027–1036
32. Mendrysa SM et al (2003) mdm2 is critical for inhibition of p53 during lymphopoiesis and the response to ionizing irradiation. Mol Cell Biol 23(2):462–472

33. Lozano G (2010) Mouse models of p53 functions. Cold Spring Harb Perspect Biol 2(4):a001115
34. Juven-Gershon T, Oren M (1999) Mdm2: the ups and downs. Mol Med 5(2):71–83
35. Jones SN et al (1995) Rescue of embryonic lethality in Mdm2-deficient mice by absence of p53. Nature 378(6553):206–208
36. Leveillard T et al (1998) MDM2 expression during mouse embryogenesis and the requirement of p53. Mech Dev 74(1–2):189–193
37. Parant J et al (2001) Rescue of embryonic lethality in Mdm4-null mice by loss of Trp53 suggests a nonoverlapping pathway with MDM2 to regulate p53. Nat Genet 29(1):92–95
38. Finch RA et al (2002) mdmx is a negative regulator of p53 activity *in vivo*. Cancer Res 62(11):3221–3225
39. Migliorini D et al (2002) Mdm4 (Mdmx) regulates p53-induced growth arrest and neuronal cell death during early embryonic mouse development. Mol Cell Biol 22(15):5527–5538
40. Marine JC, Jochemsen AG (2004) Mdmx and Mdm2: brothers in arms? Cell Cycle 3(7):900–904
41. Toledo F, Wahl GM (2006) Regulating the p53 pathway: *in vitro* hypotheses, *in vivo* veritas. Nat Rev Cancer 6(12):909–923
42. Laurie NA et al (2006) Inactivation of the p53 pathway in retinoblastoma. Nature 444(7115):61–66
43. Lenos K, Jochemsen AG (2011) Functions of MDMX in the modulation of the p53-response. J Biomed Biotechnol 2011:876173
44. Jochemsen AG (2012) Mdmx: a p53 activator? Cell Cycle 11(5):843
45. Wade M, Wahl GM (2009) Targeting Mdm2 and Mdmx in cancer therapy: better living through medicinal chemistry? Mol Cancer Res 7(1):1–11
46. Honda R, Yasuda H (2000) Activity of MDM2, a ubiquitin ligase, toward p53 or itself is dependent on the RING finger domain of the ligase. Oncogene 19(11):1473–1476
47. Fang S et al (2000) Mdm2 is a RING finger-dependent ubiquitin protein ligase for itself and p53. J Biol Chem 275(12):8945–8951
48. Woods DB, Vousden KH (2001) Regulation of p53 function. Exp Cell Res 264(1):56–66
49. Ringshausen I et al (2006) Mdm2 is critically and continuously required to suppress lethal p53 activity *in vivo*. Cancer Cell 10(6):501–514
50. Leng RP et al (2003) Pirh2, a p53-induced ubiquitin-protein ligase, promotes p53 degradation. Cell 112(6):779–791
51. Dornan D et al (2004) The ubiquitin ligase COP1 is a critical negative regulator of p53. Nature 429(6987):86–92
52. Chen D et al (2005) ARF-BP1/Mule is a critical mediator of the ARF tumor suppressor. Cell 121(7):1071–1083
53. Boutell C, Everett RD (2003) The herpes simplex virus type 1 (HSV-1) regulatory protein ICP0 interacts with and Ubiquitinates p53. J Biol Chem 278(38):36596–36602
54. Rajendra R et al (2004) Topors functions as an E3 ubiquitin ligase with specific E2 enzymes and ubiquitinates p53. J Biol Chem 279(35):36440–36444
55. Esser C, Scheffner M, Hohfeld J (2005) The chaperone-associated ubiquitin ligase CHIP is able to target p53 for proteasomal degradation. J Biol Chem 280(29):27443–27448
56. Laine A et al (2006) Regulation of p53 localization and activity by Ubc13. Mol Cell Biol 26(23):8901–8913
57. Yamasaki S et al (2007) Cytoplasmic destruction of p53 by the endoplasmic reticulum-resident ubiquitin ligase ‘Synoviolin’. EMBO J 26(1):113–122
58. Le Cam L et al (2006) E4F1 is an atypical ubiquitin ligase that modulates p53 effector functions independently of degradation. Cell 127(4):14
59. Yang W et al (2007) CARPs are ubiquitin ligases that promote MDM2-independent p53 and phospho-p53ser20 degradation. J Biol Chem 282(5):3273–3281
60. Laine A, Ronai Z (2007) Regulation of p53 localization and transcription by the HECT domain E3 ligase WWP1. Oncogene 26(10):1477–1483
61. Kruse JP, Gu W (2009) MSL2 promotes Mdm2-independent cytoplasmic localization of p53. J Biol Chem 284(5):3250–3263

62. Lee JT, Gu W (2010) The multiple levels of regulation by p53 ubiquitination. Cell Death Differ 17(1):86–92
63. Itahana K et al (2007) Targeted inactivation of Mdm2 RING finger E3 ubiquitin ligase activity in the mouse reveals mechanistic insights into p53 regulation. Cancer Cell 12(4):355–366
64. Kobet E et al (2000) MDM2 inhibits p300-mediated p53 acetylation and activation by forming a ternary complex with the two proteins. Proc Natl Acad Sci U S A 97(23):12547–12552
65. Ito A et al (2001) p300/CBP-mediated p53 acetylation is commonly induced by p53-activating agents and inhibited by MDM2. EMBO J 20(6):1331–1340
66. Jin Y et al (2004) MDM2 mediates p300/CREB-binding protein-associated factor ubiquitination and degradation. J Biol Chem 279(19):20035–20043
67. Ito A et al (2002) MDM2-HDAC1-mediated deacetylation of p53 is required for its degradation. EMBO J 21(22):6236–6245
68. Wang C et al (2005) MDM2 interaction with nuclear corepressor KAP1 contributes to p53 inactivation. EMBO J 24(18):3279–3290
69. Xirodimas DP et al (2004) Mdm2-mediated NEDD8 conjugation of p53 inhibits its transcriptional activity. Cell 118(1):83–97
70. Minsky N, Oren M (2004) The RING domain of Mdm2 mediates histone ubiquitylation and transcriptional repression. Mol Cell 16(4):631–639
71. Lai Z et al (2001) Human mdm2 mediates multiple mono-ubiquitination of p53 by a mechanism requiring enzyme isomerization. J Biol Chem 276(33):31357–31367
72. Carter S et al (2007) C-terminal modifications regulate MDM2 dissociation and nuclear export of p53. Nat Cell Biol 9(4):428–435
73. Li M et al (2003) Mono- versus polyubiquitination: differential control of p53 fate by Mdm2. Science 302(5652):1972–1975
74. Joseph TW, Zaika A, Moll UM (2003) Nuclear and cytoplasmic degradation of endogenous p53 and HDM2 occurs during down-regulation of the p53 response after multiple types of DNA damage. FASEB J 17(12):1622–1630
75. Marchenko ND et al (2007) Monoubiquitylation promotes mitochondrial p53 translocation. EMBO J 26(4):923–934
76. Lohrum MA et al (2001) C-terminal ubiquitination of p53 contributes to nuclear export. Mol Cell Biol 21(24):8521–8532
77. Feng L et al (2005) Functional analysis of the roles of posttranslational modifications at the p53 C terminus in regulating p53 stability and activity. Mol Cell Biol 25(13):5389–5395
78. Krummel KA et al (2005) The C-terminal lysines fine-tune P53 stress responses in a mouse model but are not required for stability control or transactivation. Proc Natl Acad Sci U S A 102(29):10188–10193
79. Chan WM et al (2006) Ubiquitination of p53 at multiple sites in the DNA-binding domain. Mol Cancer Res 4(1):15–25
80. Yin Y et al (2002) p53 Stability and activity is regulated by Mdm2-mediated induction of alternative p53 translation products. Nat Cell Biol 4(6):462–467
81. Inoue T et al (2001) MDM2 can promote the ubiquitination, nuclear export, and degradation of p53 in the absence of direct binding. J Biol Chem 276(48):45255–45260
82. Yap DB, Hsieh JK, Lu X (2000) Mdm2 inhibits the apoptotic function of p53 mainly by targeting it for degradation. J Biol Chem 275(47):37296–37302
83. Unger T et al (1999) Critical role for Ser20 of human p53 in the negative regulation of p53 by Mdm2. EMBO J 18(7):1805–1814
84. Marine JC et al (2006) Keeping p53 in check: essential and synergistic functions of Mdm2 and Mdm4. Cell Death Differ 13(6):927–934
85. Clegg HV, Itahana K, Zhang Y (2008) Unlocking the Mdm2-p53 loop: ubiquitin is the key. Cell Cycle 7(3):287–292
86. Wade M, Wang YV, Wahl GM (2010) The p53 orchestra: Mdm2 and Mdmx set the tone. Trends Cell Biol 20(5):299–309
87. Stommel JM, Wahl GM (2005) A new twist in the feedback loop: stress-activated MDM2 destabilization is required for p53 activation. Cell Cycle 4(3):411–417

88. Bode AM, Dong Z (2004) Post-translational modification of p53 in tumorigenesis. Nat Rev Cancer 4(10):793–805
89. Kruse JP, Gu W (2008) SnapShot: p53 posttranslational modifications. Cell 133(5):930–930. e1
90. Olsson A et al (2007) How important are post-translational modifications in p53 for selectivity in target-gene transcription and tumour suppression? Cell Death Differ 14(9):1561–1575
91. Bean LJ, Stark GR (2002) Regulation of the accumulation and function of p53 by phosphorylation of two residues within the domain that binds to Mdm2. J Biol Chem 277(3):1864–1871
92. Sakaguchi K et al (2000) Damage-mediated phosphorylation of human p53 threonine 18 through a cascade mediated by a casein 1-like kinase. Effect on Mdm2 binding. J Biol Chem 275(13):9278–9283
93. Banin S et al (1998) Enhanced phosphorylation of p53 by ATM in response to DNA damage. Science 281(5383):1674–1677
94. Canman CE, Lim DS (1998) The role of ATM in DNA damage responses and cancer. Oncogene 17(25):3301–3308
95. Khanna KK et al (1998) ATM associates with and phosphorylates p53: mapping the region of interaction. Nat Genet 20(4):398–400
96. Lakin ND, Hann BC, Jackson SP (1999) The ataxia-telangiectasia related protein ATR mediates DNA-dependent phosphorylation of p53. Oncogene 18(27):3989–3995
97. Berger M et al (2005) Mutations in proline 82 of p53 impair its activation by Pin1 and Chk2 in response to DNA damage. Mol Cell Biol 25(13):5380–5388
98. Bartek J, Falck J, Lukas J (2001) CHK2 kinase – a busy messenger. Nat Rev Mol Cell Biol 2(12):877–886
99. Meulmeester E et al (2005) ATM-mediated phosphorylations inhibit Mdmx/Mdm2 stabilization by HAUSP in favor of p53 activation. Cell Cycle 4(9):1166–1170
100. Stommel JM, Wahl GM (2004) Accelerated MDM2 auto-degradation induced by DNA-damage kinases is required for p53 activation. EMBO J 23(7):1547–1556
101. Dornan D, Hupp TR (2001) Inhibition of p53-dependent transcription by BOX-I phosphopeptide mimetics that bind to p300. EMBO Rep 2(2):139–144
102. Dumaz N, Meek DW (1999) Serine15 phosphorylation stimulates p53 transactivation but does not directly influence interaction with HDM2. EMBO J 18(24):7002–7010
103. Feng H et al (2009) Structural basis for p300 Taz2-p53 TAD1 binding and modulation by phosphorylation. Structure 17(2):202–210
104. Jenkins LM et al (2009) Two distinct motifs within the p53 transactivation domain bind to the Taz2 domain of p300 and are differentially affected by phosphorylation. Biochemistry 48(6):1244–1255
105. Lambert PF et al (1998) Phosphorylation of p53 serine 15 increases interaction with CBP. J Biol Chem 273(49):33048–33053
106. Lee CW et al (2009) Mapping the interactions of the p53 transactivation domain with the KIX domain of CBP. Biochemistry 48(10):2115–2124
107. Teufel DP, Bycroft M, Fersht AR (2009) Regulation by phosphorylation of the relative affinities of the N-terminal transactivation domains of p53 for p300 domains and Mdm2. Oncogene 28(20):2112–2118
108. Sakaguchi K et al (1998) DNA damage activates p53 through a phosphorylation-acetylation cascade. Genes Dev 12(18):2831–2841
109. MacPherson D et al (2004) Defective apoptosis and B-cell lymphomas in mice with p53 point mutation at Ser 23. EMBO J 23(18):3689–3699
110. Wu Z et al (2002) Mutation of mouse p53 Ser23 and the response to DNA damage. Mol Cell Biol 22(8):2441–2449
111. Takai H et al (2002) Chk2-deficient mice exhibit radioresistance and defective p53-mediated transcription. EMBO J 21(19):5195–5205
112. Maclaine NJ, Hupp TR (2009) The regulation of p53 by phosphorylation: a model for how distinct signals integrate into the p53 pathway. Aging (Albany, NY) 1(5):12

113. Vousden KH, Prives C (2009) Blinded by the light: the growing complexity of p53. Cell 137(3):413–431
114. Falck J et al (2001) Functional impact of concomitant versus alternative defects in the Chk2-p53 tumour suppressor pathway. Oncogene 20(39):5503–5510
115. Bartek J, Bartkova J, Lukas J (2007) DNA damage signalling guards against activated oncogenes and tumour progression. Oncogene 26(56):7773–7779
116. Chao C et al (2003) Cell type- and promoter-specific roles of Ser18 phosphorylation in regulating p53 responses. J Biol Chem 278(42):41028–41033
117. Sluss HK et al (2004) Phosphorylation of serine 18 regulates distinct p53 functions in mice. Mol Cell Biol 24(3):976–984
118. Chao C et al (2006) Ser18 and 23 phosphorylation is required for p53-dependent apoptosis and tumor suppression. EMBO J 25(11):2615–2622
119. Hay TJ, Meek DW (2000) Multiple sites of *in vivo* phosphorylation in the MDM2 oncoprotein cluster within two important functional domains. FEBS Lett 478(1–2):183–186
120. Blattner C et al (2002) Hypophosphorylation of Mdm2 augments p53 stability. Mol Cell Biol 22(17):6170–6182
121. Meek DW, Hupp TR (2010) The regulation of MDM2 by multisite phosphorylation-opportunities for molecular-based intervention to target tumours? Semin Cancer Biol 20(1):19–28
122. Kruse JP, Gu W (2009) Modes of p53 regulation. Cell 137(4):609–622
123. Meek DW, Hupp TR (2010) The regulation of MDM2 by multisite phosphorylation – opportunities for molecular-based intervention to target tumours? Semin Cancer Biol 20(1):19–28
124. Mayo LD, Turchi JJ, Berberich SJ (1997) Mdm-2 phosphorylation by DNA-dependent protein kinase prevents interaction with p53. Cancer Res 57(22):5013–5016
125. Zhang Y, Xiong Y (2001) A p53 amino-terminal nuclear export signal inhibited by DNA damage-induced phosphorylation. Science 292(5523):1910–1915
126. Sionov RV et al (2001) c-Abl regulates p53 levels under normal and stress conditions by preventing its nuclear export and ubiquitination. Mol Cell Biol 21(17):5869–5878
127. Levav-Cohen Y et al (2005) C-Abl as a modulator of p53. Biochem Biophys Res Commun 331(3):737–749
128. Goldberg Z et al (2002) Tyrosine phosphorylation of Mdm2 by c-Abl: implications for p53 regulation. EMBO J 21(14):3715–3727
129. Dias SS, Milne DM, Meek DW (2006) c-Abl phosphorylates Hdm2 at tyrosine 276 in response to DNA damage and regulates interaction with ARF. Oncogene 25(50):6666–6671
130. Khosravi R et al (1999) Rapid ATM-dependent phosphorylation of MDM2 precedes p53 accumulation in response to DNA damage. Proc Natl Acad Sci U S A 96(26):14973–14977
131. Maya R et al (2001) ATM-dependent phosphorylation of Mdm2 on serine 395: role in p53 activation by DNA damage. Genes Dev 15(9):1067–1077
132. Kharbanda S et al (1998) Determination of cell fate by c-Abl activation in the response to DNA damage. Oncogene 17(25):3309–3318
133. Zhu J, Wang JY (2004) Death by Abl: a matter of location. Curr Top Dev Biol 59:165–192
134. Cheng Q et al (2009) ATM activates p53 by regulating MDM2 oligomerization and E3 processivity. EMBO J 28(24):3857–3867
135. Cheng Q, Chen J (2010) Mechanism of p53 stabilization by ATM after DNA damage. Cell Cycle 9(3):472–478
136. Wang X, Jiang X (2012) Mdm2 and MdmX partner to regulate p53. FEBS Lett 586(10):1390–1396
137. Pei D, Zhang Y, Zheng J (2012) Regulation of p53: a collaboration between Mdm2 and Mdmx. Oncotarget 3(3):228–235
138. Shadfan M, Lopez-Pajares V, Yuan ZM (2012) MDM2 and MDMX: Alone and together in regulation of p53. Transl Cancer Res 1(2):88–89
139. Mayo LD, Donner DB (2001) A phosphatidylinositol 3-kinase/Akt pathway promotes translocation of Mdm2 from the cytoplasm to the nucleus. Proc Natl Acad Sci U S A 98(20):11598–11603
140. Mayo LD et al (2002) PTEN protects p53 from Mdm2 and sensitizes cancer cells to chemotherapy. J Biol Chem 277(7):5484–5489

141. Ogawara Y et al (2002) Akt enhances Mdm2-mediated ubiquitination and degradation of p53. J Biol Chem 277(24):21843–21850
142. Zhou BB, Elledge SJ (2000) The DNA damage response: putting checkpoints in perspective. Nature 408(6811):433–439
143. Feng J et al (2004) Stabilization of Mdm2 via decreased ubiquitination is mediated by protein kinase B/Akt-dependent phosphorylation. J Biol Chem 279(34):35510–35517
144. Meek DW, Knippschild U (2003) Posttranslational modification of MDM2. Mol Cancer Res 1(14):1017–1026
145. Testa JR, Bellacosa A (2001) AKT plays a central role in tumorigenesis. Proc Natl Acad Sci U S A 98(20):10983–10985
146. Song MS et al (2008) The tumour suppressor RASSF1A promotes MDM2 self-ubiquitination by disrupting the MDM2-DAXX-HAUSP complex. EMBO J 27(13):1863–1874
147. Levav-Cohen Y, Haupt S, Haupt Y (2005) Mdm2 in growth signaling and cancer. Growth Factors 23(3):183–192
148. Gottlieb TM et al (2002) Cross-talk between Akt, p53 and Mdm2: possible implications for the regulation of apoptosis. Oncogene 21(8):1299–1303
149. Oren M et al (2002) Regulation of p53: intricate loops and delicate balances. Ann N Y Acad Sci 973:374–383
150. Nag S et al (2013) The MDM2-p53 pathway revisited. J Biomed Res 27(4):254–271
151. Fahraeus R, Olivares-Illana V (2013) MDM2's social network. Oncogene. doi: 10.1038/onc.2013.410. [Epub ahead of print]
152. Okamoto K et al (2002) Cyclin G recruits PP2A to dephosphorylate Mdm2. Mol Cell 9(4):761–771
153. Kimura SH et al (2001) Cyclin G1 is involved in G2/M arrest in response to DNA damage and in growth control after damage recovery. Oncogene 20(25):3290–3300
154. Wu X et al (1998) The PTEN/MMAC1 tumor suppressor phosphatase functions as a negative regulator of the phosphoinositide 3-kinase/Akt pathway. Proc Natl Acad Sci U S A 95(26):15587–15591
155. Lee MH, Lozano G (2006) Regulation of the p53-MDM2 pathway by 14-3-3 sigma and other proteins. Semin Cancer Biol 16(3):225–234
156. Chalhoub N, Baker SJ (2009) PTEN and the PI3-kinase pathway in cancer. Annu Rev Pathol 4:127–150
157. Courtney KD, Corcoran RB, Engelman JA (2010) The PI3K pathway as drug target in human cancer. J Clin Oncol 28(6):9
158. Song MS, Salmena L, Pandolfi PP (2012) The functions and regulation of the PTEN tumour suppressor. Nat Rev Mol Cell Biol 13(5):283–296
159. Freeman DJ et al (2003) PTEN tumor suppressor regulates p53 protein levels and activity through phosphatase-dependent and -independent mechanisms. Cancer Cell 3(2):117–130
160. Stambolic V et al (2001) Regulation of PTEN transcription by p53. Mol Cell 8(2):317–325
161. Puc J et al (2005) Lack of PTEN sequesters CHK1 and initiates genetic instability. Cancer Cell 7(2):193–204
162. Puc J, Parsons R (2005) PTEN loss inhibits CHK1 to cause double stranded-DNA breaks in cells. Cell Cycle 4(7):927–929
163. Walker KK, Levine AJ (1996) Identification of a novel p53 functional domain that is necessary for efficient growth suppression. Proc Natl Acad Sci U S A 93(26):15335–15340
164. Sakamuro D et al (1997) The polyproline region of p53 is required to activate apoptosis but not growth arrest. Oncogene 15(8):887–898
165. Venot C et al (1998) The requirement for the p53 proline-rich functional domain for mediation of apoptosis is correlated with specific PIG3 gene transactivation and with transcriptional repression. EMBO J 17(16):4668–4679
166. Toledo F et al (2006) A mouse p53 mutant lacking the proline-rich domain rescues Mdm4 deficiency and provides insight into the Mdm2-Mdm4-p53 regulatory network. Cancer Cell 9(4):273–285
167. Zhu J et al (1999) Differential regulation of cellular target genes by p53 devoid of the PXXP motifs with impaired apoptotic activity. Oncogene 18(12):2149–2155

168. Baptiste N et al (2002) The proline-rich domain of p53 is required for cooperation with anti-neoplastic agents to promote apoptosis of tumor cells. Oncogene 21(1):9–21
169. Berger M et al (2001) A role for the polyproline domain of p53 in its regulation by Mdm2. J Biol Chem 276(6):3785–3790
170. Zilfou JT et al (2001) The corepressor mSin3a interacts with the proline-rich domain of p53 and protects p53 from proteasome-mediated degradation. Mol Cell Biol 21(12):3974–3985
171. Zacchi P et al (2002) The prolyl isomerase Pin1 reveals a mechanism to control p53 functions after genotoxic insults. Nature 419(6909):853–857
172. Zheng H et al (2002) The prolyl isomerase Pin1 is a regulator of p53 in genotoxic response. Nature 419(6909):849–853
173. Bergamaschi D et al (2006) iASPP preferentially binds p53 proline-rich region and modulates apoptotic function of codon 72-polymorphic p53. Nat Genet 38(10):1133–1141
174. Seo YR, Kelley MR, Smith ML (2002) Selenomethionine regulation of p53 by a ref1-dependent redox mechanism. Proc Natl Acad Sci U S A 99(22):14548–14553
175. Bottger A et al (1997) Design of a synthetic Mdm2-binding mini protein that activates the p53 response in vivo. Curr Biol 7(11):860–869
176. Buschmann T et al (2001) Stabilization and activation of p53 by the coactivator protein TAFII31. J Biol Chem 276(17):13852–13857
177. Bai L, Merchant JL (2001) ZBP-89 promotes growth arrest through stabilization of p53. Mol Cell Biol 21(14):4670–4683
178. Biderman L, Manley JL, Prives C (2012) Mdm2 and MdmX as regulators of gene expression. Genes Cancer 3(3–4):264–273
179. Hock A, Vousden KH (2010) Regulation of the p53 pathway by ubiquitin and related proteins. Int J Biochem Cell Biol 42(10):1618–1621
180. Matheu A, Maraver A, Serrano M (2008) The Arf/p53 pathway in cancer and aging. Cancer Res 68(15):6031–6034
181. Ozenne P et al (2010) The ARF tumor suppressor: structure, functions and status in cancer. Int J Cancer 127(10):2239–2247
182. Dominguez-Brauer C et al (2010) Tumor suppression by ARF: gatekeeper and caretaker. Cell Cycle 9(1):86–89
183. Itahana Y, Itahana K (2012) Emerging roles of mitochondrial p53 and ARF. Curr Drug Targets 13(13):1633–1640
184. Sherr CJ (2001) Parsing Ink4a/Arf: “pure” p16-null mice. Cell 106(5):531–534
185. Sherr CJ, Weber JD (2000) The ARF/p53 pathway. Curr Opin Genet Dev 10(1):94–99
186. Damalas A et al (2001) Deregulated beta-catenin induces a p53- and ARF-dependent growth arrest and cooperates with Ras in transformation. EMBO J 20(17):4912–4922
187. Llanos S et al (2001) Stabilization of p53 by p14ARF without relocation of MDM2 to the nucleolus. Nat Cell Biol 3(5):445–452
188. Honda R, Yasuda H (1999) Association of p19(ARF) with Mdm2 inhibits ubiquitin ligase activity of Mdm2 for tumor suppressor p53. EMBO J 18(1):22–27
189. Rocha S, Campbell KJ, Perkins ND (2003) p53- and Mdm2-independent repression of NF-kappa B transactivation by the ARF tumor suppressor. Mol Cell 12(1):15–25
190. Ries S et al (2000) Opposing effects of Ras on p53: transcriptional activation of mdm2 and induction of p19ARF. Cell 103(2):321–330
191. Hayon IL, Haupt Y (2002) p53: an internal investigation. Cell Cycle 1(2):111–116
192. Kamijo T et al (1997) Tumor suppression at the mouse INK4a locus mediated by the alternative reading frame product p19ARF. Cell 91(5):649–659
193. Collins CJ, Sedivy JM (2003) Involvement of the INK4a/Arf gene locus in senescence. Aging Cell 2(3):145–150
194. Ruas M, Peters G (1998) The p16INK4a/CDKN2A tumor suppressor and its relatives. Biochim Biophys Acta 1378(2):F115–F177
195. Christophorou MA et al (2006) The pathological response to DNA damage does not contribute to p53-mediated tumour suppression. Nature 443(7108):214–217

196. Efeyan A et al (2006) Tumour biology: policing of oncogene activity by p53. Nature 443(7108):159
197. Khan SH, Moritsugu J, Wahl GM (2000) Differential requirement for p19ARF in the p53-dependent arrest induced by DNA damage, microtubule disruption, and ribonucleotide depletion. Proc Natl Acad Sci U S A 97(7):3266–3271
198. Jimenez GS et al (1999) p53 regulation by post-translational modification and nuclear retention in response to diverse stresses. Oncogene 18(53):7656–7665
199. Liang SH, Clarke MF (2001) Regulation of p53 localization. Eur J Biochem 268(10):2779–2783
200. Murray-Zmijewski F, Slee EA, Lu X (2008) A complex barcode underlies the heterogeneous response of p53 to stress. Nat Rev Mol Cell Biol 9(9):702–712
201. You H, Yamamoto K, Mak TW (2006) Regulation of transactivation-independent proapoptotic activity of p53 by FOXO3a. Proc Natl Acad Sci U S A 103(24):9051–9056
202. Mihara M et al (2003) p53 has a direct apoptogenic role at the mitochondria. Mol Cell 11(3):577–590
203. Chipuk JE et al (2004) Direct activation of Bax by p53 mediates mitochondrial membrane permeabilization and apoptosis. Science 303(5660):1010–1014
204. Tomita Y et al (2006) WT p53, but not tumor-derived mutants, bind to Bcl2 via the DNA binding domain and induce mitochondrial permeabilization. J Biol Chem 281(13):8600–8606
205. Leu JI et al (2004) Mitochondrial p53 activates Bak and causes disruption of a Bak-Mcl1 complex. Nat Cell Biol 6(5):443–450
206. Chipuk JE et al (2005) PUMA couples the nuclear and cytoplasmic proapoptotic function of p53. Science 309(5741):1732–1735
207. Bernardi R, Pandolfi PP (2007) Structure, dynamics and functions of promyelocytic leukaemia nuclear bodies. Nat Rev Mol Cell Biol 8(12):1006–1016
208. de The H, Le Bras M, Lallemand-Breitenbach V (2012) The cell biology of disease: acute promyelocytic leukemia, arsenic, and PML bodies. J Cell Biol 198(1):11–21
209. Ferbeyre G et al (2000) PML is induced by oncogenic ras and promotes premature senescence. Genes Dev 14(16):2015–2027
210. Fogal V et al (2000) Regulation of p53 activity in nuclear bodies by a specific PML isoform. EMBO J 19(22):6185–6195
211. Carbone R et al (2002) PML NBs associate with the hMre11 complex and p53 at sites of irradiation induced DNA damage. Oncogene 21(11):1633–1640
212. Pearson M et al (2000) PML regulates p53 acetylation and premature senescence induced by oncogenic Ras. Nature 406(6792):207–210
213. Guo A et al (2000) The function of PML in p53-dependent apoptosis. Nat Cell Biol 2(10):730–736
214. Pearson M, Pelicci PG (2001) PML interaction with p53 and its role in apoptosis and replicative senescence. Oncogene 20(49):7250–7256
215. de Stanchina E et al (2004) PML is a direct p53 target that modulates p53 effector functions. Mol Cell 13(4):523–535
216. Langley E et al (2002) Human SIR2 deacetylates p53 and antagonizes PML/p53-induced cellular senescence. EMBO J 21(10):2383–2396
217. Vaziri H et al (2001) hSIR2(SIRT1) functions as an NAD-dependent p53 deacetylase. Cell 107(2):149–159
218. Cheng Z et al (2008) Functional characterization of TIP60 sumoylation in UV-irradiated DNA damage response. Oncogene 27(7):931–941
219. Sykes SM et al (2006) Acetylation of the p53 DNA-binding domain regulates apoptosis induction. Mol Cell 24(6):841–851
220. Tang Y et al (2006) Tip60-dependent acetylation of p53 modulates the decision between cell-cycle arrest and apoptosis. Mol Cell 24(6):827–839

221. Moller A et al (2003) PML is required for homeodomain-interacting protein kinase 2 (HIPK2)-mediated p53 phosphorylation and cell cycle arrest but is dispensable for the formation of HIPK domains. Cancer Res 63(15):4310–4314
222. D'Orazi G et al (2002) Homeodomain-interacting protein kinase-2 phosphorylates p53 at Ser 46 and mediates apoptosis. Nat Cell Biol 4(1):11–19
223. Hofmann TG et al (2002) Regulation of p53 activity by its interaction with homeodomain-interacting protein kinase-2. Nat Cell Biol 4(1):1–10
224. Louria-Hayon I et al (2003) The promyelocytic leukemia protein protects p53 from Mdm2-mediated inhibition and degradation. J Biol Chem 278(35):33134–33141
225. Alsheich-Bartok O et al (2008) PML enhances the regulation of p53 by CK1 in response to DNA damage. Oncogene 27(26):3653–3661
226. Schon O et al (2002) Molecular mechanism of the interaction between MDM2 and p53. J Mol Biol 323(3):491–501
227. Winter M et al (2004) Protein kinase CK1delta phosphorylates key sites in the acidic domain of murine double-minute clone 2 protein (MDM2) that regulate p53 turnover. Biochemistry 43(51):16356–16364
228. Yang S et al (2002) PML-dependent apoptosis after DNA damage is regulated by the checkpoint kinase hCds1/Chk2. Nat Cell Biol 4(11):865–870
229. Yang S et al (2006) Promyelocytic leukemia activates Chk2 by mediating Chk2 autophosphorylation. J Biol Chem 281(36):26645–26654
230. Wei X et al (2003) Physical and functional interactions between PML and MDM2. J Biol Chem 278(31):29288–29297
231. Zhu H, Wu L, Maki CG (2003) MDM2 and promyelocytic leukemia antagonize each other through their direct interaction with p53. J Biol Chem 278(49):49286–49292
232. Kurki S, Latonen L, Laiho M (2003) Cellular stress and DNA damage invoke temporally distinct Mdm2, p53 and PML complexes and damage-specific nuclear relocalization. J Cell Sci 116(Pt 19):3917–3925
233. Bernardi R et al (2004) PML regulates p53 stability by sequestering Mdm2 to the nucleolus. Nat Cell Biol 6(7):665–672
234. Culjkovic B et al (2006) eIF4E is a central node of an RNA regulon that governs cellular proliferation. J Cell Biol 175(3):415–426
235. Zhu N et al (2005) Transcriptional repression of the eukaryotic initiation factor 4E gene by wild type p53. Biochem Biophys Res Commun 335(4):1272–1279
236. Gostissa M et al (2004) The transcriptional repressor hDaxx potentiates p53-dependent apoptosis. J Biol Chem 279(46):48013–48023
237. Li Q et al (2007) Daxx cooperates with the Axin/HIPK2/p53 complex to induce cell death. Cancer Res 67(1):66–74
238. Cummins JM, Vogelstein B (2004) HAUSP is required for p53 destabilization. Cell Cycle 3(6):689–692
239. Tang J et al (2006) Critical role for Daxx in regulating Mdm2. Nat Cell Biol 8(8):855–862

Chapter 10
Mechanisms of Mutant p53 Stabilization in Cancer

Rebecca A. Frum and Steven R. Grossman

Abstract p53 transactivates cell cycle inhibitory, apoptosis or senescence-related genes in response to DNA damage to protect the genetic integrity of the cell. Highlighting its critical tumor suppressor functions, p53 is mutated, lost, or functionally inactivated in nearly all cancers. When mutated within its core DNA binding domain, p53's normal instability is abrogated, and oncogenic gain-of-function properties are observed accompanied by massive accumulation of steady state mutant p53 protein levels relative to the low or undetectable steady state level of wild-type (WT) p53 in normal cells. Mutation of p53 may affect its stability through a combination of mutant p53's inherent biochemical and biophysical properties as well as pathways aberrantly activated in genetically damaged cells. The increased stability of mutant p53 proteins is key to its ability to accumulate to high levels and phenotypically exhibit "gain-of-function" properties. In this chapter we will address the multifaceted ways in which intrinsic mutant p53 properties intersect with emergent properties of cancer cells to yield the stable mutant p53 phenotype.

Keywords p53 • Gain of function • Ubiquitin • Proteasome • Stability • Mdm2 • ATM • DNA damage

---

R.A. Frum • S.R. Grossman (✉)
Division of Hematology, Oncology, and Palliative Care, VCU Massey Cancer Center,
Virginia Commonwealth University, 1201 E. Marshall St., Box 980070,
Richmond, VA 23298, USA
e-mail: srgrossman@vcu.edu

S.P. Deb and S. Deb (eds.), *Mutant p53 and MDM2 in Cancer*,
Subcellular Biochemistry 85, DOI 10.1007/978-94-017-9211-0_10,

Introduction

p53 transactivates cell cycle inhibitory, apoptosis or senescence-related genes in response to DNA damage to protect the genetic integrity of the cell (reviewed in [1]). Highlighting its critical tumor suppressor functions, p53 is mutated, lost, or functionally inactivated in nearly all cancers (reviewed in [2]). When mutated within its core DNA binding domain, p53's normal instability is abrogated, and oncogenic gain-of-function properties are observed accompanied by massive accumulation of steady state mutant p53 protein levels relative to the low or undetectable steady state level of wild-type (WT) p53 in normal cells [3, 4]. Mutation of p53 may affect its stability through a combination of mutant p53's inherent biochemical and biophysical properties as well as pathways aberrantly activated in genetically damaged cells. The increased stability of mutant p53 proteins is key to its ability to accumulate to high levels and phenotypically exhibit "gain-of-function" properties (reviewed in [3]). In this chapter we will address the multifaceted ways in which intrinsic mutant p53 properties intersect with emergent properties of cancer cells to yield the stable mutant p53 phenotype.

WT p53 Stabilization and Degradation Cycle

Normal p53 Turnover in the Absence of Genotoxic Stress

Due to its growth inhibitory properties, p53 must be maintained at low levels in the absence of DNA damage, to allow normal cell cycle progression (reviewed in [5]). The Mdm2 E3 ubiquitin ligase maintains p53 instability, facilitating ubiquitination that targets it to the proteasome for degradation (reviewed in [5, 6]). In addition to Mdm2's E3 activity towards p53, p53 is also regulated by the E4 ubiquitin ligases p300/CBP that promote the polyubiquitination of p53 in the cytoplasm [7] by acting as a scaffold that assembles a polyubiquitinating complex for p53 that includes Mdm2, and accelerates the degradation of p53 by targeting it to the proteasome [8]. Mdm2 and p300/CBP, and possibly other E3/E4's [34] depending on context, thus maintain p53 as an unstable low abundance protein in the absence of genotoxic or other stress, protecting cells from p53's harmful growth suppressive or apoptotic effects when not needed.

Stabilization of WT p53 by Post-translational Modification After Genotoxic Damage

When DNA damage occurs in the cell, signaling through the DNA damage checkpoint via the ATM/ATR checkpoint kinases phosphorylates both p53 and Mdm2, and these posttranslational modifications lead to the stabilization of p53 ([9–11]; reviewed in [12]). In response to ionizing radiation, ATM has been shown to phosphorylate WT p53 on serine 15 [13, 14], and in response to UV light, ATR has also been found to phosphorylate WT p53 on serine 15 [15]. The stabilization of p53 was inhibited by mutation of serine 15 or serine 20 (a chk2 phosphorylation site; reviewed in [16]) to alanine [17, 18], suggesting this phosphorylation contributes to stabilization, though by still an unclear mechanism. In addition to modification of p53, Mdm2 is also phosphorylated after DNA damage by ATM at serine 395 [11], and this phosphorylation was suggested to contribute to the stabilization of p53 by compromising the ability of Mdm2 to promote nucleo-cytoplasmic shuttling of p53 that would normally enable p53 degradation [9]. More recently, it was found that Mdm2 forms oligomers that enable the polyubiquitination and degradation of p53 in the absence of damage [10]. Phosphorylation of Mdm2 by ATM was found to inhibit the ability of the Mdm2 RING domain to oligomerize, which suppressed the polyubiquitination of p53 [10]. The stabilization of WT p53 following DNA damage enables it to accumulate to high levels and activate a checkpoint response, inhibiting the cell cycle at the G1/S boundary via p21 induction [19] or inducing apoptosis via transactivation of downstream effectors such as Bax, PERP, PIG3, and Puma, among many other genes (reviewed in [1, 2]).

Stabilization of Mutant p53 by Physiologic Stimuli

Whereas stabilization of WT p53 enables a checkpoint response to resolve the initiating damage or induce apoptosis to protect cells from becoming tumorigenic, stabilization of mutant p53 enables it to also accumulate to high levels but to the opposite effect, promoting growth, invasive/metastatic properties, and genetic instability [3]. It has been shown that although mutant p53 is constitutively stable in cancer cells [4, 20], it does not possess inherent stability in normal cells [4], suggesting that the cancer cell provides an environment that converts mutant p53 into a more stable form.

Impact of DNA Damage, Oncogenes, Reactive Oxygen Species on Mutant p53 Stability

Given that mutant p53 is generally only stable in cancer cells, a number of studies have examined whether mutant p53 is subject to the same stability regulation by stress as WT p53, despite the usual baseline increase in stability seen for most alleles of mutant p53. Surprisingly, mutant p53 can be further stabilized by genotoxic signalling in cancer cells. Treatment with UV-C or ionizing radiation (IR) causes stabilization of WT p53, and similarly, UV-C stabilized mutant p53 stably expressed in MCF7 cells [20], and IR was shown to stabilize p53R175H protein in B cells harboring a single R175H allele [21].

The effects of oncogene activation, reactive oxygen species and the chemotherapeutic agent doxorubicin on mutant p53 were also analyzed in mouse models. Mice expressing activated Ras were crossed to mice expressing the mutant p53R172H protein [21]. Western blot analysis of p53 levels from these mice showed that the level of mutant p53 was higher in the lungs of K-Ras/p53R172H mice than in p53R172H mice that did not harbor activated K-Ras, suggesting that p53R172H is stabilized by K-Ras activation. Similar results were found using heterozygous p53R172H/+mutant mice expressing another oncogene, c-Myc [21]. Splenic lymphomas from Eμ-myc/p53R172H/+mice were found to express p53 at higher levels in comparison with WT spleens, suggesting that mutant p53R172H can be stabilized by higher expression of c-Myc. It was also found that loss of the tumor suppressor p16 led to stabilized p53R172H expression in tumors [21], suggesting that a change in the balance of expression of growth promoting or inhibiting genes affects the stability of mutant p53. In addition, it was found that induction of reactive oxygen species (ROS) also leads to stabilization of mutant p53 in vivo [21]. Another DNA damaging agent, doxorubicin, which is known to activate WT p53, was also found to stabilize mutant p53 [21]. Taken together, all of these data suggest that the pathways that are normally involved in the stabilization and activation of the WT p53 tumor response can also stabilize and activate mutant p53. The activation of p53 mutant "gain of function" phenotypic activity following this stabilization is evident in some of the cases analyzed, as gamma irradiation or expression of K-Ras, though not c-Myc expression, loss of p16 or doxorubicin treatment, were found to lead to more aggressive tumor growth or decreased survival [21].

DNA Damage-Induced Modifications of Mutant p53

Since mutant p53 is activated by similar pathways or agents as WT p53, the contribution of post-translational modifications to WT p53 may have significance for the stability of mutant p53. Previous work has suggested that phosphorylation of serine 15 or serine 20 may be required to stabilize WT p53, as mutation of these sites to alanine inhibited stabilization of p53 in vivo [17, 18] following DNA damage [22]. However, in several mutant p53 expressing cell lines, including EB2, T47D and DU145, constitutive phosphorylation of serine 15 was not observed [23]. Of note however, both MDA-MB231 and MDA-MB468 cells, which contain mutant p53, were found to have constitutively phosphorylated serine 15 [23], suggesting there may be different mechanisms for the stabilization of mutant p53 depending on cellular context, and certainly great variability in cancer cell environments that can directly impinge on differential modifications of key residues of mutant p53.

Stabilization of Mutant p53 Through Inhibition of Degradation

In addition to activation of processes that stabilize mutant p53 because of defects in the cellular environment or modifications that make mutant p53 more stable, there are also defects in the normal degradation or turnover of mutant p53 in cancer cells which lead to its longer stability and half-life.

Ubiquitination and Degradation of WT or Mutant p53

Addition of ubiquitins to either WT or mutant p53 by their respective ubiquitin ligases normally targets them for degradation. Ubiquitin is a 76 amino acid protein that is conjugated to lysines in target proteins either as monoubiquitination, which can occur as single ubiquitins on multiple sites or chains of up to three ubiquitin molecules on each site, or as polyubiquitination which is designated as chains of at least 4 ubiquitin molecules linked together on one or more lysines (reviewed in [24]). Monoubiquitination of a protein is associated with non-degradative functions such as transactivation and the DNA damage response, and is required for subsequent polyubiquitination, which does lead to targeting to the proteasome and degradation [25]. Mdm2 is an E3 ubiquitin ligase for p53 that is capable of adding single ubiquitin molecules onto 6 lysines in the C-terminus of p53 [26]. Following monoubiquitination, degradation of p53 is promoted by the addition of additional ubiquitins to the original ubiquitin monomers, creating chains of ubiquitin proteins which, when linked through ubiquitin lysine 48, target p53 for degradation [26]. Degradation of WT or mutant p53 then occurs through the 26S proteasome [23].

Several E3 ligases have been identified in addition to Mdm2 which can promote the degradation of p53, including CHIP (C-terminus of Hsp70-interacting protein) [27], Pirh2 [28], Cop1 [29], ARF-BP1 [30], Carps [31], Synoviolin [32] and TOPORS [33] (reviewed in [34]). Mdm2 and CHIP have been identified as the major E3 ubiquitin ligases for mutant p53 since destabilization of mutant p53 is partially inhibited by knockdown of Mdm2 or CHIP [23]. It has been suggested that Mdm2/WT p53 binding is necessary to enable Mdm2 to act as a ubiquitin ligase [35]; however, the ability of Mdm2 to ubiquitinate WT p53 was found to be more efficient than Mdm2's ability to ubiquitinate mutant p53. This may be a result of alternate binding regions of Mdm2 to mutant p53 [36]. It has also been found that there is largely a lack of ubiquitination of mutant p53 in cancer cells [23]. This deficiency in ubiquitination could account at least, in part, for the limited degradation deficiency of mutant p53 [23].

Regulation of Mutant p53 Stability by Cellular Chaperones

Intrinsic functional properties of mutant p53 that inhibit its ubiquitination and subsequent degradation have been identified. Mutant p53 has a conformation that is aberrant from WT p53, and this conformation has been shown to cause it to associate with the Hsp90 chaperone protein, which is overexpressed in cancer, in order to prevent aggregation of mutant p53 [23]. The conformation of mutant p53 has been shown to cause mutant p53 to bind to upregulated Hsp90, trapping it and Mdm2 in a complex, preventing Mdm2 from ubiquitinating mutant p53 [23]. The contribution of mutant p53's aberrant conformation to its stability is highlighted, since knockdown of Hsp90 in MDA-MB231 and MCF-7 cancer cells leads to destabilization of mutant p53 [23]. Additionally, treatment of MDA-MB231 cancer cells with 17AAG, which inhibits ATP binding to Hsp90, was found to release mutant p53 from Hsp90, leading to rescue of its ubiquitination and degradation [23]. Furthermore, the complex of mutant p53 with Hsp90 was proposed to also protect mutant p53 from degradation, allowing mutant p53 to accumulate to a high level to enable its gain-of-function [23]. Confirming this mechanism in the cell types tested, an inhibitor of Hsp90 led to degradation of mutant p53 by its E3 ubiquitin ligases Mdm2 and CHIP [23]. This inhibitor, suberoylanilide hydroxamic acid (SAHA), is able to inhibit stabilization of mutant p53 by Hsp90 by inhibiting HDAC6, which activates Hsp90 [23]. This finding suggests that degradation of mutant p53 by reversing the mechanisms that keep it stable may have a potential role in future cancer treatment.

In addition to Hsp90, Hsp70 has also been identified to contribute to mutant p53 stabilization [37]. By transfecting p53 R175H into non-transformed mouse embryonic fibroblasts (MEF) (Trp53–/–, Mdm2–/–) with CHIP or Mdm2 along with Hsp70, these authors showed that Hsp70 promotes the degradation of mutant p53 through CHIP but inhibits the degradation of mutant p53 by Mdm2. Similar to Hsp90, Hsp70 and Mdm2 were both found to be needed for the formation of mutant p53 aggregates in MEFs as well as cancer cell lines such as H1299 and SK-BR-3. This was proposed to be a mechanism for mutant p53 stabilization.

The Cellular Environment

In addition to mechanisms that stabilize mutant p53 through its activation or by inhibition of degradation, aberrations in the cellular environment that contribute to mutant p53 stability have also been identified. Whereas mutant p53 becomes stabilized in cancer cells, it is inherently unstable in normal cells [4]. Homozygous mutant p53 mice expressing the R172H (p53H/H) mutation were found to express p53 in 79 % of tumors whereas heterozygous mutant R172H expressing mice were positive for p53 staining in only 70 % of the primary tumors. However, immunostaining of R172H homo and heterozygous mutant R172H expressing mice indicated no staining of p53 in normal tissues of the lung, thymus, spleen, heart, bone marrow, brain, pancreas, liver, intestine and kidney [4]. These data suggest that alterations in the cancer cell environment lead to the hyperstability of mutant p53 in these cells.

To determine specific cellular alterations that stabilize mutant p53, p53H/H mice were crossed with Mdm2 null mice, and normal tissues were stained for p53 [4]. In the absence of Mdm2, mutant p53 was found to express in normal tissues from several organs, suggesting that Mdm2 suppresses expression of mutant p53 in normal tissues. Since mutant p53 was normally turned over by Mdm2 in normal cells similarly to WT p53, this suggests that WT and mutant p53 are regulated by similar mechanisms in the absence of damage. To determine the effects of DNA damage on mutant p53 in the presence and absence of Mdm2, p53H/H and p53H/H/Mdm2−/− mice were irradiated with 5 Gy to the whole body, and tissues were harvested and subjected to Western blot analysis. IR was found to stabilize mutant p53 in the spleen and thymus, and though absence of Mdm2 led to an increase in mutant p53 level, IR treatment did not increase the level further [4]. In response to IR, WT p53 levels increased following irradiation but returned to basal levels 7 h later. In contrast, mutant p53 levels were also increased following IR but remained elevated up to 15 h later [4], suggesting WT and mutant p53 are both activated by the same mechanism in response to DNA damage. However, unlike WT p53, mutant p53 did not become rapidly destabilized after resolution of checkpoint activity and repair possibly due to the impaired transactivation of Mdm2 transcription after DNA damage in cells with mutated p53 that is unable to transactivate the usual p53 target genes. These data implicate Mdm2 in a role for regulation of mutant p53 stability.

Another important component of the cell environment that is commonly inactivated in cancer was examined for its role in contributing to mutant p53 stability. p53H/H mice were crossed with p16INK4a−/− mice that retained the p19Arf allele, which shares common genetic code in exons 2 and 3 with the p16Ink4a gene. The absence of p16 was found to stabilize mutant p53 in normal tissues as well as in some tumors [4]. Other defects in the cell context were found along with mutated p53, including upregulation of cyclin D1 and Rb phosphorylation [4], suggesting further that alterations in the cellular environment may contribute to mutant p53 stabilization in cancer cells.

Transactivation Deficiency Towards Mdm2

Mdm2 is an E3 ubiquitin ligase for mutant p53 and is normally upregulated by WT p53, which is thought to lead to its degradation following resolution of the DNA damage signal. However, due to the mutation, mutant p53 is unable to transcriptionally upregulate normal WT p53 targets, including Mdm2. The inability of mutant p53 to therefore transactivate Mdm2 has been suggested to contribute to mutant p53 stability due to an insufficient level of Mdm2 [20]. Analyzing the RNA from spleens of WT, p53R172 (p53H/+) and p53R172H (p53H/H) mutant mice

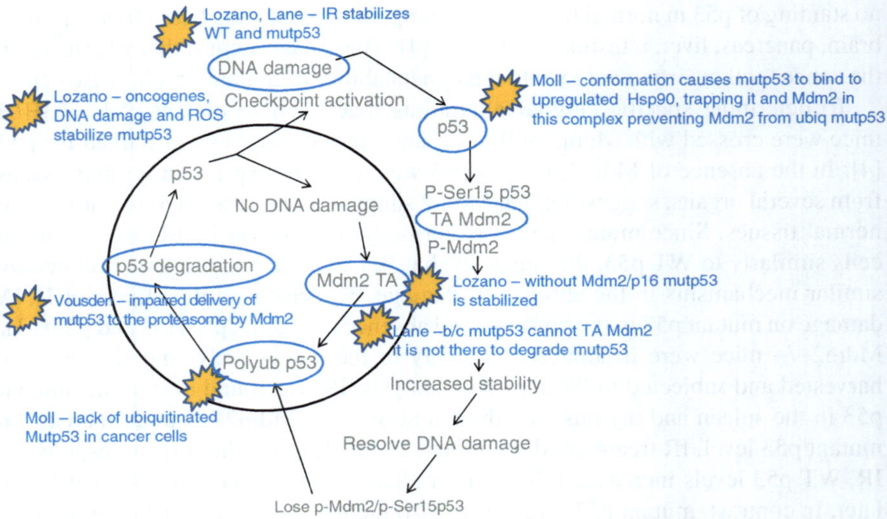

Fig. 10.1 Many of the steps that are known to normally stabilize WT p53 in the DNA damage response have been identified to contribute to the stability of mutant p53, magnifying the oncogenic potential of mutant p53, as described in the text. The general steps that have been identified that lead to mutant p53 stability are shown in circles, and a more detailed description of the contributing factors are shown as starred bullet points

following IR treatment showed that Mdm2 levels in p53H/+mice increased, but to half the extent of WT p53 mice, while p53H/H mice had no change in Mdm2 levels following DNA damage [4]. This demonstrates that mutant p53 is defective in transactivating Mdm2 following IR, suggesting that it is the inherent inability of mutant p53 to activate p53 target genes that prevents it from activating a factor that normally contributes to its destabilization when DNA repair is complete. This may result in a more prolonged upregulation of mutant p53 after stress in non-transformed cells, as opposed to WT p53. However, it is unclear if this mechanism contributes in any substantial way to constitutive mutant p53 stability seen in resting cancer cells.

Conclusions

The gain-of-function phenotype of mutant p53 is thought to depend on its increased stability and resulting elevated levels in cancer cells. Many of the steps that are known to normally stabilize WT p53 in the DNA damage response have been identified to contribute to the stability of mutant p53, magnifying the oncogenic potential of mutant p53 (Fig. 10.1). WT p53 is normally kept at low levels in the absence of DNA damage by transactivating its E3 ubiquitin ligase Mdm2, which along with its E4 p300/CBP promotes its polyubiquitination and degradation by the 26S proteasome. When there is an oncogenic challenge to a cell, WT p53 becomes stabilized by the DNA damage checkpoint to enable cells to either arrest the cell cycle to repair the damage or senesce or initiate apoptosis. Similarly to the WT p53 that is activated in response to an oncogenic challenge, factors such as oncogenes, DNA damage, and ROS have been shown to stabilize mutant p53. WT and mutant p53 both become post-translationally modified following damage, contributing to their stability. In contrast to WT p53, which gets degraded following the performance of its intended function to enable resolution of DNA damage, mutant p53 is unable to facilitate damage repair, which may keep mutant p53 hyperstabilized due to constitutively activated checkpoint pathways found in cancer cells (R.A. Frum and S.R. Grossman, unpublished observations).

Impaired polyubiquitination and subsequent defects in degradation of mutant p53 has also been reported to occur in cancer cells. In addition, the aberrant conformation of mutant p53 has been shown to contribute to its stabilization by causing it to interact with heat shock proteins, which prevents it from being degraded. The work to date illustrates that mutant p53 is regulated by the same signaling pathways as WT p53, and the same factors that contribute to WT p53 stability also stabilize mutant p53. However, the inherent properties of mutant p53 such as its conformation, or inability to transactivate WT p53 target genes, also plays a role in keeping mutant p53 stable in cancer cells, suggesting that the loss of WT function combined with the acquired functions of mutant p53 and the specific cancer cell environment, together, all play a role in the stability and oncogenic properties of mutant p53 in cancer cells.

References

1. Haupt S, Berger M, Goldberg Z, Haupt Y (2003) Apoptosis – the p53 network. J Cell Sci 116(Pt 20):4077–4085
2. Lane D (2004) Anthony Dipple Carcinogenesis Award. p53 from pathway to therapy. Carcinogenesis 25(7):1077–1081
3. Oren M, Rotter V (2010) Mutant p53 gain-of-function in cancer. Cold Spring Harb Perspect Biol 2(2):a001107
4. Terzian T, Suh YA, Iwakuma T, Post SM, Neumann M, Lang GA et al (2008) The inherent instability of mutant p53 is alleviated by Mdm2 or p16INK4a loss. Genes Dev 22(10):1337–1344
5. Moll UM, Petrenko O (2003) The MDM2-p53 interaction. Mol Cancer Res 1(14):1001–1008
6. Toledo F, Wahl GM (2006) Regulating the p53 pathway: in vitro hypotheses, in vivo veritas. Nat Rev Cancer 6(12):909–923
7. Shi D, Pop MS, Kulikov R, Love IM, Kung AL, Grossman SR (2009) CBP and p300 are cytoplasmic E4 polyubiquitin ligases for p53. Proc Natl Acad Sci U S A 106(38):16275–16280
8. Grossman SR, Perez M, Kung AL, Joseph M, Mansur C, Xiao ZX et al (1998) p300/MDM2 complexes participate in MDM2-mediated p53 degradation. Mol Cell 2(4):405–415
9. Maya R, Balass M, Kim ST, Shkedy D, Leal JF, Shifman O et al (2001) ATM-dependent phosphorylation of Mdm2 on serine 395: Role in p53 activation by DNA damage. Genes Dev 15(9):1067–1077

10. Cheng Q, Chen L, Li Z, Lane WS, Chen J (2009) ATM activates p53 by regulating MDM2 oligomerization and E3 processivity. EMBO J 28(24):3857–3867
11. Khosravi R, Maya R, Gottlieb T, Oren M, Shiloh Y, Shkedy D (1999) Rapid ATM-dependent phosphorylation of MDM2 precedes p53 accumulation in response to DNA damage. Proc Natl Acad Sci U S A 96(26):14973–14977
12. Appella E, Anderson CW (2001) Post-translational modifications and activation of p53 by genotoxic stresses. Eur J Biochem 268(10):2764–2772
13. Banin S, Moyal L, Shieh S, Taya Y, Anderson CW, Chessa L et al (1998) Enhanced phosphorylation of p53 by ATM in response to DNA damage. Science 281(5383):1674–1677
14. Canman CE, Lim DS, Cimprich KA, Taya Y, Tamai K, Sakaguchi K et al (1998) Activation of the ATM kinase by ionizing radiation and phosphorylation of p53. Science 281(5383):1677–1679
15. Tibbetts RS, Brumbaugh KM, Williams JM, Sarkaria JN, Cliby WA, Shieh SY et al (1999) A role for ATR in the DNA damage-induced phosphorylation of p53. Genes Dev 13(2):152–157
16. Stolz A, Ertych N, Bastians H (2011) Tumor suppressor CHK2: regulator of DNA damage response and mediator of chromosomal stability. Clin Cancer Res 17(3):401–405
17. Chehab NH, Malikzay A, Stavridi ES, Halazonetis TD (1999) Phosphorylation of ser-20 mediates stabilization of human p53 in response to DNA damage. Proc Natl Acad Sci U S A 96(24):13777–13782
18. Unger T, Sionov RV, Moallem E, Yee CL, Howley PM, Oren M et al (1999) Mutations in serines 15 and 20 of human p53 impair its apoptotic activity. Oncogene 18(21):3205–3212
19. Kuerbitz SJ, Plunkett BS, Walsh WV, Kastan MB (1992) Wild-type p53 is a cell cycle checkpoint determinant following irradiation. Proc Natl Acad Sci U S A 89(16):7491–7495
20. Midgley CA, Lane DP (1997) p53 protein stability in tumour cells is not determined by mutation but is dependent on Mdm2 binding. Oncogene 15(10):1179–1189
21. Suh YA, Post SM, Elizondo-Fraire AC, Maccio DR, Jackson JG, El-Naggar AK et al (2011) Multiple stress signals activate mutant p53 in vivo. Cancer Res 71(23):7168–7175
22. Chao C, Saito S, Anderson CW, Appella E, Xu Y (2000) Phosphorylation of murine p53 at ser-18 regulates the p53 responses to DNA damage. Proc Natl Acad Sci U S A 97(22):11936–11941
23. Li D, Marchenko ND, Schulz R, Fischer V, Velasco-Hernandez T, Talos F et al (2011) Functional inactivation of endogenous MDM2 and CHIP by HSP90 causes aberrant stabilization of mutant p53 in human cancer cells. Mol Cancer Res 9(5):577–588
24. Al-Hakim A, Escribano-Diaz C, Landry MC, O'Donnell L, Panier S, Szilard RK et al (2010) The ubiquitous role of ubiquitin in the DNA damage response. DNA Repair (Amst) 9(12):1229–1240
25. Myung J, Kim KB, Crews CM (2001) The ubiquitin-proteasome pathway and proteasome inhibitors. Med Res Rev 21(4):245–273
26. Lee JT, Gu W (2010) The multiple levels of regulation by p53 ubiquitination. Cell Death Differ 17(1):86–92
27. Esser C, Scheffner M, Hohfeld J (2005) The chaperone-associated ubiquitin ligase CHIP is able to target p53 for proteasomal degradation. J Biol Chem 280(29):27443–27448
28. Leng RP, Lin Y, Ma W, Wu H, Lemmers B, Chung S et al (2003) Pirh2, a p53-induced ubiquitin-protein ligase, promotes p53 degradation. Cell 112(6):779–791
29. Dornan D, Wertz I, Shimizu H, Arnott D, Frantz GD, Dowd P et al (2004) The ubiquitin ligase COP1 is a critical negative regulator of p53. Nature 429(6987):86–92
30. Chen D, Kon N, Li M, Zhang W, Qin J, Gu W (2005) ARF-BP1/mule is a critical mediator of the ARF tumor suppressor. Cell 121(7):1071–1083
31. Tang J, Qu LK, Zhang J, Wang W, Michaelson JS, Degenhardt YY et al (2006) Critical role for daxx in regulating Mdm2. Nat Cell Biol 8(8):855–862
32. Yamasaki S, Yagishita N, Sasaki T, Nakazawa M, Kato Y, Yamadera T et al (2007) Cytoplasmic destruction of p53 by the endoplasmic reticulum-resident ubiquitin ligase 'synoviolin'. EMBO J 26(1):113–122

33. Rajendra R, Malegaonkar D, Pungaliya P, Marshall H, Rasheed Z, Brownell J et al (2004) Topors functions as an E3 ubiquitin ligase with specific E2 enzymes and ubiquitinates p53. J Biol Chem 279(35):36440–36444
34. Love IM, Grossman SR (2012) It takes 15 to tango: making sense of the many ubiquitin ligases of p53. Genes Cancer 3(3–4):249–263
35. Wallace M, Worrall E, Pettersson S, Hupp TR, Ball KL (2006) Dual-site regulation of MDM2 E3-ubiquitin ligase activity. Mol Cell 23(2):251–263
36. Lukashchuk N, Vousden KH (2007) Ubiquitination and degradation of mutant p53. Mol Cell Biol 27(23):8284–8295
37. Wiech M, Olszewski MB, Tracz-Gaszewska Z, Wawrzynow B, Zylicz M, Zylicz A (2012) Molecular mechanism of mutant p53 stabilization: the role of HSP70 and MDM2. PLoS One 7(12):e51426

Chapter 11
Crosstalk Between Mdm2, p53 and HIF1-α: Distinct Responses to Oxygen Stress and Implications for Tumour Hypoxia

E. Douglas Robertson, Kostyantyn Semenchenko, and Bohdan Wasylyk

Abstract The E3 ubiquitin ligase Mdm2 regulates two transcription factors, p53 and HIF1α, which appear to be tailored towards different and specific roles within the cell, the DNA damage and hypoxia responses, respectively. However, evidence increasingly points towards the interplay between these factors being crucial for the regulation of cellular metabolism and survival in times of oxygen stress, which has particular relevance for tumour formation. Mdm2, p53 and HIF1α all respond to hypoxia, and intriguingly, have distinct roles depending on the level of hypoxia. The data from numerous studies across different conditions hint at the interplay between these key factors in cellular homeostasis. Here we try to weave these strands together, to create a picture of the complex tapestry of interactions that demonstrates the importance of the crosstalk between these key regulatory proteins during hypoxia.

Keywords MDM2 • p53 • HIF1α • Hypoxia • pVHL • PHD1-3 • GLUT • PNUTS • Protein-protein interactions • Protein degradation • Post-translational modification • Target genes • VEGF • PAI-1 • ET-1 • Metabolism • Glycolysis • Angiogenesis • Cellular homeostasis • Cell cycle control • Apoptosis • Growth-factor response

E.D. Robertson • K. Semenchenko • B. Wasylyk (✉)
Department of Functional Genomics and Cancer Biology,
Institut de Génétique et de Biologie Moléculaire et Cellulaire IGBMC,
UMR 7104 CNRS-UdS, U946 INSERM, 1 rue Laurent Fries,
BP 10142, 67404 Illkirch Cedex, France
e-mail: boh@igbmc.fr

S.P. Deb and S. Deb (eds.), *Mutant p53 and MDM2 in Cancer*,
Subcellular Biochemistry 85, DOI 10.1007/978-94-017-9211-0_11,

The Hypoxic Gradient and the Key Regulators

Normal Cells Adapt to Available Oxygen Levels

The term "hypoxia" is used to describe a reduction in the availability of oxygen to cells and tissues. In terms of human tissues, depending on the organ and the distance from blood vessels, the concentration of oxygen varies between 2 and 9 % [7]. Hypoxia then occurs with increasing severity below 2 % oxygen until the absence of oxygen and anoxia. The effect of oxygen withdrawal on cells and tissues is dramatic. At the molecular level, cells must switch from the oxygen dependent TCA cycle to the less energy efficient glycolysis pathway. In order to compensate for the resulting energy deficit, cells must increase their uptake of glucose, and so many enzymes in the glycolysis pathway are up-regulated in response to hypoxia [16].

Tumour Cells Have a Different Response to Reduced Oxygen

A common feature of tumours is that they are hypoxic [37]. As cancerous cells divide and rapidly proliferate, they use up the available oxygen and, without a proper vasculature, many tumour regions become chronically hypoxic. One may therefore suppose that cancerous cells are specially adapted to a hypoxic environment. Indeed, it was observed that tumour cells have a peculiarity known as the Warburg effect; that is, they up-regulate glucose intake (characteristic of cells in hypoxia) even under normal oxygen conditions. Increased glycolysis provides both the energy and biosynthetic needs for proliferating tumour cells [37].

The Key Component

The key regulator of hypoxia has been identified as HIF1α (hypoxia inducible factor 1 alpha), a member of the PAS protein family of transcription factors. HIF1α is degraded under normoxic conditions. Hydroxylation of key proline residues by one of three prolyl-hydroxylases (PHD1-3) targets it to the E3 ubiquitin ligase pVHL, which subsequently leads to HIF1α degradation by the proteasome [25]. In low oxygen conditions, however, the hydroxylation reaction can no longer take place, allowing HIF1α to be stabilized and translocated to the nucleus, where it can interact with its binding partner HIF1β to form a transcription factor that can regulate downstream gene transcription [25]. This stabilization allows for a rapid response to oxygen deprivation.

Other Components

However, this is not the whole story, because other factors are also involved in managing the hypoxic response, such as HIF2α, HIF3α and the ETS family member Elk3 [18, 19, 46]. Furthermore, it is increasingly clear that p53 and its ubiquitin ligase Mdm2 are also important factors in managing the cellular response to low oxygen levels. Under conditions of cellular stress, such as DNA damage, p53 is induced and acts as a transcription factor to specifically target genes involved in cell cycle control and apoptosis. Additionally, it up-regulates Mdm2, thus forming an auto-regulatory loop [53, 34, 27] whereby p53 controls its own activity. It is easy to imagine a role for p53 in protecting cells from the damaging effects of prolonged or severe hypoxia, however it is interesting to note that, as discussed here, Mdm2 also has a distinct role controlling hypoxic gene expression. Thus HIF1α, p53 and Mdm2 have individual and overlapping roles regulating shared genes, but respond to different stimuli and furthermore, all 3 physically interact in a hypoxic "ménage a trois", the balance of which is crucial in determining the response (see Fig. 11.1 for a general overview).

Losses, Gains and Alterations of MDM2, p53 and HIF1α in Hypoxia

Loss of Mdm2 in Hypoxia

Mdm2 is down regulated in hypoxia, but the evidence is sometimes conflicting and likely depends on the cell type and severity of hypoxia. After 6 h at 0.02 % oxygen, human colorectal carcinoma (RKO) cells express significantly less Mdm2 and concurrently more p53 [1]. Other studies have also shown that Mdm2 is down regulated under hypoxia in both primary and transformed cell lines [26, 59] although in some cancer cell lines (HCT116, Mum2B and PC-3) a transient induction of Mdm2 has been observed in hypoxia [38].

Different mechanisms appear to contribute to down-regulation of Mdm2 in hypoxia. Mdm2 RNA is down-regulated in primary neuronal cells, initiating after 8 h in 0 % oxygen and reaching a 10 fold reduction in mRNA after 16 h [59]. Mdm2 localization in MCF-7 cells switches from being almost exclusively cytoplasmic to nuclear under virtually anoxic conditions, without a change in the level of Mdm2. Interestingly, this translocation of Mdm2 is blocked by geldanamycin, an inhibitor of Hsp90 required for HIF1α stabilization [51]. Mdm2 is down-regulated by phosphorylation. Several kinases have been identified as being activated and phosphorylated under hypoxia, including p38. Inhibition of p38 phosphorylation in hypoxia reverses the down-regulation of Mdm2 and up-regulation of p53, suggesting that

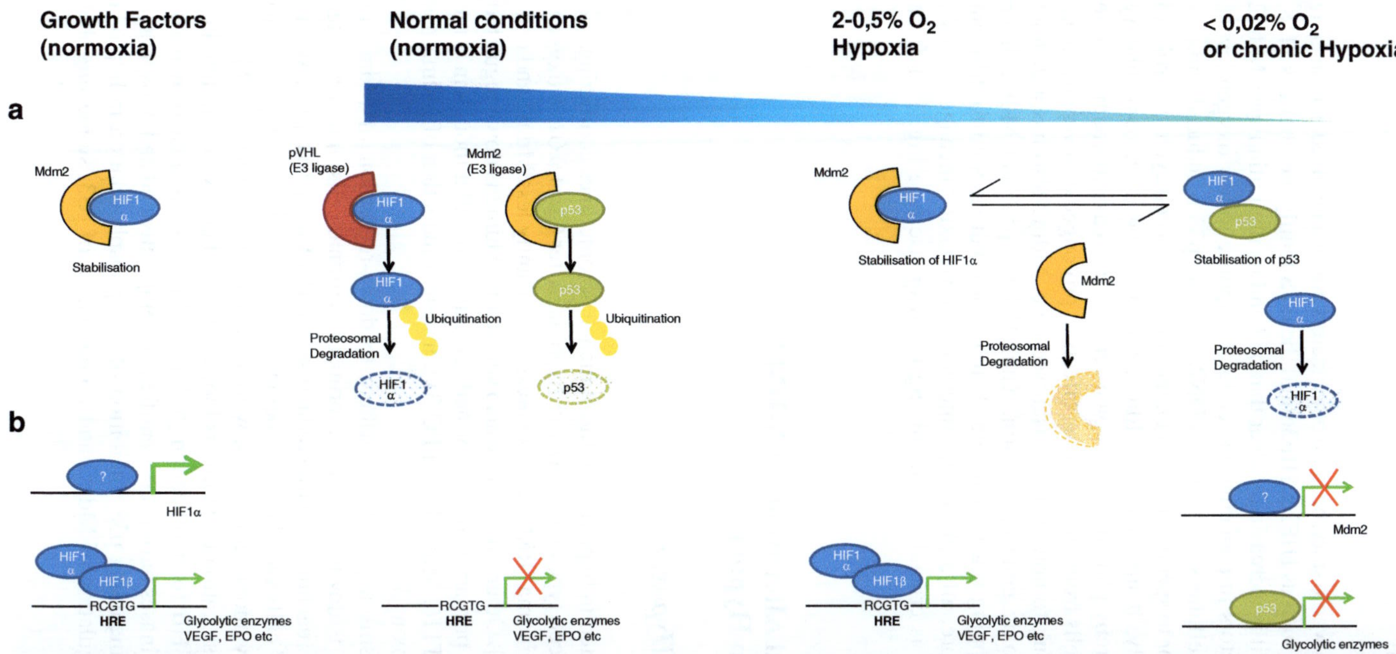

Fig. 11.1 Model of the contribution of MDM2 to the effects of hypoxia and growth factors on the stability and activities of HIF1α and p53
Parts **a** and **b** reflect changes in protein and gene expression, respectively. In normoxia, in the absence of growth factor stimulation, MDM2 interacts with p53 and pVHL with HIF1α, leading to p53 and HIF1α degradation and loss of regulation of their target genes. Growth factors stabilize MDM2, which interacts with and stabilizes HIF1α. Increased HIF1α stability and transcription of its gene leads to increased expression of HIF1 target genes. In mild hypoxia, MDM2 may contribute to the stabilization of HIF1α through protein-protein interactions. Under severe or chronic hypoxia, MDM2 mediated degradation of p53 is lost through MDM2 degradation, downregulation of transcription of the Mdm2 gene, and interference with MDM2 binding to p53 by HIF1α interactions with both MDM2 and p53. The exchange of HIF1α between Mdm2 and p53 may depend on the severity and the longevity of hypoxia. p53 terminates the metabolic response to mild hypoxia, by inhibition of many of the metabolic genes induced by HIF1α under mild hypoxia, and possibly by facilitating HIF1α degradation

p38 drives the Mdm2/p53 hypoxic response, at least in neuronal cells [59]. Yet another mechanism of hypoxic Mdm2 regulation involves hypoxia-induced PNUTS (a protein phosphatase-1 binding protein). Overexpression of PNUTS decreases Mdm2 protein but not mRNA, and inhibiting proteosomal activity reverses this effect. The mechanism of PNUTS activated Mdm2 degradation is unclear. PNUTS has no effect on the p53/Mdm2 interaction and does not itself directly interact with Mdm2 [32]. Thus, in most instances, Mdm2 appears to be down regulated in hypoxia, although it is difficult to unravel the whole picture due to the different cell types and hypoxic conditions used in the various studies.

Gain of p53 in Hypoxia

The response of p53 to hypoxia is its induction under extreme conditions. Untransformed human fibroblasts exposed to 0.02 % oxygen have increased levels of p53 after 2 h, reaching a peak at 5 h, which is maintained until 22 h. Restoring the cells to normoxic conditions results in a return to basal expression levels after 1 h [17]. Hypoxia mimics, DFX (Desferrioxamine) and $CoCl_2$, which inhibit the PHD proteins that target HIF1α for degradation, also induce p53 in murine ES cells [39]. This could indicate a role for HIF1α in p53 stabilisation, but not enough is known about other targets for the PHD proteins, which could also interact with p53. p53 is not induced by hypoxia in some cell lines (MCF-7 and HCT116 cells, 16 h at 0.05 % O_2) [39], suggesting that the mechanisms can vary depending on the cell background. The kinetics of the response of HIF1α and p53 to hypoxia appears to be slightly different, depending on both the severity and longevity of treatment. Hammond et al. [20] found that whilst HIF1α was induced at 2 % oxygen with similar kinetics to lower oxygen concentrations, p53 was only stabilised under 0.02 % hypoxia. Therefore, while p53 is induced in hypoxia like HIF1α, it is distinct in that it is not as sensitive to the depletion of oxygen.

Up-Regulation of HIF1α in Hypoxia Affects the p53/MDM2 Balance

There is considerable evidence for an interaction between HIF1α and Mdm2 that does not require p53 and leads to the stabilization of HIF1α [41]. Using *in vitro* GST pull-down assays, Chen et al. [9] were able to demonstrate that HIF1α binds strongly to Mdm2, although, in contrast to other studies, they were unable to show a similar interaction between HIF1α and p53. This suggests that the effect of hypoxia on p53 is primarily mediated by the interaction between Mdm2 and HIF1α. However, although unable to detect an *in vitro* interaction, they were able to show a p53/HIF1α interaction *in vivo* that was enhanced by increasing the levels of Mdm2 [9].

The consequence of this interaction appears to be that HIF1α can inhibit Mdm2 ubiquitination of p53, leading to its stabilization and increased transcriptional activity. In an RKO cell line expressing HPV E6, p53 accumulation results from the inhibition of E6 interaction with p53, without an effect on E6 protein levels [1]. Overexpression of HIF1α can inhibit p53 degradation caused by the overexpression of Mdm2 but not by the HPV E6 protein [9]. These data suggest that HIF1α interacts with either p53 or Mdm2, and inhibits Mdm2 regulated p53 degradation.

HIF1α has been shown to directly interact with Mdm2 in both normoxia and hypoxia in cancer cell lines, and in both $p53^{+/+}$ and $p53^{-/-}$ backgrounds [38]. Since Mdm2 is a ubiquitin ligase and HIF1α is itself targeted by the ubiquitin ligase pVHL, it could be possible that part of the function of the HIF1α/Mdm2 interaction is to target HIF1α for degradation by the proteasome. However, there is no difference in the rate of HIF1α degradation after reoxygenation in Mdm2 deficient cells. Instead, overexpressing Mdm2 actually increases HIF1α stability and its ability to induce VEGF transcription in hypoxia, although there is no effect under normoxic conditions [38]. This effect has also been observed by LaRusch et al. [29], who showed that knockdown of Mdm2 or Nutlin3 treatment (a chemical inhibitor of the p53/Mdm2 interaction) reduces expression of VEGF or EPO reporters. Additionally, they were able to identify the regions of HIF1α/Mdm2 interaction. Pull-down assays using truncated Mdm2 located the HIF1α interaction to the same region as p53. Examination of the sequence identified similarity between N-terminus of HIF1α (near to the DNA binding domain) and the p53, E2F, p73 sequences that bind to Mdm2. Mutation of the critical phenylalanine of the p53-binding motif of Mdm2 inhibits the interaction of Mdm2 with HIF1α [29]. Additionally, the HIF1α/Mdm2 interaction occurs independent of p53 but, unlike other studies, a HIF1α/p53 interaction was not detected [29]. This evidence points to HIF1α competing with p53 for binding to Mdm2. The Mdm2/HIF1α interaction could help to stabilize p53 as well as HIF1α. However, since previous data suggest that Mdm2 is down-regulated by hypoxia, the conditions under which this interaction is relevant remains to be established.

Evidence for a HIF1α/Mdm2 interaction also comes from studies with specific inhibitors of HIF1α. The chemical inhibitor YC-1 was shown to inhibit the accumulation of HIF1α protein in hypoxia (HepG2 hepatocellular carcinoma cells at 0.1 % O_2), by inhibiting both the protein synthesis and stability of HIF1α [30]. Treatment with YC-1 also down-regulates Mdm2 in hypoxia, and transfection with an Mdm2 overexpression vector rescues the effect of YC-1 on HIF1α [30]. Finally, YC-1 is able to inhibit the activity of the Mdm2 promoter in hypoxia [30]. This data suggests that YC-1 inhibition of HIF1α is mediated through its regulation of Mdm2. Similarly, Apigenin (a dietary flavonoid), also inhibits HIF1α through Mdm2 in hypoxia (1 % O_2 in an ovarian cancer cell line) [12]. Apigenin significantly affects HIF1α protein stability, but also slightly down-regulates HIF1α mRNA, and again this corresponds with down-regulation of Mdm2 [12]. Therefore, these studies with chemical inhibitors highlight the importance of Mdm2 stabilisation of HIF1α, although differences in cell lines and oxygen concentrations make it difficult to get a clear picture of when Mdm2 stabilisation of HIF1α is important.

In conclusion, it is well established that there is an interaction between Mdm2 and HIF1α that leads to the stabilization of HIF1α and does not require p53. In addition, there is also data to suggest that Mdm2 has a role beyond that of HIF in controlling hypoxic gene expression.

Mdm2 Regulates Hypoxia-Responsive Genes in Normoxia and Hypoxia

Mdm2 in Normoxia

Besides its role in hypoxia, HIF1α can have effects in normoxia, either when normal mechanisms of HIF1α degradation are disrupted, or through stabilisation by growth factors and other mechanisms [28]. Studies in normoxia reveal some of the mechanisms of Mdm2 interaction with HIF1α. In VHL defective Renal Carcinoma Cells, HIF1α and HIF2α are not degraded in normoxia and instead are constitutively expressed. Down-regulation of Mdm2 decreases expression of HIF1α and HIF2α at the mRNA level, although the effect on protein stability has not been examined [8]. The regulation of classical "hypoxic" gene targets of HIF appears to be different, suggesting a "hypoxic-like" role for Mdm2. Loss of Mdm2 results in an increase of VEGF and PAI-1 expression, and a dramatic down-regulation of ET-1. However, in the same cells, down-regulation of HIF1α has no effect, and down-regulation of HIF2α results in reduced expression of VEGF, PAI-1 and ET-1 [8]. These normoxic effects are not due to p53, but instead to increased Erk1/2 signalling. In cells treated with siRNA against Mdm2, blocking Erk1/2 phosphorylation abolishes the effect of Mdm2 knockdown on PAI-1 and VEGF [8]. These data suggest a HIF independent role for Mdm2 in the expression of hypoxic gene targets mediated by Erk1/2. However, it should be noted that these experiments were done under normoxic conditions and so it is unclear if the same would have been true in hypoxia. In addition, abnormal cancerous cells were used, and it is not clear how well the results apply to normal cells.

HIF1α can also be induced in normoxia, by stimulation of cells with growth factors [31], such as IGF-1 and GDF-15 (Growth differentiation factor 15). Whilst the mechanisms are poorly understood, they appear to be dependent on Mdm2. Bárdos et al. [5] found that transient expression of Mdm2 results in increased HIF1α expression, in agreement with other reports. IGF-1 induces both HIF1α and Mdm2 irrespective of p53 expression, however the induction of HIF1α is significantly impaired in the Mdm2/p53 double mutant. HIF1α induction occurs through the PI-3K-AKT/PKB signalling pathway, since it is blocked with a PI-3K specific inhibitor. Increased expression of HIF1α under these conditions occurs at the level of protein synthesis rather than protein stability, since inhibition of protein synthesis with cycloheximide decreases HIF1α induced by Mdm2 in normoxia but not in hypoxia [5]. PI-3K inhibition has also been shown to block HIF1α induction of VEGF [48]. A consequence of PI-3K inhibition is inhibition of Mdm2 expression and phosphorylation. Forced expression of Mdm2 is sufficient to restore expression of VEGF [48]. This provides evidence for a role for Mdm2 in the expression of HIF1α target genes. The growth factor GDF-15 can also induce HIF1α [49]. Mdm2 is significantly up-regulated after GDF-15 treatment, leading to p53 degradation. Treatment with Nutlin3 abolishes this interaction and also prevents GDF-15 induction of HIF1α and VEGF, however interactions between HIF1α and Mdm2 were not examined [49]. Therefore, there is a strong case for Mdm2 involvement in HIF1α gene regulation under conditions where HIF1α is induced by growth factors (see Fig. 11.1).

Mdm2 in Hypoxia

Mdm2 directly influences the expression of certain hypoxia-regulated genes. In one study, Mdm2 and VEGF expression in HUVEC cells were shown to correlate under hypoxia, and treatment with Nutlin3 was shown to block the accumulation of VEGF and inhibit tube-formation [35]. In another study, MEF cells expressing Mdm2 were shown to have higher levels of VEGF than those that do not express the protein [58]. In addition, Mdm2 was shown to increase in the cytoplasm and decreases in the nucleus under hypoxia in a human neuroblastoma cell line, and also to interact directly with the 3′UTR of VEGF mRNA [58].

The ability of Mdm2 to control the expression of downstream hypoxic targets as well as its interaction with HIF1α has implications for cancer therapy. Nutlin3 is a chemical inhibitor of the Mdm2/p53 interaction that has great potential for cancer therapy. By blocking this interaction, it is possible to reactivate p53 in tumour cells, leading to cell death. Nutlin3 has also been shown to block the Mdm2/HIF1α interaction [29] augmenting its anti-cancer potential because, as mentioned above, Mdm2 can stabilise HIF1α in hypoxia. Nutlin3 was found to radiosensitise prostate cancer cells, independent of p53 [50], perhaps due to this effect. Additionally, Nutlin3 was shown to inhibit HIF1α production in hypoxia, leading to reduced expression of VEGF that is independent of p53 [40]. A different analysis of the mechanism of Nutlin3 inhibition of HIF1α however, found it to be p53 dependent and VHL independent [33], suggesting that the mechanisms could be dissimilar under different conditions. Inhibition of HIF1α was found to be mediated through blocking the Mdm2/HIF1α interaction and reinforcing FIH/HIF1α inhibition [33]. Blocking the MDM2/HIF1α interaction could therefore inhibit tumour cells ability to survive in a hypoxic environment.

In conclusion, Mdm2 has a distinct hypoxic role, and is also of high importance under conditions where HIF1α is induced under normoxic conditions. While this feature is poorly understood, it is likely to be important in cancerous cells. High Mdm2 could have a double effect of inhibiting p53 tumour suppressor functions and enhancing HIF1α gene targets that control angiogenesis and metabolism.

p53 Regulates Gene Expression and Physiological Functions in Hypoxia

p53 is Important for the Hypoxic Response

There is evidence for a physiological role for p53 in managing the organismal response to low oxygen levels. Prolonged exposure of p53 knock-out and wild type mice to a hypoxic environment (8 weeks in 10 % oxygen) results not only in differences in HIF1α and VEGF expression, but also physiological changes such as thickened blood vessels [36]. Convincing evidence comes from changes in p53 in animals adapted to life in oxygen-deprived environments. Intriguing studies of the subterranean mole rat, *Spalax ehrenbergi*, have revealed that it has an altered p53. The exchange of arginine to lysine at positions 174 and 209 mirrors changes often found in human tumours and correlates with decreased expression of apoptotic genes but increased expression of Mdm2 and genes associated with cell cycle arrest and DNA repair [3]. Mdm2 is highly up-regulated under hypoxic conditions in *Spalax* in comparison to other rats, pointing to its importance in adaptation to hypoxia [4]. However, Mdm2 is up-regulated in another animal adapted to low oxygen, the red-eared slider turtle (*Trachemys scripta elegans*), although other p53 target genes are affected [57]. Apaf1, a regulator of apoptosis, is less dramatically activated in *Spalax* in comparison to rats [4]. The p53 activated DNA repair gene p53R2 is significantly induced in *Spalax* during hypoxia, however this is not the case in mice [47]. Interestingly, *Spalax* does not develop tumours, either in captivity or in the wild [47]. These studies show that in order to live in oxygen-depleted environments, complex organisms have had to alter p53, and they suggest that there is a link between p53's role in adaptation to hypoxia and cancer.

p53 Responds Differently to Hypoxia and to DNA Damage

The transcriptional activity of p53 that is induced by hypoxia is different from the activity induced by other stresses, in that it mainly represses rather than activates downstream effector genes [26]. There is at least one exception. p53 can trigger apoptosis in hypoxia by directly up-regulating Bnip3L, without affecting other target genes, such as Puma, Bax and DR5 [13]. In contrast, knockdown of p53 de-represses expression of the closely related gene, BNIP3 [14]. Binding of p53 to the promoter of BNIP3 is required for the repression and unlike BNIP3L this repression occurs in hypoxia, highlighting that up-regulation of Bnip3L is highly specific [14]. The microRNA *miR-17-92* has been identified as one of the targets for p53 repression under hypoxia (0.1 % O_2) [55]. Overexpression of this microRNA has been found to inhibit hypoxia induced apoptosis and knockdown of p53 prevents this inhibition during hypoxia, showing some of the intricacies of the hypoxic role of p53 [55].

The increase in p53 due to hypoxia appears to occur through a different pathway from the increase due to ionizing radiation. Cells expressing HPV E6 protein or infected with HPV-18 have increased levels of p53 under hypoxia but not after ionizing radiation [17]. HIF1α is present in hypoxia in both phosphorylated and unphosphorylated forms. The phosphorylated form of HIF1α preferentially binds to its dimerization partner ARNT, whereas the dephosphorylated form interacts preferentially with p53 [51]. Geldanamycin, which induces the dephosphorylated form of HIF1α, inhibits p53 induction by hypoxia, but not by DNA damaging agents, again suggesting hypoxia and DNA damage act to stabilize p53 by different pathways [51]. The significance of p53 induction under hypoxia is not entirely clear, as it is not necessary for the G1 checkpoint induced by the lack of oxygen [17].

p53 Crosstalk with HIF1α

There is evidence for crosstalk between HIF1α and p53 in hypoxia, even though p53 is not required for the induction of HIF1α under hypoxia. HIF1α is required for p53 accumulation in the presence of hypoxia mimics such as DFX. HIF1α co-precipitates with p53 under these conditions, suggesting that the interaction promotes p53 stability [2]. p53 is not induced by DFX in mouse ES cells in which HIF1α can not be induced, due to the lack of HIF1β. Similarly, p53 is not stabilized under hypoxia in a hepatoma cell line that is not able to produce HIF1α [2]. Prolonged exposure to DFX, for 48–72 h, further increases the levels of p53 and Mdm2, but reduces HIF1α. This decrease in HIF1α under prolonged hypoxia depends on p53 and is independent of Mdm2, suggesting a negative feedback system [10]. Other studies have shown that HIF1α binds to the DNA binding domain of p53 and p53 binds to the Oxygen Degradation Domain (ODD) of HIF1α [21, 44].

Additional components of the hypoxia pathway have been shown to interact with p53. pVHL is the E3 ubiquitin ligase responsible for the degradation of HIF1α following its hydroxylation in normoxia. Roe et al. [43] demonstrated that p53 can directly interact with pVHL independently of either HIF1α or Mdm2, and interestingly the regions of pVHL and p53 that interact are different from those involved in the interactions with HIF1α and Mdm2. The functions of this interaction are not only to stabilize p53 and prevent its degradation via Mdm2 but also to prevent its translocation from the nucleus [43]. Furthermore, pVHL can induce p53 acetylation and consequent transactivation, leading to the up-regulation of p53 target genes. The induction of p53 target genes is also reduced when pVHL is knocked down, showing that the endogenous protein has an active role in the p53 pathway [43]. Thus the interplay between HIF1α, p53 and their ubiquitin ligases pVHL and Mdm2 has important effects in both the induction of genes in response to hypoxia and the induction of HIF1α by growth factors.

p53 May Have Other Partners

HIF1α appears to be required for the hypoxic accumulation of p53. However, p53 is only induced in extreme hypoxic conditions, that is, 0.02 % oxygen or in the presence of a hypoxia mimic, whilst HIF1α can be stabilized between 1 and 3 % oxygen. p53 induction under hypoxia is both a function of severity and time, and if left for long enough at less severe oxygen concentrations, p53 accumulation can be observed. These differences in sensitivity to deprivation of oxygen imply that there may be other factors that are required for the p53 hypoxic response and remain to be discovered.

p53 Has Effects on Metabolism and Angiogenesis in Hypoxia

One of the key features of hypoxia at the cellular level is a switch from the TCA cycle to the oxygen independent glycolysis pathway. This requires more glucose uptake, in order to counter the energy deficit. Cancerous cells maintain this increased uptake of glucose in the presence of high oxygen levels, which is known as the Warburg effect. The up-regulation of the glucose transporters GLUT1 and GLUT4 is a key feature of the hypoxic response, and GLUT genes are regulated by HIF1α. HIF1α has been shown to regulate expression of *Glut1* and *Glut3,* in mice chrondrocytes [42]. Furthermore *Glut1* has been shown to be up-regulated in CHO cells under hypoxia but not in cells where HIF1α is not functional [52]. Interestingly, it was discovered that GLUT1 and GLUT4 are targets for p53 repression [45]. Indeed, mutation of the DNA binding domain of p53 results in up-regulation of both genes, linking p53 to the Warburg effect [45]. In addition to a role in repressing elements of the glycolysis pathway, p53 has also been shown to induce the expression of TIGAR (TP53-induced-glycolysis and apoptosis regulator). Expression of TIGAR inhibits glycolysis and protects cells from oxidative stress [6]. One of the features of the Warburg effect is the increased production from pyruvate of lactate, rather than acetyl-CoA. This switch is mediated by Pdk2, which can inhibit Pdc, the catalyst for the conversion of pyruvate into acetyl-CoA and entry into the citric acid cycle. p53 can repress Pdk2 transcription, thus loss of p53 leads to more lactate production [11]. A further example of p53's ability to restrict the use of glycolysis for energy production is that, in *p53*$^{-/-}$ MEFs, there is an increase in the rate of glucose uptake and an increase in *Glut3* expression, although there is no effect of p53 on *Glut1* or *Glut4* expression [24]. p53 can also regulate aerobic respiration, through the ability to induce transcription of GLS2 (Glutaminase 2) that converts glutamine to glutamate [22].

Angiogenesis is an important process during development, but occurs rarely in the adult. Angiogenesis is important for tumour growth since, although tumours are often adapted for hypoxic conditions, they still need to be supplied with nutrients and oxygen in order to continue expanding. Unsurprisingly then, p53 has been identified as having properties that inhibit angiogenesis. Correlations have been observed between p53 expression and VEGF expression and micro-vessel density in different tumour types [23, 15]. Indeed, loss of p53 function has been shown to lead to increased VEGF and vascularization in tumours [56]. Another mechanism by which p53 could regulate the hypoxic response in tumours is via expression of microRNAs. p53 regulates the expression of miR107, which can target the binding partner of HIF1α, HIF1β, leading to decreased hypoxic gene expression. Overexpression of this miR in mice suppresses VEGF expression and angiogenesis, and in tumours its expression inversely correlates with HIF1β [54]. Therefore, the importance for p53 as a counterbalance for HIF1α during chronic or extreme hypoxia is reasonably well established, as is its importance in preventing the activation of key angiogenic and glycolytic pathways that are essential for tumour growth.

A Need for Further Study

It is clear that there is significant crosstalk between Mdm2, HIF1α and p53, which all have important roles in the hypoxia pathway. The specific roles of Mdm2 and p53 are clearly dependent on the severity of hypoxia; however, more in depth studies are required to examine the effects of duration of hypoxia since prolonged mild hypoxia could have a similar effect to severe hypoxia. In addition, some cells are more sensitive to hypoxia than others, and given the importance of adaptation to prolonged hypoxia for cancer cells, their response may be completely abnormal. The roles of p53 and Mdm2 under hypoxia are distinctly different to those in response to DNA damaging agents. The importance of p53 for gene regulation under hypoxia is highlighted by the identification of its modification in animals specially adapted to low oxygen environments. Indeed, since it is more likely that tissues will experience hypoxia than extensive DNA damage during their existence, it could be argued that the hypoxic role of Mdm2 and p53 is the most prevalent.

Both p53 and Mdm2 are reported to be able to interact with HIF1α, although the consequences of these interactions are different. Mdm2 functions to stabilize HIF1α whilst p53, at least in one study, appears to be important for HIF1α degradation. In response to growth factors, the Mdm2 mediated stabilization of HIF1α is apparently p53 independent. The interaction between HIF1α and p53 protects p53 from Mdm2 mediated degradation. Whilst the function of HIF1α induction in response to growth factors is not as well understood as its role in hypoxia, it is apparent that Mdm2 stabilisation is crucial in this process. HIF1α and p53 also share control of many of the same genes involved in metabolism although their roles are opposing. Under severe hypoxia p53 keeps metabolism in check, whereas in the hypoxic tumour microenvironment, loss of p53 aids the cancerous cells by removing the brake from metabolism.

The weight of evidence demonstrates that both Mdm2 and p53 are important players in the HIF1α regulated hypoxia response pathways, with important implications for cancer and the development of tumours in the hypoxic microenvironment. Both p53 and Mdm2 have distinct roles, as has been identified by the numerous studies described here. However, more work needs to be done to build up a consistent story; inconsistencies likely arise from the use of different cell lines in different oxygen concentrations for different periods of time, and constructing a definitive story for Mdm2 and p53 in hypoxia will require these issues to be addressed.

Given the established importance of hypoxia in cancer progression and treatment, a clearer understanding of, at the very least, the interactions between these regulators could be expected to contribute to better therapies.

Acknowledgments E. Douglas Robertson is a recipient of an INCa (HypoNet project) post-doctoral fellowship, and Kostyantyn Semenchenko PhD fellowships from the FP7 Marie Curie ITN Cancure and the Association pour la Recherche sur le Cancer. Research in the laboratory of B. Wasylyk is supported by the CNRS, INSERM, the Association pour la Recherche sur le Cancer, the University of Strasbourg, the Institut National du Cancer (INCa), the Ligue Nationale contre le Cancer and the Conférence de Coordination Interrégionale du Grand Est de la Ligue contre le Cancer.

References

1. Alarcón R, Koumenis C, Geyer RK, Maki CG, Giaccia AJ (1999) Hypoxia induces p53 accumulation through MDM2 down-regulation and inhibition of E6-mediated degradation. Cancer Res 59(24):6046–6051
2. An WG, Kanekal M, Simon MC, Maltepe E, Blagosklonny MV, Neckers LM (1998) Stabilization of wild-type p53 by hypoxia-inducible factor 1alpha. Nature 392(6674):405–408
3. Ashur-Fabian O, Avivi A, Trakhtenbrot L, Adamsky K, Cohen M, Kajakaro G, Joel A, Amariglio N, Nevo E, Rechavi G (2004) Evolution of p53 in hypoxia-stressed Spalax mimics human tumor mutation. Proc Natl Acad Sci U S A 101(33):12236–12241
4. Band M, Ashur-Fabian O, Avivi A (2010) The expression of p53-target genes in the hypoxia-tolerant subterranean mole-rat is hypoxia-dependent and similar to expression patterns in solid tumors. Cell Cycle 9(16):3347–3352
5. Bárdos JI, Chau NM, Ashcroft M (2004) Growth factor-mediated induction of HDM2 positively regulates hypoxia-inducible factor 1α expression. Mol Cell Biol 24(7):2905–2914
6. Bensaad K, Tsuruta A, Selak MA, Vidal MNC, Nakano K, Bartrons R, Gottlieb E, Vousden KH (2006) TIGAR, a p53-inducible regulator of glycolysis and apoptosis. Cell 126(1):107–120
7. Bertout JA, Patel SA, Simon MC (2008) The impact of O2 availability on human cancer. Nat Rev Cancer 8(12):967–975
8. Carroll VA, Ashcroft M (2008) Regulation of angiogenic factors by HDM2 in renal cell carcinoma. Cancer Res 68(2):545–552
9. Chen D, Li M, Luo J, Gu W (2003) Direct interactions between HIF-1α and Mdm2 modulate p53 function. J Biol Chem 278(16):13595–13598
10. Choy M-K, Movassagh M, Bennett MR, Foo RSY (2010) PKB/Akt activation inhibits p53-mediated HIF1A degradation that is independent of MDM2. J Cell Physiol 222(3):635–639. doi:10.1002/jcp.21980
11. Contractor T, Harris CR (2012) p53 negatively regulates transcription of the pyruvate dehydrogenase kinase Pdk2. Cancer Res 72(2):560–567
12. Fang J, Xia C, Cao Z, Zheng JZ, Reed E, Jiang BH (2005) Apigenin inhibits VEGF and HIF-1 expression via PI3K/AKT/p70S6K1 and HDM2/p53 pathways. FASEB J 19(3):342–353

13. Fei P, Wang W, Kim SH, Wang S, Burns TF, Sax JK, Buzzai M, Dicker DT, McKenna WG, Bernhard EJ, El-Deiry WS (2004) Bnip3L is induced by p53 under hypoxia, and its knockdown promotes tumor growth. Cancer Cell 6(6):597–609
14. Feng X, Liu X, Zhang W, Xiao W (2011) P53 directly suppresses BNIP3 expression to protect against hypoxia-induced cell death. EMBO J 30(16):3397–3415
15. Gasparini G, Weidner N, Maluta S, Pozza F, Boracchi P, Mezzetti M, Testolin A, Bevilacqua P (1993) Intratumoral microvessel density and p53 protein: correlation with metastasis in head-and-neck squamous-cell carcinoma. Int J Cancer 55(5):739–744
16. Goda N, Kanai M (2012) Hypoxia-inducible factors and their roles in energy metabolism. Int J Hematol 95(5):457–463
17. Graeber TG, Peterson JF, Tsai M, Monica K, Fornace AJ Jr, Giaccia AJ (1994) Hypoxia induces accumulation of p53 protein, but activation of a G1- phase checkpoint by low-oxygen conditions is independent of p53 status. Mol Cell Biol 14(9):6264–6277
18. Gross C, Buchwalter G, Dubois-Pot H, Cler E, Zheng H, Wasylyk B (2007) The ternary complex factor net is downregulated by hypoxia and regulates hypoxia- responsive genes. Mol Cell Biol 27(11):4133–4141
19. Gross C, Dubois-Pot H, Wasylyk B (2008) The ternary complex factor Net/Elk-3 participates in the transcriptional response to hypoxia and regulates HIF-1α. Oncogene 27(9):1333–1341
20. Hammond EM, Denko NC, Dorie MJ, Abraham RT, Giaccia AJ (2002) Hypoxia links ATR and p53 through replication arrest. Mol Cell Biol 22(6):1834–1843
21. Hansson LO, Friedler A, Freund S, Rüdiger S, Fersht AR (2002) Two sequence motifs from HIF-1α bind to the DNA-binding site of p53. Proc Natl Acad Sci USA 99(16):10305–10309. doi:10.1073/pnas.122347199
22. Hu W, Zhang C, Wu R, Sun Y, Levine A, Feng Z (2010) Glutaminase 2, a novel p53 target gene regulating energy metabolism and antioxidant function. Proc Natl Acad Sci U S A 107(16): 7455–7460
23. Kang SM, Maeda K, Onoda N, Chung YS, Nakata B, Nishiguchi Y, Sowa M (1997) Combined analysis of p53 and vascular endothelial growth factor expression in colorectal carcinoma for determination of tumor vascularity and liver metastasis. Int J Cancer 74(5):502–507
24. Kawauchi K, Araki K, Tobiume K, Tanaka N (2008) p53 regulates glucose metabolism through an IKK-NF-κB pathway and inhibits cell transformation. Nat Cell Biol 10(5):611–618
25. Ke Q, Costa M (2006) Hypoxia-inducible factor-1 (HIF-1). Mol Pharmacol 70(5):1469–1480
26. Koumenis C, Alarcon R, Hammond E, Sutphin P, Hoffman W, Murphy M, Derr J, Taya Y, Lowe SW, Kastan M, Giaccia A (2001) Regulation of p53 by hypoxia: dissociation of transcriptional repression and apoptosis from p53-dependent transactivation. Mol Cell Biol 21(4):1297–1310
27. Kubbutat MHG, Jones SN, Vousden KH (1997) Regulation of p53 stability by Mdm2. Nature 387(6630):299–303
28. Kuschel A, Simon P, Tug S (2012) Functional regulation of HIF-1alpha under normoxia – is there more than post-translational regulation? J Cell Physiol 227(2):514–524. doi:10.1002/jcp.22798
29. LaRusch GA, Jackson MW, Dunbar JD, Warren RS, Donner DB, Mayo LD (2007) Nutlin3 blocks vascular endothelial growth factor induction by preventing the interaction between hypoxia inducible factor 1α and Hdm2. Cancer Res 67(2):450–454. doi:10.1158/0008-5472.can-06-2710
30. Lau CK, Yang ZF, Lam CT, Tam KH, Poon RTP, Fan ST (2006) Suppression of hypoxia inducible factor-1α (HIF-1α) by YC-1 is dependent on murine double minute 2 (Mdm2). Biochem Biophys Res Commun 348(4):1443–1448
31. Laughner E, Taghavi P, Chiles K, Mahon PC, Semenza GL (2001) HER2 (neu) signaling increases the rate of hypoxia-inducible factor 1α (HIF-1α) synthesis: novel mechanism for HIF-1-mediated vascular endothelial growth factor expression. Mol Cell Biol 21(12):3995–4004
32. Lee SJ, Lim CJ, Min JK, Lee JK, Kim YM, Lee JY, Won MH, Kwon YG (2007) Protein phosphatase 1 nuclear targeting subunit is a hypoxia inducible gene: its role in post-translational modification of p53 and MDM2. Cell Death Differ 14(6):1106–1116

33. Lee YM, Lim JH, Chun YS, Moon HE, Lee MK, Huang LE, Park JW (2009) Nutlin-3, an Hdm2 antagonist, inhibits tumor adaptation to hypoxia by stimulating the FIH-mediated inactivation of HIF-1α. Carcinogenesis 30(10):1768–1775
34. Lin J, Chen J, Elenbaas B, Levine AJ (1994) Several hydrophobic amino acids in the p53 amino-terminal domain are required for transcriptional activation, binding to mdm-2 and the adenovirus 5 E1B 55-kD protein. Genes Dev 8(10):1235–1246
35. Ma J, Xue Y, Cui W, Li Y, Zhao Q, Ye W, Zheng J, Cheng Y, Ma Y, Li S, Han T, Miao L, Yao L, Zhang J, Liu W (2012) Ras homolog gene family, member A promotes p53 degradation and vascular endothelial growth factor-dependent angiogenesis through an interaction with murine double minute 2 under hypoxic conditions. Cancer 118(17):4105–4116
36. Mizuno S, Bogaard HJ, Kraskauskas D, Alhussaini A, Gomez-Arroyo J, Voelkel NF, Ishizaki T (2011) P53 gene deficiency promotes hypoxia-induced pulmonary hypertension and vascular remodeling in mice. Am J Physiol Lung Cell Mol Physiol 300(5):L753–L761
37. Mucaj V, Shay JES, Simon MC (2012) Effects of hypoxia and HIFs on cancer metabolism. Int J Hematol 95(5):464–470
38. Nieminen AL, Qanungo S, Schneider EA, Jiang BH, Agani FH (2005) Mdm2 and HIF-1α interaction in tumor cells during hypoxia. J Cell Physiol 204(2):364–369
39. Pan Y, Oprysko PR, Asham AM, Koch CJ, Simon MC (2004) p53 cannot be induced by hypoxia alone but responds to the hypoxic microenvironment. Oncogene 23(29):4975–4983
40. Patterson DM, Gao D, Trahan DN, Johnson BA, Ludwig A, Barbieri E, Chen Z, Diaz-Miron J, Vassilev L, Shohet JM, Kim ES (2011) Effect of MDM2 and vascular endothelial growth factor inhibition on tumor angiogenesis and metastasis in neuroblastoma. Angiogenesis 14(3):255–266
41. Ravi R, Mookerjee B, Bhujwalla ZM, Sutter CH, Artemov D, Zeng Q, Dillehay LE, Madan A, Semenza GL, Bedi A (2000) Regulation of tumor angiogenesis by p53-induced degradation of hypoxia- inducible factor 1α. Genes Dev 14(1):34–44
42. Ren BF, Deng LF, Wang J, Zhu YP, Wei L, Zhou Q (2008) Hypoxia regulation of facilitated glucose transporter-1 and glucose transporter-3 in mouse chondrocytes mediated by HIF-1α. Joint Bone Spine 75(2):176–181
43. Roe J-S, Kim H, Lee S-M, Kim S-T, Cho E-J, Youn H-D (2006) p53 stabilization and transactivation by a von Hippel-Lindau protein. Mol Cell 22(3):395–405. doi:10.1016/j.molcel.2006.04.006
44. Sánchez-Puig N, Veprintsev DB, Fersht AR (2005) Binding of natively unfolded HIF-1α ODD domain to p53. Mol Cell 17(1):11–21. doi:http://dx.doi.org/10.1016/j.molcel.2004.11.019
45. Schwartzenberg-Bar-Yoseph F, Armoni M, Karnieli E (2004) The tumor suppressor p53 down-regulates glucose transporters GLUT1 and GLUT4 gene expression. Cancer Res 64(7):2627–2633
46. Serchov T, Dubois-Pot-Schneider H, Charlot C, Rösl F, Wasylyk B (2010) Involvement of net and Hif1α in distinct yet intricately linked hypoxia-induced signaling pathways. J Biol Chem 285(28):21223–21232
47. Shams I, Malik A, Manov I, Joel A, Band M, Avivi A (2013) Transcription pattern of p53-targeted DNA repair genes in the hypoxia-tolerant subterranean mole rat Spalax. J Mol Biol 425:1111–1118
48. Skinner HD, Zheng JZ, Fang J, Agani F, Jiang BH (2004) Vascular endothelial growth factor transcriptional activation is mediated by hypoxia-inducible factor 1α, HDM2, and p70S6K1 in response to phosphatidylinositol 3-kinase/AKT signaling. J Biol Chem 279(44):45643–45651
49. Song H, Yin D, Liu Z (2012) GDF-15 promotes angiogenesis through modulating p53/HIF-1α signaling pathway in hypoxic human umbilical vein endothelial cells. Mol Biol Rep 39(4):4017–4022
50. Supiot S, Hill RP, Bristow RG (2008) Nutlin-3 radiosensitizes hypoxic prostate cancer cells independent of p53. Mol Cancer Ther 7(4):993–999
51. Suzuki H, Tomida A, Tsuruo T (2001) Dephosphorylated hypoxia-inducible factor 1α as a mediator of p53-dependent apoptosis during hypoxia. Oncogene 20(41):5779–5788

52. Wood SM, Wiesener MS, Yeates KM, Okada N, Pugh CW, Maxwell PH, Ratcliffe PJ (1998) Selection and analysis of a mutant cell line defective in the hypoxia- inducible factor-1 α-subunit (HIF-1α). Characterization of HIF-1α- dependent and -independent hypoxia-inducible gene expression. J Biol Chem 273(14):8360–8368
53. Wu X, Bayle JH, Olson D, Levine AJ (1993) The p53-mdm-2 autoregulatory feedback loop. Genes Dev 7(7A):1126–1132
54. Yamakuchi M, Lotterman CD, Bao C, Hruban RH, Karim B, Mendell JT, Huso D, Lowenstein CJ (2010) P53-induced microRNA-107 inhibits HIF-1 and tumor angiogenesis. Proc Natl Acad Sci U S A 107(14):6334–6339
55. Yan HL, Xue G, Mei Q, Wang YZ, Ding FX, Liu MF, Lu MH, Tang Y, Yu HY, Sun SH (2009) Repression of the miR-17-92 cluster by p53 has an important function in hypoxia-induced apoptosis. EMBO J 28(18):2719–2732
56. Yoshioka Y, Shimizu S, Ito T, Taniguchi M, Nomura M, Nishida T, Sawa Y (2012) p53 inhibits vascular endothelial growth factor expression in solid tumor. J Surg Res 174(2):291–297, http://dx.doi.org/10.1016/j.jss.2010.12.028
57. Zhang J, Biggar KK, Storey KB (2013) Regulation of p53 by reversible post-transcriptional and post-translational mechanisms in liver and skeletal muscle of an anoxia tolerant turtle, Trachemys scripta elegans. Gene 513(1):147–155
58. Zhou S, Gu L, He J, Zhang H, Zhou M (2011) MDM2 regulates vascular endothelial growth factor mRNA stabilization in hypoxia. Mol Cell Biol 31(24):4928–4937
59. Zhu Y, Mao XO, Sun Y, Xia Z, Greenberg DA (2002) p38 mitogen-activated protein kinase mediates hypoxic regulation of Mdm2 and p53 in neurons. J Biol Chem 277(25):22909–22914

Chapter 12
MDM2 Overexpression, Activation of Signaling Networks, and Cell Proliferation

Swati Palit Deb, Shilpa Singh, and Sumitra Deb

Abstract Frequent overexpression of MDM2 in human cancers suggests that the protein confers a survival advantage to cancer cells. However, overexpression of MDM2 in normal cells seems to restrict cell proliferation. This review discusses the cell growth regulatory functions of MDM2 in normal and genetically defective cells to assess how cancer cells evade the growth-restricting consequence of MDM2 overexpression. Similar to oncoproteins that induce a DNA damage response and oncogene induced senescence in non-transformed cells, MDM2 induces G1-arrest and intra-S phase checkpoint responses that control untimely DNA replication in the face of genetic challenges.

Keywords MDM2 • Oncogenesis • p53 • Cell cycle • Unscheduled DNA replication • Akt signaling

MDM2, the Oncoprotein

It has been more than two decades since the human homologue of the mouse double minute-2 (*mdm2*) gene was discovered as an oncogene. The gene was found in amplified double minute chromosomes in a spontaneously transformed and tumorigenic cell line known as 3T3DM. In addition, artificial amplification of *mdm2* genomic DNA in murine cells induced tumorigenesis. These observations suggested the oncogenic consequences of *mdm2* gene amplification [1], and were followed by the findings that the *mdm2* gene is amplified and overexpressed in 17 of 47 sarcomas, and 5 out of 5 sarcomas with *mdm2* gene amplification did not have a p53 gene mutation commonly found in human tumors [2]. These findings, along with the

S.P. Deb (✉) • S. Singh • S. Deb
Department of Biochemistry and Molecular Biology, and the Massey Cancer Center, Virginia Commonwealth University, Richmond, VA 23298, USA
e-mail: spdeb@vcu.edu

S.P. Deb and S. Deb (eds.), *Mutant p53 and MDM2 in Cancer*, Subcellular Biochemistry 85, DOI 10.1007/978-94-017-9211-0_12,

observation that MDM2 complexes with wild-type p53 [3], led to the conclusion that the oncogenic function of MDM2 is due to its ability to interact with WT p53 and inhibit its ability to activate transcription of its target genes. It was thought that, since MDM2 acts by inactivating p53, MDM2 overexpression in tumors carrying p53 mutations does not confer any further selective advantage.

Subsequently, MDM2 was found to be overexpressed in various human tumors, either due to gene amplification and gain of copy number, or due to enhanced expression at the level of protein or RNA (reviewed in [4–6]). A single nucleotide gene polymorphism (snp) in the promoter region of MDM2 (snp 309) was found to enhance MDM2 expression by enhancing the affinity of the promoter sequence for the transcription factor sp1 [6]. The gene polymorphism has been associated with MDM2 expression and a higher risk of cancer in some studies [7].

Although amplification of the MDM2 gene was originally observed in soft tissue sarcomas with WT p53, occurrence of p53 mutation and MDM2 overexpression are not mutually exclusive in human cancers [4, 5, 8]. In many studies, MDM2 overexpression was observed along with p53 mutation [9–12], while several groups reported MDM2 overexpression without gene amplification (reviewed in [4]). Furthermore SNP T309G polymorphism of the MDM2 promoter region, known to enhance MDM2 expression, has been reported in cancers with p53 mutations [13, 14]. These evidences indicate that cancers bearing p53 mutations may also overexpress MDM2. Whether these two oncogenic events cooperate with each other in establishing oncogenic phenotypes or are selected in a cell type- or tumor-specific manner remains to be established. Since both gene aberrations are considered to be early events during oncogenesis, and occur frequently in various cancers [15], present evidences do not exclude the possibility that their joint appearance could be co-incidental. However, accumulation of the two proteins in a certain tumor may alter the biochemical nature of the tumor, its growth characteristics as well as clinical outcome depending on the occurrence of other gene mutations in the same tumor. Some studies have reported a worse prognosis for tumors carrying p53 mutation and overexpressing MDM2 [9], and this observation was supported by studies in transgenic mice that show a higher incidence of tumor formation in p53-knockout mice overexpressing MDM2 compared to MDM2 transgenic or p53-knockout mice [16].

Function of MDM2 in Non-cancerous Cells

The presence of MDM2 during embryogenesis is essential to control accumulation and growth suppressive effects of WT p53 [17, 18]. Thus, the embryonic lethality posed by MDM2 deficiency could be rescued by deletion of WT p53.

The p53-independent functions of MDM2 have met with skepticism due to the fact that studies in mouse models did not show an overt phenotype in p53–/– MDM2–/– mice compared to p53–/– mice. However, the knockout mouse models suffer from the limitation that MDM2 is a WT p53-inducible gene [19–21], and therefore compared to normal mice, p53–/– mice express lower levels of MDM2, which cannot be induced under stress. Thus, phenotype comparison of p53–/–:MDM2–/– and p53–/– mice may not show significant biological consequences. It is essential to generate knock-in MDM2 mice expressing MDM2 mutants precisely inactive in the function under scrutiny. Such models for ubiquitin ligase functions of MDM2 have been generated [22, 23], and have been used to understand the consequence of the p53-degrading function of MDM2. Generation of similar mouse models would be important to determine the significance of p53-independent functions of MDM2. However, as discussed below, in cell culture studies MDM2 has been found to regulate crucial cell cycle regulatory genes in normal cells [24–26]. Functional redundancy of crucial regulatory genes in higher eukaryotes may defend a major phenotype alteration due to a single gene inactivation by the absence of MDM2.

Interaction of MDM2 with Known Growth Regulators

One of the indicators of biological or oncogenic functions of MDM2 would be the proteins it interacts with. Consistent with its crucial role in cell growth regulation, MDM2 interacts with a large number of proteins [27]. Many of these interactions have been related to the ability of MDM2 to regulate WT p53 levels or other ubiquitin ligase substrates, while the consequences of some of the interactions are p53-independent.

Interaction with p53

Since its discovery, MDM2 is known to interact with WT p53 and inhibit p53-mediated transcriptional activation [3, 28]. The p53-interaction domain of MDM2 was needed for its ability to inhibit WT p53-mediated transcriptional activation [28–31]. Subsequently, it was discovered that MDM2 degrades WT p53 due to its ubiquitin ligase function, and thus controls its accumulation in cell [32–34]. In the current literature, the p53 degrading function of MDM2 is considered to be the primary biological function of MDM2 both in normal and cancer cells. Interaction of MDM2 with WT p53 has been studied extensively. Earlier studies implicated 100 amino acid residues at the N-terminus of MDM2 in interacting with p53 [35, 36]. Later reports suggest that binding of p53 with MDM2 induces a conformational change in WT p53 leading to secondary binding of the central acidic domain of MDM2 to the DNA-binding domain of p53 [37–40]. Consistent with these reports, the central acidic domain of MDM2 has been reported to interact with central domain of tumor-derived mutants of p53 [41]. This observation argues for the possibility that conformation change due to mutation in the DNA-binding domain of p53 facilitates its interaction with MDM2. These studies signify that functions of

MDM2 that regulate WT or mutant p53-mediated transcription are context dependent, and respond to the need of the moment. Thus, phosphorylation and/or acetylation of p53 [40, 42–44] or MDM2 [39, 45–49] at various sites induced by UV or ionizing radiation, DNA intercalating agents, or nucleotide analogues would protect their mutual interaction, accumulating both proteins for respective functions. Several excellent reviews are available that discuss the E3 ubiquitin ligase functions of MDM2 [50–52].

Interaction with Other Growth Regulators

MDM2 interacts with several growth or cell cycle regulators, such as retinoblastoma susceptibility gene product RB [53], E2F1/DP1 transcription factor [54], DNA replication proteins (DNA polymerase ε and η) [55–57] and transcription factors such as sp1 [58, 59], p65RelA [60] or RE-1 silencing transcription factor [26]. As reviewed recently [27], MDM2 is also known to interact with the cyclin kinase inhibitor p21, the catalytic subunit TERT of telomerase, Runt related transcription factor 3, RUNX3 and with histones H2A and H2B, resulting in their ubiquitination or degradation, while its interaction with other factors such as the growth suppressor p14/p19, and ribosomal proteins (L5, L11, L23, L26) regulates MDM2-mediated ubiquitination.

Interaction with Retinoblastoma Susceptibility Gene Product RB and Transcription Factor E2F

Since RB and E2F1 are crucial S phase regulatory factors, their interaction with MDM2 may have implications for cell cycle regulation. Experiments in cultured cells suggest several consequences of these interactions. Binding of MDM2 with RB interferes with the anti-apoptotic function of MDM2 [61], whereas overexpression of RB in cultured cells unmasks p53-mediated apoptosis even in the presence of elevated MDM2 levels. This study also reported that RB does not interfere with the ability of MDM2 to inhibit transactivation by p53. Instead, it prevents MDM2-mediated p53 degradation, and rescues the trans-repression function of p53. MDM2 also induces proteasome-mediated ubiquitin independent degradation of RB [62].

Overexpression studies in human cancer cell lines have revealed that MDM2 stimulates the transactivation function of E2F/DP1 complex through direct interaction [54]. Several mechanisms have been proposed for MDM2-mediated activation of E2F1. MDM2-mediated degradation of RB has been proposed to cause release of E2F1, thereby activating its transcriptional function (reviewed in [63]). It has been reported that in cancer cells containing WT p53, MDM2 inhibits WT p53 transactivation, reducing p21 expression, which in turn activates cyclin dependent kinases to induce RB phosphorylation and release E2F1 to become transcriptionally active [64]. It is likely that a combination of these mechanisms contribute to MDM2-mediated upregulation of E2F1/DP1 activity. However, studies in mouse models demonstrate that targeted expression of MDM2 in mouse mammary gland leads to accumulation of epithelial cells in S phase, and absence or increase in E2F1 expression does not alter this phenotype [65, 66], suggesting that MDM2-mediated upregulation of E2F1 activity is not involved in the observed alterations in cell cycle.

Regulating the Cell Cycle

Given the oncogenic properties of MDM2, overexpression of MDM2 would be expected to impart a growth advantage to host cells. Instead, transient MDM2 overexpression from its cDNA establishes sharp G1-arrest in non-transformed cells [67]. Transformed cells derived from human tumors show partial resistance to MDM2-mediated G1-arrest, showing a slower rate of cell cycle transition [67]. Consistently, non-transformed cells do not show stable overexpression of MDM2 from its cDNA suggesting that cells overexpressing MDM2 do not multiply. This study identified the domains of MDM2 required for G1-arrest, deletion of which abrogates MDM2-mediated G1-arrest, allows stable expression of MDM2 in non-transformed cells, and unmasks its tumorigenic property.

The G1-arrest function of MDM2 does not explain why MDM2 is overexpressed in cancer cells. The quest for the presence of an oncogenic form of MDM2 revealed several splice variants with deletion or alteration in the growth suppressor domains [68]. However, in most cases, predominant expression of unmutated full-length MDM2 was found [69]. Full-length MDM2 cDNA isolated from human cancer cells such as OsACL (osteosarcoma) or MCF-7 (breast cancer) cells efficiently establishes G1-arrest in human diploid cells such as WI38 or MRC5 [69]. This suggests that cancer cells that overexpress MDM2 escape the MDM2-induced G1-arrest function due to abnormalities in downstream cellular genes needed to establish MDM2-mediated G1-arrest. Thus, these mutations allow the cells to escape MDM2-mediated G1-arrest, while retaining or unmasking its tumorigenic functions. Consistently, it is possible to stably express MDM2 in many (but not all) cell lines derived from human tumors [67].

Yet, several laboratories have generated transgenic mice or drosophila with general or targeted expression of MDM2. The observed phenotype of MDM2 transgene expression varies widely. Transgenic mice expressing MDM2 from the *mdm2* gene under the control of its own promoter showed an increase in spontaneous tumor formation at a rate slower than in p53–/– mice, with tumor formation of 50 % heterozygous transgenic mice in 84 weeks and homozygous transgenic mice in 81 weeks, as opposed to 20 weeks by p53-null mice [16]. Furthermore, the MDM2 transgenic mice showed increased sarcoma formation. These observations suggest tumorigenic consequences of MDM2 overexpression, and implicate both p53-dependent and independent tumor forming abilities of MDM2.

Targeted overexpression of MDM2 in mouse mammary gland from a bovine β lactoglobulin promoter induced during gestation and lactation [65] resulted in inhibition of mammary gland development, but did not show signs of apoptosis. This property was found to be p53-independent. In spite of inhibition of mammary gland development and fewer numbers of epithelial cells present in transgenic mammary glands, the epithelial cells were active in incorporating the nucleotide analogue BrdU, suggesting accumulation of cells in S phase. Consistently, the cells showed a lack of cell division. However, this study also reports increase in tumor formation in the retired breeders of MDM2 transgenic mice. Although the mechanism of tumor formation in these mice is not clear, the observation suggests a possibility that even in the absence of induction, basal MDM2 expression from beta lactoglobulin promoter may induce tumor formation. Alternatively, aberrant DNA replication during MDM2 expression may cause gene abnormality, and some of this abnormal cells may acquire selective growth advantage later in life. However this alternative possibility would argue for a "hit and run" mechanism not observed in the case other oncogenes.

Expression of MDM2 from its cDNA in wing imaginal discs of Drosophila resulted in either gnarled or blistered wings along with a dramatic increase in apoptotic cells. In the same study, MDM2 overexpression in the eye imaginal disc showed either a small or rough eye phenotype with no signs of apoptosis. Furthermore, no increase in cell proliferation was observed in either case [70].

In a separate study [71], transgenic mice overexpressing MDM2 from pCMV promoter showed skin abnormality (desquamation and hyperkeratosis) in new born mice, thickening of epidermis and altered expression of differentiation markers, increased numbers of cells synthesizing DNA and undergoing apoptosis in adult mice, but no significant increase in tumorigenesis. A more detailed study by this group [72] reported that targeted MDM2 overexpression in the granular layer of epidermis using a human K14 promoter generates similar changes as described above, but with an increased incidence of chemical carcinogen-induced tumor formation.

The variation in the phenotypes in the transgenic models as discussed above not only suggests a cell-type specific and context specific consequence of MDM2 expression, but also involvement of MDM2 in multiple crucial cellular pathways. For example, expression of MDM2 in mouse skin [71] or wing of Drosophila [70] induced apoptosis, but not in mammary epithelial cells [65]. Frequency of tumor formation was higher in MDM2 transgenic mice generated by Jones et al. compared to targeted expression in mammary gland [65] or the granular layer of the epidermis [72].

One obvious difference in the studies described above is the promoter that drove MDM2 expression, and the second difference is the MDM2 coding templates. Jones et al. [16] used the normal MDM2 promoter and genomic sequence. Given that MDM2 expression is induced by stress or ionizing radiation by WT p53, one would expect that this construct would maintain the normal cellular regulatory events and MDM2 transcripts, except that in these mice the levels of MDM2 would be higher than that in non-transgenic mice. Although cells isolated from these transgenic mice show MDM2-mediated checkpoint arrest [25], this mechanism is utilized by the cells to maintain timely DNA replication. The frequency and latency period of tumor formation observed in these mice argues for the possibility that a prolonged increase in MDM2 expression limits the availability of WT p53 and therefore, over time, selects for cells containing mutations. It would be interesting to determine if splice variants of MDM2 lacking the G1-arrest domains are also overexpressed in this system, thus facilitating tumor formation.

Targeted overexpression of MDM2 in mouse mammary epithelial cells was achieved by using a promoter that was induced by hormones during pregnancy and lactation. As in the case of the transgenic mice developed by Jones et al., MDM2 overexpression would be predicted to lower WT p53 levels in mouse mammary epithelial cells. However, inhibition of mammary gland development and the presence of fewer epithelial cells suggest that a surge of MDM2 expression prevented proliferation of mouse mammary epithelial cells, causing their accumulation in the S phase. This observation is consistent with the report that elevated levels of MDM2 induce intra-S phase checkpoint arrest [25]. Similarly, expression of MDM2 from pCMV or GAL4 responsive HSP70 TATA promoters (UAS-T) did not show oncogenic effects of MDM2. Thus, the results of these studies show both oncogenic and growth suppression properties of MDM2 [65, 70], perhaps by different mechanism.

Although presumably MDM2 is overexpressed from its own promoter in human tumors, the expression of MDM2 is often deregulated by various mechanisms. Therefore, overexpression of MDM2 may select for cells with additional mutations to overcome MDM2-mediated growth arrest and checkpoint regulation. These mechanisms are discussed in the following sections.

Controlling Expression of Cyclins

Expression of Cyclin A

The mechanistic basis of the dichotomy in the consequence of MDM2 expression on cell growth and cell cycle regulation is not clear. The cell cycle regulatory function of MDM2 is consistent with the fact that MDM2 interacts with several proteins, such as RB or the transcription factor E2F, that are crucial for cell cycle regulation. Consistent with its ability to induce G1-arrest in apparently normal diploid cells [67], MDM2 inhibits expression of cyclin A, which is required for cellular DNA replication [24]. The inhibition is at the level of transcription, and requires factors (WT p53, the cyclin kinase inhibitor p16 and the transcription factor BRG1) that are needed for timely expression of cyclin A. Silencing of any one of these proteins disables the ability of MDM2 to inhibit cyclin A expression [24]. These observations suggest that MDM2 controls the timely expression of cyclin A during the G1 to S phase transition. If this pathway is deregulated due to genetic defects, MDM2 loses its control over cyclin A expression.

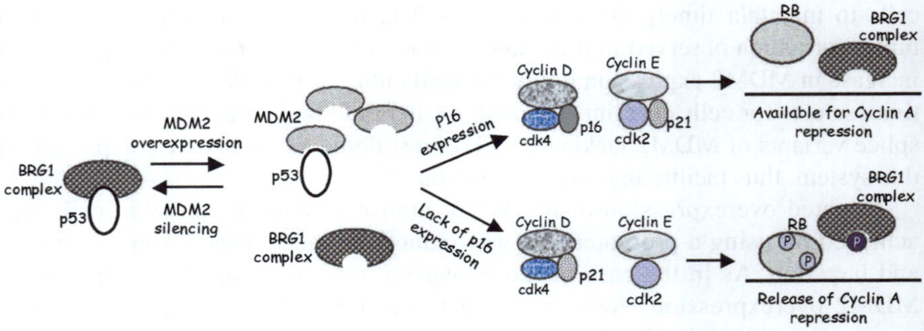

Fig. 12.1 Proposed mechanism by which MDM2 restricts timely cyclin A expression [24]

Analysis of the sequence requirement revealed that MDM2 requires amino acid residues 58 to 109 for inhibition of cyclin A [24]. Although this domain overlaps with the p53-interaction domain, the first 58 amino acid residues of MDM2 are required for MDM2-p53 interaction but dispensable for inhibition of cyclin A [24, 29, 31]. Thus, a stable MDM2-p53 interaction is not a requirement for cyclin A down-regulation. MDM2 harbors a SWIB (SWI/SNF complex B) domain, homologous to a conserved domain of chromosome remodeling factor BRG1-associated factor, structurally capable of binding p53 at the MDM2-binding site [73]. BRG1 complexed with associated factors represses cyclin A expression. WT p53 recruits this BRG1 complex to up-regulate p21 expression [74, 75]. MDM2 overexpression releases this complex from WT p53, thereby making it available to inhibit cyclin A expression. The transcriptional repression of cyclin A expression by BRG1 is released by cyclin E, which ensures timely expression of cyclin A. In the absence of the cdk4 inhibitor p16, activated cyclin D-cdk4 complex releases cyclin E-cdk2 activity to phosphorylate RB and BRG1, thus releasing repression of cyclin A [76, 77]. Consistent with this model, Giono and Manfredi [78] have reported that silencing of MDM2 expression leads to an increase in the levels of WT p53 and its target cyclin kinase inhibitor p21, and interaction of p21 with cyclin E-cdk2 complex, but it does not lead to G1-arrest. Since p53-BRG1 interaction would not be challenged in the absence of MDM2, p53 would recruit BRG1 to p21 promoter, and therefore BRG1 would not be available to repress cyclin A expression (Fig. 12.1).

As expected, therefore, the cyclin A inhibitory domain of MDM2 is capable of inducing G1-arrest in apparently normal diploid cells such as WI38 or MRC5, but not in cancer cells such as H1299 or NIH 3 T3 that lack WTp53, BRG1 or p16 [24]. Overexpression of cyclin A in apparently normal human diploid cells rescues G1-arrest mediated by the cyclin A inhibitory domain of MDM2 [69]. Since inactivation of p53, BRG1 or p16 are frequent events in human cancer [79–81], MDM2-mediated inhibition of cyclin A expression and G1-arrest are desensitized in cancer cells. Consistently, overexpression of cyclin A is a frequent event in cancer cells [69, 82, 83].

Transient transfection and promoter analysis in cultured cells have revealed that MDM2 is capable of up-regulating cyclin A promoter activity, implicating that a resulting increase in cyclin A expression would hasten DNA replication and, consequently, cell proliferation [84]. Later studies have shown that the ability of MDM2 to upregulate cyclin A is unmasked in cells with genetic defects that deregulate cyclin A expression [24, 25]. Although an increase in cyclin A expression alone does not unleash DNA replication, it may contribute to the oncogenic properties of MDM2.

Expression of Cyclin D

Consistent with its ability to upregulate cyclin A expression in the absence of p53, studies in lung cells isolated from p53-null MDM2 transgenic mice or human cancer cells expressing MDM2 from an inducible promoter show that induction of MDM2 expression increases expression of both cyclin D2 transcript and protein [25]. Conversely, knockdown of MDM2 reduces cyclin D2 expression in lung cells derived from p53-null MDM2 transgenic mice.

How MDM2 activates cyclin D2 expression is yet to be determined. MDM2 harbors the structural features of transcriptional regulators, including an acidic activation domain, Zn-finger motifs and a basic region. MDM2 interacts with the TATA-binding protein (TBP) [85] and TATA-associated factor TAFII250 [84, 86] and controls expression of several genes [12, 58, 87, 88]. Therefore, MDM2 may control cyclin D2 expression due to its ability to regulate transcription by interacting directly with transcription factors on the cyclin D2 promoter.

Cyclin D2 is a member of the family of D-type cyclins. D-type cyclins, (cyclin D1, D2 and D3) are expressed in a cell-type specific manner and are functionally redundant. After mitosis, cells make a decision in the early G1 phase whether to enter or exit from the cell cycle to multiply or differentiate, respectively. D-type cyclins are thus synthesized early in the G1 phase and commit the cells to a round of cell cycle. Synthesis of cyclin D is thus growth factor dependent. Following expression of D-type cyclins, after a point in the G1 phase (the restriction point) cell cycle progression becomes growth factor independent [89]. Accordingly, cyclin D2 is expressed in response to mitogenic signals such as growth factors [90]. Therefore, MDM2 may activate cyclin D2 expression through its ability to activate mitogenic signaling [26].

Controlling the Akt Signaling Pathway

The growth factor dependent activation of D-type cyclins requires the PI3-Kinase/Akt signaling pathway [91–93]. The Akt signaling pathway transmits signals from membrane bound receptors to regulate cell proliferation, survival and motility [94]. Mutation in genes that encode regulators of the Akt signaling pathway has been found in many cancers, and these mutations have been correlated with poor prognosis [95–97]. MDM2 interacts with several components of the PI3-kinase/Akt signaling pathway. Activation of PI3-kinase or Akt phosphorylation leads to phosphorylation of MDM2 at serine 166 and 186. Furthermore, MDM2 interacts with Akt [98]. Phosphorylation of MDM2 promotes its nuclear entry [99]. According to the current model, nuclear localization of MDM2 leads to p53-binding and degradation. MDM2 also physically interacts with Akt2 and the translation elongation factor EF1α, which is known to activate the PI4-kinase [100].

Interestingly, MDM2 activates the Akt signaling pathway [26]. Studies in lung cells isolated from MDM2 transgenic mice with a WT or p53-null background, and human cancer cell lines with WT or mutant p53, show that upregulation of MDM2 expression activates, while its knock-down diminishes, Akt phosphorylation. Primary mouse embryo fibroblast (MEF) cells from normal mice show higher efficiency of phosphorylation of an AKT reporter plasmid compared to p53–/–:MDM2–/– MEF cells. Re-introduction of MDM2 in p53–/–:MDM2–/– MEF rescues the Akt phosphorylating activity. MDM2 also induces phosphorylation of a downstream target of Akt signaling, glycogen synthase kinase GSK3β, at Ser 9 and activates PI3-kinase activity by repressing the expression of the regulatory subunit p85 of PI3-kinase.

MDM2-mediated upregulation of cyclin D2, therefore, could be a consequence of enhanced PI3K/Akt signaling. PI3-kinase is a central regulator of cyclin D2 expression. It may activate cyclin D2 by cooperating with the transcription factor MYC [101, 102], by inactivating the transcription factor Foxo3a [103], or inactivating GSK3β by phosphorylation at Ser 9 [93]. Consistently, inhibition of PI3-kinase activity inhibits increase in cyclin D2 expression mediated by MDM2 [25]. It is yet to be determined whether inhibition of PI3-kinase activity inhibits MDM2-mediated cyclin D2 expression by inhibiting phosphorylation of MDM2 at Ser 166 and Ser 186, preventing its nuclear localization.

Controlling DNA Replication and Checkpoint Response

As mentioned earlier, transient overexpression of MDM2 inhibits cyclin A expression in apparently normal diploid cells. Since cyclin A is required for initiation of DNA replication, inhibition of cyclin A expression in these cells leads to inhibition of DNA replication and consequently G1 arrest [24]. However, MDM2 does not inhibit cyclin A expression in cells that are defective in timely expression of cyclin A, yet reduces the frequency of actively replicating cells, suggesting that MDM2 inhibits DNA replication in these cells [69]. Studies in cultured lung cells from p53-null MDM2 transgenic mice show that elevated levels of MDM2 increase cyclin A expression in the absence of p53 and leads to early S phase entry. However in spite of early S phase entry, the lung cells from p53-null MDM2 transgenic mice show fewer DNA replication origin firing events compared to lung cells from p53-null mice at the onset of S phase [25]. Fiber analysis of replicating DNA from H1299 cells expressing MDM2 from an inducible promoter showed similar inhibition of origin firing [25].

The inhibition of DNA replication origin firing by MDM2 at the onset of S phase can be rescued by treatment with caffeine, an inhibitor of ATM and ATR kinases. Consistently, MDM2 induces chk1 phosphorylation at the onset of S phase [25]. Checkpoint kinases are known to restrict origin firing during normal unperturbed S phase progression in response to single stranded DNA exposed at replication forks [104–108]. A time course analysis of origin firing and induction of chk1 phosphorylation in synchronized cells revealed that MDM2 specifically inhibits origin firing and induces chk1 phosphorylation at the onset of S phase, considerably delaying S phase progression. p53-null cells displays induction of chk1 phosphorylation and restricted origin firing at a later time point [25].

Induction of the intra-S phase checkpoint response is known to inhibit origin firing by checkpoint-dependent phosphorylation and accumulation of mixed lineage lymphoma (MLL) histone methyl transferase, which induces trimethylation of histone H3 at lysine 4 (H3K4), thereby preventing CDC45 loading and delaying DNA replication [109]. Accordingly, elevated levels of MDM2 cause accumulation of MLL histone methyl transferase and checkpoint dependent trimethylation of H3K4, implying that induction of chk1 phosphorylation by MDM2 inhibits firing of DNA replication origins at the onset of S phase through accumulation of MLL and consequent H3K4 methylation [25].

In metazoan cells, cyclin dependent kinases and cdc7 kinases phosphorylate the minichromosome maintenance (MCM) proteins to activate MCM helicases, thus inducing melting of double stranded DNA at DNA replication origins [106, 110]. Inhibition of this activity abrogates the checkpoint response induced by MDM2. Thus, MDM2 upregulates expression of cyclin D2 and, consequently, cyclin A enhancing activation of origins (origin melting) and hastening S phase entry. However, this activity generates an early chk1 phosphorylation in response to single stranded DNA exposed at the replication fork, and inhibits further origin firing.

The current literature reports that several oncogenes overexpressed in various cancers have growth restricting properties in non-transformed cells [111, 112]. Oncogenic Ras and Myc induce DNA damage and checkpoint responses [113–115], and Raf-1 induces cell cycle arrest and senescence [116]. Signs of oncogene induced senescence have been detected in pre-malignant lesions of human and animal tumors [117]. Oncogene mediated growth suppression is considered to be a safety mechanism to prevent untimely cell proliferation. Inhibition of origin firing and chk1 phosphorylation induced by MDM2 is a novel cellular defense mechanism that prevents unscheduled origin firing and DNA replication.

These observations signify that a compromised checkpoint response would induce untimely origin firing by MDM2. ATR and Chk1 kinases are required for cell survival [118], and inactivation of these kinases is not a frequent event in oncogenesis. However, mutations in the downstream checkpoint pathways, such as translocation of N-terminal MLL histone methyl transferase, are common in leukemia, lung and other cancers [109, 119]. These MLL fusion proteins act as dominant negative mutants of the WT protein, inactivating their functions and compromising the S phase checkpoint [109]. Overexpression or amplification of MDM2 is also a frequent event in leukemia and lung cancer [5, 120]. Co-occurrence of MLL translocation and MDM2 overexpression would lead to unscheduled origin firing, which is known to induce gene abnormalities [107, 121, 122].

Controlling DNA Polymerases

MDM2 is known to interact with more than one DNA polymerase. Using a yeast two-hybrid system, Vlakovic et al. detected that MDM2 interacts with the C-terminus of the catalytic subunit of DNA polymerase ε [57]. Using purified proteins, this group also found that MDM2 stimulates the catalytic activity of this enzyme *in vitro* [55]. DNA polymerase ε functions as the primary leading strand replicase during replication of the eukaryotic genome [123]. This enzyme is also known to translocate the CMG (CDC45, MCM2-7 and GINS or Go-Ichi-Nii-San) complex for progression of the CMG helicases [124]. Hence, stimulation of this enzyme by MDM2 in eukaryotic cells may accelerate leading strand synthesis.

MDM2 has been shown to interact with DNA polymerase η known to be involved in translesion DNA synthesis [56]. Using transient transfection of human lung cancer cell lines, Jung et al. have shown that MDM2 interacts with DNA polymerase η and promotes its polyubiquitination independent of the presence of WT p53. Knockdown of MDM2 increases expression of DNA polymerase η and its localization on DNA. Transient knockdown of MDM2 also reduces sensitivity of UV induced cell death. This study, therefore, implies that degradation of DNA polymerase η by MDM2 would reduce DNA repair after DNA damage causing cell death (Fig. 12.2).

Genome Abnormalities

In eukaryotic cells, during the G1/S transition and at different times during S phase, a fraction of licensed replication origins are fired in a temporally regulated manner [105, 125]. Firing of replication origins is regulated by checkpoint kinases during normal unperturbed S phase. As mentioned above, unscheduled origin firing is known to induce gene abnormalities [107, 121, 122]. Consistent with the observation that MDM2 induces unscheduled origin firing [25], MDM2 has been implicated in the generation of gene abnormalities [126]. Metaphase chromosome analysis of splenocytes from aging mice by Lushnikova et al. [127] showed increased chromosome abnormalities (chromosome and chromatid break, fusion and aneuploidy) with age. Similar analysis of splenocytes from littermate MDM2 transgenic mice reveals increases in chromosome fusion, aneuploidy and polyploidy, but not chromosome breakage. This study also reports that MDM2 overexpression leads to gain in one or more chromosome with age, rather than loss. It is interesting to note that

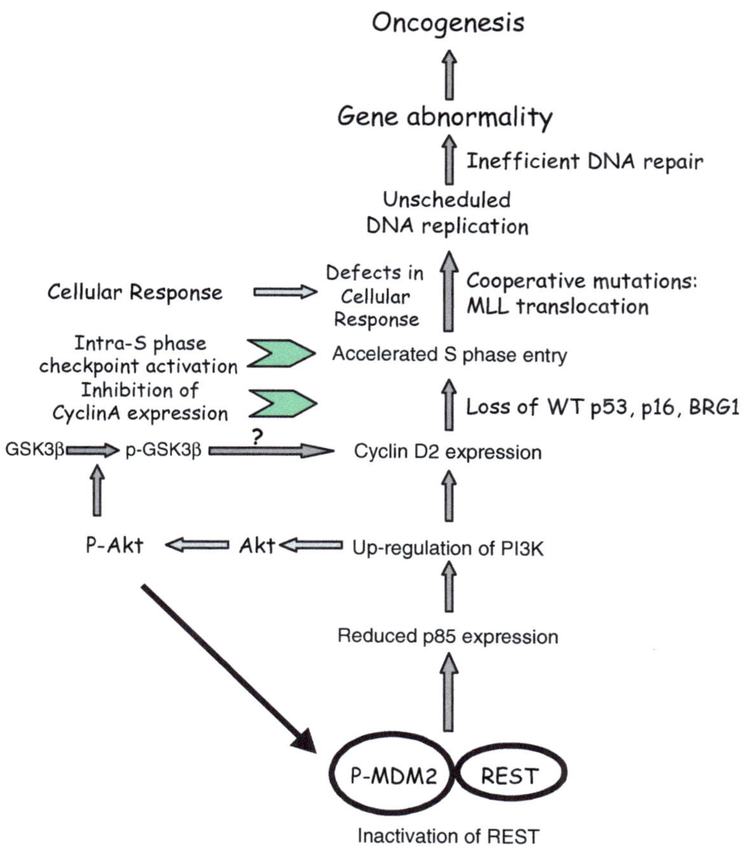

Fig. 12.2 Proposed pathway of MDM2-mediated oncogenesis

chromosome fusion, inter- or intra- chromosomal aberrations, and chromosome gain are generated from perturbations in DNA replication [128, 129].

Bouska et al. have reported that introduction of a retroviral expression vector increases chromosome and chromatid breaks in cultured cells lacking p53, presumably due to viral integration, while increased MDM2 expression using a retroviral expression vector led to an increase in chromosome and chromatid breaks and delayed DNA double strand break repair [130]. This effect has been related to interaction between Nijmegen breakage syndrome protein, Nbs1. Nbs1 is a component of Mre11/Rad50/Nbs1 DNA double strand break repair complex. The gene encoding this protein is mutated in patients exhibiting Nijmegen breakage syndrome due to compromised ability to repair DNA breaks. The study reports that MDM2 increased the number of DNA breaks introduced by retroviruses or gamma radiation.

Although the pattern of genome abnormalities introduced by MDM2 in the two studies varied, the common theme in the two reports is that these abnormalities require an additional gene defect, either acquired through aging or by chromosome breaks introduced as a result of retroviral integration. These reports are consistent with the observation by Ganguli et al. [72] that MDM2 increases tumor formation by chemical carcinogens, implying that introduction of gene damage by chemical carcinogens is essential for oncogenic functions of MDM2.

In summary, involvement of MDM2 in a multitude of crucial cell growth regulatory processes implies that functional deregulation of this protein could be a challenge for timely progression of DNA replication, DNA repair and maintenance of genomic integrity. Consistently, there are multiple levels of cellular defense mechanisms that can be invoked to resist the cell proliferative consequences of MDM2 overexpression, but which are violated by cancer cells. However, these observations also indicate the crucial involvement of MDM2 during normal cell growth. Since MDM2 is induced in response to genetic challenges, the protein perhaps participates in initiating timely DNA replication during normal cell cycle or when induced.

References

1. Fakharzadeh SS, Trusko SP, George DL (1991) Tumorigenic potential associated with enhanced expression of a gene that is amplified in a mouse tumor cell line. EMBO J 10:1565–1569
2. Oliner JD, Kinzler KW, Meltzer PS, George DL, Vogelstein B (1992) Amplification of a gene encoding a p53-associated protein in human sarcomas. Nature 358:80–83
3. Momand J, Zambetti GP, Olson DC, George D, Levine AJ (1992) The mdm-2 oncogene product forms a complex with the p53 protein and inhibits p53-mediated transactivation. Cell 69:1237–1245
4. Senturk E, Manfredi JJ (2012) Mdm2 and tumorigenesis: evolving theories and unsolved mysteries. Genes Cancer 3:192–198
5. Onel K, Cordon-Cardo C (2004) MDM2 and prognosis. Mol Cancer Res 2:1–8
6. Bond GL, Hu W, Levine AJ (2005) MDM2 is a central node in the p53 pathway: 12 years and counting. Curr Cancer Drug Targets 5:3–8
7. Bond GL, Levine AJ (2007) A single nucleotide polymorphism in the p53 pathway interacts with gender, environmental stresses and tumor genetics to influence cancer in humans. Oncogene 26:1317–1323
8. Bond GL, Hu W, Levine A (2005) A single nucleotide polymorphism in the MDM2 gene: from a molecular and cellular explanation to clinical effect. Cancer Res 65:5481–5484
9. Cordon-Cardo C, Latres E, Drobnjak M, Oliva MR, Pollack D, Woodruff JM, Marechal V, Chen J, Brennan MF, Levine AJ (1994) Molecular abnormalities of mdm2 and p53 genes in adult soft tissue sarcomas. Cancer Res 54:794–799
10. Ko JL, Cheng YW, Chang SL, Su JM, Chen CY, Lee H (2000) MDM2 mRNA expression is a favorable prognostic factor in non-small-cell lung cancer. Int J Cancer J Int du Cancer 89:265–270
11. Miwa S, Uchida C, Kitagawa K, Hattori T, Oda T, Sugimura H, Yasuda H, Nakamura H, Chida K, Kitagawa M (2006) Mdm2-mediated pRB downregulation is involved in carcinogenesis in a p53-independent manner. Biochem Biophys Res Commun 340:54–61

12. Vaughan C, Mohanraj L, Singh S, Dumur CI, Ramamoorthy M, Garrett CT, Windle B, Yeudall WA, Deb S, Deb SP (2011) Human oncoprotein MDM2 up-regulates expression of NF-kappaB2 precursor p100 conferring a survival advantage to lung cells. Genes Cancer 2:943–955
13. Faur N, Araud L, Laroche-Clary A, Kanno J, Toutain J, Yamori T, Robert J, Le Morvan V (2009) The association between the T309G polymorphism of the MDM2 gene and sensitivity to anticancer drug is dependent on the p53 mutational status in cellular models. Br J Cancer 101:350–356
14. Lind H, Zienolddiny S, Ekstrom PO, Skaug V, Haugen A (2006) Association of a functional polymorphism in the promoter of the MDM2 gene with risk of nonsmall cell lung cancer. Int J Cancer J Int du Cancer 119:718–721
15. Ding L, Getz G, Wheeler DA, Mardis ER, McLellan MD, Cibulskis K, Sougnez C, Greulich H, Muzny DM, Morgan MB et al (2008) Somatic mutations affect key pathways in lung adenocarcinoma. Nature 455:1069–1075
16. Jones SN, Hancock AR, Vogel H, Donehower LA, Bradley A (1998) Overexpression of Mdm2 in mice reveals a p53-independent role for Mdm2 in tumorigenesis. Proc Natl Acad Sci U S A 95:15608–15612
17. Jones SN, Roe AE, Donehower LA, Bradley A (1995) Rescue of embryonic lethality in Mdm2-deficient mice by absence of p53. Nature 378:206–208
18. Montes de Oca Luna R, Wagner DS, Lozano G (1995) Rescue of early embryonic lethality in mdm2-deficient mice by deletion of p53. Nature 378:203–206
19. Barak Y, Gottlieb E, Juven-Gershon T, Oren M (1994) Regulation of mdm2 expression by p53: alternative promoters produce transcripts with nonidentical translation potential. Genes Dev 8:1739–1749
20. Barak Y, Juven T, Haffner R, Oren M (1993) Mdm2 expression is induced by wild type p53 activity. EMBO J 12:461–468
21. Juven T, Barak Y, Zauberman A, George DL, Oren M (1993) Wild type p53 can mediate sequence-specific transactivation of an internal promoter within the mdm2 gene. Oncogene 8:3411–3416
22. Gannon HS, Woda BA, Jones SN (2012) ATM phosphorylation of Mdm2 Ser394 regulates the amplitude and duration of the DNA damage response in mice. Cancer Cell 21:668–679
23. Itahana K, Mao H, Jin A, Itahana Y, Clegg HV, Lindstrom MS, Bhat KP, Godfrey VL, Evan GI, Zhang Y (2007) Targeted inactivation of Mdm2 RING finger E3 ubiquitin ligase activity in the mouse reveals mechanistic insights into p53 regulation. Cancer Cell 12:355–366
24. Frum R, Ramamoorthy M, Mohanraj L, Deb S, Deb SP (2009) MDM2 controls the timely expression of cyclin A to regulate the cell cycle. Mol Cancer Res 7:1253–1267
25. Frum RA, Singh S, Vaughan C, Mukhopadhyay ND, Grossman SR, Windle B, Deb S, Deb SP (2014) The human oncoprotein MDM2 induces replication stress eliciting early intra-S phase checkpoint response and inhibition of DNA replication origin firing. Nucleic Acids Res 42:926–940
26. Singh S, Ramamoorthy M, Vaughan C, Yeudall WA, Deb S, Palit Deb S (2013) Human oncoprotein MDM2 activates the Akt signaling pathway through an interaction with the repressor element-1 silencing transcription factor conferring a survival advantage to cancer cells. Cell Death Differ 20:558–566
27. Fahraeus R, Olivares-Illana V (2013) MDM2's social network. Oncogene. doi:10.1038/onc.2013.410 [Epub ahead of print]
28. Oliner JD, Pietenpol JA, Thiagalingam S, Gyuris J, Kinzler KW, Vogelstein B (1993) Oncoprotein MDM2 conceals the activation domain of tumour suppressor p53. Nature 362:857–860
29. Brown DR, Deb S, Munoz RM, Subler MA, Deb SP (1993) The tumor suppressor p53 and the oncoprotein simian virus 40 T antigen bind to overlapping domains on the MDM2 protein. Mol Cell Biol 13:6849–6857
30. Haines DS, Landers JE, Engle LJ, George DL (1994) Physical and functional interaction between wild-type p53 and mdm2 proteins. Mol Cell Biol 14:1171–1178

31. Leng P, Brown DR, Shivakumar CV, Deb S, Deb SP (1995) N-terminal 130 amino acids of MDM2 are sufficient to inhibit p53-mediated transcriptional activation. Oncogene 10:1275–1282
32. Haupt Y, Maya R, Kazaz A, Oren M (1997) Mdm2 promotes the rapid degradation of p53. Nature 387:296–299
33. Honda R, Tanaka H, Yasuda H (1997) Oncoprotein MDM2 is a ubiquitin ligase E3 for tumor suppressor p53. FEBS Lett 420:25–27
34. Kubbutat MH, Jones SN, Vousden KH (1997) Regulation of p53 stability by Mdm2. Nature 387:299–303
35. Deb SP (2002) Function and dysfunction of the human oncoprotein MDM2. Front Biosci J Virtual Libr 7:d235–d243
36. Iwakuma T, Lozano G (2003) MDM2, an introduction. Mol Cancer Res 1:993–1000
37. Zauberman A, Barak Y, Ragimov N, Levy N, Oren M (1993) Sequence-specific DNA binding by p53: identification of target sites and lack of binding to p53–MDM2 complexes. EMBO J 12:2799–2808
38. White DE, Talbott KE, Arva NC, Bargonetti J (2006) Mouse double minute 2 associates with chromatin in the presence of p53 and is released to facilitate activation of transcription. Cancer Res 66:3463–3470
39. Wang X, Taplick J, Geva N, Oren M (2004) Inhibition of p53 degradation by Mdm2 acetylation. FEBS Lett 561:195–201
40. Cross B, Chen L, Cheng Q, Li B, Yuan ZM, Chen J (2011) Inhibition of p53 DNA binding function by the MDM2 protein acidic domain. J Biol Chem 286:16018–16029
41. Brown DR, Deb D, Frum R, Hickes L, Munoz R, Deb S, Deb SP (2001) The human oncoprotein MDM2 uses distinct strategies to inhibit transcriptional activation mediated by the wild-type p53 and its tumor-derived mutants. Int J Oncol 18:449–459
42. Sakaguchi K, Herrera JE, Saito S, Miki T, Bustin M, Vassilev A, Anderson CW, Appella E (1998) DNA damage activates p53 through a phosphorylation-acetylation cascade. Genes Dev 12:2831–2841
43. Kastan MB, Lim DS (2000) The many substrates and functions of ATM. Nat Rev Mol Cell Biol 1:179–186
44. Tang Y, Zhao W, Chen Y, Zhao Y, Gu W (2008) Acetylation is indispensable for p53 activation. Cell 133:612–626
45. Khosravi R, Maya R, Gottlieb T, Oren M, Shiloh Y, Shkedy D (1999) Rapid ATM-dependent phosphorylation of MDM2 precedes p53 accumulation in response to DNA damage. Proc Natl Acad Sci U S A 96:14973–14977
46. Maya R, Balass M, Kim ST, Shkedy D, Leal JF, Shifman O, Moas M, Buschmann T, Ronai Z, Shiloh Y et al (2001) ATM-dependent phosphorylation of Mdm2 on serine 395: role in p53 activation by DNA damage. Genes Dev 15:1067–1077
47. Dias SS, Milne DM, Meek DW (2006) c-Abl phosphorylates Hdm2 at tyrosine 276 in response to DNA damage and regulates interaction with ARF. Oncogene 25:6666–6671
48. Cheng Q, Chen L, Li Z, Lane WS, Chen J (2009) ATM activates p53 by regulating MDM2 oligomerization and E3 processivity. EMBO J 28:3857–3867
49. de Toledo SM, Azzam EI, Dahlberg WK, Gooding TB, Little JB (2000) ATM complexes with HDM2 and promotes its rapid phosphorylation in a p53-independent manner in normal and tumor human cells exposed to ionizing radiation. Oncogene 19:6185–6193
50. Love IM, Grossman SR (2012) It takes 15 to tango: making sense of the many ubiquitin ligases of p53. Genes Cancer 3:249–263
51. Manfredi JJ (2010) The Mdm2-p53 relationship evolves: Mdm2 swings both ways as an oncogene and a tumor suppressor. Genes Dev 24:1580–1589
52. Ponnuswamy A, Hupp T, Fahraeus R (2012) Concepts in MDM2 signaling: allosteric regulation and feedback loops. Genes Cancer 3:291–297
53. Xiao ZX, Chen J, Levine AJ, Modjtahedi N, Xing J, Sellers WR, Livingston DM (1995) Interaction between the retinoblastoma protein and the oncoprotein MDM2. Nature 375:694–698

54. Martin K, Trouche D, Hagemeier C, Sorensen TS, La Thangue NB, Kouzarides T (1995) Stimulation of E2F1/DP1 transcriptional activity by MDM2 oncoprotein. Nature 375:691–694
55. Asahara H, Li Y, Fuss J, Haines DS, Vlatkovic N, Boyd MT, Linn S (2003) Stimulation of human DNA polymerase epsilon by MDM2. Nucleic Acids Res 31:2451–2459
56. Jung YS, Qian Y, Chen X (2012) DNA polymerase eta is targeted by Mdm2 for polyubiquitination and proteasomal degradation in response to ultraviolet irradiation. DNA Repair 11:177–184
57. Vlatkovic N, Guerrera S, Li Y, Linn S, Haines DS, Boyd MT (2000) MDM2 interacts with the C-terminus of the catalytic subunit of DNA polymerase epsilon. Nucleic Acids Res 28:3581–3586
58. Gu L, Findley HW, Zhou M (2002) MDM2 induces NF-kappaB/p65 expression transcriptionally through Sp1-binding sites: a novel, p53-independent role of MDM2 in doxorubicin resistance in acute lymphoblastic leukemia. Blood 99:3367–3375
59. Johnson-Pais T, Degnin C, Thayer MJ (2001) pRB induces Sp1 activity by relieving inhibition mediated by MDM2. Proc Natl Acad Sci U S A 98:2211–2216
60. Heyne K, Winter C, Gerten F, Schmidt C, Roemer K (2013) A novel mechanism of crosstalk between the p53 and NFkappaB pathways: MDM2 binds and inhibits p65RelA. Cell Cycle 12:2479–2492
61. Hsieh JK, Chan FS, O'Connor DJ, Mittnacht S, Zhong S, Lu X (1999) RB regulates the stability and the apoptotic function of p53 via MDM2. Mol Cell 3:181–193
62. Sdek P, Ying H, Chang DL, Qiu W, Zheng H, Touitou R, Allday MJ, Xiao ZX (2005) MDM2 promotes proteasome-dependent ubiquitin-independent degradation of retinoblastoma protein. Mol Cell 20:699–708
63. Polager S, Ginsberg D (2009) p53 and E2f: partners in life and death. Nat Rev Cancer 9:738–748
64. Wunderlich M, Berberich SJ (2002) Mdm2 inhibition of p53 induces E2F1 transactivation via p21. Oncogene 21:4414–4421
65. Lundgren K, Montes de Oca Luna R, McNeill YB, Emerick EP, Spencer B, Barfield CR, Lozano G, Rosenberg MP, Finlay CA (1997) Targeted expression of MDM2 uncouples S phase from mitosis and inhibits mammary gland development independent of p53. Genes Dev 11:714–725
66. Reinke V, Bortner DM, Amelse LL, Lundgren K, Rosenberg MP, Finlay CA, Lozano G (1999) Overproduction of MDM2 in vivo disrupts S phase independent of E2F1. Cell Growth Differ Mol Biol J Am Assoc Cancer Res 10:147–154
67. Brown DR, Thomas CA, Deb SP (1998) The human oncoprotein MDM2 arrests the cell cycle: elimination of its cell-cycle-inhibitory function induces tumorigenesis. EMBO J 17:2513–2525
68. Bartel F, Harris LC, Wurl P, Taubert H (2004) MDM2 and its splice variant messenger RNAs: expression in tumors and down-regulation using antisense oligonucleotides. Mol Cancer Res 2:29–35
69. Zhou R, Frum R, Deb S, Deb SP (2005) The growth arrest function of the human oncoprotein mouse double minute-2 is disabled by downstream mutation in cancer cells. Cancer Res 65:1839–1848
70. Folberg-Blum A, Sapir A, Shilo BZ, Oren M (2002) Overexpression of mouse Mdm2 induces developmental phenotypes in Drosophila. Oncogene 21:2413–2417
71. Alkhalaf M, Ganguli G, Messaddeq N, Le Meur M, Wasylyk B (1999) MDM2 overexpression generates a skin phenotype in both wild type and p53 null mice. Oncogene 18:1419–1434
72. Ganguli G, Abecassis J, Wasylyk B (2000) MDM2 induces hyperplasia and premalignant lesions when expressed in the basal layer of the epidermis. EMBO J 19:5135–5147
73. Bennett-Lovsey R, Hart SE, Shirai H, Mizuguchi K (2002) The SWIB and the MDM2 domains are homologous and share a common fold. Bioinformatics 18:626–630

74. Lee D, Kim JW, Seo T, Hwang SG, Choi EJ, Choe J (2002) SWI/SNF complex interacts with tumor suppressor p53 and is necessary for the activation of p53-mediated transcription. J Biol Chem 277:22330–22337
75. Wang M, Gu C, Qi T, Tang W, Wang L, Wang S, Zeng X (2007) BAF53 interacts with p53 and functions in p53-mediated p21-gene transcription. J Biochem 142:613–620
76. Zhang HS, Gavin M, Dahiya A, Postigo AA, Ma D, Luo RX, Harbour JW, Dean DC (2000) Exit from G1 and S phase of the cell cycle is regulated by repressor complexes containing HDAC-Rb-hSWI/SNF and Rb-hSWI/SNF. Cell 101:79–89
77. Kang H, Cui K, Zhao K (2004) BRG1 controls the activity of the retinoblastoma protein via regulation of p21CIP1/WAF1/SDI. Mol Cell Biol 24:1188–1199
78. Giono LE, Manfredi JJ (2007) Mdm2 is required for inhibition of Cdk2 activity by p21, thereby contributing to p53-dependent cell cycle arrest. Mol Cell Biol 27:4166–4178
79. Medina PP, Carretero J, Ballestar E, Angulo B, Lopez-Rios F, Esteller M, Sanchez-Cespedes M (2005) Transcriptional targets of the chromatin-remodelling factor SMARCA4/BRG1 in lung cancer cells. Hum Mol Genet 14:973–982
80. Reisman DN, Sciarrotta J, Wang W, Funkhouser WK, Weissman BE (2003) Loss of BRG1/BRM in human lung cancer cell lines and primary lung cancers: correlation with poor prognosis. Cancer Res 63:560–566
81. Ruas M, Peters G (1998) The p16INK4a/CDKN2A tumor suppressor and its relatives. Biochim Biophys Acta 1378:F115–F177
82. Dutta A, Chandra R, Leiter LM, Lester S (1995) Cyclins as markers of tumor proliferation: immunocytochemical studies in breast cancer. Proc Natl Acad Sci U S A 92:5386–5390
83. Keyomarsi K, Pardee AB (1993) Redundant cyclin overexpression and gene amplification in breast cancer cells. Proc Natl Acad Sci U S A 90:1112–1116
84. Leveillard T, Wasylyk B (1997) The MDM2 C-terminal region binds to TAFII250 and is required for MDM2 regulation of the cyclin A promoter. J Biol Chem 272: 30651–30661
85. Leng P, Brown DR, Deb S, Deb SP (1995) Human oncoprotein MDM2 interacts with the TATA-binding protein in vitro and in vivo. Int J Oncol 6:251–259
86. Thut CJ, Goodrich JA, Tjian R (1997) Repression of p53-mediated transcription by MDM2: a dual mechanism. Genes Dev 11:1974–1986
87. Minsky N, Oren M (2004) The RING domain of Mdm2 mediates histone ubiquitylation and transcriptional repression. Mol Cell 16:631–639
88. Zhao J, Bilsland A, Jackson K, Keith WN (2005) MDM2 negatively regulates the human telomerase RNA gene promoter. BMC Cancer 5:6
89. Pardee AB (1974) A restriction point for control of normal animal cell proliferation. Proc Natl Acad Sci U S A 71:1286–1290
90. Mullany LK, White P, Hanse EA, Nelsen CJ, Goggin MM, Mullany JE, Anttila CK, Greenbaum LE, Kaestner KH, Albrecht JH (2008) Distinct proliferative and transcriptional effects of the D-type cyclins in vivo. Cell Cycle 7:2215–2224
91. Diao L, Chen YG (2007) PTEN, a general negative regulator of cyclin D expression. Cell Res 17:291–292
92. Banerji L, Glassford J, Lea NC, Thomas NS, Klaus GG, Lam EW (2001) BCR signals target p27(Kip1) and cyclin D2 via the PI3-K signalling pathway to mediate cell cycle arrest and apoptosis of WEHI 231 B cells. Oncogene 20:7352–7367
93. Huang W, Chang HY, Fei T, Wu H, Chen YG (2007) GSK3 beta mediates suppression of cyclin D2 expression by tumor suppressor PTEN. Oncogene 26:2471–2482
94. Fresno Vara JA, Casado E, de Castro J, Cejas P, Belda-Iniesta C, Gonzalez-Baron M (2004) PI3K/Akt signalling pathway and cancer. Cancer Treat Rev 30:193–204
95. Keniry M, Parsons R (2008) The role of PTEN signaling perturbations in cancer and in targeted therapy. Oncogene 27:5477–5485
96. Steelman LS, Chappell WH, Abrams SL, Kempf RC, Long J, Laidler P, Mijatovic S, Maksimovic-Ivanic D, Stivala F, Mazzarino MC et al (2011) Roles of the Raf/MEK/ERK and

PI3K/PTEN/Akt/mTOR pathways in controlling growth and sensitivity to therapy-implications for cancer and aging. Aging (Albany NY) 3:192–222
97. Westbrook TF, Martin ES, Schlabach MR, Leng Y, Liang AC, Feng B, Zhao JJ, Roberts TM, Mandel G, Hannon GJ et al (2005) A genetic screen for candidate tumor suppressors identifies REST. Cell 121:837–848
98. Ogawara Y, Kishishita S, Obata T, Isazawa Y, Suzuki T, Tanaka K, Masuyama N, Gotoh Y (2002) Akt enhances Mdm2-mediated ubiquitination and degradation of p53. J Biol Chem 277:21843–21850
99. Mayo LD, Donner DB (2001) A phosphatidylinositol 3-kinase/Akt pathway promotes translocation of Mdm2 from the cytoplasm to the nucleus. Proc Natl Acad Sci U S A 98:11598–11603
100. Frum R, Busby SA, Ramamoorthy M, Deb S, Shabanowitz J, Hunt DF, Deb SP (2007) HDM2-binding partners: interaction with translation elongation factor EF1alpha. J Proteome Res 6:1410–1417
101. Bouchard C, Marquardt J, Bras A, Medema RH, Eilers M (2004) Myc-induced proliferation and transformation require Akt-mediated phosphorylation of FoxO proteins. EMBO J 23:2830–2840
102. Bouchard C, Thieke K, Maier A, Saffrich R, Hanley-Hyde J, Ansorge W, Reed S, Sicinski P, Bartek J, Eilers M (1999) Direct induction of cyclin D2 by Myc contributes to cell cycle progression and sequestration of p27. EMBO J 18:5321–5333
103. Fernandez de Mattos S, Essafi A, Soeiro I, Pietersen AM, Birkenkamp KU, Edwards CS, Martino A, Nelson BH, Francis JM, Jones MC et al (2004) FoxO3a and BCR-ABL regulate cyclin D2 transcription through a STAT5/BCL6-dependent mechanism. Mol Cell Biol 24:10058–10071
104. Ben-Yehoyada M, Gautier J, Dupre A (2007) The DNA damage response during an unperturbed S-phase. DNA Repair 6:914–922
105. Grallert B, Boye E (2008) The multiple facets of the intra-S checkpoint. Cell Cycle 7:2315–2320
106. Nakanishi M, Katsuno Y, Niida H, Murakami H, Shimada M (2010) Chk1-cyclin A/Cdk1 axis regulates origin firing programs in mammals. Chromosom Res Int J Mol Supramol Evol Asp Chromosom Biol 18:103–113
107. Sorensen CS, Syljuasen RG (2012) Safeguarding genome integrity: the checkpoint kinases ATR, CHK1 and WEE1 restrain CDK activity during normal DNA replication. Nucleic Acids Res 40:477–486
108. Willis N, Rhind N (2009) Regulation of DNA replication by the S-phase DNA damage checkpoint. Cell Div 4:13
109. Liu H, Takeda S, Kumar R, Westergard TD, Brown EJ, Pandita TK, Cheng EH, Hsieh JJ (2010) Phosphorylation of MLL by ATR is required for execution of mammalian S-phase checkpoint. Nature 467:343–346
110. Bell SP, Dutta A (2002) DNA replication in eukaryotic cells. Annu Rev Biochem 71:333–374
111. Di Micco R, Fumagalli M, D'Adda di Fagagna F (2007) Breaking news: high-speed race ends in arrest–how oncogenes induce senescence. Trends Cell Biol 17:529–536
112. Yaswen P, Campisi J (2007) Oncogene-induced senescence pathways weave an intricate tapestry. Cell 128:233–234
113. Di Micco R, Fumagalli M, Cicalese A, Piccinin S, Gasparini P, Luise C, Schurra C, Garre M, Nuciforo PG, Bensimon A et al (2006) Oncogene-induced senescence is a DNA damage response triggered by DNA hyper-replication. Nature 444:638–642
114. Herold S, Herkert B, Eilers M (2009) Facilitating replication under stress: an oncogenic function of MYC? Nat Rev Cancer 9:441–444
115. Lopez-Contreras AJ, Gutierrez-Martinez P, Specks J, Rodrigo-Perez S, Fernandez-Capetillo O (2012) An extra allele of Chk1 limits oncogene-induced replicative stress and promotes transformation. J Exp Med 209:455–461

116. Zhu J, Woods D, McMahon M, Bishop JM (1998) Senescence of human fibroblasts induced by oncogenic Raf. Genes Dev 12:2997–3007
117. Collado M, Serrano M (2010) Senescence in tumours: evidence from mice and humans. Nat Rev Cancer 10:51–57
118. Segurado M, Tercero JA (2009) The S-phase checkpoint: targeting the replication fork. Biol Cell Under Auspice Eur Cell Biol Organ 101:617–627
119. de Boer J, Walf-Vorderwulbecke V, Williams O (2013) In focus: MLL-rearranged leukemia. Leukemia 27:1224–1228
120. Bueso-Ramos CE, Yang Y, deLeon E, McCown P, Stass SA, Albitar M (1993) The human MDM-2 oncogene is overexpressed in leukemias. Blood 82:2617–2623
121. Branzei D, Foiani M (2010) Maintaining genome stability at the replication fork. Nat Rev Mol Cell Biol 11:208–219
122. Errico A, Costanzo V (2012) Mechanisms of replication fork protection: a safeguard for genome stability. Crit Rev Biochem Mol Biol 47:222–235
123. Miyabe I, Kunkel TA, Carr AM (2011) The major roles of DNA polymerases epsilon and delta at the eukaryotic replication fork are evolutionarily conserved. PLoS Genet 7:e1002407
124. Handa T, Kanke M, Takahashi TS, Nakagawa T, Masukata H (2012) DNA polymerization-independent functions of DNA polymerase epsilon in assembly and progression of the replisome in fission yeast. Mol Biol Cell 23:3240–3253
125. Machida YJ, Hamlin JL, Dutta A (2005) Right place, right time, and only once: replication initiation in metazoans. Cell 123:13–24
126. Bouska A, Eischen CM (2009) Mdm2 affects genome stability independent of p53. Cancer Res 69:1697–1701
127. Lushnikova T, Bouska A, Odvody J, Dupont WD, Eischen CM (2011) Aging mice have increased chromosome instability that is exacerbated by elevated Mdm2 expression. Oncogene 30:4622–4631
128. Carr AM, Paek AL, Weinert T (2011) DNA replication: failures and inverted fusions. Semin Cell Dev Biol 22:866–874
129. Mizuno K, Miyabe I, Schalbetter SA, Carr AM, Murray JM (2013) Recombination-restarted replication makes inverted chromosome fusions at inverted repeats. Nature 493:246–249
130. Bouska A, Lushnikova T, Plaza S, Eischen CM (2008) Mdm2 promotes genetic instability and transformation independent of p53. Mol Cell Biol 28:4862–4874

Chapter 13
p53-Independent Effects of Mdm2

Stephen Bohlman and James J. Manfredi

Abstract Mdm2 is best known as the primary negative regulator of p53, but a growing body of evidence suggests that Mdm2 also has a number of functions independent of its role in regulating p53. Although these functions are not yet well-characterized, they have been implicated in regulating of a number of cellular processes, including cell-cycle control, apoptosis, differentiation, genome stability, and transcription, among others. It appears that Mdm2 exerts these functions through a surprisingly wide variety of mechanisms. For example, it has been shown that Mdm2 can ubiquitinate alternative targets, can stimulate the activity of transcription factors, and can directly bind to mRNA to regulate its stability. Dysregulation of p53-independent functions could be responsible for the oncogenic properties of Mdm2 seen even in the absence of p53, and may explain why approximately 10 % of human tumors overexpress Mdm2 instead of inactivating p53 through other mechanisms. As the p53-independent functions of Mdm2 present novel targets for potential therapeutic interventions, fully characterizing these cellular and pathogenic roles of Mdm2 will be important in the study of tumor biology and the treatment of cancer.

Keywords Mdm2 • p53 • Oncogenesis • Ubiquitination

Mdm2 (murine double minute 2 homolog) is best known for its role as a negative regulator of the tumor-suppressor p53. Due to the ability of p53 to induce cell-cycle arrest and apoptosis, tight regulation of this protein is necessary for normal cellular

S. Bohlman • J.J. Manfredi (✉)
Department of Oncological Sciences, The Graduate School of Biomedical Sciences Icahn School of Medicine at Mount Sinai, New York, NY 10029, USA
e-mail: james.manfredi@mssm.edu

S.P. Deb and S. Deb (eds.), *Mutant p53 and MDM2 in Cancer*,
Subcellular Biochemistry 85, DOI 10.1007/978-94-017-9211-0_13,

growth and development. The primary way this regulation is accomplished is through interaction with Mdm2 [26]. Mdm2 is a E3 ubiquitin ligase that ubiquitinates p53 and targets it for proteasomal degradation [15, 17, 49]. Mdm2 is also able to directly bind the N-terminus of p53, inhibiting its activity as a transcription factor for anti-proliferative genes [39]. Mdm2 itself is a transcriptional target of p53, as the *Mdm2* gene possesses a sequence-specific recognition site for p53 within an internal promoter [23, 51]. In this manner, a negative-feedback loop is established, whereby p53 transactivates its own inhibitor. In support of such a regulatory loop, studies of p53 kinetics have shown that elevated levels of p53 following DNA damage decrease in a series of dampened oscillations [29].

In response to various stress signals, the Mdm2-p53 interaction is disrupted. Genotoxic stress causes p53 to undergo phosphorylation and acetylation at a number of residues [1, 54]. However, studies have shown that these modifications are not inherently critical to p53 transactivation function, and are only necessary to counter Mdm2 regulation [13, 58, 61]. This suggests that the primary function of these post-translational modifications are the disruption of the Mdm2-p53 interaction. A number of other mechanisms for regulating this have been characterized. Of note is the inhibition of Mdm2 by either p14^{ARF} in response to oncogenic signaling [53] or several ribosomal proteins in response to ribosomal stress [72].

The importance of the Mdm2-p53 relationship to normal cellular development was dramatically illustrated by the homozygous deletion of *Mdm2* in mice, which results in lethality at the blastocyst stage due to aberrant apoptosis. Strikingly, this phenotype can be completely rescued by the concomitant deletion of p53 [20, 41]. Recently, it has been demonstrated that although p53 transactivation of Mdm2 is necessary for a normal DNA damage response, p53-independent basal expression levels of Mdm2 are sufficient for regulation of p53 in most tissues under homeostatic conditions. This was demonstrated by the use of a transgenic mouse model in which basal levels of Mdm2 are expressed from its p53-independent promoter, but has lost p53-inducible Mdm2 expression [38, 47].

Disruption of the p53 pathway is critical for the development of cancer, as demonstrated by the fact that over 50 % of human tumors contain p53 mutations [16]. Therefore, it is unsurprising that Mdm2 overexpression has been also been implicated in tumorigenesis as an alternative method of inactivating p53 [30]. In one such study, *Mdm2* was found to be overexpressed in 7 % of 3,889 human tumor samples [40]. The highest frequencies of Mdm2 overexpression were found in soft-tissue sarcomas (20 %) and osteosarcomas (16 %), while other tumor types, such as leukemias, lymphomas, and pancreatic carcinomas, showed little Mdm2 overexpression. Importantly, there is a negative association between amplification of Mdm2 and mutation of p53 [45]. This association holds potential clinical implications in the possibility of developing therapeutics that restore p53 function through the inhibition of Mdm2. Indeed, this is the rationale that underlies the investigation of Nutlin-3, a small-molecule inhibitor of the Mdm2/p53 interaction, as a potential therapy in cancers that retain functional p53 [27, 52].

Mdm2 and Cancer

Consistent with its expected role in p53 regulation, ubiquitous overexpression of Mdm2 in transgenic mice predisposes them to spontaneous tumor formation. Surprisingly, however, when Mdm2 is overexpressed in mice in a p53-null background, an increased incidence of sarcomas is observed relative to p53-null mice alone [21]. This suggests that Mdm2 overexpression can also promote tumorigenesis through p53-independent mechanisms. This finding corroborated earlier reports that Mdm2 is able to transform cells *in vitro*, independent of p53 [3]. The increased rate of sarcomas in these mice also parallels the high rate of Mdm2 amplification in human sarcomas. Targeted overexpression of Mdm2 to mammary gland tissue led to the production of polyploid mammary epithelial cells in both p53+/+ and p53−/− backgrounds, suggesting a p53-independent role for Mdm2 in the regulation of DNA synthesis and cell-cycle progression [34].

In addition to mutations and changes in expression levels, gene activity can also be regulated through alternative splicing of mRNA. Studies of Mdm2 mRNA splicing showed multiple different-sized transcripts and protein isoforms that vary in their ability to bind to p53 [3, 14]. Analysis of Mdm2 mRNA in human ovarian and bladder cancer samples showed the presence of alternative and aberrant splice variants not found in normal tissue. Interestingly, four out of the five identified alternative transcripts contained partial deletions of the p53-binding domain, and expression of these isoforms *in vitro* confirmed their inability to interact with p53. However, all of the alternative transcripts were able to transform NIH 3 T3 cells, indicating an oncogenic property for these transcripts independent of p53 [55]. Similar transcripts have also been identified in glioblastomas, breast carcinomas, pediatric rhabdomyosarcomas, and oral squamous cell carcinomas [2, 37, 33, 50]. In total, over 40 different alternative splice variants have been identified in human tumors, most of which lack the p53-binding domain. Although the exact function of these splice variants is unknown, the fact that multiple splice variants are associated with specific tumor types, and that most of these splice variants do not interact with p53, suggest that these aberrant isoforms may play a p53-independent role in promoting tumorigenesis.

Although initial studies suggested that overexpression of Mdm2 and mutation of p53 were mutually exclusive in human tumor samples, this was later found to not be the case in certain tumor types. Although the majority of soft-tissue sarcomas contain one or the other, a significant proportion of tumor samples exhibit both an overexpression of Mdm2 and mutation of p53. Interestingly, it was found that there is a significant correlation between tumors containing changes in both proteins and poor patient prognosis [7]. Another confirmatory study showed that the effect on patient survival of Mdm2/p53 cooverexpression was greater than the additive effects of each independently [67]. Additionally, high levels of Mdm2 mRNA expression are correlated with an earlier age of onset of soft-tissue sarcomas [60]. Altogether, these findings point to Mdm2 having p53-independent effects in the process of tumorigenesis.

p53-Independent Effects of Mdm2

How might such p53-independent effects occur? Although the regulation of p53 degradation is by far the most extensively characterized function of Mdm2, a number of other functions have been described.

Since Mdm2 is an E3 ubiquitin ligase, a straightforward hypothesis would be that Mdm2 ubiquitinates other proteins in addition to p53, targeting them for proteasomal degradation as well. Indeed, Mdm2 has been shown to target a number of other proteins involved in cell-cycle regulation and apoptosis.

The Ras/Raf/MEK/ERK pathway couples cell-surface receptor signals to transcription factors that regulate proliferation and differentiation, and mutations or alterations in this pathway are found in many human cancers [10]. When this pathway is activated, ERK phosphorylates Foxo3A, a transcription factor for cell-cycle regulatory proteins. In this phosphorylated form, Foxo3A becomes a target for ubiquitination by MDM2, promoting its degradation. As Foxo3 acts as a tumor suppressor, Mdm2-mediated down-regulation of Foxo3 could play a role in tumorigenesis in response to oncogenic growth factor signaling [70].

Additionally, Mdm2 has been shown to ubiquitinate the cell-adhesion protein E-cadherin and target it for degradation via the 26S proteasome [69]. E-cadherin has an extensively characterized role in the epithelial-to-mesenchymal transition that occurs when tumors initiate metastasis. Downregulation of E-cadherin causes a loss of cell polarity and cell-cell adhesion, and promotes cell motility and invasiveness; these in turn lead to invasion of the blood stream and metastasis. Interestingly, Mdm2 overexpression occurs more frequently in metastatic and recurrent tumors than it does in primary tumors [9, 28]. Overexpression of Mdm2 is also associated with a poor patient prognosis in a number of cancers [19, 24]. Thus, Mdm2 overexpression could potentially promote tumor invasion and metastasis through the downregulation of E-cadherin. However, whether or not this interaction is important in tumor progression has yet to be confirmed.

Paradoxically, Mdm2 has also been shown to target the transcription factor Slug for degradation. Slug (also known as SNAI1), is a member of the Snail family of transcriptional repressors. It is a key promoter of the epithelial-to-mesenchymal transition, which it stimulates by repressing the transcription of E-cadherin. This was shown to occur through a p53-Mdm2-Slug complex; interestingly, mutant and transcriptionally-inactive p53 inactivates the Mdm2-mediated degradation of Slug [64]. Although Mdm2 is normally thought of as an oncogene, its degradation of an invasion-promoting protein suggests that under certain circumstances, Mdm2 can also act as a tumor suppressor.

Mdm2 has also been shown to interact with the retinoblastoma protein (pRb), a tumor suppressor gene that, like p53, has a major role in cell-cycle inhibition and apoptosis. Unlike the Mdm2-p53 interaction, however, the Mdm2-pRb interaction is not mediated by ubiquitin ligation and degradation. In cells, Mdm2 forms a complex with pRb and disrupts the G1/S checkpoint by preventing pRb from binding to and inactivating certain members of the family of E2F transcription factors [68]. Interestingly, it has also been demonstrated that Mdm2 forms a trimeric pRb-Mdm2-p53 complex, with Mdm2 acting as a linker between these two transcription factors. Binding of pRb to Mdm2 does not inhibit p53 binding; this stands in contrast to other known regulators of the Mdm2-p53 interaction, such as p14^{ARF}, which act by binding to Mdm2 and inhibiting p53 binding. The pRb-Mdm2 interaction is able to inhibit Mdm2-mediated p53 degradation, but does not remove the ability of Mdm2 to inhibit p53 transactivation of its target genes [18]. The finding that pRb impacts the apoptotic ability, but not the transcriptional activity of p53 suggests that the apoptotic effects of p53 may be independent of transactivation [71]. The discovery of cross-talk between the p53 and pRb pathways is particularly exciting, as these are arguably the two most important pathways in the prevention of tumorigenesis.

Mdm2 also impacts the pRb pathway through its interaction with E2F1 and DP1, transcription factors that heterodimerize and activate genes involved in the G1/S-phase transition. E2F1 contains a series of amino acids that are homologous to the activation domain of p53. In contrast to its negative regulation of p53, Mdm2 stimulates the transcriptional activity of E2F1/DP1, promoting progression into S phase [36]. In addition, Mdm2 increases degradation of the E2F1/DP1 complex, and can prevent p53-null cells from entering E2F-mediated apoptosis [32]. Thus, the Mdm2-E2F interaction both promotes cell growth through increased gene transcription and prevents cell death by inhibiting apoptosis, two activities that are hallmarks of tumorigenesis. However, whether or not the effect of Mdm2 on E2F1 is completely p53-independent is unclear. One study has suggested that this effect is mediated through inhibition of p53-dependent transcription of p21, the resultant increase in cyclin-dependent kinase activity causing phosphorylation/inactivation of pRb, which in turn stimulates E2F1 activity [12, 66]. These effects are not necessarily mutually exclusive; it is possible that this occurs through both p53-dependent and -independent mechanisms. The nature of the Mdm2-E2F1 relationship and its effect on cellular growth and tumorigenesis warrants further investigation.

The RING finger domain of Mdm2, which is responsible for its E3 ubiquitin ligase activity, also serves as a binding site for the closely related protein MdmX [59]. Mice lacking MdmX have similarity to Mdm2-knockout mice, in that both have an embryonic-lethal phenotype, although MdmX-null lethality occurs at a later stage of development and is associated with cell cycle arrest. Both phenotypes can be rescued by the loss of p53 [35, 48]. MdmX, despite possessing homology to the RING finger domain of Mdm2, does not possess E3-ubiquitin ligase activity. MdmX also represses p53, but does not do so through p53 degradation [25]. It has been suggested that Mdm2 homo-oligomers have different functions than Mdm2-MdmX heterodimers and can modify each other's function and regulation. The growing evidence surrounding the Mdm2-MdmX-p53 triumvirate suggests a complex relationship between these proteins that modifies their functions, modifications, and stabilities [62]. The interaction between Mdm2 and MdmX, and what role it plays in the regulation of p53depdendent and -independent effects of Mdm2, remains an active area of investigation.

Mdm2 and Genome Instability

Recently, increasing evidence has been found that suggests that Mdm2 overexpression can cause genome instability in a p53-independent manner. As already noted, targeted expression of Mdm2 to the mammary gland in transgenic mice gave the surprising phenotype of mammary epithelial cells that were hypertrophic and polyploid, indicating that these cells had undergone multiple rounds of S phase without mitosis. This effect was found to be p53-independent, as the same phenotype was observed in both wild-type and p53-knockout backgrounds [34].

The Eischen laboratory has reported an interesting interaction between Mdm2 and Nbs1 (also called Nibrin or NBN), a member of the Mre11/Rad50/Nbs1 complex (more commonly referred to as MRN) that functions in initiating DNA double-strand break repair and activating cell-cycle checkpoints. Nbs1 is thought to localize the complex to double-stranded DNA breaks and play a role in activation of ATM signaling. Mdm2 overexpression in p53-null cells induced chromosome breakage and delayed DNA double-stranded break repair, but not in cells with a mutated form of Nbs1. Through mutational analysis of their respective binding domains, it was demonstrated that Mdm2 directly interacts with Nbs1 and inhibits its function, leading to a delayed DNA-damage response, possibly through reduction in ATM signaling. When this Mdm2-Nbs1 interaction is disrupted, the rate of DNA damage repair is restored [4].

Genomic instability is a common characteristic of many cancers [43]. Previous studies of Mdm2 overexpression had revealed that elevated Mdm2 levels led to increased chromosome/chromatid breaks, centrosome hyperamplification, and aneuploidy [6, 63]. At that time, however, it was assumed these effects were mediated by Mdm2 repression of p53, as loss of p53 can also promote genome instability [8, 11, 31]. However, recent evidence, such as the demonstration of Mdm2 inhibition of Nbs1, argues that Mdm2 increases genetic instability directly and in a manner independent from the instability caused through inhibition of p53. This novel function of Mdm2 could have implications for cancer treatment, as it presents a new target interaction for chemotherapeutic drugs, as well as the appropriate clinical use of Mdm2 inhibitors such as Nutlin [4].

Mdm2, Transcription, and Translation

Although its primary role is involved the degradation of proteins, one unexpected function of Mdm2 that has been described is Mdm2-dependent regulation of translation. Mdm2 is known to regulate the translation of p53 via two different mechanisms. First, Mdm2 has been shown to directly interact with p53-encoding mRNA and impact its translation. Binding of p53-mRNA to the RING finger domain of Mdm2 stimulates translation of p53, while simultaneously inhibiting the E3 ligase ability of Mdm2 [42]. Fascinatingly, this suggests that while Mdm2 negatively regulates p53 on the protein level, it is a positive regulator of p53 on the mRNA level.

In agreement with this, it has also been found that "silent" mutations of p53 – that is, nucleotide changes of the gene encoding p53 that do not change its amino acid sequence – have an impact on levels of p53 activity. Presumably, a silent mutation could alter the secondary structure of mRNA, thus altering its regulation by mRNA-binding proteins. This suggests that Mdm2-mRNA binding is an alternative method through which Mdm2 regulates p53 activity [5].

Secondly, Mdm2 has been shown to ubiquitinate and target the ribosomal protein RPL26 for degradation. RPL26 plays a critical role in the translation of p53-encoding mRNA following DNA damage [57]. Under normal conditions, Mdm2 targets RPL26 for degradation, secondarily prohibiting translation of p53. In response to DNA damage, the Mdm2 inhibition of RPL26 is attenuated, causing a rise in p53 translation [44].

Both of these mechanisms have been described as alternative methods by which Mdm2 regulates the activity of p53. In addition, they suggest additional mechanisms through which Mdm2 could regulate other cellular processes as well. Indeed, it has been recently been shown that Mdm2 binds to and stabilizes the mRNA encoding Slug. In p53-null cells, Mdm2 stabilization of Slug mRNA caused an increase in the amount of Slug protein and induced Slug-dependent effects, such as repression of E-cadherin and increased invasiveness [22]. The direct binding of Mdm2 to mRNA and altering its transcription is a novel mechanism by which Mdm2 could have tumorigenic effects independent of p53.

Future Directions

In conclusion, Mdm2 has been shown to have a number of different functions independent of its role as a regulator of p53. Through the demonstration of alternate ubiquitination targets, its effects on genome stability, and interactions with mRNA and ribosomal proteins, it is becoming increasingly clear that Mdm2 regulates cellular processes on a number of different levels. Although these p53-independent functions are not fully understood, they may contribute to the role of Mdm2 in oncogenesis. As Mdm2 is implicated in a significant portion of human tumors, these functions present novel targets for potential clinical therapies. In order to provide a clinical benefit, however, the exact nature of these p53-independent functions will need to be further characterized.

Based on current knowledge, several avenues of research need to be pursued. For one, over 40 different splice variants of the Mdm2 mRNA have been isolated from normal and tumor cells [3]. Studying the potentially distinct biological roles of these isoforms is likely to reveal more about the various functions of Mdm2. The finding that some human tumors express Mdm2 splice variants that lack the p53-binding domain suggests that these splice variants may contribute to oncogenesis through p53-independent effects [55]. One interesting area of further study would be to identify the effects of expressing these distinct splice variants in targeted tissues to determine what role they have in tumor development.

Additionally, given the variety of splice variants of Mdm2, a more rigorous analysis of Mdm2 overexpression in human cancers is warranted. Simple measurement of gene amplification or mRNA expression levels may not be truly reflective of the actual protein levels of Mdm2 and its variant forms in human tumors. A more detailed analysis that takes into account the existence of distinct isoforms will need to be performed in order to determine what isoforms are expressed in specific tumor types. Mdm2 overexpression seems to be associated with better prognosis in some tissue types, and worse prognosis in others [46]. However, these analyses have, for the most part, used methods that do not distinguish between splice variants. Comparing the expression of distinct isoforms in tumor samples with clinical outcome data could reveal whether or not Mdm2 isoforms have an effect on patient prognosis and potentially explain this paradoxical finding.

Nutlin-3 has been investigated for use as an agent that disrupts the Mdm2-p53 interaction. However, as a small-molecule inhibitor of Mdm2, it also has the potential to inhibit or alter p53-independent effects of Mdm2 as well. It has been shown that Nutlin-3 can increase cell toxicity following DNA damage in p53-null prostate cancer cells, where it acts as a radiosensitizer [56]. Recently, Nutlin-3 has also been shown to inhibit the epithelial-to-mesenchymal transition in p53-null cells by interfering with the TGF-β1-SmadSnail/Slug axis [65]. The exact mechanisms by which Nutlin-3 exerts p53-independent effects remain unclear. Such findings have important clinical relevance: traditional thinking would indicate that drugs such as Nutlin would only be clinically useful in cancers that retain wild-type p53. However, if Mdm2 inhibitors are found to interfere with the tumor-promoting, p53-independent effects of Mdm2, they could potentially be useful as chemotherapeutic agents in a much wider spectrum of tumors.

Given the central role of p53 in tumor suppression, it is unsurprising that most research on Mdm2 has been focused on its regulation of p53. Nevertheless, a growing body of evidence suggests that Mdm2 has a number of other p53-independent functions, both in normal cellular biology and tumorigenesis. The exact mechanism of these functions, under which conditions they occur, and their significance to oncogenesis have yet to be determined. As Mdm2 emerges as an important player in tumor development in its own right, fully characterizing its p53-independent functions will certainly be important in the study of tumor biology and the development of new therapies for the treatment of cancer.

References

1. Appella E, Anderson CW (2001) Post-translational modifications and activation of p53 by genotoxic stresses. Eur J Biochem 268:2764–2772
2. Bartel F, Taylor AC, Taubert H, Harris LC (2001) Novel mdm2 splice variants identified in pediatric rhabdomyosarcoma tumors and cell lines. Oncol Res 12:451–457
3. Bartel F, Taubert H, Harris LC (2002) Alternative and aberrant splicing of MDM2 mRNA in human cancer. Cancer Cell 2:9–15

4. Bouska A, Lushnikova T, Plaza S, Eischen CM (2008) Mdm2 promotes genetic instability and transformation independent of p53. Mol Cell Biol 28:4862–4874
5. Candeias MM, Malbert-Colas L, Powell DJ, Daskalogianni C, Maslon MM, Naski N, Bourougaa K, Calvo F, Fåhraeus R (2008) P53 mRNA controls p53 activity by managing Mdm2 functions. Nat Cell Biol 10:1098–1105
6. Carroll PEP, Okuda MM, Horn HFH, Biddinger PP, Stambrook PJP, Gleich LLL, Li YQY, Tarapore PP, Fukasawa KK (1999) Centrosome hyperamplification in human cancer: chromosome instability induced by p53 mutation and/or Mdm2 overexpression. Oncogene 18:1935–1944
7. Cordon-Cardo C, Latres E, Drobnjak M, Oliva MR, Pollack D, Woodruff JM, Marechal V, Chen J, Brennan MF, Levine AJ (1994) Molecular abnormalities of mdm2 and p53 genes in adult soft tissue sarcomas. Cancer Res 54:794–799
8. Dalton WBW, Yu BB, Yang VWV (2010) p53 suppresses structural chromosome instability after mitotic arrest in human cells. Oncogene 29:1929–1940
9. Datta MW, Macri E, Signoretti S, Renshaw AA, Loda M (2001) Transition from in situ to invasive testicular germ cell neoplasia is associated with the loss of p21 and gain of mdm-2 expression. Mod Pathol 14:437–442
10. Davies H, Bignell GR, Cox C, Stephens P, Edkins S, Clegg S, Teague J, Woffendin H, Garnett MJ, Bottomley W et al (2002) Mutations of the BRAF gene in human cancer. Nature 417:949–954
11. Donehower LA, Godley LA, Aldaz CM, Pyle R, Shi YP, Pinkel D, Gray J, Bradley A, Medina D, Varmus HE (1995) Deficiency of p53 accelerates mammary tumorigenesis in Wnt-1 transgenic mice and promotes chromosomal instability. Genes Dev 9:882–895
12. Ganguli G, Wasylyk B (2003) p53-independent functions of MDM2. Mol Cancer Res 1:1027–1035
13. Giono LE, Manfredi JJ (2007) Mdm2 is required for inhibition of Cdk2 activity by p21, thereby contributing to p53-dependent cell cycle arrest. Mol Cell Biol 27:4166–4178
14. Haines DS, Landers JE, Engle LJ, George DL (1994) Physical and functional interaction between wild-type p53 and mdm2 proteins. Mol Cell Biol 14:1171–1178
15. Haupt Y, Maya R, Kazaz A, Oren M (1997) Mdm2 promotes the rapid degradation of p53. Nature 387:296–299
16. Hollstein M, Sidransky D, Vogelstein B, Harris CC (1991) p53 mutations in human cancers. Science 253:49–53
17. Honda R, Tanaka H, Yasuda H (1997) Oncoprotein MDM2 is a ubiquitin ligase E3 for tumor suppressor p53. FEBS Lett 420:25–27
18. Hsieh JK, Chan F, O'Connor DJ, Mittnacht S (1999) RB regulates the stability and the apoptotic function of p53 via MDM2. Mol Cell 3:181–193
19. Ikeguchi M, Ueda T, Fukuda K, Yamaguchi K-I, Tsujitani S-I, Kaibara N (2002) Expression of the murine double minute gene 2 oncoprotein in esophageal squamous cell carcinoma as a novel marker for lack of response to chemoradiotreatment. Am J Clin Oncol 25:454–459
20. Jones SN, Roe AE, Donehower LA, Bradley A (1995) Rescue of embryonic lethality in Mdm2-deficient mice by absence of p53. Nature 378:206–208
21. Jones SN, Hancock AR, Vogel H, Donehower LA, Bradley A (1998) Overexpression of Mdm2 in mice reveals a p53-independent role for Mdm2 in tumorigenesis. Proc Natl Acad Sci 95:15608–15612
22. Jung C-H, Kim J, Park JK, Hwang S-G, Moon S-K, Kim W-J, Um H-D (2013) Mdm2 increases cellular invasiveness by binding to and stabilizing the Slug mRNA. Cancer Lett 335:270–277
23. Juven T, Barak Y, Zauberman A, George DL, Oren M (1993) Wild type p53 can mediate sequence-specific transactivation of an internal promoter within the mdm2 gene. Oncogene 8:3411–3416
24. Korkolopoulou P, Christodoulou P, Kapralos P, Exarchakos M, Bisbiroula A, Hadjiyannakis M, Georgountzos C, Thomas-Tsagli E (1997) The role of p53, MDM2 and c-erb B-2 oncoproteins, epidermal growth factor receptor and proliferation markers in the prognosis of urinary bladder cancer. Pathol Res Pract 193:767–775
25. Kruse J-P, Gu W (2009) Modes of p53 regulation. Cell 137:609–622

26. Kubbutat MH, Jones SN, Vousden KH (1997) Regulation of p53 stability by Mdm2. Nature 387:299–303
27. Künkele A, De Preter K, Heukamp L, Thor T, Pajtler KW, Hartmann W, Mittelbronn M, Grotzer MA, Deubzer HE, Speleman F et al (2012) Pharmacological activation of the p53 pathway by nutlin-3 exerts anti-tumoral effects in medulloblastomas. Neuro-Oncology 14:859–869
28. Ladanyi M, Cha C, Lewis R, Jhanwar SC, Huvos AG, Healey JH (1993) MDM2 gene amplification in metastatic osteosarcoma. Cancer Res 53:16–18
29. Lahav G (2008) Oscillations by the p53-Mdm2 feedback loop. Adv Exp Med Biol 641:28–38
30. Leach FS, Tokino T, Meltzer P, Burrell M, Oliner JD, Smith S, Hill DE, Sidransky D, Kinzler KW, Vogelstein B (1993) p53 mutation and MDM2 amplification in human soft tissue sarcomas. Cancer Res 53:2231–2234
31. Lengauer C, Kinzler KW, Vogelstein B (1998) Genetic instabilities in human cancers. Nature 396:643–649
32. Loughran O, La Thangue NB (2000) Apoptotic and growth-promoting activity of E2F modulated by MDM2. Mol Cell Biol 20:2186–2197
33. Lukas J, Gao DQ, Keshmeshian M, Wen WH, Tsao-Wei D, Rosenberg S, Press MF (2001) Alternative and aberrant messenger RNA splicing of the mdm2 oncogene in invasive breast cancer. Cancer Res 61:3212–3219
34. Lundgren K, Montes de Oca Luna R, McNeill YB, Emerick EP, Spencer B, Barfield CR, Lozano G, Rosenberg MP, Finlay CA (1997) Targeted expression of MDM2 uncouples S phase from mitosis and inhibits mammary gland development independent of p53. Genes Dev 11:714–725
35. Marine J-C, Francoz S, Maetens M, Wahl G, Toledo F, Lozano G (2006) Keeping p53 in check: essential and synergistic functions of Mdm2 and Mdm4. Cell Death Differ 13:927–934
36. Martin K, Trouche D, Hagemeier C, Sørensen TS, La Thangue NB, Kouzarides T (1995) Stimulation of E2F1/DP1 transcriptional activity by MDM2 oncoprotein. Nature 375:691–694
37. Matsumoto R, Tada M, Nozaki M, Zhang CL, Sawamura Y, Abe H (1998) Short alternative splice transcripts of the mdm2 oncogene correlate to malignancy in human astrocytic neoplasms. Cancer Res 58:609–613
38. Mendrysa SMS, Perry MEM (2000) The p53 tumor suppressor protein does not regulate expression of its own inhibitor, MDM2, except under conditions of stress. Mol Cell Biol 20:2023–2030
39. Momand J, Zambetti GP, Olson DC, George D, Levine AJ (1992) The mdm-2 oncogene product forms a complex with the p53 protein and inhibits p53-mediated transactivation. Cell 69:1237–1245
40. Momand J, Jung DD, Wilczynski SS, Niland JJ (1998) The MDM2 gene amplification database. Nucleic Acids Res 26:3453–3459. http://eutils.ncbi.nlm.nih.gov/entrez/eutils/elink.fcgi?dbfrom=pubmed&id=9671804&retmode=ref&cmd=prlinks
41. Montes de Oca Luna R, Wagner DS, Lozano G (1995) Rescue of early embryonic lethality in mdm2-deficient mice by deletion of p53. Nature 378:203–206
42. Naski N, Gajjar M, Bourougaa K, Malbert-Colas L, Fåhraeus R, Candeias M (2009) The p53 mRNA-Mdm2 interaction. Cell Cycle 8:31–34
43. Negrini S, Gorgoulis VG, Halazonetis TD (2010) Genomic instability – an evolving hallmark of cancer. Nat Rev Mol Cell Biol 11:220–228
44. Ofir-Rosenfeld Y, Boggs K, Michael D, Kastan MB, Oren M (2008) Mdm2 regulates p53 mRNA translation through inhibitory interactions with ribosomal protein L26. Mol Cell 32:180–189
45. Oliner JD, Kinzler KW, Meltzer PS, George DL, Vogelstein B (1992) Amplification of a gene encoding a p53-associated protein in human sarcomas. Nature 358: 80–83. http://eutils.ncbi.nlm.nih.gov/entrez/eutils/elink.fcgi?dbfrom=pubmed&id=1614537&retmode=ref&cmd=prlinks
46. Onel K, Cordon-Cardo C (2004) MDM2 and prognosis. Mol Cancer Res 2:1–8

47. Pant V, Xiong S, Jackson JG, Post SM, Abbas HA, Quintás-Cardama A, Hamir AN, Lozano G (2013) The p53-Mdm2 feedback loop protects against DNA damage by inhibiting p53 activity but is dispensable for p53 stability, development, and longevity. Genes Dev 27:1857–1867
48. Parant JJ, Chavez-Reyes AA, Little NAN, Yan WW, Reinke VV, Jochemsen AGA, Lozano GG (2001) Rescue of embryonic lethality in Mdm4-null mice by loss of Trp53 suggests a nonoverlapping pathway with MDM2 to regulate p53. Nat Genet 29:92–95
49. Ponnuswamy AA, Hupp TT, Fåhraeus RR (2012) Concepts in MDM2 signaling: allosteric regulation and feedback loops. Genes Cancer 3:291–297
50. Sam KK, Gan CP, Yee PS, Chong CE, Lim KP, Karen-Ng LP, Chang WS, Nathan S, Rahman ZAA, Ismail SM et al (2012) Novel MDM2 splice variants identified from oral squamous cell carcinoma. Oral Oncol 48:1128–1135
51. Saucedo LJL, Carstens BPB, Seavey SES, Albee LDL, Perry MEM (1998) Regulation of transcriptional activation of mdm2 gene by p53 in response to UV radiation. Cell Growth Differ 9:119–130
52. Secchiero P, Bosco R, Celeghini C, Zauli G (2011) Recent advances in the therapeutic perspectives of Nutlin-3. Curr Pharm Des 17:569–577
53. Sherr CJ (2006) Divorcing ARF and p53: an unsettled case. Nat Rev Cancer 6:663–673
54. Shieh SY, Ikeda M, Taya Y, Prives C (1997) DNA damage-induced phosphorylation of p53 alleviates inhibition by MDM2. Cell 91:325–334
55. Sigalas I, Calvert AH, Anderson JJ, Neal DE, Lunec J (1996) Alternatively spliced mdm2 transcripts with loss of p53 binding domain sequences: transforming ability and frequent detection in human cancer. Nat Med 2:912–917
56. Supiot SS, Hill RPR, Bristow RGR (2008) Nutlin-3 radiosensitizes hypoxic prostate cancer cells independent of p53. Mol Cancer Ther 7:993–999
57. Takagi M, Absalon MJ, McLure KG, Kastan MB (2005) Regulation of p53 translation and induction after DNA damage by ribosomal protein L26 and nucleolin. Cell 123:49–63
58. Tang Y, Zhao W, Chen Y, Zhao Y, Gu W (2008) Acetylation is indispensable for p53 activation. Cell 133:612–626
59. Tanimura SS, Ohtsuka SS, Mitsui KK, Shirouzu KK, Yoshimura AA, Ohtsubo MM (1999) MDM2 interacts with MDMX through their RING finger domains. FEBS Lett 447:5–9
60. Taubert H, Bartel F, Greither T, Bache M, Kappler M, Köhler T, Böhnke A, Lautenschläger C, Schmidt H, Holzhausen H-J et al (2008) Association of HDM2 transcript levels with age of onset and prognosis in soft tissue sarcomas. Mol Cancer Res 6:1575–1581
61. Vassilev LT, Vu BT, Graves B, Carvajal D, Podlaski F, Filipovic Z, Kong N, Kammlott U, Lukacs C, Klein C et al (2004) In vivo activation of the p53 pathway by small-molecule antagonists of MDM2. Science 303:844–848
62. Wade M, Wang YV, Wahl GM (2010) The p53 orchestra: Mdm2 and Mdmx set the tone. Trends Cell Biol 20:299–309
63. Wang P, Lushnikova T, Odvody J, Greiner TC, Jones SN, Eischen CM (2008) Elevated Mdm2 expression induces chromosomal instability and confers a survival and growth advantage to B cells. Oncogene 27:1590–1598
64. Wang S-P, Wang W-L, Chang Y-L, Wu C-T, Chao Y-C, Kao S-H, Yuan A, Lin C-W, Yang S-C, Chan W-K et al (2009) p53 controls cancer cell invasion by inducing the MDM2-mediated degradation of Slug. Nat Cell Biol 11:694–704
65. Wu Y, Fu Y, Zheng L, Lin G, Ma J, Lou J, Zhu H, He Q, Yang B (2014) Nutlin-3 inhibits epithelial-mesenchymal transition by interfering with canonical transforming growth factor β1-Smad-Snail/Slug axis. Cancer Lett 342:82–91
66. Wunderlich M, Berberich SJ (2002) Mdm2 inhibition of p53 induces E2F1 transactivation via p21. Oncogene 21:4414–4421
67. Würl P, Meye A, Schmidt H, Lautenschläger C, Kalthoff H, Rath FW, Taubert H (1998) High prognostic significance of Mdm2/p53 co-overexpression in soft tissue sarcomas of the extremities. Oncogene 16:1183–1185

68. Xiao ZX, Chen J, Levine AJ, Modjtahedi N, Xing J, Sellers WR, Livingston DM (1995) Interaction between the retinoblastoma protein and the oncoprotein MDM2. Nature 375:694–698
69. Yang J-Y, Zong CS, Xia W, Wei Y, Ali-Seyed M, Li Z, Broglio K, Berry DA, Hung M-C (2006) MDM2 promotes cell motility and invasiveness by regulating E-cadherin degradation. Mol Cell Biol 26:7269–7282
70. Yang J-YJ, Zong CSC, Xia WW, Yamaguchi HH, Ding QQ, Xie XX, Lang J-YJ, Lai C-CC, Chang C-JC, Huang W-CW et al (2008) ERK promotes tumorigenesis by inhibiting FOXO3a via MDM2-mediated degradation. Nat Cell Biol 10:138–148
71. Yap DB, Hsieh JK, Chan FS, Lu X (1999) mdm2: a bridge over the two tumour suppressors, p53 and Rb. Oncogene 18:7681–7689
72. Zhang Y, Lu H (2009) Signaling to p53: ribosomal proteins find their way. Cancer Cell 16:369–377

Chapter 14
Splice Variants of MDM2 in Oncogenesis

Melissa Rosso, Danielle E. Okoro, and Jill Bargonetti

Abstract Many types of human cancers overexpress MDM2 protein. A common characteristic among these cancers is an associated increase in *mdm2* splice variants. Provided here is a comprehensive list, based on a literature review, of over 70 *mdm2* variants. These variants are grouped according to in-frame versus out-of-frame status and their potential (or ability) to be translated into isoform proteins. We describe the putative functions for these *mdm2* splice variant mRNAs, as well as the mechanistic drivers associated with increased *mdm2* transcription and splicing. The paradoxical signal transduction functions of the most commonly studied variants *mdm2-a,-b* and *-c* are addressed for their outcomes in the presence and absence of wild-type p53. These outcomes vary from tumor promotion to growth arrest. Finally, we present issues in the detection of endogenous MDM2 protein and how many of the antibodies commonly used to detect MDM2 do not present a full picture of the cellular representation of the isoform proteins. This review provides a focusing lens for individuals interested in learning about the complexities of *mdm2* mRNAs and their protein isoforms as well as the roles MDM2 isoforms may play in cancer progression.

Keywords MDM2 • Splicing

Introduction

Many human cancers over-express MDM2 and the *mdm2* gene locus produces a diverse array of *mdm2* splice variants [1, 2]. Alternative splicing is predominantly co-transcriptional [3] with approximately 6.3 alternatively spliced transcripts

M. Rosso • D.E. Okoro • J. Bargonetti (✉)
Department of Biological Sciences, Hunter College, The City University of New York, 695 Park Ave, New York, NY 10065, USA

The Graduate Center of the City University of New York, The City University of New York, 695 Park Ave, New York, NY 10065, USA
e-mail: Bargonetti@genectr.hunter.cuny.edu

S.P. Deb and S. Deb (eds.), *Mutant p53 and MDM2 in Cancer*, Subcellular Biochemistry 85, DOI 10.1007/978-94-017-9211-0_14,

occurring per human gene [4]. The coordination of splicing with transcription highlights the importance of alternative splicing in signal transduction. Unlike the average human gene locus, the *mdm2* gene gives rise to more than the average 6.3 alternative spliced transcripts. Increased transcription and splicing of *mdm2 is* associated with increased tumorigenesis [1]. However, it is unclear exactly what biological functions the alternative spliced *mdm2* transcripts contribute to tumorigenesis. We reviewed the literature describing *mdm2* splice variants associated with oncogenesis and found that at least 72 have been described. It has not been determined how many of these splice variants express protein in cancer cells.

The association of high MDM2 expression with tumorigenic potential was identified in 1991 in the Donna George laboratory [5]. In 1996, the Lunec group detected increased expression of *mdm2* splice variants in multiple tumor types [6]. The alternative spliced transcripts were named *mdm2*-a, *mdm2*-b, *mdm2*-c, *mdm2*-d and *mdm2*-e in order to contrast them with the full-length version called *mdm2-fl* [6]. Since 1996 the list of *mdm2* alternative spliced variants related to oncogenesis has vastly increased (see Table 14.1). We will focus on the three most commonly found, and best studied, exon skipped transcripts that can be translated into protein [23].

Table 14.1 Known MDM2 splice variant transcripts

Group	Mdm2 variant	Transcript size	In frame	Amino acid number	Predicted protein size (kDa)	Apparent protein size (kDa)	Reference
A	Hdm365	365	Nuclear				[7]
	MYO-2	262	No				[8]
	MYO-3	269	No				[8]
	MYO-5	302	No				[8]
	MYO-6	313	No				[8]
	MYO-7	337	No				[8]
	MYO-10	371	No				[8]
	MYO-13	440	No				[8]
	MYO-14	485	No				[8]
	MYO-16	490	No				[8]
	MYO-17	514	No				[8]
	MYO-19	544	No				[8]
	MYO-21	575	No				[8]
	MYO-22	578	No				[8]
	MYO-27	724	No				[8]
	MYO-28	736	No				[8]
	MYO-29	763	No				[8]
	MYO-30	791	No				[8]
	MYO-31	893	No				[8]
	MYO-33	1,385	No				[8]
	Hdm2-1007	1,007	No				[9]
	Mdm2-FB26		No				[10]
	Mdm2-FB28		No				[10]

(continued)

Table 14.1 (continued)

Group	Mdm2 variant	Transcript size	In frame	Amino acid number	Predicted protein size (kDa)	Apparent protein size (kDa)	Reference
	Mdm2-FB30		No				[10]
	Mdm2-FB55		No				[10]
	Mdm2-Var1		No				[10, 11]
	Mdm2-219	219	No				[12, 13]
	Mdm2-254	254	No				[12]
	Mdm2-H	386	No				[14]
	Mdm2-LN229a	259	No				[15]
	Mdm2-LN229b	195	No				[15]
	Mdm2-LN18	234	No				[15]
	Mdm2-G116	201	No				[15]
	Mdm2-G150	211	No				[15]
B	MYO-1	252	Yes	84	9.6		[8]
	MYO-4	285	Yes	95	10.9		[8]
	MYO-8	351	Yes	117	13.4		[8]
	MYO-9	360	Yes	120	13.8		[8]
	MYO-11	405	Yes	135	15.5		[8]
	MYO-12	405	Yes	135	15.5		[8]
	MYO-18	486	Yes	162	55.8		[8]
	MYO-20	546	Yes	182	20.9		[8]
	MYO-23	621	Yes	207	23.7		[8]
	MYO-24	654	Yes	218	25		[8]
	MYO-25	660	Yes	220	25.2		[8]
	MYO-26	717	Yes	239	27.4		[8]
	MYO-32	1,095	Yes	365	41.9		[8]
	Hdm2-1338	1,338	Yes	446	50.7		[9]
	Hdm2-1200	1,200	Yes	400	45.5		[9]
	P2-Mdm2-10	1,338	Yes	446	50.7		[16]
	P2-Mdm2-C1	906	Yes	302	34.6		[16]
	Mdm2-N1_40	522	Yes	174	20		[17]
	Mdm2-KB2	732	Yes	243	27.9		[17]
	Mdm2-KB3	219	Yes	73	8.4		[17]
	Mdm2-JN1	207	Yes	69	7.9		[17]
	Mdm2-DS2	364	Yes				[17]
	Mdm2-DS3	294	Yes	98	11.2		[17]
	Mdm2-IS1	456	Yes	152	17.4		[17]
	Mdm2-PM2	393	Yes	131	15		[17]
	Mdm2-EU2	297	Yes	99	11.4		[17]
	Mdm2-281	281	Yes	94	10.8		[12]
	Mdm2-DelE	327	Yes	109	12.5		[13]
	Mdm2-DelF	303	Yes	101	11.6		[13]
	Mdm2-FB29	1,074	Yes	358	41.1		[10]
	Mdm2-FB25	750	Yes	250	28.7		[10]
	Mdm2-Var2		Yes				[11]

(continued)

Table 14.1 (continued)

Group	Mdm2 variant	Transcript size	In frame	Amino acid number	Predicted protein size (kDa)	Apparent protein size (kDa)	Reference
C	Mdm2-A (Alt2)	941	Yes	295	33.8	75	[6, 18–20]
	Mdm2-A1	798	Yes	270	22.6	55	[19, 21]
	Mdm2-B (Alt1)	707	Yes	217	24.9	48	[6, 12, 14, 17–20, 22]
	Mdm2-C (Alt3)	1,016	Yes	321	36.6	85	[6, 14, 18–20]
	Mdm2-D	449	Yes	132	15.1	30	[6, 19, 20]
	Mdm2-E	303	Yes	102	11.7	16	[6, 19, 20]
	Mdm2-F	1,391	Yes	468	53.7	85	[2, 14, 19]
	Mdm2-G	1,361	Yes	444	50.9	85	[2, 14, 21]
	Mdm2-FL	1,526	Yes	491	55	90	[6]

Table 14.2 Common MDM2 transcripts associated with cancers

Cancer type	MDM2-A/ALT2 (Exons 4–9)	MDM2-B/ALT1 (Exons 4–11)	MDM2-C/ALT3 (Exons 5–9)	Reference
Colorectal	No	Yes	No	[9]
Ovarian	Yes	Yes	Yes	[6]
Bladder	Yes	Yes	Yes	[6]
Leukemia	Yes	Yes	Yes	[6]
Breast carcinoma	Yes	Yes	Yes	[12, 13, 22, 24, 25]
Soft tissue sarcoma	Yes	Yes	Yes	[1, 10, 17]
Hodgkin's lymphoma	Yes	Yes	Yes	[22, 26]
Glioblastoma	Yes	Yes	Yes	[1, 15, 20]
Liposarcoma	Yes	Yes	Yes	[1]
Lung carcinoma	Yes	Yes	Yes	[1, 27]
Oral squamous carcinoma	No	Yes	Yes	[8]
Burkitt's lymphoma	Not tested	Not tested	Yes	[16]
Osteosarcoma	Not tested	Not tested	Yes	[16]
Pediatric rhabdomyosarcoma	Yes	Yes	Yes	[10]

The most commonly detected isoforms are *mdm2*-a, *mdm2*-b and *mdm2*-c. These three *mdm2* splice variants are found in human leukemia, soft tissue sarcoma, Hodgkin's lymphoma, glioblastoma, rhabdomyosarcoma, liposarcoma, and many different carcinomas including ovarian, breast, bladder, lung and oral squamous cell carcinoma (see Table 14.2). It is highly likely that the protein products MDM2-A, MDM2-B and MDM2-C contribute to the diversity of the human cancer proteome [23]. In addition to *mdm2* spliced variants potentiating cancer proteome diversity, alternative spliced *mdm2* products have also been associated with a small nuclear RNA processed form called *hdm365* [7]. This suggests that some human *mdm2* transcripts may possess RNA-based functions.

Transcripts of Human *mdm2* in Cancer

Two different promoters, P1 and P2, control transcription of the *mdm2* gene, giving rise to two different mRNA messages that encode MDM2-FL [28]. The *mdm2* gene has 12 exons and the P1 promoter drives transcription from upstream of exon 1 and coordinates the splicing out of exon 2. The P1 promoter is responsible for basal *mdm2* transcription and is controlled in part by NF-κB binding sites [29]. The P2 promoter drives transcription from upstream of exon 2 and is controlled by numerous transcription factors. Directly adjacent to the P2 dependent promoter are two binding sites for the transcription factor p53 and in response to stress p53 activates *mdm2* transcription [28, 30]. Other transcription factor binding sites adjacent to the P2 promoter include the Ets/Ap-1, E-box, RXR and Smad binding sites and GC boxes (reviewed in [31]). The Ras signaling pathway can stimulate *mdm2* transcription via the Ap-1/Ets sites [32]. TGF-β signaling stimulates transcription via Smad 2/3 transcription factors binding to the Smad binding sites and MYCN in neuroblastoma cells binds to the E-box near the P2 promoter [33, 34]. The GC boxes are the regions that bind the Sp1 transcription factor. A single nucleotide polymorphism in the GC box region, at position 309, that changes a T to G increases the affinity for Sp1 binding to drive *mdm2* transcription [35]. Patients who are homozygous G/G SNP 309 in the *mdm2* gene have increased susceptibility to multiple cancers [35, 36]. In addition, the tissue-specific RXRγ transcription factor and binding region in retinoblastoma cells can activate P2 dependent *mdm2* expression [37]. High levels of estrogen also activate *mdm2* by activating P2 promoter transcription [38, 39]. While promoter usage has not been shown to be a factor in the alternative splicing of *mdm2* transcripts, the robust signaling of oncogenes present in cancers often drives P2 *mdm2* oncogene mediated transcription [40, 41]. We hypothesize that this increased transcription from the P2 promoter changes the *mdm2* splice variant isoform ratio.

In 2002, Bartel, Taubert and Harris summarized the existence of over 40 different human tumor associated *mdm2* splice variants [2]. At that time the list of distinctive *mdm2* mRNAs was as follows: *mdm2*-fl, *mdm2*-a, *mdm2*-b, *mdm2*-c, *mdm2*-d, *mdm2*-e, *mdm2*-a1, *mdm2*-kb2, *mdm2*-kb3, *mdm2*-jn1, *mdm2*-ds2, *mdm2*-ds3, *mdm2*-is1, *mdm2*-gk1, *mdm2*-pm2, *mdm2*-eu2, *mdm2*-bl, *mdm2*-n, *mdm2*-fb25, *mdm2*-fb26, *mdm2*-fb28, *mdm2*-fb29, *mdm2*-fb30, *mdm2*-fb55, *mdm2*-281 bp, *mdm2*-219 bp, *mdm2*-254 bp, *mdm2*-f, *mdm2*-g, *mdm2*-h, *mdm2*-ln229a, *mdm2-ln229b*, *mdm2*-ln18, *mdm2*-g116, *mdm2*-g150, *mdm2*-var2, *mdm2*-var1, *mdm2-delF*, *mdm2*-delE, and *mdm2*-fb60 (see Table 14.1). These transcripts are found in human cancers, but until recently, no corresponding endogenous protein products had been detected. Many of the *mdm2* splice variant transcripts produced by exon skipping lack the coding region for the p53 interacting domain [2]. The most common variants associated with many different types of human cancers (and missing the p53 interacting domain) are: *mdm2*-a (lacking exons 4–9), *mdm2*-b (lacking exons 4–11), and *mdm2*-c (lacking exons 5–9) (see Table 14.2). All three of these have the potential to endogenously encode oncogenic proteins (see Table 14.3) (reviewed in [23]).

Table 14.3 Biological activity of MDM2-A, MDM2-B, and MDM2-C

Biological/ biochemical activity	MDM2-A	MDM2-B	MDM2-C	References
Increases tumor formation in mice	Yes	Yes	Not Tested	[42, 43]
Alters tumor spectrum in transgenic mice	Yes	Yes	Not Tested	[42, 44]
Overexpression is incompatible with normal mouse development, or viability after birth, in a wt p53 background	Yes	Yes	Not Tested	[42, 43]
Transformation of NIH 3 T3 cells	Yes	Yes	Yes	[6]
Overexpression inhibits proliferation of cells with wt p53	Yes	Yes	No	[26, 43, 45]
Increases proliferation of some cell lines with wt p53	No	Yes	Yes	[26, 42, 45]
Overexpression increases proliferation of cells lacking wt p53	Yes	Yes	Yes	[42, 44, 45]
Upregulates p21, Cyclin D1 and Cyclin E	Yes	Yes	Not Tested	[26, 43]
Inhibits apoptotic signaling by upregulating p65 RelA	Not Tested	Yes	Not Tested	[42]
Binds Mdm2-FL	Yes	Yes	Yes	[24, 43, 45, 46]

Alternatively spliced *mdm2* transcripts in human cancer continue to be detected. The list has increased in number beyond the previously identified 40 (see Table 14.1). The most recent additions to the list of *mdm2* splice variants come from a study of oral squamous cell carcinoma (OSCC). This study shows that *mdm2* splice variants associate with increased likelihood to form OSCC [8] as *mdm2* splice variants are detected in 89 % of oral squamous cell carcinoma [40]. Four splice variants in OSCC are the previously identified *mdm2*-b, *mdm2*-c, *mdm2*-pm2 and *mdm2*-eu2 (with *mdm2*-b found most often). Interestingly, 26 *mdm2* OSCC variants are novel isoforms. Those found to be in-frame range in size from 252 bp to 1,095 bp, and were named: MYO-1, MYO-4, MYO-8, MYO-9, MYO-11, MYO-12, MYO-18, MYO-20, MYO-23, MYO-24, MYO-25 and MYO-32. What is consistent for the in-frame oral cancer variants is that they retain the MDM2 ring-finger binding domain. A significant number of OSCC *mdm2* transcripts are out-of-frame. These range in size from 262 bp to 1,385 bp and were named: MYO-2, MYO-3, MYO-5, MYO-6, MYO-7, MYO-10, MYO-13, MYO-14, MYO-17, MYO-19, MYO-21, MYO-27, MYO-28, MYO-29, MYO-30, MYO-31 and MYO-33.

At least 72 *mdm2* alternative spliced transcripts have been identified in human cancers. This number of 72 includes those 40 compiled in 2002 [2], the OSCC transcripts [8], two novel transcripts that we documented that are driven from the

P2 promoter (P2*mdm2*-10 and P2*mdm2*-C1 [16]) and the RNA-based functional form, *hdm365* [7]. It is likely that the tally of 72 variable *mdm2* transcripts is an underestimate because they continue to be identified. Moreover, while many *mdm2* alternatively spliced transcripts have been detected, the identification of endogenous MDM2 splice variant polypeptides is still lacking. This is partially due to a lack of specific antibodies to detect them. However, the fact that many of the alternative and aberrantly spliced *mdm2* messages are not competent to encode protein suggests that some *mdm2* splice variants might function as regulatory RNA molecules.

The ENCODE project consortium guidelines for functional elements of the genome demonstrates that only a small percentage of the genome (2.9 %) covers areas of protein-coding exons. Furthermore, 62 % of the genome represents RNA molecules with only 5.5 % accounted for in protein-annotated regions [4]. Therefore, the majority of the functional RNA molecules encoded by the human genome represent non-coding regions and for *mdm2* transcripts may indicate a major RNA-based function.

Out-of-Frame Versus In-Frame *mdm2* Transcripts

There are over 70 known splice variant transcripts and they represent alternatively and aberrantly spliced mRNAs (see Table 14.1). Alternatively spliced *mdm2* transcripts are those that result due to exon-exon splicing and give rise, more often than not, to in-frame transcripts with the potential to produce protein [2]. Aberrantly spliced transcripts represent those that result due to the use of cryptic internal splice sites within the *mdm2* exon or intron sequences [2]. Aberrant *mdm2* splicing produces transcripts that are mainly out-of-frame and these do not have the potential to generate protein.

Of the known *mdm2* transcripts, approximately 46 % do not encode protein and all but one of these is spliced out-of-frame to the full-length *mdm2* transcript (Table 14.1, group A). Group A represents this subset of numerous *mdm2* transcripts generated in human cells. One major product in this group is the *hdm365* transcript (in Table 14.1) that potentially has an RNA-based function [7]. This transcript is initiated from the P2 promoter of *mdm2* and retains exons 2, 3, 4 and 5 [7]. The *hdm365* transcript resides in the nucleus and is located at sites of *mdm2* transcription [7]. This localization suggests a role for this *mdm2* transcript in splicing or regulation of the *mdm2* mRNA message.

The *mdm2* transcripts that are assumed, but not proven, to encode protein make up approximately 41.7 % of the identified *mdm2* transcripts (Table 14.1, group B). Group B represents both alternatively and aberrantly spliced *mdm2* transcripts. It is not clear if these transcripts form protein in the cell, as the tools to properly identify each potential MDM2 protein isoform need to be developed.

The final category of *mdm2* transcripts accounts for 12.5 % of the known *mdm2* transcripts. They have been confirmed by in vitro translation assays to encode MDM2 protein isoforms (Table 14.1, group C). Interestingly, only MDM2-FL,

MDM2-A, MDM2-B, MDM2-C, MDM2-D and MDM2-E have been shown to have a biological function in vitro or in vivo [6, 18, 24, 26, 42–44]. Furthermore, none of these MDM2 protein isoforms except for the full-length MDM2 (MDM2-FL) have been detected as expressed endogenously in cancer cells. Although there are high levels of *mdm2* transcripts found in cancers, the level of transcripts do not correlate with high MDM2 protein levels [12, 17]. The reason for this may be due to the absence of proper antibody epitope recognition since antibodies detect some, but not all, MDM2 isoforms within the background of MDM2-FL.

Full-length MDM2, translated from exons 3–12, possesses both oncogenic and tumor suppressive properties [31]. Translation of the MDM2 protein begins in exon 3 and P2-derived transcripts are more efficiently translated than P1-derived transcripts [30]. Some oncogenic properties of MDM2 come from the ability of the protein to interact with the tumor suppressor p53 and target it for proteasome-mediated degradation [47]. However, MDM2 also interacts with the p53 mRNA and increases the translation of p53 protein [48]. This apparent paradox for MDM2 function is increased in complexity by the fact that some *mdm2* splice variants have the capacity to encode polypeptides that lack portions of the p53 interacting domain [2, 31]. Therefore, some of this paradoxical behavior may be explained by determining the functions of specific MDM2 splice variant isoforms.

Mechanisms That Drive Alternative Splicing of *mdm2* Transcripts in Cancer

It is common to find a loss of splicing fidelity in cancer cells [49]. The mechanisms responsible for changes in splicing in cancer continue to emerge. Evidence attributes some of these changes to variations in *cis*-regulatory elements, sequences within the RNA which effect splice-site usage and recognition [50]. Many Serine/Arginine rich (SR) and heterogeneous ribonucleoprotein (hnRNP) splicing factor proteins are up-regulated in cancers and these *trans*- acting splicing factors can increase splicing events [49, 51, 52]. The oncogene c-MYC drives upregulation of specific splicing factors including polypyrimidine-tract binding protein (PTB) and hnRNP A1 and A2 (reviewed in [51–53]). With oncogenes driving alternative splicing, it is not surprising that alternatively spliced transcripts of *mdm2* are found in many different cancers (see Tables 14.1 and 14.2). The *mdm2* splice variants *mdm2-a, mdm2*-b and *mdm2*-c result from exon skipping. This exon skipping occurs because some *mdm2* introns have a defective polypyrimidine tract, a *cis*-regulatory element important for splicing factor binding and 3′ splice site recognition [21]. The splice variants *mdm2*-d and *mdm2*–e on the other hand result from an aberrant splicing mechanism that does not use the normal exon-intron boundaries [21]. Interestingly, some known aberrantly spliced *mdm2* transcripts have a common splicing pattern due to sequences of high homology in the *mdm2* transcript that serve as cryptic splice donor and acceptor sites for splicing factor binding [2].

Alternative splicing of *mdm2* transcripts and transcription from the P2 promoter are also driven by genotoxic stress conditions such as cisplatin or ultraviolet radiation [22, 25]. Some splice variants produced under genotoxic stress conditions, like *mdm2*-b, are seen at high frequency in cancers [40]. A conserved *cis*-regulatory element in intron 11 of the *mdm2* gene promotes this stress-induced regulation of *mdm2* splicing [40]. Stress-induced splicing, in particular that seen with cisplatin treatment, induces co-transcriptional *mdm2* exon skipping through disruption of the EWS-YB1 interaction [41]. EWS is a protein that interacts with the RBP7 subunit of RNA pol II and YB1 interacts with the spliceosome [54, 55]. The stress-induced cotranscriptional exon skipping of *mdm2* produces *mdm2* variants missing the p53 interaction domains. Therefore, exon skipping may help to promote a more robust p53 response by inhibiting the production of MDM2 that interacts with p53 [41].

The Biological Functions of Ectopically Expressed MDM2-A, MDM2-B, and MDM2-C

The biological outcomes of ectopically expressed MDM2-A, MDM2-B, and MDM2-C range from growth activation to growth inhibition under different circumstances (see Table 14.3). The variable outcomes are associated with the presence or absence of wild-type p53 protein. For example, if wild-type p53 is expressed then MDM2-A transgenic homozygous mouse pups die of unknown causes shortly after birth [43]. The only mice that survive with MDM2-A are hemizygous [43]. However, in a *p53*-null background homozygous mice survive and the expression of MDM2-A alters the tumor spectrum of transgenic *p53*-null mice toward increased T-cell lymphomagenesis [44]. Additionally, *p53* heterozygous mice crossed with MDM2-A expressing transgenic mice develop aggressive mammary tumors [44]. Furthermore, the expression of MDM2-A in *p53*-null mouse embryo fibroblasts (MEFs) promotes cell transformation [44]. The exogenous expression of MDM2-A in wild-type MEFs inhibits cell growth. This inhibition of cell growth correlates with an increase in p53 transcriptional activity and high p21 protein levels [43]. Similarly, in the immortalized primary BJ fibroblast cell line ectopic expression of MDM2-A up-regulates p21 along with Cyclin D1 and Cyclin E [26]. Exogenously expressed MDM2-A interacts with endogenous MDM2-FL and activates wild-type p53 activity thus explaining some of the differences seen in a *p53*-null background [43].

Similar to MDM2-A, exogenous expression of MDM2-B also has differential outcomes in the presence or absence of wild-type p53. The exogenous expression of MDM2-B in transgenic mice is not compatible with normal development [42]. Only when MDM2-B is expressed under a promoter with limited tissue expression are mice able to survive. The transfection of the *mdm2*-b into NIH/3T3 cells increases cell proliferation and transformation capabilities [42]. Interestingly, the expression of MDM2-B in NIH/3T3 cells interferes with the induction of apoptosis without affecting p53 stability or activity and is linked to an increase of p65 RelA protein levels [42]. Surviving MDM2-B transgenic mice with tissue specific expression have increased tumorigenesis that correlates with this increase in p65 protein levels [42]. MDM2-B expression also increases cell proliferation in p53-null, ARF-null and Rb-null MEFs, therefore indicating a p53-independent mechanism of action [42]. However, other studies show exogenously expressed MDM2-B interacts with MDM2-FL protein localizing MDM2-FL to the cytoplasm in numerous cell lines to allow wild-type p53 protein to be activated [24, 46]. Ectopic expression of MDM2-B also up-regulates p21 expression in immortalized BJ fibroblasts correlating with an inhibition of cell proliferation [26].

Our laboratory works on the MDM2-C splice variant. Unpublished studies from our laboratory were presented at the 2011 MDM2 Workshop in New York City and were recently published [45]. We have designed a specific antibody toward MDM2-C to detect the endogenous MDM2-C protein isoform and we have explored the biological functions of MDM2-C. Exogenous expression of MDM2-C in the presence or absence of p53 in H1299 lung carcinoma cells showed increased colony formation as compared to MDM2-FL or vector control [45]. Therefore, like MDM2-A and MDM2-B, MDM2-C shows a p53-independent transformation function. Furthermore, the transfection of *mdm2*-c in the presence or absence of *p53* into H1299 cells increased colony formation, indicated by transforming ability. The co-transfection of *mdm2*-c and *p53* into H1299 cells did not significantly decrease p53 transcriptional activity or change p53 protein levels and MDM2-C was also able to interact with MDM2-FL [45]. Therefore, unlike MDM2-A and MDM2-B, MDM2-C does not increase the activity of wild-type p53. An in vivo mouse model has yet to be carried out for MDM2-C. Until this is done, we will not know the full biological functions of MDM2-C.

Detection of Endogenous MDM2

There are a number of MDM2 specific antibodies that detect the endogenous MDM2 protein in cancer cells and cancer tissues (reviewed in [23]). These MDM2 antibodies recognize epitopes of multiple MDM2 domains including the amino terminus, the central region, and carboxyl terminus of the protein. However, the MDM2 antibodies utilized to determine MDM2 protein levels in cancers are often to the central region. Therefore, they are not appropriate to detect the majority of MDM2 splice variant isoforms. This is especially true since the main antibody used in immunohistochemistry of cancer tissues for MDM2 protein expression is IF2 (Ab-1). The epitope of recognition for the IF2 antibody lies within amino acids 26–169, which represents the p53-binding domain of the MDM2 protein and spans exons 4 and 5 [19, 56]. Therefore, using the IF2 antibody (or any other antibody to a region deleted by a splicing event) will not show a true representation of the levels of MDM2 protein present in the cancer tissue.

Work to examine the expression of endogenous MDM2 splice variant protein isoforms is being carried out in our laboratory. We generated a rabbit polyclonal antibody to the MDM2-C isoform. The MDM2-C rabbit polyclonal antibody specifically detects MDM2-C, and not MDM2-FL, expressed by an in vitro translation system [45]. This MDM2-C specific antibody also detects endogenously expressed MDM2-C in cancer cell lines and cancer tissues. To our knowledge, we are the first group to generate an antibody specific for an MDM2 spliced variant protein isoform. The *mdm2*-c transcript is the third most common *mdm2* transcript found in cancer cells and tissues [1]. Therefore, detection of the MDM2-C protein isoform may provide a new cancer biomarker. MDM2 endogenous expression undoubtedly results from a mixture of *mdm2* transcripts such as *mdm2*-a, *mdm2*-b, and *mdm2*-c. The proteins expressed from these transcripts are all potential cancer biomarkers. It is important that these biomarkers be detected with the proper MDM2 antibodies that are specific for various isoforms. The use of antibodies to MDM2 in the clinic have led to the conclusion that cancers with high levels of spliced variant transcripts have less MDM2 protein [57]. In actuality, not detecting MDM2 protein in breast cancers with *mdm2* splice variant transcripts is a false negative [57]. Future research needs to make use of MDM2 splice-variant specific antibodies, or antibodies to either the extreme amino or carboxyl terminus of MDM2, in order to evaluate the true nature of MDM2 protein expression in cancer.

Summary

The diverse array of *mdm2* splice variants in human cancers suggests they have functional significance and can serve as cancer biomarkers. To date, MDM2 protein biomarker studies have been carried out with antibodies that give false negative results for the accumulation of MDM2 isoforms lacking central regions of the polypeptide. Future MDM2 biomarker studies must be carried out with consideration given to detecting multiple isoforms. In order to detect multiple MDM2 isoforms, antibody reagents must recognize either the amino or carboxyl terminus of MDM2 because as shown in Table 14.1 most *mdm2* splice variants retain these regions. Alternatively, future MDM2 biomarker studies could make use of mixtures of antibodies with specificity to the MDM2 amino and carboxyl termini as well as the specific amino acid splice junction residues for focused splice variants. Recommendations for future MDM2 biomarker studies should combine new methods for the detection of *mdm2* splice variant RNA messages along with the detection of multiple MDM2 isoform proteins. This is because the MDM2 polypeptides and RNA sequences may cooperate in the transformation process. The oncogenic MDM2 pathway is a central node in cancer progression that may make use of many isoforms of the MDM2 protein and *mdm2* RNA and future research should center on this exciting oncogenic hub.

References

1. Jeyaraj S, O'Brien DM, Chandler DS (2009) MDM2 and MDM4 splicing: an integral part of the cancer spliceome. Front Biosci 14:2647–2656, doi:3402 [pii]
2. Bartel F, Taubert H, Harris LC (2002) Alternative and aberrant splicing of MDM2 mRNA in human cancer. Cancer Cell 2(1):9–15
3. Tilgner H, Knowles DG, Johnson R, Davis CA, Chakrabortty S, Djebali S, Curado J, Snyder M, Gingeras TR, Guigo R (2012) Deep sequencing of subcellular RNA fractions shows splicing to be predominantly co-transcriptional in the human genome but inefficient for lncRNAs. Genome Res 22(9):1616–1625. doi:10.1101/gr.134445.111
4. Dunham I, Kundaje A, Aldred SF, Collins PJ, Davis CA, Doyle F, Epstein CB, Frietze S, Harrow J, Kaul R, Khatun J, Lajoie BR, Landt SG, Lee BK, Pauli F, Rosenbloom KR, Sabo P, Safi A, Sanyal A, Shoresh N, Simon JM, Song L, Trinklein ND, Altshuler RC, Birney E, Brown JB, Cheng C, Djebali S, Dong X, Ernst J, Furey TS, Gerstein M, Giardine B, Greven M, Hardison RC, Harris RS, Herrero J, Hoffman MM, Iyer S, Kelllis M, Kheradpour P, Lassman T, Li Q, Lin X, Marinov GK, Merkel A, Mortazavi A, Parker SC, Reddy TE, Rozowsky J, Schlesinger F, Thurman RE, Wang J, Ward LD, Whitfield TW, Wilder SP, Wu W, Xi HS, Yip KY, Zhuang J, Bernstein BE, Green ED, Gunter C, Snyder M, Pazin MJ, Lowdon RF, Dillon LA, Adams LB, Kelly CJ, Zhang J, Wexler JR, Good PJ, Feingold EA, Crawford GE, Dekker J, Elinitski L, Farnham PJ, Giddings MC, Gingeras TR, Guigo R, Hubbard TJ, Kellis M, Kent WJ, Lieb JD, Margulies EH, Myers RM, Starnatoyannopoulos JA, Tennebaum SA, Weng Z, White KP, Wold B, Yu Y, Wrobel J, Risk BA, Gunawardena HP, Kuiper HC, Maier CW, Xie L, Chen X, Mikkelsen TS, Gillespie S, Goren A, Ram O, Zhang X, Wang L, Issner R, Coyne MJ, Durham T, Ku M, Truong T, Eaton ML, Dobin A, Lassmann T, Tanzer A, Lagarde J, Lin W, Xue C, Williams BA, Zaleski C, Roder M, Kokocinski F, Abdelhamid RF, Alioto T, Antoshechkin I, Baer MT, Batut P, Bell I, Bell K, Chakrabortty S, Chrast J, Curado J, Derrien T, Drenkow J, Dumais E, Dumais J, Duttagupta R, Fastuca M, Fejes-Toth K, Ferreira P, Foissac S, Fullwood MJ, Gao H, Gonzalez D, Gordon A, Howald C, Jha S, Johnson R, Kapranov P, King B, Kingswood C, Li G, Luo OJ, Park E, Preall JB, Presaud K, Ribeca P, Robyr D, Ruan X, Sammeth M, Sandu KS, Schaeffer L, See LH, Shahab A, Skancke J, Suzuki AM, Takahashi H, Tilgner H, Trout D, Walters N, Wang H, Hayashizaki Y, Reymond A, Antonarakis SE, Hannon GJ, Ruan Y, Carninci P, Sloan CA, Learned K, Malladi VS, Wong MC, Barber GP, Cline MS, Dreszer TR, Heitner SG, Karolchik D, Kirkup VM, Meyer LR, Long JC, Maddren M, Raney BJ, Grasfeder LL, Giresi PG, Battenhouse A, Sheffield NC, Showers KA, London D, Bhinge AA, Shestak C, Schaner MR, Kim SK, Zhang ZZ, Mieczkowski PA, Mieczkowska JO, Liu Z, McDaniell RM, Ni Y, Rashid NU, Kim MJ, Adar S, Zhang Z, Wang T, Winter D, Keefe D, Iyer VR, Sandhu KS, Zheng M, Wang P, Gertz J, Vielmetter J, Partridge EC, Varley KE, Gasper C, Bansal A, Pepke S, Jain P, Amrhein H, Bowling KM, Anaya M, Cross MK, Muratet MA, Newberry KM, McCue K, Nesmith AS, Fisher-Aylor KI, Pusey B, DeSalvo G, Parker SL, Balasubramanian S, Davis NS, Meadows SK, Eggleston T, Newberry JS, Levy SE, Absher DM, Wong WH, Blow MJ, Visel A, Pennachio LA, Elnitski L, Petrykowska HM, Abyzov A, Aken B, Barrell D, Barson G, Berry A, Bignell A, Boychenko V, Bussotti G, Davidson C, Despacio-Reyes G, Diekhans M, Ezkurdia I, Frankish A, Gilbert J, Gonzalez JM, Griffiths E, Harte R, Hendrix DA, Hunt T, Jungreis I, Kay M, Khurana E, Leng J, Lin MF, Loveland J, Lu Z, Manthravadi D, Mariotti M, Mudge J, Mukherjee G, Notredame C, Pei B, Rodriguez JM, Saunders G, Sboner A, Searle S, Sisu C, Snow C, Steward C, Tapanari E, Tress ML, van Baren MJ, Washieti S, Wilming L, Zadissa A, Zhengdong Z, Brent M, Haussler D, Valencia A, Raymond A, Addleman N, Alexander RP, Auerbach RK, Bettinger K, Bhardwaj N, Boyle AP, Cao AR, Cayting P, Charos A, Cheng Y, Eastman C, Euskirchen G, Fleming JD, Grubert F, Habegger L, Hariharan M, Harmanci A, Iyenger S, Jin VX, Karczewski KJ, Kasowski M, Lacroute P, Lam H, Larnarre-Vincent N, Lian J, Lindahl-Allen M, Min R, Miotto B, Monahan H, Moqtaderi Z, Mu XJ, O'Geen H, Ouyang Z, Patacsil D, Raha D, Ramirez L, Reed B, Shi M, Slifer T, Witt H, Wu L, Xu X, Yan KK, Yang X, Struhl K, Weissman SM, Tenebaum SA, Penalva LO, Karmakar S, Bhanvadia RR, Choudhury A, Domanus M, Ma L,

Moran J, Victorsen A, Auer T, Centarin L, Eichenlaub M, Gruhl F, Heerman S, Hoeckendorf B, Inoue D, Kellner T, Kirchmaier S, Mueller C, Reinhardt R, Schertel L, Schneider S, Sinn R, Wittbrodt B, Wittbrodt J, Jain G, Balasundaram G, Bates DL, Byron R, Canfield TK, Diegel MJ, Dunn D, Ebersol AK, Frum T, Garg K, Gist E, Hansen RS, Boatman L, Haugen E, Humbert R, Johnson AK, Johnson EM, Kutyavin TM, Lee K, Lotakis D, Maurano MT, Neph SJ, Neri FV, Nguyen ED, Qu H, Reynolds AP, Roach V, Rynes E, Sanchez ME, Sandstrom RS, Shafer AO, Stergachis AB, Thomas S, Vernot B, Vierstra J, Vong S, Weaver MA, Yan Y, Zhang M, Akey JA, Bender M, Dorschner MO, Groudine M, MacCoss MJ, Navas P, Stamatoyannopoulos G, Stamatoyannopoulos JA, Beal K, Brazma A, Flicek P, Johnson N, Lukk M, Luscombe NM, Sobral D, Vaquerizas JM, Batzoglou S, Sidow A, Hussami N, Kyriazopoulou-Panagiotopoulou S, Libbrecht MW, Schaub MA, Miller W, Bickel PJ, Banfai B, Boley NP, Huang H, Li JJ, Noble WS, Bilmes JA, Buske OJ, Sahu AO, Kharchenko PV, Park PJ, Baker D, Taylor J, Lochovsky L (2012) An integrated encyclopedia of DNA elements in the human genome. Nature 489(7414):57–74. doi:10.1038/nature11247

5. Fakharzadeh SS, Trusko SP, George DL (1991) Tumorigenic potential associated with enhanced expression of a gene that is amplified in a mouse tumor cell line. Embo J 10(6):1565–1569
6. Sigalas I, Calvert AH, Anderson JJ, Neal DE, Lunec J (1996) Alternatively spliced mdm2 transcripts with loss of p53 binding domain sequences: transforming ability and frequent detection in human cancer. Nat Med 2(8):912–917
7. Bartl S, Ban J, Weninger H, Jug G, Kovar H (2003) A small nuclear RNA, hdm365, is the major processing product of the human mdm2 gene. Nucleic Acids Res 31(4):1136–1147
8. Sam KK, Gan CP, Yee PS, Chong CE, Lim KP, Karen-Ng LP, Chang WS, Nathan S, Rahman ZA, Ismail SM, Cheong SC (2012) Novel MDM2 splice variants identified from oral squamous cell carcinoma. Oral Oncol 48(11):1128–1135. doi:10.1016/j.oraloncology.2012.05.016
9. Yu Z, Zhang B, Cui B, Wang Y, Han P, Wang X (2012) Identification of spliced variants of the proto-oncogene HDM2 in colorectal cancer. Cancer 118(4):1110–1118. doi:10.1002/cncr.26330
10. Bartel F, Taylor AC, Taubert H, Harris LC (2001) Novel mdm2 splice variants identified in pediatric rhabdomyosarcoma tumors and cell lines. Oncol Res 12(11–12):451–457
11. Schlott T, Nagel H, Laskawi R, Eiffert H, Droese M (2001) Genetic analysis of the human oncoprotein MDM2 in benign and malignant tumors of the salivary gland. Pathobiology 69(2):67–76
12. Lukas J, Gao DQ, Keshmeshian M, Wen WH, Tsao-Wei D, Rosenberg S, Press MF (2001) Alternative and aberrant messenger RNA splicing of the mdm2 oncogene in invasive breast cancer. Cancer Res 61(7):3212–3219
13. Hori M, Shimazaki J, Inagawa S, Itabashi M (2000) Alternatively spliced MDM2 transcripts in human breast cancer in relation to tumor necrosis and lymph node involvement. Pathol Int 50(10):786–792
14. Tamborini E, Della Torre G, Lavarino C, Azzarelli A, Carpinelli P, Pierotti MA, Pilotti S (2001) Analysis of the molecular species generated by MDM2 gene amplification in liposarcomas. Int J Cancer 92(6):790–796. doi:10.1002/ijc.1271
15. Kraus A, Neff F, Behn M, Schuermann M, Muenkel K, Schlegel J (1999) Expression of alternatively spliced mdm2 transcripts correlates with stabilized wild-type p53 protein in human glioblastoma cells. Int J Cancer 80(6):930–934
16. Arva NC, Talbott KE, Okoro DR, Brekman A, Qiu WG, Bargonetti J (2008) Disruption of the p53-Mdm2 complex by Nutlin-3 reveals different cancer cell phenotypes. Ethn Dis 18(2 Suppl 2): S2-1–8
17. Bartel F, Meye A, Wurl P, Kappler M, Bache M, Lautenschlager C, Grunbaum U, Schmidt H, Taubert H (2001) Amplification of the MDM2 gene, but not expression of splice variants of MDM2 MRNA, is associated with prognosis in soft tissue sarcoma. Int J Cancer 95(3):168–175
18. Evans SC, Viswanathan M, Grier JD, Narayana M, El-Naggar AK, Lozano G (2001) An alternatively spliced HDM2 product increases p53 activity by inhibiting HDM2. Oncogene 20(30):4041–4049. doi:10.1038/sj.onc.1204533
19. Anderson JJ, Challen C, Atkins H, Suaeyun R, Crosier S, Lunec J (2007) MDM2 RNA binding is blocked by novel monoclonal antibody h-MDM2-F4-14. Int J Oncol 31(3):545–555

20. Matsumoto R, Tada M, Nozaki M, Zhang CL, Sawamura Y, Abe H (1998) Short alternative splice transcripts of the mdm2 oncogene correlate to malignancy in human astrocytic neoplasms. Cancer Res 58(4):609–613
21. Liang H, Atkins H, Abdel-Fattah R, Jones SN, Lunec J (2004) Genomic organisation of the human MDM2 oncogene and relationship to its alternatively spliced mRNAs. Gene 338(2):217–223
22. Chandler DS, Singh RK, Caldwell LC, Bitler JL, Lozano G (2006) Genotoxic stress induces coordinately regulated alternative splicing of the p53 modulators MDM2 and MDM4. Cancer Res 66(19):9502–9508. doi:10.1158/0008-5472.CAN-05-4271
23. Okoro DR, Rosso M, Bargonetti J (2012) Splicing up mdm2 for cancer proteome diversity. Genes Cancer 3(3–4):311–319. doi:10.1177/1947601912455323
24. Dias C, Liu Y, Yau A, Westrick L, Evans S (2006) Regulation of hdm2 by stress-induced hdm2alt1 in tumor and nontumorigenic cell lines correlating with p53 stability. Cancer Res 66(19):9467–9473
25. Dias CS, Liu Y, Yau A, Westrick L, Evans SC (2006) Regulation of hdm2 by stress-induced hdm2alt1 in tumor and nontumorigenic cell lines correlating with p53 stability. Cancer Res 66(19):9467–9473. doi:10.1158/0008-5472.CAN-05-3013
26. Sanchez-Aguilera A, Garcia JF, Sanchez-Beato M, Piris MA (2006) Hodgkin's lymphoma cells express alternatively spliced forms of HDM2 with multiple effects on cell cycle control. Oncogene 25(18):2565–2574. doi:10.1038/sj.onc.1209282
27. Weng MW, Lai JC, Hsu CP, Yu KY, Chen CY, Lin TS, Lai WW, Lee H, Ko JL (2005) Alternative splicing of MDM2 mRNA in lung carcinomas and lung cell lines. Environ Mol Mutagen 46(1):1–11. doi:10.1002/em.20118
28. Zauberman A, Flusberg D, Haupt Y, Barak Y, Oren M (1995) A functional p53-responsive intronic promoter is contained within the human mdm2 gene. Nucleic Acids Res 23(14):2584–2592
29. Busuttil V, Droin N, McCormick L, Bernassola F, Candi E, Melino G, Green DR (2010) NF-kappaB inhibits T-cell activation-induced, p73-dependent cell death by induction of MDM2. Proc Natl Acad Sci U S A 107(42):18061–18066. doi:10.1073/pnas.1006163107
30. Barak Y, Gottlieb E, Juven-Gershon T, Oren M (1994) Regulation of mdm2 expression by p53: alternative promoters produce transcripts with nonidentical translation potential. Genes Dev 8:1739–1749
31. Manfredi J (2010) The Mdm2-p53 relationship evolves: Mdm2 swings both ways as an oncogene and a tumor suppressor. Genes Dev 24(15):1580–1589, 24/15/1580 [pii]. doi:10.1101/gad.1941710
32. Ries S, Biederer C, Woods D, Shifman O, Shirasawa S, Sasazuki T, McMahon M, Oren M, McCormick F (2000) Opposing effects of Ras on p53: transcriptional activation of mdm2 and induction of p19ARF. Cell 103(2):321–330
33. Slack A, Lozano G, Shohet JM (2005) MDM2 as MYCN transcriptional target: implications for neuroblastoma pathogenesis. Cancer letters 228(1–2):21–27, S0304-3835(05)00350-2 [pii]. doi:10.1016/j.canlet.2005.01.050
34. Araki S, Eitel JA, Batuello CN, Bijangi-Vishehsaraei K, Xie XJ, Danielpour D, Pollok KE, Boothman DA, Mayo LD (2010) TGF-beta1-induced expression of human Mdm2 correlates with late-stage metastatic breast cancer. J Clin Invest 120(1):290–302. doi:10.1172/JCI39194
35. Bond G, Hu W, Bond E, Robins H, Lutzker S, Arva N, Bargonetti J, Bartel F, Taubert H, Wuerl P, Onel K, Yip L, Hwang S, Strong L, Lozano G, Levine A (2004) A single nucleotide polymorphism in the MDM2 promoter attenuates the p53 tumor suppressor pathway and accelerates tumor formation in humans. Cell 119(5):591–602
36. Bond GL, Hirshfield KM, Kirchhoff T, Alexe G, Bond EE, Robins H, Bartel F, Taubert H, Wuerl P, Hait W, Toppmeyer D, Offit K, Levine AJ (2006) MDM2 SNP309 accelerates tumor formation in a gender-specific and hormone-dependent manner. Cancer Res 66(10):5104–5110
37. Xu X, Fang Y, Lee T, Forrest D, Gregory-Evans C, Almeida D, Liu A, Jhanwar S, Abramson D, Cobrinik D (2009) Retinoblastoma has properties of a cone precursor tumor and depends upon cone-specific MDM2 signaling. Cell 137(6):1018–1031, S0092-8674(09)00400-0 [pii]. doi:10.1016/j.cell.2009.03.051

38. Brekman A, Singh KE, Polotskaia A, Kundu N, Bargonetti J (2011) A p53-independent role of Mdm2 in estrogen-mediated activation of breast cancer cell proliferation. Breast Cancer Res: BCR 13(1):R3. doi:10.1186/bcr2804
39. Phelps M, Darley M, Primrose JN, Blaydes JP (2003) p53-independent activation of the hdm2-P2 promoter through multiple transcription factor response elements results in elevated hdm2 expression in estrogen receptor alpha-positive breast cancer cells. Cancer Res 63(10):2616–2623
40. Singh RK, Tapia-Santos A, Bebee TW, Chandler DS (2009) Conserved sequences in the final intron of MDM2 are essential for the regulation of alternative splicing of MDM2 in response to stress. Exp Cell Res 315(19):3419–3432. doi:10.1016/j.yexcr.2009.07.017
41. Dutertre M, Sanchez G, De Cian MC, Barbier J, Dardenne E, Gratadou L, Dujardin G, Le Jossic-Corcos C, Corcos L, Auboeuf D (2010) Cotranscriptional exon skipping in the genotoxic stress response. Nat Struct Mol Biol 17(11):1358–1366. doi:10.1038/nsmb.1912
42. Steinman HA, Burstein E, Lengner C, Gosselin J, Pihan G, Duckett CS, Jones SN (2004) An alternative splice form of Mdm2 induces p53-independent cell growth and tumorigenesis. J Biol Chem 279(6):4877–4886
43. Volk EL, Schuster K, Nemeth KM, Fan L, Harris LC (2009) MDM2-A, a common Mdm2 splice variant, causes perinatal lethality, reduced longevity and enhanced senescence. Dis Model Mech 2(1–2):47–55. doi:10.1242/dmm.000992
44. Volk EL, Fan L, Schuster K, Rehg JE, Harris LC (2009) The MDM2-a splice variant of MDM2 alters transformation in vitro and the tumor spectrum in both Arf- and p53-null models of tumorigenesis. Mol Cancer Res 7(6):863–869. doi:10.1158/1541-7786.MCR-08-0418
45 Okoro D, Arva N, Gao C, Polotskaia A, Puente C, Rosso M, Bargonetti J (2013) Endogenous human MDM2-C is highly expressed in human cancers and functions as a p53-independent growth activator. PLoS One 8(10):e77643
46. Evans S, Viswanathan M, Grier J, Narayana M, El-Naggar A, Lozano G (2001) An alternatively spliced HDM2 product increases p53 activity by inhibiting HDM2. Oncogene 20(30):4041–4049
47. Haupt Y, Maya R, Kazaz A, Oren M (1997) Mdm2 promotes the rapid degradation of p53. Nature 387:296–299
48. Naski N, Gajjar M, Bourougaa K, Malbert-Colas L, Fahraeus R, Candeias MM (2009) The p53 mRNA-Mdm2 interaction. Cell Cycle 8(1):31–34
49. Ghigna C, Valacca C, Biamonti G (2008) Alternative splicing and tumor progression. Curr Genomics 9(8):556–570. doi:10.2174/138920208786847971
50. Kalnina Z, Zayakin P, Silina K, Line A (2005) Alterations of pre-mRNA splicing in cancer. Genes Chromosomes Cancer 42(4):342–357. doi:10.1002/gcc.20156
51. David CJ, Manley JL (2010) Alternative pre-mRNA splicing regulation in cancer: pathways and programs unhinged. Genes Dev 24(21):2343–2364. doi:10.1101/gad.1973010
52. Grosso AR, Martins S, Carmo-Fonseca M (2008) The emerging role of splicing factors in cancer. EMBO Rep 9(11):1087–1093. doi:10.1038/embor.2008.189
53. David CJ, Chen M, Assanah M, Canoll P, Manley JL (2010) HnRNP proteins controlled by c-Myc deregulate pyruvate kinase mRNA splicing in cancer. Nature 463(7279):364–368. doi:10.1038/nature08697
54. Chansky HA, Hu M, Hickstein DD, Yang L (2001) Oncogenic TLS/ERG and EWS/Fli-1 fusion proteins inhibit RNA splicing mediated by YB-1 protein. Cancer Res 61(9):3586–3590
55. Petermann R, Mossier BM, Aryee DN, Khazak V, Golemis EA, Kovar H (1998) Oncogenic EWS-Fli1 interacts with hsRPB7, a subunit of human RNA polymerase II. Oncogene 17(5):603–610. doi:10.1038/sj.onc.1201964
56. Leach FS, Tokino T, Meltzer P, Burrell M, Oliner JD, Smith S, Hill DE, Sidransky D, Kinzler KW, Vogelstein B (1993) p53 Mutation and MDM2 amplification in human soft tissue sarcomas. Cancer Res 53(10 Suppl):2231–2234
57. Hori M, Shimazaki J, Inagawa S, Itabashi M (2002) Overexpression of MDM2 oncoprotein correlates with possession of estrogen receptor alpha and lack of MDM2 mRNA splice variants in human breast cancer. Breast Cancer Res Treat 71(1):77–83

Chapter 15
Mdm2 and MdmX Involvement in Human Cancer

Steven J. Berberich

Abstract Discovered in 1987 and 1997 respectively, Mdm2 and MdmX represent two critical cellular regulators of the p53 tumor suppressor. This chapter reviews each from initial discovery to our current understanding of their deregulation in human cancer with a focus on how each regulator impacts p53 function. While p53 independent activities of Mdm2 and MdmX are noted the reader is directed to other reviews on this topic. The chapter concludes with an examination of the various mechanisms of Mdm-deregulation and an assessment of the current therapeutic approaches to target Mdm2 and MdmX overexpression.

Keywords Mdm2 • MdmX • p53 • E3 ligase • Gene amplification • Alternative splicing

Introduction

The p53 tumor suppressor gene is the most frequently inactivated gene in human cancers. The p53 protein is post-transcriptionally induced in response to a variety of stresses internal and external to the cell and functions predominately as a transcription factor. p53 activates genes and miRNAs that lead to cell cycle arrest, metabolism, cellular senescence and apoptosis [53]. However early on p53 inactivation in human tumors was viewed almost exclusively in terms of gene mutations. Based on numerous studies it has been determined that 50 % of human cancers harbor p53 mutations, predominately missense mutations in its DNA binding domain [39]. These mutations inhibit p53 function leading to the inability to block the transmission of DNA damage to future progeny thus leading to enhanced tumorigenesis.

S.J. Berberich, Ph.D. (✉)
Biochemistry and Molecular Biology, Wright State University,
3640 Colonel Glenn Hwy, Dayton, OH 45435, USA
e-mail: steven.berberich@wright.edu

S.P. Deb and S. Deb (eds.), *Mutant p53 and MDM2 in Cancer*,
Subcellular Biochemistry 85, DOI 10.1007/978-94-017-9211-0_15,

So what about the other 50 % of human tumors that do not harbor mutant p53 genes? Is inactivation of the p53 signaling pathway in these tumors unnecessary? One formative step towards answering these questions was discovering that the p53 protein associated with a cellular protein of 90 kDa named the murine double minute 2 (Mdm2) protein [71].

Mdm2 Discovery

Interested in understanding the origin and function of amplified DNA sequences found on extra-chromosomal elements known as double-minute (DM) chromosomes, Dr. Donna George's laboratory is credited with first discovering the mdm-2 gene as one of several genes cloned from DM chromosomes in a spontaneously transformed line of mouse 3T3 cells [13]. Using a differential hybridization approach a cDNA library cloned into lambda was separately screened (11,000 recombinant plaques) with radiolabeled probes constructed from 3T3-DM and A31 mRNA. Forty-eight plaques showed stronger signals from the 3T3-DM probes of which 7 clones were determined to be true amplified sequences based on Southern blot analysis of 3T3-DM DNA. Subcloning of the cDNAs from these seven clones and a series of cross-hybridization experiments confirmed that they fell into two classes, named mdm-1 and mdm-2.

Consistent with DNA amplification, Northern blot analysis confirmed that both mdm-1 and mdm2 genes were overexpressed in 3T3-DM cells relative to other mouse fibroblast cell lines. The group went on to demonstrate that these two genes were enriched in double minute chromosomes, were localized to chromosome 10 (region C1-C3) and were conserved in other species. Later studies focusing on the mdm-1 gene uncovered alternative spliced mRNAs encoding potentially three distinct protein products with unknown function [91]. A third DM-linked amplified gene mdm-3 was uncovered using field inversion electrophoresis and chromosome walking. To test whether any of these genes possessed transforming capacity the George laboratory created expression vectors of gene genomic clone (mdm-1, mdm-2 and mdm-3) and transfected them into immortalized NIH3T3 or Rat2 fibroblasts and derived clones based on selection for antibiotic resistance genes present on the transfected cosmids. These cell lines were subsequently shown to overexpress their respective mdm-gene. However when these cells were injected into athymic (nude) mice, only the cells overexpressing mdm-2 produced tumors [26]. This was the first evidence that the mdm-2 gene was oncogenic when overexpressed. An analysis of the amino acid sequence of the Mdm2 protein led the authors to suggest that this nuclear protein may function as a DNA binding transcription factor. While they were correct in identifying that the metal binding and acidic amino acid domains were important and that Mdm2 possessed a nuclear localization sequence, they were unaware that Mdm2 was actually a modulator of one of the most critical DNA binding transcription factors in human cancer, p53.

Mdm2 Functions

Mdm2 moved to the epicenter of cancer research when Dr. Arnold Levine's laboratory discovered that a novel 90 kDa cellular protein associating with p53 was in fact Mdm2. Levine's group demonstrated that Mdm2's association with p53 inhibited p53 transactivation [71]. The Vogelstein laboratory cloned the human mdm-2 gene and reported that human Mdm2 associated with p53 *in vitro* and that the human mdm-2 gene was amplification (hdm-2) in a third of 47 human sarcomas [75]. As will be discussed later, this Hdm-2 overexpression correlates with tumors possessing wild-type p53, a result first reported in MethA tumor cells [76], thus providing a method by which human tumors inactive p53 cell signaling. These publications placed Mdm2/Hdm2 as the first identified cellular regulator of the p53 tumor suppressor.

Shortly after the Mdm2:p53 association was discovered three groups reported that p53 directly transactivates the mdm-2 gene. Moshe Oren's group was the first to discover increases in Mdm2 binding to p53 when a cell line harboring a temperature sensitive p53 allele was shifted to a temperature that converted p53 to its wild-type conformation [2]. They went on to show this increase in complex formation resulted from increased mdm-2 mRNA levels and based on kinetic studies suggested that wild-type p53 transactivated the mdm-2 gene. A few months later the Oren laboratory provided unequivocal evidence that p53 directly transactivated the mdm-2 promoter from a p53 DNA binding site located downstream from mdm-2 exon 1 [45]. The Levine laboratory was the first to demonstrate that in cell harboring wild-type p53 DNA damage could trigger p53 to upregulate the mdm-2 gene creating what is now well known as the p53-Mdm2 autoregulatory feedback loop [80].

Crystal structures using the first 109 amino acids of Mdm2 and 15 amino acids of p53 show that the Mdm2 creates a deep hydrophobic cleft in which the p53 amino terminal alpha-helix binds thereby masking the p53 transactivation domain [50]. However the interaction of Mdm2 with p53 goes beyond simple inhibition of transactivation. Oren and Vousden laboratories were the first to report that Mdm2 triggered a reduction in the steady state levels of p53 [37, 49]. p53 destabilization was dependent on the ability of Mdm2 to directly associate with p53 but also required a region of the Mdm2 carboxyl-terminal region. Finally, Mdm2 was capable of lowering mutant p53 levels but had no impact on p53 transactivation. Less than 6 months later using recombinant forms of p53 and Mdm2, Honda et al. performed *in vitro* assays to demonstrate that Mdm2 functioned as an E3 ubiquitin ligase for p53. They went on to show that Mdm2 C464 was essential for Mdm2's E3 ligase activity towards p53.

Taken together these three publications provided an explanation for how the p53-Mdm2 negative feedback loop might function in normal cells to maintain p53 levels low, especially following a genotoxic or cell stress. While this pathway is central to Mdm2's role in tumorigenesis, recent data suggests that other functions of Mdm2, independent of its regulation of p53, may play equally important roles in tumorigenesis. The p53-independent activities of Mdm2 are not discussed in this review but can be found in the following recent reviews [57, 68, 88].

Mdm2 Overexpression in Cell Culture

The discovery of an amplified mdm-2 gene in double minute chromosomes of a spontaneously transformed mouse fibroblast cell line and subsequent demonstration that immortalized fibroblasts overexpressing Mdm2 could form tumors in immuno-compromised mice suggested that the mdm-2 gene was a proto-oncogene. However those studies and more classic cell transformation assays by Cathy Findlay using primary rat embryo fibroblasts demonstrating mdm2 could cooperate with an activated ras gene to transform cells employed expression vectors containing the mdm-2 genomic clone [28]. As predicted, the expression of wild-type p53 blocked the transformation of mdm-2 and ras. Surprisingly, other groups have had varied results with Mdm2 overexpression in cell culture when employing mdm-2 cDNA expression vectors. Interestingly attempts to overexpress a full-length mdm-2 cDNA in primary or immortalized cells failed to lead stably overexpressing Mdm2 cells [12]. These findings appeared to correlate with an inhibition of G1 to S phase transition and deletion mutants of Mdm2 implicated two domains in the central region of the protein. Growth inhibitory effects were also reported with mdm-2 splice variants expressing only the C-terminal region of Mdm2 [18]. The lack of a transforming capability with full-length Mdm2 cDNAs suggest that either a particular mdm-2 splice variant possesses the transforming activity or a specific genetic background is required for cells in culture to elicit the transforming effects of Mdm2 overexpression (Table 15.1).

Table 15.1 Mouse models of Mdm2 or MdmX overexpression

Gene	Findings	Citation
mdm2 cDNA	Targeted to the mammary gland	[61]
	Inhibited mammary development	
	Increased mammary tumors	
Amplified mdm2 genomic	Increased spontaneous tumors	[43]
	Increase in sarcomas relative to tumors in p53 null mice	
	Sarcoma increase was independent of p53	
mdmx transgene	Accelerated tumorigenesis in p53 null background with different tumor spectrum	[103]
	Increase in spontaneous sarcomas compared to p53 null	
HA-tagged mdmX	No increase in spontaneous tumors	[20]
	No increase in Myc-induced tumors	
	Homozygous mice demonstrated embryonic lethality	

Mouse Models of Mdm2 Loss

While cell culture studies set the foundation of the p53: Mdm2 negative feedback loop it was mouse knockout studies of Mdm2 that established Mdm2's critical role in regulating p53, at least during mouse embryogenesis. The laboratories of Gigi Lozano and Allen Bradley both attempted to create mdm-2 knockout mice [44, 73]. Both groups employed targeting vectors aimed at the exons encoding the carboxyl-terminal region of the Mdm2 although both reported the inability to detect RNA [44] or small Mdm2 proteins [73] consistent with the knockouts effectively eliminating Mdm2 protein production. From over 250 progeny screened from mdm-2 heterozygote crosses in both laboratories no mdm-2 –/– mice were observed suggesting that loss of Mdm2 resulted in embryonic lethality. Using PCR analysis of genomic DNA isolated from yolk sacks of mdm2 –/– embryos, the Bradley laboratory concluded that mdm2 –/– embryos die before gestational day 7.5. Coincidentally, this is the time during embryogenesis where Mdm2 and p53 are ubiquitously expressed. The Lozano laboratory explored embryos during various days of gestation and reported 22 % showed empty deciduae concluding that Mdm2 loss triggered embryonic lethality around gestational day 5.5. To test whether the embryonic lethality was due to the loss of Mdm2's ability to regulate p53 or some other p53 independent function of Mdm2 both groups crossed mdm-2 +/– mice with p53 –/– mice, mated the compound heterozygotes and then analyzed the progeny. mdm-2 –/–; p53 –/– mice were observed at frequencies leading both groups to conclude that the absence of p53 could completely rescue the embryonic lethality seen in Mdm2 deficient mice.

The unchecked p53 activity in these early Mdm2 null embryos suggested that p53 signaling was resulting in either complete G1 arrest or potentially p53 dependent apoptosis. Given that p53 activation of the cyclin dependent kinase inhibitor p21 is the most likely p53-effector of cell cycle arrest [24] the Lozano laboratory tested whether loss of mdm-2 could be rescued by crossing the mice to p21 –/– mice. In 46 mice born from mdm-2 heterozygotes crossed to p21 null mice, none were double null [72]. Later, they would test this question using mouse embryo fibroblasts from p53 null and p53/mdm-2 double null embryos. By transducing these cells with a temperature sensitive p53 allele they were able to show that shifting the cells to induce wild-type p53 in the double null cells led to apoptosis [23]. These early studies supported the model where Mdm2 was required for constitutive degradation of p53 during embryogenesis.

To explore the role of Mdm2 loss in specific tissues various groups created mdm-2 conditional knockout alleles using different Cre-recombinase systems. To examine the role of Mdm2 in cardiomyocyte development, Lozano's laboratory created a conditional allele and after demonstrating with CMV-Cre that its knockout produced an embryonic lethality crossed their mdm-2 conditional mice with mice expressing Cre under the control of the alpha myosin heavy chain promoter [34]. While all mice were phenotypically normal at gestational day 9, the mice lacking Mdm2 died at E13.5 and showed no evidence of a heart. As anticipated, this effect was dependent upon p53 as mice lacking p53 and the loss of the mdm-2 conditional allele did not demonstrate this lethality. Using a neuronal specific Cre recombinase under the control of a Nestin enhancer the Lozano group explored how the loss of Mdm2 impacted neuronal cell development. Mice lacking Mdm2 specifically in the central nervous system resulted in hydranecephaly at E12.5 resulting in neonatal lethality that was rescueable by elimination of p53 [104]. Taking the approach of using a p53 knock-in allele, the Marine laboratory was able to target re-expression of endogenous p53 in p53/mdm2 double null mice [29]. The re-expression of p53 using the nestin-Cre mice in p53/Mdm2 null cells confirmed that loss of Mdm2 was critical to maintain neuronal progenitors [29]. Overall, it is clear that Mdm2 plays an essential role in p53 regulation *in vivo*.

Mdm2 Has a Brother, MdmX

Using a radioactive p53 protein to screen a mouse embryo fibroblast expression library, Aart Jochemsen's laboratory reported a novel p53-binding protein with homology to Mdm2 [90]. They originally named this novel gene MdmX, but mouse knockout studies renamed the gene Mdm4. While both gene names are used in the literature it is important to note that the mdmX gene has nothing to do with murine double minute genes and was named based on its homology to Mdm2. MdmX possesses significant structural homology with Mdm2 in the amino terminal p53 DNA binding domain, central acidic and zinc finger domains and the carboxyl terminal Ring finger domain (Fig. 15.1, shaded areas) [100]. Mdm2 possesses nuclear localization and nuclear export sequences that impart the ability of the protein to shuttle in and out of the nucleus. Interestingly the cytoplasmic localization of Mdm2 provides a regulatory mechanism to control Mdm2's ability to regulate p53. In response to mitogen signaling AKT phosphorylation of Mdm2 triggers nuclear localization [67]. In contrast, MdmX lacks detectable NLS or NES

Fig. 15.1 Comparison of the Mdm2 and MdmX proteins. *Patterned boxes* represent protein domains with homology between the two proteins. *NLS* nuclear localization sequence, *NES* nuclear export sequence, *NoLS* nucleolar localization sequence

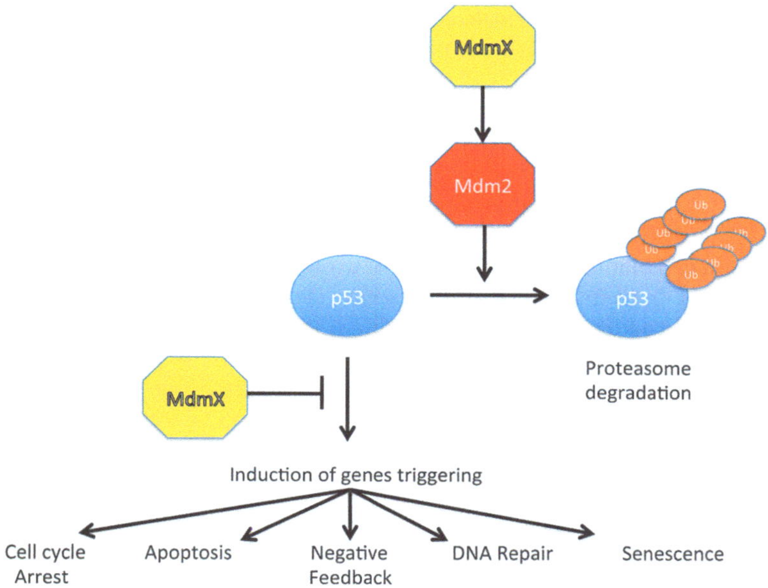

Fig. 15.2 Overview of the Mdm2-MdmX-p53 signaling pathway

motifs and appears to reside cytoplasmic. Both proteins have been shown to form homodimers through associations in their Ring Finger domains. The human homology of MdmX, HdmX also associates with p53 and inhibits its transactivation [89]. It is located on chromosome 12.14 a chromosomal region showing gene amplification in several human tumors (see below, MdmX gene amplification).

While MdmX can form homodimers like Mdm2, MdmX lacks detectable E3 ligase activity in its Ring finger domains and thus MdmX alone cannot degrade p53 [42, 93]. Interestingly using a yeast two hybrid assay Mdm2 and MdmX were shown to heterodimerize [94]. Heterodimerization required the Ring Finger domains of both proteins. Finally while MdmX is a very stable protein (half- life over 24 h), Mdm2 is a rapidly degraded protein (half-life 30 min).

The ability of MdmX and Mdm2 to heterodimerize can decrease or increase Mdm2's E3 activity towards p53 depending on its abundance [59]. Heterodimerization also triggers ubiquitin-mediated degradation of HdmX and MdmX [21, 46, 77]. It appears that following DNA damage, MdmX becomes downregulated through its association with and subsequent ubiquitination by Mdm2. An Mdm2 dependent proteasome degradation of MdmX was also seen upon ARF induction. The findings in total demonstrate that p53 activation can occur through loss of MdmX levels resulting from Mdm2 being retargeting its E3 ligase activity towards MdmX.

While MdmX alone does not impact p53 stability [92] it is able to bind to p53 with affinities comparable to Mdm2 [11] and thus inhibit p53 transactivation [42]. Thus the current model of MdmX function shows it possesses distinct impacts on p53 and Mdm2 (Fig. 15.2). Several excellent reviews provide more details into the unique interactions, cellular localization and post-translational regulations of Mdm2, MdmX and p53 as well as details of the growing number of p53-independent activities ascribed to MdmX and Mdm2 [57, 63, 88, 99].

Unlike mdm2 gene regulation, mdmX gene expression appears to be constitutively expressed throughout the cell cycle [41], with promoter activity through the ERK mitogenic signaling [32] or under certain conditions by p53 [81]. Until recently, evidence suggested that p53 did not activate the mdmX gene directly [32, 41]. However Jochemsen's laboratory has provided evidence suggesting that hdmX possesses a p53 binding site downstream of exon 1 and thus can be induced by wild-type p53 [81]. In contrast, several studies have shown that mdmX mRNA levels are down regulated under DNA damage conditions through miR34a regulation [62, 64]. Given that miR34a is a p53 activated gene [7, 16, 38] these two groups appear to be observing alternative affects of p53 in relation to mdmX gene regulation. Clearly more studies will be required to unravel this aspect of p53 gene regulation.

MdmX Overexpression in Cell Culture

Unlike the differing results seen with mdm2 overexpression (genomic vs. cDNA clone) forced overexpression of the full-length mdmX cDNA via retroviral transduction led to immortalization of mouse embryo fibroblasts and coinfection with activated Ras led to increased colony formation and growth in soft agar, two cell-based indicators of transformation [19]. In fact, the injection of these MdmX + Ras^{V12} cells into immunocompromised mice resulted in tumor formation. Similar transformation results were observed using HdmX and activated Ras in hTERT transduced human fibroblasts [55].

One unique aspect of endogenous MdmX appears to be its ability to suppress multipolar mitosis and transformation in hyperploid p53 deficient cells. MdmX loss in p53 deficient cells and mice appears to enhance cell transformation by inducing multipolar mitosis and subsequent chromosomal loss [65, 66]. This suggests that MdmX suppresses tumorigenesis in contrast to studies below that implicate its overexpression as playing a causative role in human tumorigenesis. Since this unique activity is observed in the absence of p53 it points to a p53-independent activity of MdmX.

Mouse Models of MdmX Loss

With the discovery of MdmX as a negative regulator of p53 in cell culture the next logical question was to ascertain how loss of MdmX impacted mouse embryogenesis. Using a classical knockout approach Lozano's laboratory removed mdmX exons 3–5 encompassing amino acids 27–96 and observed mdmX null embryos die at gestational day 7.5–8.5 [78]. The abnormal embryos from MdmX null mice appeared to show loss of cell proliferation with no detectable apoptosis, in stark contrast to the findings with Mdm2 null embryos. Consistent with loss of mdm-2, MdmX null mice were developmentally normal when crossed with mice lacking p53. Although the timing of embryonic lethality was slightly later (gestational days 10–12), similar results were observed in two other laboratories using gene-trap ES cells to knockout mdmX [27, 69].

Targeted knockouts of MdmX uncovered results that suggested synergistic effects with Mdm2 and others that suggest MdmX possesses distinct regulatory activities in certain cellular contexts. In the CNS, the loss of MdmX targeted a lethality that was later than that seen with similar Mdm2 knockouts [104]. However the loss of both Mdms resulted in an earlier more severe CNS phenotype suggesting that both proteins function in regulating p53 function in the CNS. Marine's laboratory [29] observed a similar synergistic effect in neuronal cells and suggested that the two proteins both function to regulate p53 activity in proliferating and postmitotic neuronal cells in non-redundant ways, consistent with the observation that loss of either Mdm2 or MdmX is embryonic lethal suggesting that the remaining member is unable to compensate for the loss of the other. In contrast to the embryonic lethality observed when Mdm2 was knocked out of embryonic cardiomyocytes or adult smooth muscle, loss of MdmX led to modest or no detectable cellular defects, respectively [6, 34]. Similarly using a Villin-Cre transgene, mice lacking MdmX in their intestinal epithelium display normal morphology with increased p53-apoptosis in the proliferative compartment of the epithelium [96].

In a completely different approach, loss of MdmX was studied by crossing mdmX heterozygotes with a p53ER^{TAM} knockin mouse where p53 can be activated in specific tissues through treatment with 4-hydroxyl-tamoxifen [30]. Interestingly when p53 is activated by tamoxifen treatment the adult mice lacking MdmX survive the transient p53 induction with all observed abnormal pathologies reversed upon loss of p53. These findings suggest that approaches to eliminate MdmX overexpression in human cancers may have less detrimental effects on normal tissue compared to targeting Mdm2.

Mdm Mouse Models of Overexpression

Towards addressing the tumorigenic properties of Mdm2 and MdmX various laboratories created mouse models of Mdm overexpression. The Finlay and Lozano laboratories targeted Mdm2 overexpression to the mammary gland using a mdm2 cDNA cloned into the ovine milk protein promoter beta-lactoglobulin (BLG) vector [61]. Targeted overexpression of Mdm2 during pregnancy and lactation resulted in defective mammary gland development. The defect appeared to result from multiple rounds of DNA replication without completed mitosis. Mammary tumors developed between 14 and 18 months in two different transgenic Mdm2 mouse strains. In another study Mdm2 was globally using a cosmid containing the mdm2 genomic clone under the control of its own promoter. This construct was electroporated into mouse ES cells. ES cells overexpressing Mdm2 due to genomic amplification were identified and then used to create an Mdm2 transgenic mice [43]. These Mdm2 overexpressing mice were susceptible to spontaneous tumorigenesis albeit at a lower level than p53 null mice. Tumors were detected in 100 % of the mice with a spectrum of tumors that showed a higher proportion of sarcomas than seen with p53 null mice. The change in tumor phenotype may suggest that Mdm2 overexpression plays a p53-independent role in tumor initiation. Taken together these studies confirm the Mdm2 overexpression impacts tumorigenesis *in vivo*.

In contrast to Mdm2, transgenic mdmX overexpressing mice have given conflicting results. Using a construct where the mdmX cDNA was placed under the control of a chicken Actin promoter containing a CMV enhancer three mdmX transgenic mouse strains with varying expression were established [103]. The mice demonstrated accelerated tumorigenesis and like mdm2 transgenic mice, displayed a tumor spectrum with a high percentage of sarcomas, distinct from the tumor spectrum seen with p53 null mice. When mdmX was overexpressed in mice heterozygous for p53 an accelerated tumorigenesis was observed with tumors maintaining the wild-type p53 allele. Taken together these findings suggested that mdmX was a bona fide oncogene. In contrast, De Clercq et al. [20] used an HA-tagged mdmX vector to create conditional transgenic mdmX mice. While the mice show MdmX overexpression they failed to demonstrate an expected increase in spontaneous, radiation-induced or Myc-induced tumors. In fact, crosses failed to produce homozygous MdmX transgenic mice. The embryos appear to demonstrate vascular defects beginning at gestational day 12.5. It is unclear the basis for the difference in findings between these two studies but they suggest that the context under which MdmX overexpression occurs may prove critical for its oncogenic activity.

Mdm2 and MdmX Expression in Human Tumors

Consistent with large body of *in vitro* and *in vivo* evidence showing that Mdm2 and MdmX play non-overlapping roles regulating p53 activity and the 50 % of human tumors harboring wild-type p53, numerous groups set out to test whether Mdm2 and MdmX overexpression was detected in human tumors. Pubmed lists well over 1,000 publications examining Mdm2 deregulation in human cancer. Below is an overview of the various methods of Mdm2 deregulation.

Mdm2 Gene Amplification

In 1998 Momand et al. undertook a review of the literature to explore spectrum of Mdm2 abnormalities in human tumors taking specific note of the distribution of Mdm2 amplification and p53 mutation. Examining 28 human tumors they reported that Mdm2 gene amplification was observed in approximately 20 % soft tissue tumors (e.g. lipomas, Ewing sarcomas), brain tumors (2–8 %), carcinomas (0–13 %) and sarcomas (13–30 %) [70]. The simultaneous screening for p53 mutations in several of they reported that mdm2 amplification and p53 mutations were mutually exclusive in human tumors. A report by Cordon-Cardo et al. [17] suggesting that Mdm2 overexpression was not always seen in tumors harboring mdm-2 gene amplification led Patterson et al., to examine 129 soft tissue tumors for both mdm2 gene amplification and overexpression of Mdm2 protein. They observed that in every tumor harboring mdm-2 gene amplification, Mdm2 protein levels were also elevated [79]. Interestingly they observed ten tumors with elevated Mdm2 protein without any detectable mdm2 gene amplification. Findings like this led to the examination of human tumors for other mechanisms to overexpress Mdm2 protein in human cancer.

Mdm2 Increased Transcription/Alternative Splicing

The mdm2 gene possesses two promoters and it has been reported that more than 40 mRNA variants have been identified [3, 74]. Some variants result from alternative splicing while others arise from aberrant splicing. One of the better-studied and most frequently detected Mdm2 isoform is Mdm2-B or Alt1. Mdm2-B is unable to bind to p53 but is capable of binding full length Mdm2, triggering cytoplasmic sequestration of the protein. Evans et al. reported that the binding of Mdm2-B to Mdm2 increased wild-type p53 activity [25]. Several of these Mdm2 isoforms appear to have tumor promoting capabilities independent of p53. Since levels of these various isoforms may affect function, more work is required to decipher the complexity of mdm2 gene splicing and tumorigenesis.

One significant discovery in the area of mdm2 gene transcription was a T to G single-nucleotide polymorphism (SNP309) located within the mdm2 P2 promoter. This SNP creates an additional SP1 transcription factor binding site leading to a higher affinity of the SP1 transcription factor for this binding site, resulting in higher levels of mdm2 mRNA and an attenuation of the p53 pathway. This modest increase in mdm2 mRNA and protein is correlated with a gender-specific increase in accelerated tumor formation [8–10]. Consistent with these findings a SNP309 knock-in mouse model demonstrated increased Mdm2 expression and accelerated spontaneous tumorigenesis [82].

Mdm2 Enhanced Translation

Choriocarcinoma cell lines were the first reported to express elevated Mdm2 protein levels through an enhanced translation mechanism [52]. Subsequent studies from the George laboratory uncovered that the enhanced translation resulted from mdm2 transcripts possessing a distinct 5′ UTR [51]. Later, Mdm2 enhanced translation

Table 15.2 MdmX overexpression in human cancers

Tumor type	Detection	%	#	Citation
Glioblastomas	Gene amplification	7		[14]
Hepatoblastomas	Gene amplification	7	4 of 56	[1]
Malignant gliomas	Gene amplification	4	5 of 125	[85]
				[86]
Retinoblastomas	Gene Amplification	47	3 of 7	[54]
	+mRNA+IHC	65	36 of 56	
Soft tissue sarcoma	Gene amplification	17	11 of 66	[5]
Urothelial cell carcinoma	Gene amplification	17	16 of 98	[98]
Papillary thyroid carcinomas	HdmX-S mRNA	21	12 of 57	[83]
Soft tissue sarcoma	HdmX-S mRNA	14	8 of 57	[4]
				[5]
Retinoblastomas	MdmX SNP			[22]
Breast cancer	mRNA	19	41 of 218	[19]
Colon cancer	mRNA	19	5 of 27	[19]
		49		[32]
Lung cancer	mRNA	18	16 of 88	[19]
Retinoblastoma	mRNA and Protein	20	2 of 10	[35]
Mantle cell lymphomas	Nuclear HdmX	95	18 of 19	[58]
	HdmX-S protein			
Acute lymphocytic leukemias	Protein	80	39 of 49	[36]
Cutaneous melanomas	Protein	65	35 of 54	[31]
Head/Neck squamous carcinomas	Protein	50		[95]

was reported in Burkitt's lymphoma cells [15] and several B-cell leukemias and lymphomas [102]. It have been noted by others in the field that the overall percentage of human tumors harboring increased Mdm2 protein expression is likely significantly underestimated due in part to poor antibodies and early studies that did not consider two to four fold increases as significant [68]. Understanding the mechanism for this enhanced translation may lead to new therapeutic approaches.

MdmX Gene Amplification

Similar to Mdm2, initial studies of MdmX deregulation in human tumors focused on gene amplification studies. While many tumor types that possess mdm2 gene amplification also show mdmX gene amplification (e.g. glioblastomas, soft tissue sarcomas), some MdmX specific deregulations in human tumors have been reported (Table 15.2). Laurie et al. uncovered one such MdmX specific deregulation when they discovered that retinoblastomas, human tumors of the retinoblasts that are initiated by mutations in the retinoblastoma (Rb) gene, harbor inactivation of the p53 signaling pathway through MdmX overexpression (Table 15.2, [54]). Owing to different pathways to MdmX overexpression the percentage of mdmX gene amplification (47 %) did not capture all of the cases of MdmX overexpression at the mRNA and protein level were considered (65 %, Table 15.2). Melanomas are another human tumor where MdmX deregulation is observed at a high frequency (65 %). Interestingly p53 mutations are a rare occurrence in melanomas suggesting that MdmX elevation is playing a critical role in overcoming p53 tumor suppression. Consistent with MdmX playing a role in blocking p53 Marine's laboratory showed that inhibiting the MdmX:p53 interaction in melanoma cells made them highly sensitive to cytotoxic chemotherapy and inhibitors of BRAF, an effective target in melanomas harboring oncogenic mutant BRAF(V600E) protein [31]. As better reagents become available a better picture of MdmX deregulation in human tumors will develop.

MdmX Transcription/Alternative Splicing

Less is known about the transcriptional upregulation of mdmX in human tumors. Human colon tumors harboring elevated phosphorylated ELK demonstrated a two-fold increase in mdmX transcripts suggesting a role for ELK signaling in tumor development (Table 15.2, [32]). Additionally several mdmX splice variants have been reported. The first mdmX splice variant discovered (MdmX-S or HdmX-S) encodes a truncated protein (114 amino acids) that retains p53 binding activity and is overexpressed in several sarcomas and carcinomas [4, 5, 83]. These findings suggest that HdmX-S may represent a novel biomarker in human cancer [56]. Studies examining the existence of SNPs near the mdmX gene that correlate with cancer has not been as fruitful as mdm2 SNP309 however a recent study of retinoblastomas suggest that MDM4 rs116197192G was significantly higher in RB patients implying it may increase the risk of acquiring retinoblastomas (Table 15.2, [22]). More studies will be required to confirm these findings.

MdmX Protein Expression

Thus far it does not appear that enhanced translation is a mechanism leading to MdmX overexpression in human tumors. However clinical analysis of MdmX protein levels has proven useful in identifying MdmX deregulation in acute lymphocytic leukemias, melanomas and squamous carcinomas (Table 15.2). The results in melanomas have particular clinical relevance since MdmX overexpression appears to be a key factor in aggressive melanomas where p53 mutations are extremely rare [31]. As mentioned above these findings suggest that targeting MdmX in melanoma tumors will be provide a therapeutic benefit to apply to current anti-melanoma therapies.

Targeting Mdm2 and MdmX in Human Cancers

Several excellent reviews have been written on the targeting of Mdm2 and MdmX in human cancers [57, 63, 99]. The approaches currently being tested can be classified into four areas. The first and most active area of research is to target the Mdm:p53 interaction. Foremost in this category of compounds is a class of drugs known as Nutlins [97]. These small molecules antagonize the Mdm2:p53 interaction activating p53 in a non-genotoxic manner. One Nutlin derivative RG7112 is currently in Phase I clinical trials. While Nutlins have proven effective against Mdm2:p53 interactions they surprisingly do not antagonize MdmX:p53 interactions [40, 48, 101]. The second approach would be to target MdmX. To date the work has been limited to compounds like SJ-172550 [84] and RO-5963 [33] that target MdmX:p53 complexes. Clearly, this is an area worthy of additional study.

The third approach has been to target Mdm2 E3 ligase activity however the first round of compounds appear at higher concentrations or with increased potency to exhibit p53-independent activity such as inhibiting other E3 ligases suggesting their suitability in clinical trails maybe limited [87, 105]. Finally the fourth approach has to been to target downstream of the Mdm2 overexpression stage and target the Mdm2:p53 association with the proteasome. Again, early indications suggest that targeting this point in the pathway will lead to p53-independent activities [47].

Concluding Remarks

The discovery and subsequent characterization of Mdm2 and MdmX deregulation in human cancers has broadened the importance of studying p53 signaling and continuing to translate what is found at the research bench to the cancer patient. Human cancers possessing overexpressed Mdm2 or MdmX and wild-type p53 represent prime targets for small molecules the impinge upon the Mdm:p53 pathway. What research has also uncovered is that MdmX and Mdm2 overexpression likely have a broader impact than targeting p53 thus care must be taken with biomarker screening as these compounds reach clinical trials. What is certain is the 30 plus years of p53 research have taught us more about a cellular protein associated with large T antigen than anyone could have expected [60]. The next 30 years offer the potential for real improvements as we move to a more personalized cancer treatment regiment where Mdm2 and MdmX will most certainly be critical targets.

References

1. Arai Y et al (2010) Genome-wide analysis of allelic imbalances reveals 4q deletions as a poor prognostic factor and MDM4 amplification at 1q32.1 in hepatoblastoma. Genes Chromosom Cancer 49(7):596–609

2. Barak Y et al (1993) mdm2 expression is induced by wild type p53 activity. EMBO J 12(2):461–468
3. Bartel F et al (2004) MDM2 and its splice variant messenger RNAs: expression in tumors and down-regulation using antisense oligonucleotides. Mol Cancer Res 2(1):29–35
4. Bartel F et al (2004) HDMX amplification and high levels of HDMX-S splice variant are correlated with a poor prognosis in soft tissue sarcomas. Verh Dtsch Ges Pathol 88:199–206
5. Bartel F et al (2005) Significance of HDMX-S (or MDM4) mRNA splice variant overexpression and HDMX gene amplification on primary soft tissue sarcoma prognosis. Int J Cancer 117(3):469–475
6. Boesten LS et al (2006) Mdm2, but not Mdm4, protects terminally differentiated smooth muscle cells from p53-mediated caspase-3-independent cell death. Cell Death Differ 13(12):2089–2098
7. Bommer GT et al (2007) p53-mediated activation of miRNA34 candidate tumor-suppressor genes. Curr Biol 17(15):1298–1307
8. Bond GL et al (2006) MDM2 SNP309 accelerates tumor formation in a gender-specific and hormone-dependent manner. Cancer Res 66(10):5104–5110
9. Bond GL et al (2004) A single nucleotide polymorphism in the MDM2 promoter attenuates the p53 tumor suppressor pathway and accelerates tumor formation in humans. Cell 119(5):591–602
10. Bond GL et al (2005) A single nucleotide polymorphism in the MDM2 gene: from a molecular and cellular explanation to clinical effect. Cancer Res 65(13):5481–5484
11. Bottger V et al (1999) Comparative study of the p53-mdm2 and p53-MDMX interfaces. Oncogene 18:189–199
12. Brown DR et al (1998) The human oncoprotein MDM2 arrests the cell cycle: elimination of its cell-cycle-inhibitory function induces tumorigenesis. EMBO J 17(9):2513–2525
13. Cahilly-Snyder L et al (1987) Molecular analysis and chromosomal mapping of amplified genes isolated from a transformed mouse 3T3 cell line. Somat Cell Mol Genet 13(3):235–244
14. Cancer Genome Atlas Research Network (2008) Comprehensive genomic characterization defines human glioblastoma genes and core pathways. Nature 455(7216):1061–1068
15. Capoulade C et al (1998) Overexpression of MDM2, due to enhanced translation, results in inactivation of wild-type p53 in Burkitt's lymphoma cells. Oncogene 16(12):1603–1610
16. Chang TC et al (2007) Transactivation of miR-34a by p53 broadly influences gene expression and promotes apoptosis. Mol Cell 26(5):745–752
17. Cordon CC et al (1994) Molecular abnormalities of mdm2 and p53 genes in adult soft tissue. Cancer Res 54(3):794–799
18. Dang J et al (2002) The RING domain of Mdm2 can inhibit cell proliferation. Cancer Res 62(4):1222–1230
19. Danovi D et al (2004) Amplification of Mdmx (or Mdm4) directly contributes to tumor formation by inhibiting p53 tumor suppressor activity. Mol Cell Biol 24(13):5835–5843
20. De Clercq S et al (2010) Widespread overexpression of epitope-tagged Mdm4 does not accelerate tumor formation in vivo. Mol Cell Biol 30(22):5394–5405
21. de Graaf P et al (2003) Hdmx protein stability is regulated by the ubiquitin ligase activity of Mdm2. J Biol Chem 278(40):38315–38324
22. de Oliveira Reis AH et al (2012) Influence of MDM2 and MDM4 on development and survival in hereditary retinoblastoma. Pediatr Blood Cancer 59(1):39–43
23. de Rozieres S et al (2000) The loss of mdm2 induces p53-mediated apoptosis. Oncogene 19(13):1691–1697
24. el-Deiry WS et al (1993) WAF1, a potential mediator of p53 tumor suppression. Cell 75(4):817–825
25. Evans SC et al (2001) An alternatively spliced HDM2 product increases p53 activity by inhibiting HDM2. Oncogene 20(30):4041–4049
26. Fakharzadeh SS et al (1991) Tumorigenic potential associated with enhanced expression of a gene that is amplified in a mouse tumor cell line. EMBO J 10(6):1565–1569

27. Finch RA et al (2002) mdmx is a negative regulator of p53 activity in vivo. Cancer Res 62(11):3221–3225
28. Finlay CA (1993) The mdm-2 oncogene can overcome wild-type p53 suppression of transformed cell growth. Mol Cell Biol 13(1):301–306
29. Francoz S et al (2006) Mdm4 and Mdm2 cooperate to inhibit p53 activity in proliferating and quiescent cells in vivo. Proc Natl Acad Sci U S A 103(9):3232–3237
30. Garcia D et al (2011) Validation of MdmX as a therapeutic target for reactivating p53 in tumors. Genes Dev 25(16):1746–1757
31. Gembarska A et al (2012) MDM4 is a key therapeutic target in cutaneous melanoma. Nat Med 18:1239–1247
32. Gilkes DM et al (2008) Regulation of MDMX expression by mitogenic signaling. Mol Cell Biol 28(6):1999–2010
33. Graves B et al (2012) Activation of the p53 pathway by small-molecule-induced MDM2 and MDMX dimerization. Proc Natl Acad Sci U S A 109(29):11788–11793
34. Grier JD et al (2006) Tissue-specific differences of p53 inhibition by Mdm2 and Mdm4. Mol Cell Biol 26(1):192–198
35. Guo Y et al (2008) Expression of p14ARF, MDM2, and MDM4 in human retinoblastoma. Biochem Biophys Res Commun 375(1):1–5
36. Han X et al (2007) HDM4 (HDMX) is widely expressed in adult pre-B acute lymphoblastic leukemia and is a potential therapeutic target. Mod Pathol 20(1):54–62
37. Haupt Y et al (1997) Mdm2 promotes the rapid degradation of p53. Nature 387(6630):296–299
38. He L et al (2007) A microRNA component of the p53 tumour suppressor network. Nature 447(7148):1130–1134
39. Hollstein M et al (1994) Database of p53 gene somatic mutations in human tumors and cell lines. Nucleic Acids Res 22(17):3551–3555
40. Hu B et al (2006) MDMX overexpression prevents P53 activation by the MDM2 inhibitor nutlin. J Biol Chem 281:33030–33035
41. Jackson MW, Berberich SJ (1999) Constitutive mdmx expression during cell growth, differentiation, and DNA damage. DNA Cell Biol 18(9):693–700
42. Jackson MW, Berberich SJ (2000) MdmX protects p53 from Mdm2-mediated degradation. Mol Cell Biol 20(3):1001–1007
43. Jones SN et al (1998) Overexpression of Mdm2 in mice reveals a p53-independent role for Mdm2 in tumorigenesis. Proc Natl Acad Sci U S A 95(26):15608–15612
44. Jones SN et al (1995) Rescue of embryonic lethality in Mdm2-deficient mice by absence of p53. Nature 378:206–208
45. Juven T et al (1993) Wild type p53 can mediate sequence-specific transactivation of an internal promoter within the mdm2 gene. Oncogene 8(12):3411–3416
46. Kawai H et al (2003) DNA damage-induced MDMX degradation is mediated by MDM2. J Biol Chem 278(46):45946–45953
47. Kojima K et al (2010) The novel tryptamine derivative JNJ-26854165 induces wild-type p53- and E2F1-mediated apoptosis in acute myeloid and lymphoid leukemias. Mol Cancer Ther 9(9):2545–2557
48. Kranz D, Dobbelstein M (2006) Nongenotoxic p53 activation protects cells against S-phase-specific chemotherapy. Cancer Res 66(21):10274–10280
49. Kubbutat MH et al (1997) Regulation of p53 stability by Mdm2. Nature 387(6630):299–303
50. Kussie PH et al (1996) Structure of the MDM2 oncoprotein bound to the p53 tumor suppressor transactivation domain. Science 274(5289):948–953
51. Landers JE et al (1997) Translational enhancement of mdm2 oncogene expression in human tumor cells containing a stabilized wild-type p53 protein. Cancer Res 57(16):3562
52. Landers JE et al (1994) Enhanced translation: a novel mechanism of mdm2 oncogene overexpression identified in human tumor cells. Oncogene 9(9):2745–2750
53. Lane D, Levine A (2010) p53 research: the past thirty years and the next thirty years. Cold Spring Harb Perspect Biol 2(12):a000893

54. Laurie NA et al (2006) Inactivation of the p53 pathway in retinoblastoma. Nature 444(7115):61–66
55. Lenos K et al (2011) Oncogenic functions of hMDMX in in vitro transformation of primary human fibroblasts and embryonic retinoblasts. Mol Cancer 10:111
56. Lenos K et al (2012) Alternate splicing of the p53 inhibitor HDMX offers a superior prognostic biomarker than p53 mutation in human cancer. Cancer Res 72(16):4074–4084
57. Li Q, Lozano G (2012) Molecular pathways: targeting Mdm2 and Mdm4 in cancer therapy. Clin Cancer Res 19:34–41
58. Liang M et al (2010) HDM4 is overexpressed in mantle cell lymphoma and its inhibition induces p21 expression and apoptosis. Mod Pathol 23(3):381–391
59. Linares LK et al (2003) HdmX stimulates Hdm2-mediated ubiquitination and degradation of p53. Proc Natl Acad Sci U S A 100(21):12009–12014
60. Linzer DIH, Levine AJ (1979) Characterization of a 54 K dalton cellular SV40 tumor antigen present in SV40-transformed cells and uninfected embryonal carcinoma cells. Cell 17:43–52
61. Lundgren K et al (1997) Targeted expression of MDM2 uncouples S phase from mitosis and inhibits mammary gland development independent of p53. Genes Dev 11(6):714–725
62. Mandke P et al (2012) MicroRNA-34a modulates MDM4 expression via a target site in the open reading frame. PLoS One 7(8):e42034
63. Marine JC (2011) MDM2 and MDMX in cancer and development. Curr Top Dev Biol 94:45–75
64. Markey M, Berberich SJ (2008) Full-length hdmX transcripts decrease following genotoxic stress. Oncogene 27(52):6657–6666
65. Matijasevic Z et al (2008) MdmX regulates transformation and chromosomal stability in p53-deficient cells. Cell Cycle 7(19):2967–2973
66. Matijasevic Z et al (2008) MdmX promotes bipolar mitosis to suppress transformation and tumorigenesis in p53-deficient cells and mice. Mol Cell Biol 28(4):1265–1273
67. Mayo LD, Donner DB (2001) A phosphatidylinositol 3-kinase/Akt pathway promotes translocation of Mdm2 from the cytoplasm to the nucleus. Proc Natl Acad Sci U S A 98(20):11598–11603
68. Melo AN, Eischen CM (2012) Protecting the genome from mdm2 and mdmx. Genes Cancer 3(3–4):283–290
69. Migliorini D et al (2002) Mdm4 (Mdmx) regulates p53-induced growth arrest and neuronal cell death during early embryonic mouse development. Mol Cell Biol 22(15):5527–5538
70. Momand J et al (1998) The MDM2 gene amplification database. Nucleic Acids Res 26(15):3453–3459
71. Momand J et al (1992) The mdm-2 oncogene product forms a complex with the p53 protein and inhibits p53-mediated transactivation. Cell 69(7):1237–1245
72. Montes de Oca Luna R et al (1997) Deletion of p21 cannot substitute for p53 loss in rescue of mdm2 null lethality. Nat Genet 16(3):336–337
73. Montes de Oca Luna R et al (1995) Rescue of early embryonic lethality in mdm2-deficient mice by deletion of p53. Nature 378:203–206
74. Okoro DR et al (2012) Splicing up mdm2 for cancer proteome diversity. Genes Cancer 3(3–4):311–319
75. Oliner JD et al (1992) Amplification of a gene encoding a p53-associated protein in human sarcomas. Nature 358(6381):80–83
76. Otto A, Deppert W (1993) Upregulation of mdm-2 expression in meth A tumor cells tolerating wild-type p53. Oncogene 8:2591–2603
77. Pan Y, Chen J (2003) MDM2 promotes ubiquitination and degradation of MDMX. Mol Cell Biol 23(15):5113–5121
78. Parant J et al (2001) Rescue of embryonic lethality in Mdm4-null mice by loss of Trp53 suggests a nonoverlapping pathway with MDM2 to regulate p53. Nat Genet 29(1):92–95
79. Patterson H et al (1997) Amplification and over-expression of the MDM2 gene in human soft tissue tumours. Sarcoma 1(1):17–22

80. Perry ME et al (1993) The mdm-2 gene is induced in response to UV light in a p53-dependent. Proc Natl Acad Sci U S A 90(24):11623–11627
81. Phillips A et al (2010) HDMX-L is expressed from a functional p53-responsive promoter in the first intron of the HDMX gene and participates in an autoregulatory feedback loop to control p53 activity. J Biol Chem 285(38):29111–29127
82. Post SM et al (2010) A high-frequency regulatory polymorphism in the p53 pathway accelerates tumor development. Cancer Cell 18(3):220–230
83. Prodosmo A et al (2008) Analysis of human MDM4 variants in papillary thyroid carcinomas reveals new potential markers of cancer properties. J Mol Med (Berl) 86(5):585–596
84. Reed D et al (2010) Identification and characterization of the first small molecule inhibitor of MDMX. J Biol Chem 285(14):10786–10796
85. Riemenschneider MJ et al (1999) Amplification and overexpression of the MDM4 (MDMX) gene from 1q32 in a subset of malignant gliomas without TP53 mutation or MDM2 amplification. Cancer Res 59(24):6091–6096
86. Riemenschneider MJ et al (2003) Refined mapping of 1q32 amplicons in malignant gliomas confirms MDM4 as the main amplification target. Int J Cancer 104(6):752–757
87. Roxburgh P et al (2012) Small molecules that bind the Mdm2 RING stabilize and activate p53. Carcinogenesis 33(4):791–798
88. Senturk E, Manfredi JJ (2012) Mdm2 and tumorigenesis: evolving theories and unsolved mysteries. Genes Cancer 3(3–4):192–198
89. Shvarts A et al (1997) Isolation and identification of the human homolog of a new p53-binding protein, Mdmx. Genomics 43(1):34
90. Shvarts A et al (1996) MDMX: a novel p53-binding protein with some functional properties of MDM2. EMBO J 15(19):5349–5357
91. Snyder LC, Trusko SP, Freeman N, Eshleman JR, Fakharzadeh SS, George DL (1988) A gene amplified in a transformed mouse cell line undergoes complex transcriptional processing and encodes a nuclear protein. J Biol Chem 263:17150–17158
92. Stad R et al (2001) Mdmx stabilizes p53 and Mdm2 via two distinct mechanisms. EMBO Rep 2(11):1029–1034
93. Stad R et al (2000) Hdmx stabilizes Mdm2 and p53. J Biol Chem 275(36):28039–28044
94. Tanimura S et al (1999) MDM2 interacts with MDMX through their RING finger domains [In Process Citation]. FEBS Lett 447(1):5–9
95. Valentin-Vega YA et al (2007) High levels of the p53 inhibitor MDM4 in head and neck squamous carcinomas. Hum Pathol 38(10):1553–1562
96. Valentin-Vega YA et al (2009) Mdm4 loss in the intestinal epithelium leads to compartmentalized cell death but no tissue abnormalities. Differentiation 77(5):442–449
97. Vassilev LT et al (2004) In vivo activation of the p53 pathway by small-molecule antagonists of MDM2. Science 303(5659):844–848
98. Veerakumarasivam A et al (2008) High-resolution array-based comparative genomic hybridization of bladder cancers identifies mouse double minute 4 (MDM4) as an amplification target exclusive of MDM2 and TP53. Clin Cancer Res 14(9):2527–2534
99. Wade M et al (2012) MDM2, MDMX and p53 in oncogenesis and cancer therapy. Nat Rev Cancer 13(2):83–96
100. Wade M et al (2010) The p53 orchestra: Mdm2 and Mdmx set the tone. Trends Cell Biol 20(5):299–309
101. Wade M et al (2006) Hdmx modulates the outcome of p53 activation in human tumor cells. J Biol Chem 281(44):33036–33044
102. Watanabe T et al (1994) The MDM2 oncogene overexpression in chronic lymphocytic leukemia and low-grade lymphoma of B-cell origin. Blood 84(9):3158–3165
103. Xiong S et al (2010) Spontaneous tumorigenesis in mice overexpressing the p53-negative regulator Mdm4. Cancer Res 70(18):7148–7154
104. Xiong S et al (2006) Synergistic roles of Mdm2 and Mdm4 for p53 inhibition in central nervous system development. Proc Natl Acad Sci U S A 103(9):3226–3231
105. Yang Y et al (2005) Small molecule inhibitors of HDM2 ubiquitin ligase activity stabilize and activate p53 in cells. Cancer Cell 7(6):547–559

Chapter 16
Targeting p53-MDM2-MDMX Loop for Cancer Therapy

Qi Zhang, Shelya X. Zeng, and Hua Lu

Abstract The tumor suppressor p53 plays a central role in anti-tumorigenesis and cancer therapy. It has been described as "the guardian of the genome", because it is essential for conserving genomic stability by preventing mutation, and its mutation and inactivation are highly related to all human cancers. Two important p53 regulators, MDM2 and MDMX, inactivate p53 by directly inhibiting its transcriptional activity and mediating its ubiquitination in a feedback fashion, as their genes are also the transcriptional targets of p53. On account of the importance of the p53-MDM2-MDMX loop in the initiation and development of wild type p53-containing tumors, intensive studies over the past decade have been aiming to identify small molecules or peptides that could specifically target individual protein molecules of this pathway for developing better anti-cancer therapeutics. In this chapter, we review the approaches for screening and discovering efficient and selective MDM2 inhibitors with emphasis on the most advanced synthetic small molecules that interfere with the p53-MDM2 interaction and are currently on Phase I clinical trials. Other therapeutically useful strategies targeting this loop, which potentially improve the prospects of cancer therapy and prevention, will also be discussed briefly.

Keywords p53 • MDM2 • MDMX • Drug discovery • Drug design • Drug development • Cancer therapy

Q. Zhang • S.X. Zeng • H. Lu (✉)
Department of Biochemistry & Molecular Biology and Tulane Cancer Center,
Tulane University School of Medicine, 1430 Tulane Ave, Louisiana, LA 70112, USA
e-mail: hlu2@tulane.edu

S.P. Deb and S. Deb (eds.), *Mutant p53 and MDM2 in Cancer*,
Subcellular Biochemistry 85, DOI 10.1007/978-94-017-9211-0_16,

Introduction

The p53-MDM2-MDMX-Loop

The tumor suppressor p53 is inarguably the most recognized and studied protein involving human cancers. Its vital importance in preventing human cancer development and progression is simply reflected by the fact that mutations of its gene TP53 are detected in approximately 50 % of all types of human cancers, and the functions and stability of the p53 protein are often abrogated via posttranslational mechanisms in the rest of human cancers that harbor wild type TP53 [1–3]. Cancers often deactivate p53, because it can trigger cell growth arrest, apoptosis, autophagy, and/or senescence, which are detrimental to cancer cells [4, 5], and impede cell migration, metabolism, and/or angiogenesis, which are favorable to cancer cell progression and metastasis [5]. These physiological functions of p53 are executed primarily through its transcription-dependent and independent activities [5]. However, because these functions are also deleterious to normally growing stem cells and developing tissues [6], p53 is tightly monitored by two closely related proteins called MDM2 (sometime called HDM2 for its human analog) [7–9] and MDMX (also known as MDM4) [10] in higher eukaryotes [11].

MDM2 and MDMX execute their oncogenic activity mainly by negatively regulating the stability and activity of the p53 protein in a feedback fashion (Fig. 16.1). They work together to block the transcriptional activity of p53 [5, 8, 9, 12] and to mediate p53 rapid degradation via ubiquitin-dependent proteolysis [13, 14], as MDM2 possesses an E3 ubiquitin ligase activity [15], and p53 stimulates MDM2 and MDMX mRNA expression [7, 9, 16–18]. This dual action of MDM2 and MDMX on p53 leads to the barely detectable level and activity of p53 in most normal mammalian cells or tissues. MDM2 and MDMX can also inhibit p53 independently of each other. Often, MDMX negates p53 transcriptional activity, while MDM2 can inhibit both of the p53 protein stability and activity [19]. Hence, in order to activate p53, eukaryotic cells have developed mechanisms to block this negative feedback regulation in response to a variety of cellular, genotoxic, or non-genotoxic stresses [20–22]. These mechanisms include posttranslational modifications of either p53 or MDM2/MDMX, such as acetylation [23], phosphorylation [24–27], and protein-protein interactions, such as ribosomal proteins-MDM2 interaction, or Arf-MDM2 interaction [20, 28], ultimately leading to p53 activation that prevents cells from undergoing transformation and neoplasia. Interestingly, two different modifications, acetylation and ubiquitylation, often occur at a similar set of lysine residues within p53, and thus are mutually exclusive. For example, acetylation of p53 by p300/CBP prevents its degradation by MDM2 and activates its activity whereas MDM2 inhibits p53 acetylation by p300/CBP [29–31]. Conversely, deacetylation of p53 by an NAD-dependent deacetylase, SIRT1 [32–34], or a class I histone deacetylase, HDAC1 [35], favors MDM2-mediated p53 degradation, leading to p53 inactivation. Remarkably, cancers often take advantages of this feedback loop to promote their own growth, as human breast cancers,

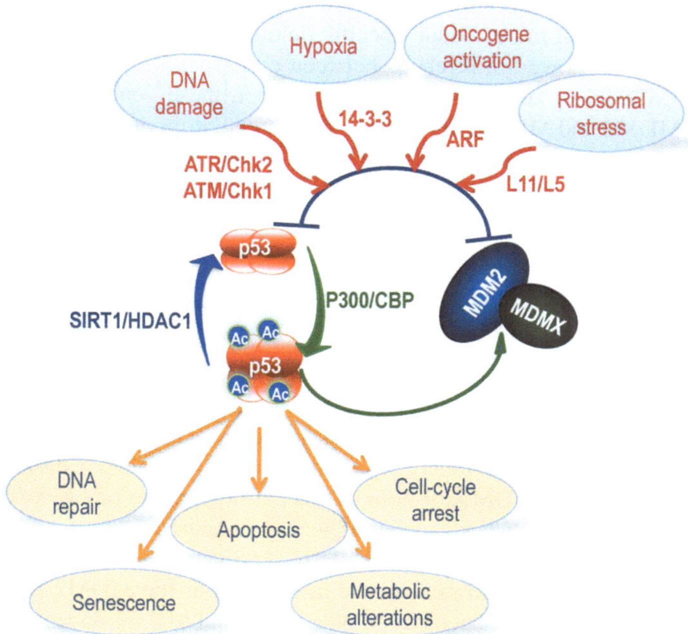

Fig. 16.1 The p53-MDM2-MDMX feedback loop. Two p53 suppressors, MDM2 and MDMX, which are highly expressed in tumors, often work together as one complex to inactivate p53 by mediating its ubiquitination and degradation as well as to directly inhibit p53 transcriptional activity in a feedback fashion. This feedback regulation is however untied through various mechanisms in response to a variety of stress signals, including DNA damage, hypoxia, oncogene activation and ribosomal stress, leading to p53 activation and consequently cellular responses, such as DNA repair, cell-cycle arrest, senescence, apoptosis and metabolic alterations. On mechanism is through p53 acetylation by p300/CBP and deacetylation by SIRT1 or HDAC1. Deacetylation of p53 by SIRT1 facilitates MDM2/MDMX-mediated p53 degradation, while p53 acetylation by p300/CBP prevents p53 ubiquitination by MDM2/MDMX, thus stabilizing p53 in response to these stress signals

osteosarcomas, lymphomas, leukemia or melanoma express high levels of MDM2 or MDMX through distinct mechanisms without p53 mutation [17, 36]. Also, the high level of deacetylases is often detected in cancers [37–40]. Therefore, it is likely that deacetylases might play a role in maintaining p53 in a deacetylated status in cancer cells, consequently facilitating MDM2/MDMX-mediated degradation.

The physiological relevance of the MDM2/MDMX-p53 regulation has been also demonstrated in mouse genetic studies [22, 41, 42]. For example, embryonic lethality of either MDM2 or MDMX-null mice can be completely rescued by the simultaneous deletion of the TP53 gene. In addition, compared to wild-type adult mice, genetically engineered mice expressing reduced levels of MDM2 and MDMX are small in size, have reduced organ weight, and are radiosensitive [43]. The p53 dependence of these phenotypes has been shown by reversing the phenotypes when crossed with p53-null mice. Tissue-specific deletion of either the MDM2 gene or the MDMX gene showed differences between cell types for their dependency on MDM2 or MDMX to keep p53 in check. Together, these genetic studies demonstrate that MDM2 and MDMX are critical for the regulation of p53 functions during embryonic development as well as in adult mice, and the changes of MDM2 and MDMX levels can dictate abnormality and tumorigenesis.

Targeting the p53-MDM2-MDMX Loop for Cancer Therapy

The highly cancer-pertinent and well-defined p53-MDM2-MDMX pathway offers multiple molecular targets for screening small molecules as potential therapies for wild type p53-harboring cancers. Over the past decade, a number of new small molecules in addition to the previously reported Nutlin-3 [44] have been identified to target either MDM2 or MDMX or both in this pathway, some of which as listed in Table 16.1 have been further advanced to the stage of clinical trials.

For example, a small molecule named JNJ-26854165 was shown to bind the RING domain of MDM2 and prevent the interaction of the MDM2-p53 complex with the proteasome and has been put on phase I clinical trial for advanced or refractory tumor [45, 46]. Also, the other three small compounds, called HLI98, MPD and MEL, respectively, were discovered to inhibit the E3 ubiquitin ligase activity of HDM2, preventing p53 degradation [47–49]. Furthermore, NSC279287, NSC66811 and terphenyl compounds are small molecules that can also disrupt the MDM2-p53 interaction [50–52]. Additional small molecule inhibitors of the MDM2-p53 interaction have been revealed later on, such as TDP521252, TDP665759, PXN727, PXN822 and isoindolinones, which are currently under pre-clinical development [53–57]. By contrast, the two molecules SJ172550 and XI-006 were recently identified to restore WT p53 activity by targeting MDMX. SJ172550 interferes with the MDMX-p53 interaction [58, 59], while XI-006 inhibits MDMX transcription [60]. Interestingly, the indolyl hydantoin class of compounds, RO-2443 and RO-5963, demonstrated as dual inhibitors of the MDM2/MDMX-p53 binding display the potential for further chemical optimization to achieve more desirable pharmacological characteristics with improved potency [61]. More interestingly, naturally-derived molecules have been found to inhibit the p53-MDM2 interaction. Some of them have been shown to decrease MDM2 expression and activity in vitro and in vivo. These newly identified natural MDM2 inhibitors include a plethora of diverse chemical frameworks, ranging from flavonoids, steroids, and sesquiterpenes to alkaloids. Their complex and unique molecular architectures could stimulate further development of synthetic analogs in the near future (see review for more information in [62]). Remarkably, although several advanced synthetic small molecule inhibitors, such as RG7112, MI-773, CGM097, and MK-8242, which interfere with p53-MDM2 interactions based on imidazoline Nutlin, spirooxindole MI-219 and undisclosed scaffolds,

Table 16.1 Inhibitors targeting MDM2-MDMX loop discovered by different strategies

Small molecule	Status	ID	Scaffold	Discover methods	K_i or IC_{50}	In vivo test	Reference
M2-p53 interaction[a]	Preclinical	Nutlin-3[c]	Imidazolines	HTS, SPR assay	~0.1 μM	Retinoblastoma	[44]
	Phase I	RG7112 (RO5045337)			18 nM	Advanced malignancies, except leukemia	[100]
M2-p53 interaction	Preclinical	MI63, MI-219	Spirooxindoles	Structure-based design	~1–50 nM	SJSA-1 xenograft	[94]
	Phase I	MI-773 (SAR405838)				Advanced cancer	[63]
M2-p53 interaction		NSC333003[c]	Benzothiazoles	Virtual screening	~20 μM		[150]
M2-p53 interaction	Preclinical	TPD222669	Benzodiazepines	HTS ThermoFlor	0.08 μM	A375 xenograft	[151]
M2-p53 interaction	Preclinical	YH265	Pyrazoles	HCS, Biosensor	1.8 μM	NCI60 cancer cells	[83, 84] WO2011106650
M2-p53 interaction	Phase I	CGM097 (Novartis)			<50 nM	Selected advanced and refractory solid tumors	[63]
M2-p53 interaction	Phase I	MK-8242 (Merck)	Piperidines				[63]
M2-p53 interaction	Preclinical	PXN727 (Priaxon)	Isoquinoline-1-ones, pyrrolinones	Virtual screening	8 nM–1 μM	LNCaP xenograft	http://www.priaxon.com/content/_docs/doc1264519272209.pdf

(continued)

Table 16.1 (continued)

Small molecule	Status	ID	Scaffold	Discover methods	K_i or IC_{50}	In vivo test	Reference
M2-p53 interaction			Tetra-substituted heteroaryls	Structure-based design	1 nM–50 μM		[63]
M2-p53 interaction		[c]	Substituted pyrazoles	Structure-based design	M2: ~0.02 μM; MX: ~10 μM		
M2-p53 interaction		[c]	Indole-2-carboxylic acids	Structure-based design	130–400 nM		
M2-p53 interaction			Indolinones	Structure-based design	1.2 nM-3.7 nM		
M2-p53 interaction			Piperidines	Structure-based design	0.02–0.27 μM		
M2-p53 interaction	Preclinical	AM-8553	Piperidinones	Structure-based design	150 nM	SJSA xenograft	[96]
M2-p53 interaction	Preclinical		Pyrrolidone	Virtual screening, FP assay	0.26 μM	A549 xenograft	[152]
MX-p53 interaction[b]		SJ-172550[c]	Imidazolines	HTS, FP assay	~1 μM	Retinoblastoma cells	[59]
	Preclinical	SJ000558295		HTS, FP	M2:9.1 μM; MX: ~9.0 μM		[63]
	Preclinical	SJ000558304		HTS, FP	M2: >150 nM; MX: ~2 μM		
MX-p53 interaction		XI-006	Benzofuroxans	HCS, luciferase reporter assay		MCF-7 cells	[60]
Dual inhibitor	Preclinical	RO-2443, RO-5963[c]	Indolyl Hydantoins	HTS, TR-FRET Assay	M2 and MX: <50 nM	MCF7, HCT, RKO cells	[61]

Dual inhibitor?	Phase I	RO5503781	Indolyl Hydantoins?			Soft tissue sarcoma; leukemia	[103] NCT01462175
Dual inhibitor		Novartis	Isoquinolinones	TR-FRET	M2: 0.8 nM; MX: 1.36 μM		[63]
Dual inhibitor			Imidazoles, imidazolines	Virtual screening, MCR	M2: 15 nM; MX: 1 μM	HCT116 cells	[65]
M E3 ligase inhibitor	Preclinical	MEL23, MEL24		Cell-based assay	7.5 μM, 9.2 μM	U2OS, HCT116 et al.	[49]
M2 E3 ligase inhibitor	Preclinical	MPD		Cell-based assay	1–50 μM		[48]
M2 E3 ligase inhibitor	Preclinical	HLI98		Cell-based assay	~50 μM		[47]
M2-proteasome interaction	Phase I	JNJ-26854165	Cyclic alkyl amines	Cell-based assay	>0.5 μM	Advanced stage solid tumors	[45] WO2008132175
p53 N terminal binder?	Preclinical	RITA		Cell-based assay			[153, 154]
Peptidic compound							
Dual inhibitor	Preclinical	SAH-p53-8		Phage library	M2: 55 nM; MX: 2.3 nM	Cutaneous melanoma	[87, 88, 114]
Peptidic compound							
Dual inhibitor		β53-16		Phage library	M2: ~50 nM; MX: >100 nM		[155]

(continued)

Table 16.1 (continued)

Small molecule	Status	ID	Scaffold	Discover methods	K_i or IC_{50}	In vivo test	Reference
Peptide							
Dual inhibitor		(D)PMI-gamma[c]	DWWPLAFEALLR	Mirror image phage display	M2 and MX: <50 nM	U87 xenograft	[79]
Peptide							
Dual inhibitor		PDI[c]	LTFRYWCIRQLS	Phage library	220 pM		[80]

[a]M2 indicates MDM2
[b]MX indicates MDMX
[c]3D co-crystal structures are available via PDB protein data bank

have entered Phase I clinical trials, the game of drug screening for small molecules that target this pathway will not be ending, as more patents and patent applications related to MDM2 and/or MDMX inhibitors and more deals among biotech/pharmco companies/universities and institutions have been either submitted or undergoing over the past several years (see review articles in [63–65] for more information). In addition to the aforementioned direct inhibitors, indirectly interrupting the MDM2-p53 negative feedback loop by enhancing p53 acetylation, which prevents MDM2-mediated degradation of p53 [66, 67], has also been explored for the development of anti-cancer molecule-target therapy. For instance, several studies demonstrated that attaining p53 acetylation through inhibition of its deacetylase SIRT1 [68–70] or activation of the acetyltransferase p300 [71] by small molecules leads to the inhibition of tumor growth.

These exciting studies not only consolidate the p53-MDM2-MDMX pathway as a valid target, but also provide multiple candidates for their future development into clinically applicable anti-cancer drugs. Here, we review the strategies that have been employed to identify and discover MDM2/MDMX-targeted inhibitors, including high-throughput screening (HTS) using biochemical, physicochemical and/or cell-based assays, combinatorial chemistry, and computational aided drug design. We will also provide the status of small molecules that are currently in clinical trials. Finally, we will discuss other potential and therapeutically useful approaches targeting the p53-MDM2-MDMX loop.

Discovery of Inhibitors Targeting the p53-MDM2-MDMX Loop

As briefly described above, numerous small molecules, such as synthetic small-molecules, small peptides, peptidomimetics, or natural products, have been revealed to target the p53-MDM2-MDMX pathway. These molecules are identified via a variety of drug discovery strategies that have been undertaken to block p53-MDM2/MDMX interactions, or modulate MDM2's E3 ubiquitin ligase activity, and/or inhibit MDM2-mediated ubiquitination of p53 (Table 16.1). However, the main strategies are high-throughput screening (HTS) using biochemical, physicochemical and/or cell-based assays, combinatorial chemistry and computational aided drug design including structure-based rational drug design and virtual screening. Because of the complexity of the p53 pathway as a drug target, combining different drug screening or design approaches, such as crystal structure-guided molecular design, biochemical assays and cell-based phenotypic approaches, would be more successful in developing more effective and selective drugs targeting this pathway. In this section, we provide a detailed description of several frequently used strategies as follows (Fig. 16.2).

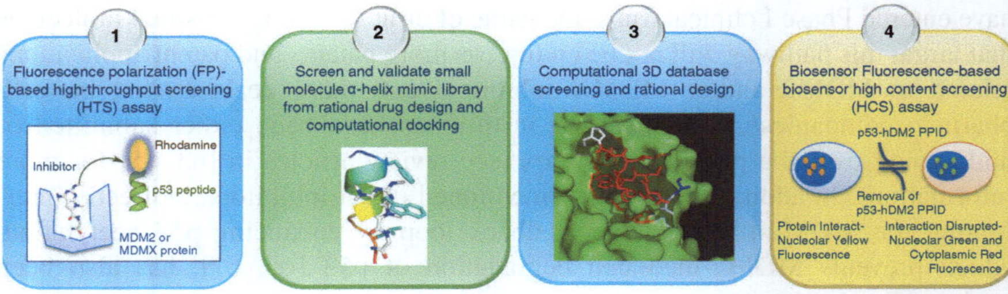

Fig. 16.2 Drug discovery strategies for identification of antagonists of MDM2 and MDMX. Some representative strategies for Drug discovery by searching MDM2 and/or MDMX antagonists are shown here. For details, see the description in sections "High-throughput Screening", "α-Helix Mimetic Based Chemistry or Combinational Chemistry", and "Computational Aided Drug Design"

High-Throughput Screening

A number of screening strategies are based on the available 3D structures of either the MDM2-p53 or MDMX-p53 complex, as the X-ray crystal structure of human MDM2 (aa, 17–125) bound by a 15-residue p53 peptide (aa, 15–29) (PDB: 1YCR) and the structure of human MDMX (aa, 1–140) with p53 peptide (PDB: 2Z5X, 2Z5T) revealed crucial binding information in these complexes [72]. Although designing non-peptidic small-molecule inhibitors disrupting the binding of p53-MDM2/MDMX has been challenging, the structures provide a good starting point for developing affinity-based assays for high throughput screenings in order to identify lead compounds. Such screening approaches include surface plasmon resonance, fluorescence polarization, and fluorescence resonance energy transfer as well as fluorescence-correlation spectroscopy experiments. A secondary NMR-based and isothermal calorimetric approach is often used to further confirm the identified molecules and to determine the (apparent) dissociation constant (K_D) between p53 and MDM2 or MDMX in the presence of the molecules or the affinity between the proteins and the molecules. Following these steps, identified compounds are routinely tested for their ability to inhibit proliferation of wild type p53-, but not mutant p53- or null, containing cells in order to determine if their cellular effects are due to their specific inhibition of MDM2 or MDMX. Now let us walk you through each of these comprehensive approaches below.

Surface Plasmon Resonance

The first potent small molecule inhibitor of MDM2 capable of activating p53 in cells and in vivo was cis-imidazoline or Nutlin as reported by the Hoffman Roche team led by Vassilev, L. T. in 2004, 8 years after the first published p53-MDM2 crystal structure [44, 72]. It was discovered by screening a diverse library of synthetic compounds using surface plasmon resonance (SPR). The SPR assays were performed on a Biacore Series S Sensor chip CM5 derived from immobilization of a Penta-His antibody for capture of the His-tagged p53 protein (aa, 1–312). The assays were conducted with MDM2 fragments (aa, 25–108). Nutlins were one class of lead structures identified and optimized for potency and selectivity. These compounds displaced the recombinant p53 protein from its complex with MDM2 at inhibitory concentration (IC_{50}) values in the 100–300 nM range. Nutlin-1 and Nutlin-2 are racemic mixtures, and Nutlin-3a is an active enantiomer isolated from racemic Nutlin-3. The IC_{50} of Nutlin-3a is 90 nM, which is 150 times more potent than the enantiomer-b of Nutlin-3. X-ray crystal structure of the MDM2-Nutlin-2 complex showed that Nutlin binds to the p53 binding site in MDM2 [PDB: 1RV1]. Nutlin-2 mimics the interaction between the p53 peptide and MDM2 with a bromophenyl moiety sitting deeply in the Trp23 pocket and the other bromophenyl group occupying the Leu26 pocket. The ethyl ether side chain is directed towards the Phe19 pocket. Computational modeling of Nutlin-3 with MDMX suggested that Nutlin-3 might also block the MDMX-p53 interaction. However, Nutlin does not bind to MDMX as efficiently as to MDM2 due to structural differences in the p53-binding pockets of the MDM2 and MDMX proteins [73]. Consistent with these observations, Nutlin-3a binds MDMX with at least a 40-fold weaker equilibrium binding constant (K_d) than that for MDM2 [61]. Recent model simulations also indicate that the inherent plasticity of MDM2 is higher than that of MDMX, enabling it to bind both p53 and Nutlin. The less flexible MDMX interacts with the more mobile p53, because the peptide can adapt different conformations to dock to MDMX, albeit with a reduced affinity. Nutlin, however, is rigid and hence can only interact with MDMX with low affinity [74]. This series of studies sets up the first successful example of targeting the MDM2-p53 interaction for anti-cancer drug discovery.

Fluorescence Polarization

However, MDMX inhibitors were not reported until more than half a decade later when a high throughput screen campaign of a library consisting of around 300,000 chemicals resulted in the identification of SJ-172550 by using a fluorescence polarization (FP) assay [59]. The FP assay was carried out using FITC labeled p53 peptide (a.a.15–29) and MDM2-GST (a.a. 1–188) or MDMX-GST (a.a. 1–185). The specificity of this FP assay was confirmed by the competitive displacement of unlabeled p53 peptides and Nutlin. Over 1,150 active compounds from the FP assay were further subjected to a cytotoxicity cell-based assay using retinoblastoma cells with MDMX amplification. Further studies of SJ-172550 using p53-deficient (SJmRbl-8) cells and the BJ cells, a human foreskin fibroblast cell line, as an additional control to estimate general cytotoxicity of the compounds, indicated its p53-dependent cytotoxicity [59]. It was through this series of screening that this compound was finally selected as the most potent compound. Interestingly, it contains a substituted imidazoline group similar to that in Nutlins. Although this compound could effectively kill retinoblastoma cells, a series of biochemical and structural modeling studies suggested that SJ-172550 alkylated Cys76 of MDMX protein and could lock MDMX into a conformation to block p53 binding. Disappointingly, this covalent binding hinders the further development of SJ-172550 [58]. Regardless of this obstacle, the same research group reported two novel small molecules, SJ000558295 (IC_{50}: MDM2 = 9.1 μM; MDMX = 9.0 μM) and SJ000558304 (IC_{50}: MDM2 = 14.4 μM; MDMX = 9.0 μM), as MDM2 and MDMX inhibitors, as shown in a recent international PCT application (WO2012045018). Among the two, SJ000558304 was with a reportedly improved binding affinity for MDMX over MDM2 in contrast to Nutlin-3. The affinity of these compounds to MDMX was further confirmed by nuclear magnetic resonance (NMR), monitoring the compound-induced NMR chemical shift perturbations. These compounds displayed desired pharmacokinetic properties and may be eventually developed into a therapy for pediatric cancers and various adult tumors that overexpress MDMX or have similar genetic lesions, though more studies are necessary to achieve this clinical goal.

Time Resolved-Fluorescence Resonance Energy Transfer

Because MDM2 inhibitors do not inhibit the p53-MDMX interaction, and their effectiveness can be compromised in tumors overexpressing MDMX, developing dual MDM2/MDMX antagonists is necessary to restore p53 apoptotic activity in the presence of high levels of MDMX and may offer a more effective therapeutic modality for MDMX-overexpressing cancers. To reach this goal, Hoffmann-La Roche identified indolyl hydantoins as dual MDM2/MDMX antagonists from a diverse small-molecule library using a Time Resolved-Fluorescence Resonance Energy Transfer assay (TR-FRET) [61]. The TR-FRET assay was performed using GST-tagged MDM2 or GST-MDMX and biotinylated p53 peptide, europium (Eu)-conjugated streptavidin and allophycocyanin (APC)-conjugated anti-GST antibody. Binding signals were monitored by reading excitation at 340 nm and emission fluorescence at 615 nm and 665 nm. A series of indolyl hydantoin compounds emerged as potent, dual MDM2/MDMX antagonists. The binding potency and binding mode were further measured and verified by Trp fluorescence quenching, ITC, NMR and X-ray crystallography (PDB: 3U15, 3VBG). For example, RO-2443 showed a remarkably similar inhibitory activity against both MDM2 (IC_{50} = 33 nM) and MDMX (IC_{50} = 41 nM) binding to p53. For its size, RO-2443 appeared highly potent and likely to fit into, at most, two of the three sub-pockets on the surface of MDMX or MDM2. These pockets were defined by the original structure of a p53 peptide bound to MDM2, which showed three key residues of the p53 peptide, Phe19, Trp23, and Leu26, important for MDM2 binding. Crystal structures of MDMX bound to RO-2443 revealed that the inhibitor blocks the interaction of p53 by inducing the formation of dimeric complexes of MDMX. Further modifications of RO-2443 in part to improve its physicochemical properties resulted in RO-5963. As expected, RO-5963 restored p53 transcriptional activity and overcame the apoptotic resistance of MDMX-overexpressing cell line, SJSA-X, to Nutlin-3. Therefore, this type of MDM2/MDMX dual inhibitors could be potentially developed into more potent anti-cancer drugs, though more studies are apparently necessary to translate them into clinical therapy.

Fluorescence Correlation Spectroscopy-Based Assay

The researchers from the University of Tokyo identified new MDMX inhibitors after screening almost 40,000 commercially available compounds by employing a fluorescence correlation spectroscopy (FCS)-based high-throughput screening (HTS) assay [75]. In this assay, an MDMX fragment (aa 26–106) was fused to a green fluorescent protein (GFP), and a p53 peptide was labeled with a fluorescence quencher. The MDMX-p53 interaction was evaluated by detecting the quenching of the fluorescence of GFP; the inhibition of the interaction was detected by the recovery of GFP fluorescence. Two hundred and fifty five compounds were selected after the first screening, and six of these compounds were further confirmed by SPR binding assay to have IC_{50} values less than 5 μM. The hit compound **1** increased p53 and p21 levels and triggered apoptosis in wild type p53-containing MV4-11 leukemia cells and more selectively killed MV4-11 compared to H1299 p53-null and WI38 normal cells. This study presents another new compound that can inhibit MDMX-p53 binding and activate p53 as a promising anti-cancer drug candidate.

ThermoFluor Microcalorimetry

3DP (3-Dimensional Pharmaceuticals) (Yardley, PA, USA), which was later merged with Johnson & Johnson, reported a MDM2 complex containing a benzodiazepinedione (BDP) based inhibitor (PDB: 1T4E) [76, 77]. The scaffold was found by HTS with the temperature-dependent protein-unfolding assay using the ThermoFluor microcalorimetry technology. In this technology, fluorescent dyes were used to monitor protein unfolding as a function of temperature for the identification of compounds that bind to MDM2. Detection of compound binding to MDM2 was measured by the resultant increase in thermal stability. Thermal stability was quantified as a change in midpoint transition temperature in the presence of the compound at a single concentration. The sensitivity of this assay was verified by the shift in T_m on addition of peptides known to bind to MDM2, with higher-affinity peptides generating larger shifts. From the initial screening, 1,216 compounds out of 338,000 compounds from combinatorial libraries were selected, which included 116 compounds belonging to a benzodiazepinedione library. The affinity of the selected compounds was determined using an FP-based assay to monitor the binding of p53 peptide to MDM2. TDP222669 with a benzodiazepinedione scaffold and a carboxylic acid group was proved to be the most potent inhibitor with a binding affinity of 80 nM. The crystal structure of its complex with an MDM2 fragment demonstrated that it occupies the same pocket as the peptide side chains Phe19, Trp23, and Leu26 of p53 in the MDM2 binding cleft. The MDM2 interactions with the inhibitor are largely nonspecific van der Waals contacts and, similar to the p53 peptide; the bound conformation of the inhibitor is amphipathic. Unfortunately, despite the initial tremendous effort, this series of compounds series seems to have been abandoned, most likely because of their insufficient drug-like properties.

Screening of Phage Display Library

Although most of the screening against the MDM2/MDMX-p53 interaction have been based on biochemical and biophysical assays as discussed above, a high-affinity peptide inhibitor of p53 interaction with MDM2 and MDMX was also identified by screening a duodecimal peptide library displayed on M13 phage using site-specific biotinylated p53-binding domains of human MDM2 and MDMX [78]. The peptide inhibitor (TSFAEYWNLLSP), termed PMI, bound to MDM2 and MDMX with low nanomolar affinities-approximately 2 orders of magnitude stronger than the wild-type p53 peptide of the same length (K_d: 3.6 nM for MDM2; 4.15 nM for MDMX). The crystal structure of MDM2 or MDMX in complex with PMI was solved at 1.6 Å resolution (PDB: 3EQS, 3EQY). An extensive, tightened intramolecular H-bonding network in the PMI-bound complex was identified by comparative structural analysis. The H-bonding network contributed to the conformational stability of this complex, thus enhancing binding to MDM2 and MDMX proteins. Importantly, the C-terminal residue Pro of PMI induced the formation of a hydrophobic cleft in MDMX previously unseen in the structures of p53-bound MDM2 or MDMX [78].

Small peptides are often capable of efficiently blocking protein interactions with high affinity and supreme specificity. However, their poor in vivo stability and membrane permeability impede the thriving of peptide-based therapeutics. To overcome their instability in vivo, aided by native chemical ligation and mirror image phage display, several proteolysis-resistant D-peptide inhibitors termed D-PMI-α, β, γ were identified to bind MDM2 with K_d values of 50–200 nM [79]. Mirror image phage display is a straightforward technique to identify proteolysis-resistant D-peptide ligands of a native protein through phage library screening against the D enantiomer of the L target. The screening of the duodecimal peptide phage library was carried out against N79K-biotin-d-$^{25-109}$MDM2 immobilized on streptavidin-agarose resin. Bound phage particles were competitively eluted with 1 mM D-$^{15-29}$p53, and subsequently amplified in host strain Escherichia coli ER2738. The D-peptides from the screening was confirmed by SPR and FP assays, and structural studies coupled with mutational analysis also verified the mode of action of the D-peptide as an MDM2-dependent p53 activator (PDB: 3IWY). Despite being resistant to proteolysis, both (D) PMI-alpha and (D) PMI-gamma failed to actively traverse the cell membrane and, when conjugated to a cationic cell-penetrating peptide, were indiscriminately cytotoxic independently of p53 status. This issue was solved when encapsulated in liposomes decorated with an integrin-targeting cyclic-RGD peptide. By using this approach, (D) PMI-alpha was found to exert potent p53-dependent growth inhibitory activity against human glioblastoma in cell cultures and nude mouse xenograft models. The findings validate D-peptide antagonists of MDM2 as a class of p53 activators for targeted molecular therapy of malignant neoplasms harboring WT p53 and elevated levels of MDM2. Remarkably, this group led by Lu, W has continued to modify this type of D-peptides and recently reported a superactive D-peptide antagonist of MDM2, termed D-PMI-δ (K_d = 220 pM) [80]. Their X-ray crystallographic studies (PDB: 3EQS, 3LNJ) validated D-PMI-δ as an exceedingly potent inhibitor of the p53-MDM2 interaction, which could be a highly attractive anti-cancer drug candidate. The field will be excited to see its future success as the first peptide drug for cancer therapy.

In Vitro Autoubiquitination and MDM2-Catalyzed p53 Ubiquitination Assay

Beside the aforementioned N-terminal p53-binding pocket of MDM2 or MDMX, the C-terminal RING finger domain with an intrinsic E3 ubiquitin ligase activity of the MDM2 protein has also been utilized as a target for anti-cancer drug screening. There have been at least two types of high throughput assays that are aimed at the discovery of inhibitors that might modulate ligase activity of MDM2. One of them measures the ubiquitination of p53 by MDM2, and the other measures the self-ubiquitination of MDM2. The Vousden team first identified MDM2 ubiquitin ligase inhibitors, 7-nitro-5-deazaflavin compounds (HLI98s), through an HTS of small-molecule libraries of 10,000 compounds using an in vitro MDM2 autoubiquitination assay [47]. HLI98 compounds specifically inhibited the RING finger domain of MDM2, and not the regions that interact with p53. When primary human fibroblasts were treated with HLI98 compounds, both p53 and MDM2 levels increased. Ubiquitylated p53 was not detected, which is consistent with the ability of the compounds to inhibit ubiquitylation instead of proteasome function. HLI98 compounds showed some selectivity for MDM2 compared with other RING finger E3s in cell lines. The compounds did not stabilize p53 in the absence of MDM2 in mouse embryonic fibroblasts (MEFs), indicating that they do not inhibit other reported E3s that target p53 for degradation, such as PIRH2 and COP1. Although the compound stabilized p53 and MDM2, and activated p53-dependent transcription and apoptosis, it also had p53-independent cytotoxicity. Furthermore, as expected, the compound worked much better in cancer cells containing wild-type p53 than in those containing mutant p53, because targeting MDM2 should, in theory, have little or no effect on human cancers with mutant p53. However, in vivo antitumor activity of HLI98, using human xenograft models has not been reported. Therefore, the anti-cancer potential of this compound remains to be further investigated.

A novel electrochemiluminescent assay system was developed to monitor the inhibition of MDM2's E3 ubiquitin ligase activity for a high-throughput screening of natural product extracts [81]. This assay was used to screen more than 144,000 natural product extracts. Interestingly, sempervirine was identified to inhibit MDM2 auto-ubiquitination and MDM2-mediated p53 degradation, and to induce p53 level in cancer cells. Consequently, this natural product preferentially induced apoptosis in transformed cells expressing wild-type p53, suggesting that it could be a potential lead natural product for anticancer therapeutics. In addition to this, three new alkaloids, isolissoclinotoxin B, diplamine B, and lissoclinidine B, have also been identified from Lissoclinum cf. badium. Lissoclinidine B was found to inhibit ubiquitylation and degradation of p53 and selectively kill transformed cells harboring wild-type p53, suggesting this compound could also be used for the development of new anti-cancer treatments.

Additionally, Murray MF et al. developed a TR-FRET assay for measuring the ubiquitination of p53 by MDM2 and executed an HTS campaign with >600,000 compounds [82]. An assay for MDM2 autoubiquitination was also developed using the same TR-FRET format to find compounds selective for p53 ubiquitination versus MDM2 autoubiquitination. The most selective compound identified displayed an IC_{50} of 8 μM in the MDM2/p53 assay and no discernible inhibition up to 100 μM in the autoubiquitination assay. These studies aimed at targeting the RING finger domain of MDM2 could be promising, though much more remain to be done before an ideal candidate with a more specific inhibitory activity toward MDM2 could be eventually identified.

Cell-Based Auto-ubiquitination Assay

It has been believed that MDM2 can regulate itself in cells through a so-called auto-ubiquitination and auto-degradation mechanism. Based on this notion, Herman, A. G. et al. developed a high-throughput cellular MDM2 auto-ubiquitination assay to discover a class of small-molecule MDM2-MDMX ligase inhibitors [49]. In this assay, the compounds were screened using wild-type MDM2 or mutant MDM2 (C464A)-luciferase cell lines. The C464A mutation disrupts a metal-binding site in the RING domain, thereby eliminating the MDM2 E3 ligase activity. Compounds that specifically increase the luminescence of MDM2 (wt)-luciferase fusion protein rather than MDM2 (C464A) would likely inhibit MDM2 E3 ligase activity or proteasomal degradation of MDM2. Two of such compounds, MEL23 and MEL24, were identified. They were found effectively to inhibit the E3 ligase activity of the MDM2-MDMX hetero-complex, thereby inhibiting MDM2 and p53 ubiquitination in cells, reducing viability of cells with wild-type p53, and synergizing with DNA-damaging agents to cause cell death. Also, the activity of MEL compounds was specific to MDM2, and independent of p53 transcription. As MDM2-MDMX ligase inhibitors, MEL compounds may be used as molecular tools to validate novel targets of MDM2-MDMX and to inhibit MDM2's E3 ligases, which could be beneficial in diverse applications, as it remains to be determined if they could be druggable candidates or not.

Fluorescence-Based Biosensor High Content Screening Assay

Fluorescence-based biosensor high content screening (HCS) is a cell-based assay that was developed recently by a research team from University of Pittsburgh. Compared to biochemical assay-based HTS screening, an HTS application in living cells by a biosensor assay can be used to only identify the "drug like" compounds that are able to block the p53-MDM2 interaction in the nucleus of cancer cells [83]. In this novel imaging-based biosensor HCS assay, the recombinant adenovirus biosensors used to express the N-terminal domains of p53 and MDM2 were fused to green or red fluorescent proteins (GRP or RFP) and co-infected in U2OS cells. By monitoring the p53-MDM2 interactions through changes in the subcellular localization of the MDM2 component of the biosensor and analyzing the multiparameter data from images of the 3 fluorescent channels, this team was able to identify and eliminate compounds that were cytotoxic or fluorescent artifacts [83]. They identified compounds with methylbenzo-naphthyridin-5-amine (MBNA) after screening a collection of over 220,000 compounds using this novel biosensor HCS assay. The MBNAs were proved to induce p53 protein levels, increase the expression of p53 target genes, cause G1 arrest during the cell cycle, induce apoptosis, and inhibit cell proliferation with an IC_{50} ~ 4μM in p53-WT HCT116 cells. However, unlike Nutlin-3, MBNAs also increased the percentage of apoptosis in p53 null cells and exhibited similar potencies for growth inhibition in isogenic cell lines null of p53 or p21. Therefore, these compounds might target other cellular proteins and have p53-independent cytotoxicity.

Reported by the same group, compounds containing the substituted pyrazole scaffold were also identified by employing this biosensor HCS assay [84]. The potency of the best compound YH265 in the biosensor assay was determined at 1.8 μM (WO2011106650). The affinity (K_i) of the selected compounds toward MDM2 and MDMX was further demonstrated in vitro FP assays. The affinity toward MDM2 and MDMX ranged from 0.02 to 10 μM. The ability of these compounds to dissociate MDM2-p53 or MDMX-p53 complexes was confirmed based on fluorescent polarization analysis, AIDA NMR and HSQC NMR analysis. Selected compounds were also tested on the NCI60 cell panel, and the results showed that the compounds could arrest the growth of various cancer cells in culture. The crystal structure of a compound analogous to YH265, but based on an imidazole rather than pyrazole scaffold, bound to MDMX has been solved (PDB: MDMX: 3LBJ, MDM2: 3LBK) and represents the only MDMX small molecule co-crystal structure so far [see a recent review of available structural information on MDM2/X-inhibitor interaction for more details [77, 85, 86]]. The YH265 analogs could be hot candidates for further development, although this set of cell-based biosensor screening might yield more candidate compounds that target the MDM2-MDMX-p53 loop.

α-Helix Mimetic Based Chemistry or Combinational Chemistry

HTS is not the only strategy that has been employed to identify small molecules or peptides that might target the MDM2-MDMX-p53 loop. Other approaches, such as rational design to imitate the structure of the N-terminal MDM2/MDMX-binding domain of p53, have also been explored. For instance, using a chemical approach termed "hydrocarbon stapling", Federico Bernal et al. designed Stabilized Alpha Helix of p53 (SAH-p53) [17, 87, 88] peptides based on the peptide sequence of the p53 transactivation domain α-helix [87, 88]. They replaced natural amino acids at positions S20 and P27 with synthetic olefinic residues and generated the structurally reinforcing hydrocarbon staple by olefin metathesis. Other residues within the MDM2/MDMX-binding domain of p53 not required for MDM2/MDMX interaction were also modified to improve peptide solubility and uptake. The newly designed SAH-p53 peptide was found to preferably bind to MDMX compared to MDM2 (K_d for MDMX: 2.3 ± 0.2 nM; K_d for MDM2: 55 ± 11 nM) and subsequently reactivate the p53 pathway and suppress tumor growth by targeting MDMX. Also this novel MDMX-binding peptide restored the sensitivity of Nutlin in cancer cells with high levels of MDMX expression and in JEG-3 xenograft mice with little toxicity.

Similarly, a peptidomimetic strategy was also employed to synthesize small molecules with a peptidomimetic 1,4-thienodiazepine-2, 5-dione scaffold as an α-helix mimetic of the MDM2-binding peptide of p53 to disturb the p53-MDM2 interaction [89]. This strategy takes advantage of an Ugi-deprotection-cyclization reaction that has been exploited in combinatorial chemistry because it combines four separate components to make one scaffold, providing an easy access to create diversity around a single scaffold. A small library of 18 diverse thienodiazepine-2, 5-diones selected from a large virtual library was prepared in one pot by solution phase synthesis via an Ugi-deprotection-cyclization reaction. The compounds were found to antagonize the p53-MDM2 interaction in an FP assay, exhibiting a dose dependent effect to compete with a p53-like peptide. Also, in an NMR competition assay, two compounds were found to dissociate the MDM2-p53 complex with K_d values of 30 ± 20 μM and 10 ± 6 μM, respectively [89]. Further studies are needed to show its clinical application for cancer therapy.

Recently, a library of 900 compounds based on a pyrrolopyrimidine scaffold as an α-helix mimetic, was prepared by using solid phase parallel synthesis in hope to discover small molecules able to disrupt the interaction between p53 and MDMX/MDM2 [90]. The structural basis for their pyrrolopyrimidine-based molecules is Hamilton's terphenyls, which are among the most prominent and most-imitated scaffolds in the field. However, this work is much more than a simple variation of the terphenyl scaffold. It rises above many other imitators through a clever set of features, including the use of a scaffold known by medicinal chemists to have favorable aqueous solubility and cell permeability, a unique and simple synthesis route that is immediately amenable to library generation, an FP screening approach for inhibitor discovery rather than reliance on rational design. Two compounds from the library were identified as sub-micromolar p53-MDMX and p53-MDM2 inhibitors in the FP binding assay, a relatively novel profile, especially for a small molecule. Their activity was also shown in cells for dose-dependent increase of p53 and triggering of apoptosis. Although the properties of these pyrrolopyrimidine derivatives established them as promising lead compounds for further structural elaboration toward p53-MDM2/MDMX inhibitors with improved drug-like attributes, much more need to be done to verify the pharmacological activity of these small molecules and to further develop them into a possible anti-cancer therapy. This approach might also be useful for targeting other protein-protein interactions.

Finally, a fragment-based strategy, involving "multicomponent reaction chemistry" (MCR), identified imidazolines as dual MDM2/MDMX inhibitors [91]. The crystal structures of p53-MDM2 (PDB: 1YCR) and imidazole antagonist PB12 bound to MDM2 (PDB: 3LBK) were used as templates to identify a fragmentation/anchor. The anchor was imported into a database containing MCR-accessible scaffolds to generate a virtual library of compounds, which subsequently were docked into the binding pocket of the target protein. Results from docking then were used to select compounds for synthesis and complementary screening by an NMR-based binding assay. Building upon the success of the imidazole and using multicomponent reactions, compounds with imidazoline scaffolds were identified with low micromolar activity in HCT116 cancer cells, claiming dual inhibition of the p53-MDM2 as well as p53-MDMX protein interactions with a $K_d < 1$ μM (US20080280769) [63]. With alternative structures to Nutlins, which require a rather lengthy synthetic route with more than eight individual steps, imidazolines are accessible via a straightforward multicomponent reaction in just one or two steps.

In summary, using the p53 peptide imitating strategy, several promising mimetic peptides or small molecules that could inhibit the interaction of MDM2 and/or MDMX with p53 have been discovered. Although none of them has yet been developed to the stage of preclinical studies, as more studies are necessary to demonstrate their druggable potential, this approach provides solid evidence for the proof of principle and would eventually offer some promising candidate molecules for anti-cancer drug development.

Computational Aided Drug Design

Medicinal chemistry is not merely limited to the mimetic-based design of inhibitors, but it also exploits advances in bioinformatics. Two major approaches in medicinal chemistry developed by taking advantage of information technology (IT)-based computational power are the straightforward in silico compound-selection (virtual screening) and structure-based de novo design. These tools have also been applied to the design of optimal MDM2/MDMX inhibitors. The compounds were designed to generally mimic the interactions provided by the important amino acid side chains, such as Phe19, Trp23 and Leu26, within the p53 peptide, to disrupt the p53-MDM2 and/or p53-MDMX bindings. In addition, it is believed that Leu22 is important for the p53-MDM2/X complex, and it could assist in the development of molecules with higher binding affinity than the p53 TAD peptide [92]. Based on structural analysis, the p53 residues Phe19 and Trp23 interact in a similar way to MDM2 and MDMX, but not Leu26. The interaction contributions by the MDM2 residues Leu54, His96 and Ile99 are different from the equivalent MDMX residues Met53, Pro95 and Leu98, which could account for the differential binding of the p53 peptide to MDM2/X.

Structure-Based de novo Design

Using a structure-based de novo design strategy, the Wang laboratory at the University of Michigan discovered the spiro-oxindoles MI-63 and MI-219 that are more potent and selective than Nutlin. In a similar fashion to the Nutlins, spiro-oxindoles bind MDM2 by mimicking the interactions of crucial hydrophobic residues (Phe19, Trp23 and Leu26) in the p53 peptide [93, 94]. The design started from a search for chemical moieties that can mimic the interaction of Trp23, the most critical for binding to MDM2. The results showed that oxindole can nicely mimic the side chain of Trp23 for interaction with MDM2. Further substructure search identified the natural alkaloids such as spirotryprostatin A and alstonisine, which both contain a spirooxindole substructure. Computational modeling studies predicted the oxindole can closely mimic the Trp23 side chain in p53 in both hydrogen-bonding formation and hydrophobic interactions with MDM2, and the spiropyrrolidine ring provides a rigid scaffold from which two hydrophobic groups can be projected to mimic the side chain of Phe19 and Leu26. Initial compounds were designed with different hydrophobic groups and different stereochemistry, and they were docked into the MDM2 binding cleft using the GOLD program. X-ray structure of the MDM2-p53 complex, mutagenesis and alanine scanning of p53 peptides suggested that a fourth residue, Leu22, also appears to play an important role in the overall interaction between p53 and MDM2. Structure-based optimization to capture the additional interaction between Leu22 in p53 and MDM2 yielded MI-63, and improvements to ensure sufficient water solubility ultimately yielded MI-219 with desirable pharmacological properties, such as 55 % oral bioavailability in mice. MI-219 was greater than 10,000-fold more selective for MDM2 over its closely related homolog MDMX (MDM2: K_i: 5 nM). Consistent with the high binding affinity for MDM2 and disruption of the MDM2-p53 complex, spiro-oxindoles induced accumulation of p53 in p53 wild-type cancer cells, and it showed antitumor activity in human tumour xenograft models. Extensive modifications have been made on spirooxindoles and MI-147 is one of the most potent of this class of compounds, with a K_i of 0.6 nM in an FP-based competitive binding assay [95]. Additional modifications yielded new compounds with high affinities to MDM2 and improved pharmacokinetic properties. One such compound in this class, MI-773, has completed Investigational New Drug (IND) -enabling studies and currently Sanofi-Aventis is recruiting participants for Phase I clinical trials with MI-773

[63] (reference spirooxindole SAR405838). This series of studies sets a perfect example for de novo drug design if the 3D structure of a protein or protein-protein complex is available.

In another study reported by Rew, Y. et al. from Amgen, a new class of potent MDM2-p53 inhibitors bearing a piperidinone scaffold has been successfully designed and optimized using the structure-based de novo design strategy [96, 97]. On the basis of the analysis of known MDM2 inhibitors, assisted by computer modeling, SAR studies and crystal structure (PDB: 4ERE), the affinity of these compounds for MDM2 was improved through conformational control of both the piperidinone ring and the appended N-alkyl substituent. Optimization afforded AM-8553, with a K_D of 0.4 nM, especially with excellent pharmacokinetic properties. AM-8553 has an oral bioavailability of 100 % in rats and 12 % in mice. On the basis of its low human hepatocyte intrinsic clearance, AM-8553 was projected to have a long human half-life (>12 h). It is effective in inhibition of tumor growth in a dose-dependent manner and is capable of achieving partial tumor regression (R = 27 %) against SJSA-1 xenograft tumors in mice at 200 mg/kg once daily dosing [96].

Virtual Screening

Sulphonamide I (NSC279287) is the first example that was identified by screening the NCBI database using a computational pharmacophore model of MDM2 binding [50]. The HINT molecular modeling program was used to generate predictive QSAR regression equations for p53-MDM2 inhibition based on the template of the interaction between p53-based peptide inhibitors and MDM2, which led to the development of the pharmacophore model to mimic the portions of p53 necessary to bind to MDM2. In an ELISA-based p53-MDM2 binding assay, sulphonamide I showed an IC_{50} of approximately 32 μM compared to a value of 13 μM for the p53 peptide (16QETFSDLWKLLP27). The efficacy of NSC279287 to induce a p53 response was verified in a p53 reporter gene assay using MDM2-over-expressing osteosarcoma cells.

Another approach combined pharmacophore and structure-based screening that used computational database screening of a subset of the NCI database of 150,000 compounds, identifying 354 potential MDM2 inhibitors. A 3D-pharmacophore model was derived from the X-ray crystal structure of the p53 peptide associated with MDM2, with several known small-molecule inhibitors. The pharmacophore model consisted of three elements that mimic the three key hydrophobic binding residues in p53 (Phe19, Trp23, Leu26), together with three associated distance constraints. Computational docking was performed using the GOLD program to dock each hit to the p53-binding site in MDM2. Their binding affinities were ranked using Chemscore and X-score. In a fluorescence-based binding assay, the quinolinole NSC66811 was identified to inhibit MDM2 with a significantly lower K_i (120 nM) than the natural p53 peptide [51].

Molecular dynamics accounting for protein flexibility was applied to a pharmacophore model derived from the MDM2 structure. Out of 35,000 small-molecule compounds, computational analysis identified five non-peptidic MDM2 inhibitors with novel scaffolds. Based on fluorescence polarization binding assays, the most potent of them displayed a K_i of 110 nM [98]. However, besides conveying results on biochemical experiments, there were no reports on their effect at the cellular level.

A strategy to find compounds that interact with new targets based on existing knowledge of ligand-target interactions has been employed based on chemogenomics knowledge-based methodologies, which relies on homology and conserved residues that are involved in molecular recognition. Based on MDM2-p53 interactions as template, this approach was applied to the MDM4-p53 complex, where combination of virtual screening strategies resulted in unique compounds with varying degrees of selectivity for both MDM2-p53 and MDM4-p53 systems [99].

Development of Inhibitors Targeting the p53-MDM2-MDMX Loop

Until now several of the aforementioned MDM-p53 targeted inhibitors that directly activate wild-type p53 in cancer cells have been successfully developed for the clinical trials (Table 16.1), but additional clinical data is necessary to verify if these non-genotoxic inhibitors can effectively activate p53 and improve the clinical efficacy in patients carrying wild-type p53. Other strategies that target the p53-MDM2-MDMX loop other than disrupting the MDM2-p53 interactions have also been considered, but encounter major challenges.

Clinical Candidates of MDM2-p53 Inhibitors

RG7112 and RO5503781 (Hoffmann-La Roche)

Driven by the success of Nutlin (see "Surface Plasmon Resonance"), Hoffmann-La Roche has undertaken an extensive optimization of lead compounds that culminated in the selection of RG7112 (RO5045337), a Nutlin-3 derivative, and RO5503781, whose structure has not been disclosed, for clinical application. RG7112 is the first MDM2 inhibitor that advances to clinical trials. It is a more potent binder of MDM2 compared to Nutlin (K_d = 10.7 vs 90 nM), but similarly to Nutlin, it is inactive against MDMX [100]. The crystal structure of the RG7112-MDM2 complex revealed that RG7112 binds in the same fashion as Nutlin to the p53 pocket of MDM2 by mimicking three critical p53 amino acid residues Trp23, Leu26 and Phe19 [PDB: 4IPF]. The compound differs from Nutlin by the introduction of 4, 5-dimethyl substitution on the imidazoline ring. The substitutions possibly add greater structural rigidity to the imidazoline scaffold, which might block metabolic conversion to the inactive imidazole form. In December of 2007, Hoffmann-La Roche initiated Phase I studies of RG7112, and until now, five clinical studies in patients with liposarcomas (before debulking surgery), solid tumors, and soft tissue sarcomas, advanced solid tumors (http://www.clinicaltrials.gov; NCT01677780, NCT01143740, NCT01164033, NCT00559533, NCT00623870) have been completed. The clinical biomarker studies in 20 patients with liposarcomas (a frequently HDM2-amplified tumor) confirmed the ability of RG7112 to activate p53 and its major functions, cell cycle arrest and apoptosis, in human tumors [101]. The clinical results provided not only proof of mechanism, but also evidence for the clinical activity of the single agent RG7112. In phase I leukemia trial of RG7112, among 116 patients treated with RG7112 for 10 days followed by 18 days of rest, 6 AML patients with complete remission (CR), 5 of 31 (16 %) at the MTD (the maximum tolerated dose), and additional patients showed significant decreases in blasts [102]. Notably, RG7112 caused >10 % adverse events related to hematological toxicities. Currently, Hoffmann-La Roche is recruiting patients participating in previous cancer studies for extension studies of RG7112 (NCT01677780). Clinical studies of RG7112 in combination with doxorubicin in patients with soft tissue sarcoma, or in combination with cytarabine in patients with acute myelogenous leukemia (NCT01605526 and NCT01635296) are currently underway. Another compound RO5503781 is also entering phase I studies for advanced malignancies, except leukemia (NCT01462175), and in combination with cytarabine for patients with acute myelogenous leukemia (AML) in 2013 (NCT01773408). Although it has been almost a decade since Nutlin was discovered to take these Nutlin-derived inhibitors of MDM2-p53 binding to their phase I or II clinical trials, these progresses are still quite impressive, offering a big promise for targeting the MDM2-p53 loop as a way to develop effective and non-genotoxic anti-cancer therapy.

SAR405838 (Sanofi-Aventis)

Another anti-MDM2-p53-binding small molecule that has also been developed into phase I clinical trial is a derivative of Spirooxindoles, which represent one of the most promising chemical scaffolds for MDM inhibitors (see "Structure-Based de novo Design") and are actively being developed by several companies, including Hoffmann-La Roche, Sanofi-Aventis and Daiichi Sankyo. The first candidate SAR405838 is an analog of MI-773 and currently entering the phase I trial for patients with advanced cancer prescreening with wild-type p53 (NCT01636479) [63, 103]. Spirooxindoles MI-series were initially designed by the Wang group at the University of Michigan, further developed by Ascenta Therapeutics and licensed by Sanofi-Aventis. As discussed above, initially, MI-219 was identified in 2008 as a potent, specific, cell-permeable, and orally active MDM2 inhibitor through structure based de novo design. However, due to its high dose required, short half-life and quick metabolism, MI-219 is not an ideal candidate for the clinical development. Further optimization of MI-219 led to a more potent analogue MI-319 [104]. This compound exhibited potent activity against follicular lymphoma that retains wild-type p53 both in vitro and in vivo. In addition, MI-319 in combination with cisplatin induced cell growth inhibition and apoptosis in pancreatic cancer cells irrespective of their p53 mutational status. An even more potent analogue of MI-319, MI-773, was later claimed in 2012. This series of Spirooxindole compounds has a more complicated stero-chemistry compared to Nutlin. Thus, the newly developed optimal enantiomer, MI-77301, displayed 10 times higher binding affinity against MDM2 than did MI-773 ($K_d = 62$ vs 8.2 nM). The antitumor activity of MI-77301 was more pronounced in a set of wild type p53 xenograft models than MI-773, including SJSA-1 osteosarcoma, human prostate, melanoma, colorectal tumor, LNCAP human prostate tumor and human acute lymphoblastic leukemia (WO2012065022), although there is a less pronounced difference between the two compounds in vitro cell-based cytotoxicity assays. Apparently, this group of anti-MDM2-p53-binding small molecules is also appealing in terms of their clinical development into anti-cancer therapy.

CGM097 (Novartis)

Novartis, another pharmaceutical giant, has also joined the race to develop anti-cancer drugs that can specifically target the MDM2/MDMX-p53 loop. Very recently, this company sponsored a Phase I trial of a small molecule called CGM097 in patients with advanced solid tumor with wild type p53 (http://clinicaltrials.gov/show/NCT01760525). Ninety-two individuals will be orally dosed with CGM097. Trial dates were designed from 1st March 2013 to 1st June 2016. Although the structure of CGM097 is not disclosed, this compound appears to have a dual inhibitory activity toward both MDM2 and MDMX to different extents, as Novartis has recently showed several small molecular scaffolds as inhibitors of MDM2 and/or MDMX in patents, such as 3-imidazolylindoles, substituted dihydroimidazole derivatives, tetra-substituted heteroaryl compounds and substituted isoquinolinones and quinazolinones [63]. The most potent compound among the 3-imidazolylindoles, prepared by a van Leusen multicomponent reaction, has an IC_{50} of 15 nM for MDM2 in an FP binding assay and an IC_{50} of 1.32 μM for MDMX in a TR-FRET binding assay. Compounds derived from dihydroimidazole demonstrated IC_{50} values from 70 to 2 μM. This scaffold closely resembles Roche's Nutlins. The potency (IC_{50}) of disclosed tetra-substituted heteroaryl and isoquinolinones and quinazolinones as obtained from TR-FRET based assay ranging in the low nanomolar for MDM2, but in the low micromolar for MDMX. The cancer drug discovery field will keep its eyes wide open for the birth of this MDM2/MDMX dual inhibitor as an anti-cancer therapy.

MK-8242 (Merck)

The competition became more intensive when Merck recently claimed the US patent 7884107 on February 2011, based on a provisional application from 2006. This patent describes a substituted piperidine as a specific MDM2 inhibitor with an IC_{50}

value of 0.02 μM as determined by an FP assay. In the same year, Merck launched two Phase I trials (NCT01451437 and NCT01463696) with MK-8242, an MDM2 inhibitor of undisclosed structure in advanced solid tumor, alone and in combination with cytarabine in participants with acute myelogenous leukemia. Hopefully, these clinical trials together with those trials as described above will eventually lead to some effective therapies for human cancers that harbor wild type p53 by targeting the p53-MDM2-MDMX loop.

Other Therapeutically Useful Approaches Targeting the p53-MDM2-MDMX Loop

Modulation of MDM2 and MDMX Expression

In addition to disrupting the interaction between MDM2/MDMX and p53, other strategies have also been explored for the possibility of developing anti-cancer therapy by negatively affecting the p53-MDM2-MDMX loop. Because MDM2 and MDMX are p53 target genes, disruption of MDM2/MDMX-p53 interaction would lead to elevated MDM2 or MDMX levels and thus might be less effective in cancer cells that express high levels of MDM2 and/or MDMX. Therefore, down-regulation of MDM2 or MDMX protein in cancer cells is a straightforward approach and has been proved to activate the p53 pathway and inhibit tumor growth using small interfering RNA (siRNA), short hairpin RNA (shRNA) or miRNA in tumor xenografts in nude mice [16, 105, 106]. Unfortunately, siRNA therapy is hampered by the issues of delivery and cellular uptake, and thus it is less feasible. However, interestingly, a benzofuroxan derivative (NSC207895, XI006) that selectively inhibits MDMX expression has been identified through a reporter-based screening [60]. This small molecule caused p53-dependent trans-activation of pro-apoptotic genes in MCF-7 cells. NSC207895 was shown to exhibit the additive effect with Nutlin-3a in activating p53, inducing apoptosis and decreasing the viability of MCF-7 cells. NSC207895 seemed to repress the MDMX promoter activity and decrease MDMX transcription, although the underlying molecular mechanism of this promoter-specific targeting has not yet been deciphered. In addition, NSC207895 was highly correlated to known DNA-damaging agents, such as methyl methanesulphonate (MMS) and camptothecin, in a cross-species chemogenomic profiling screen. As DNA-damaging agents also induce MDMX degradation, it is possible that the effects of NSC207895 on MDMX protein levels and p53 activation might involve more than the repression of MDMX transcription. Besides this small molecule, several naturally derived compounds have also been shown to exert their anticancer activities by inhibiting MDM2 expression, independent of p53, such as Genistein, gambogic acid, curcumin, and berberine (disruption of MDM2-DAXX-HAUSP complex). Since this natural product field has been expanding dramatically, we refer readers to a comprehensive review on natural products that block MDM2 expression in a p53-dependent or -independent manner [62].

Interfering with MDM2 Interactions with Other Proteins

MDM2 regulates the stability of p53 protein by not only directly interacting with it and mediating its ubiquitylation, but also associating with different subunits of the 26S proteasome, such as S2, S4, S5a, S6a, S6b, to facilitate p53 proteasomal turnover. Thus targeting the interaction between MDM2 and the 26S proteasomal subunits is also an attractive approach. Indeed, a novel tryptamine derivative, orally bioavailable small molecule JNJ-26854165, was invented by Forschungszentrum Karlsruhe and Janssen Pharmaceutical using a high-throughput screening approach based on the characterization of MDM2-proteasome interaction (WO2008132175) [45, 46, 107]. JNJ-26854165 specifically bound to the ring domain of MDM2 and inhibited the binding of the MDM2-p53 complex to the proteasome, consequently blocking the degradation of p53. However, in addition to p53, degradation of other human MDM2 client proteins may be inhibited through disturbance of the ubiquitin-proteasome proteolysis (UPS)-pathway. Preclinical studies of this small molecule have revealed induction of apoptosis and anti-proliferation in p53 wild-type and p53-mutant cancer cells and a general genotoxic effect in various tumor models including NCI-H1373 non-small cell lung cancer xenografts, PC-3M orthotopic prostate and p53 mutant HT-29 colon xenografts, and U87 glioblastoma xenografts, indicating that JNJ-26854165 is not p53-specific. To date, a phase I clinical study of JNJ-26854165 (NCT00676910) to determine safety and dosing in patients with advanced stage or refractory solid tumors has been completed. Although JNJ-26854165 was initially well tolerated at clinically effective doses in patients, its cardiotoxic effects and MDM2-independent mechanism were subsequently observed. Hence this compound is no longer in competition.

It has been shown that Arf, a tumor suppressor protein, can bind to MDM2 and enhance the degradation of MDM2 and MDMX, consequently activating p53 [108–110]. This Arf activity may represent an efficient strategy for therapeutics. Indeed, a group at the St. Jude Children's Research Hospital characterized the interaction domains of Arf and MDM2 [111–114]. They further exploited the possibility of identifying and/or designing compounds that mimic, inhibit and/or enhance the effect of Arf on MDM2. Although this research is still at its embryonic stage, it might produce surprising and promising results useful for anti-cancer drug discovery.

Recently, several ribosomal proteins (RP), such as RPL11, RPL5, RPL23, RPS7, RPS14, RPL26, RPS27, RPS3, and other nucleolar proteins have been identified as native MDM2 and/or MDMX inhibitors in response to ribosomal stress. This type of stress often occurs upon perturbation of ribosomal biogenesis caused by chemicals, nutrient deprivation, DNA damaging agents, or genetic alterations [115]. Interestingly, these ribosomal proteins upon this stress can uncouple p53 from its key negative regulator MDM2, consequently leading to p53 activation and protecting cells from tumor formation. Biochemical characterization by several groups including our lab [116, 117] and particularly genetic studies by the Zhang group at University of North Carolina [117–119] have demonstrated the specific interaction of RPL11 with the Zinc domain of MDM2, which is particularly important for p53 activation in response to ribosomal stress and whose defect would lead to cancer development. Specifically, cancer-related zinc finger cysteine mutations of MDM2 can disrupt the binding of MDM2 with RPL11 and RPL5. Furthermore, our biochemical studies uncovered several non-cysteine amino acids within the Zinc finger of MDM2 and basic amino acids in the MDM2 binding domain of RPL11, which are important for their specific interactions [116]. These findings identify the Zinc finger domain of MDM2 as a new and potential target site for future anticancer drug discovery.

SIRT1 Inhibitors

SIRT1, a NAD-dependent deacetylase, has attracted much attention due to its relation to cellular longevity [120, 121] and its role in negating p53 activity and stability [33], but it is also highly expressed in cancers [122]. It would readily maintain the tumor suppressor p53 in a deacetylated and inactive status in cancer cells, consequently favoring its MDM2/MDMX-mediated degradation [123]. The role of the HIC-1-SIRT1-p53 loop in lung cancer development has been confirmed in animal models [124]. These studies offer an opportunity for developing SIRT1 inhibitors as a potential anti-cancer therapy because they would not affect p53 function in normal cells and thus be less toxic to these cells. Indeed, there are several sirtuin inhibitors with proven activity in preclinical cancer models. The current state of the art in research on SIRT1 inhibitors has been extensively reviewed by other scientists [125–127], and thus we will not further replicate the review here. Among these SIRT1 inhibitors, Inauhzin (INZ), as a newly discovered SIRT1-specific inhibitor in our lab, can re-activate p53 in several wild type p53-containing cancer cells without genotoxic, and suppresses tumor growth by targeting SIRT1 and activating p53 [69]. Moreover, INZ is more potent, but much less toxic to normal cells or tissues than Tenovin [70] and other SIRT1 inhibitors [69]. Amazingly, INZ can sensitize the anti-cancer effect of chemotherapy and Nutlin-3 as tested in colon and lung cancer models [128, 129]. Thus, this small molecule presents as a promising contender for a molecule-targeted anti-cancer therapy that indirectly targets the MDM2-p53 pathway. We are currently undertaking systematic optimization [130], new target identification [131], and intensive preclinical studies of this compound, in hoping to eventually develop it into clinical trials in the near future.

Combination Strategies

Drug resistance and toxicity have been the two major obstacles for effective chemotherapy against cancer. To overcome and minimize the emergence of resistance caused by single-agent cancer therapy and to achieve maximal therapeutic response as well as to reduce the toxicity of genotoxic chemotherapy, combinations of different types of chemotherapeutic and molecule-targeted agents have been a common practice in clinic. This strategy has also been applied to the clinical trials of MDM2 or MDMX inhibitors, as combination of non-genotoxic p53 activators or MDM2 antagonists with clinically used chemotherapeutic agents [132], such as cisplatin, doxorubicin or cytarabine, has been shown to augment efficacy in the preclinical studies [133–135] and are currently being tested in several Phase I clinical trials (ClinicalTrials.gov: NCT01773408, NCT01605526 and NCT01605526). However, the radiosensitivity of the hematopoietic system of mice to p53 activation suggests that such an approach would potentially exacerbate the toxicity that is associated with MDM2 antagonists. Thus, combining non-genotoxic specific molecule-targeted therapeutic agents with p53 agonists could be a safer alternative. Moreover, this approach targeting more than one highly cancer-related molecule of one or multiple cellular signaling pathway(s) in cancers would be more important and effective for cancer treatment. A number of laboratories have been working on combinatorial therapy with the MDM2 antagonists, as shown in Table 16.2. Combinatorial therapy in a systematic manner seems more effective than empirically determined combination regimens, as p53 mutations confer absolute resistance or de novo p53-mutated multi-drug resistance emerges to MDM2 antagonists [135, 136], and more than 50 % of human cancers carry p53 mutants [137]. Conversely, the association of mutated Flt3 (Flt-ITD) with heightened sensitivity to MDM2 inhibitors was found in AML patients; BRAF mutation was associated with MDM2 inhibitor Nutlin 3a sensitivity [138], suggesting that therapeutic blockage of Flt3 using Flt3 inhibitor or targeting BRAF as well as the p53 pathway could generate additive or synergistic effects. Indeed, co-treatment of melanomas with the p53-activating stapled peptide SAH-p53-8 and BRAF inhibitor PLX4032 significantly enhanced cytotoxicity when compared with single-agent treatment [17]. Apparently, the development of a combinatorial anti-cancer therapy with anti-MDM2 or anti-MDMX inhibitors holds an encouraging future in this field.

Summary and Outlook

There is a great impetus in finding new types of chemotherapeutics to combat the drastically growing cancers with ever increasing number of cases of resistance to existing anti-cancer drugs. Remarkable progress in the past decade has been made in the discovery of novel chemotherapeutic drugs that selectively activate p53 to trigger rapid elimination of tumors by targeting the p53-MDM2-MDMX loop. As discussed above, a number of small-molecule drugs that inhibit the interaction of MDM2-p53 and block p53 degradation have been successfully developed through comprehensive medicinal chemistry, and some of them are currently in clinical trials. Experimental therapeutics and preclinical studies have demonstrated that MDM2 antagonists can be used alone or in combination regimens with cytotoxic drugs or other molecule-targeted drugs. The clinical results of MDM2 antagonists, such as RG7112, analogs of Nutlins, provide not only the proof of the concept, but also the usefulness of this therapeutic strategy for the treatment of human cancers in the near future. Combined screening of biomarkers, such as p53 status, MDM2 or

Table 16.2 Combination of non-genotoxic p53 activators or MDM2 antagonists with targeted therapeutic agents

Drug 1	Target 1	Drug 2	Target 2	In vitro evidence	In vivo evidence	Status	Reference
Nutlin	M2-p53 interaction	Inauhzin	SIRT1 inhibitor	H460 and HCT116 cells	HCT116 xenografts	Preclinical	[128]
Nutlin-3	M2-p53 interaction	Vorinostat, NaB, MS-275, Apicidin	HDAC1 inhibitor	A549 and A2780		Preclinical	[156]
Nutlin-3	M2-p53 interaction	Sorafenib	Multi-kinase inhibitor	RCC, acute myeloid leukemic			[157–159]
Nutlin	M2-p53 interaction	17AAG	Hsp90 inhibitor	RKO, AGS, HCT116 MCF7, U2OS	Xenografts?		[160]
Nutlin-3	M2-p53 interaction	Dasatinib	Multi-kinase inhibitor	Primary B-CLL cell & cell line			[161]
Nutlin-3	M2-p53 interaction	Imatinib	BCR/ABL tyrosine kinase inhibitor	B210, IL-3			[162]
Nutlin-3	M2-p53 interaction	VX-680	Aurora kinase inhibitors	A549, HCT116			[163]
MI-219	M2-p53 interaction	$ZnCl_2$, TPEN	Zinc chelator	HCT116, MCF-7			[164]
Nutlin-3	M2-p53 interaction	1396-11, XIAP antisense oligonucleotide (ASO)	XIAP inhibitor	OCI-AML3, MOLM-13, HL-60		Phase I clinical trial	[165]
Nutlin-3	M2-p53 interaction	Bortezomib	Proteasome inhibitor	MCL, HCT116, U2OS			[166–170]
Nutlin-3	M2-p53 interaction	Velcade	Proteasome inhibitor	Myeloma (MM)			[171]
Nutlin-3	M2-p53 interaction	Valproic acid (VPA)	HDAC1 inhibitors	AML cell lines	MOLM-13 orthotopic and xenograft		[172]

(continued)

Table 16.2 (continued)

Drug 1	Target 1	Drug 2	Target 2	In vitro evidence	In vivo evidence	Status	Reference
Nutlin-3a	M2-p53 interaction	ABT-737	Bcl-2 protein inhibitor	OCI-AML-3 and MOLM-13			[173]
Nutlin-3	M2-p53 interaction	MK-0457	Aurora kinase inhibitors	OCI-AML-3 and MOLM-13			[174]
Nutlin	M2-p53 interaction	PI-103	PI3K/mTOR inhibitor	AML cells		Preclinical	[175]
Nutlin-3	M2-p53 interaction	Mithramycin	Sp1 DNA binding inhibitor	Gynecologic cancer cells	Xenografts	Preclinical	[176]
Nutlin-3	M2-p53 interaction	5-FU cisplatin	Suicide inhibitor, crosslinking of DNA	Gastric cancer cell, OVCa cells	Xenografts		[177, 178]
Paclitaxel (PTX)	AKT-M2-p53	Beta-lapachone (LPC)	(Topo I)	Retinoblastoma (Y79 cells)			[179]
Nutlin-3a	M2-p53 interaction	Roscovitine and DRB	CDK inhibitors	ARN8 cells			[180]
Nutlin-3	M2-p53 interaction	SiMDMX	MDMX	A549, U2OS, SJSA. MCF-7, JEG3			[181]
Nutlin-3	M2-p53 interaction	Doxil	Intercalating DNA	RKO, HT-29			[182]
MI-63	M2-p53 interaction	Gemcitabine	RNR inhibitor	MCL cell	Xenografts	Preclinical	[183]

MDMX gene amplification or other additional markers, would effectively select patients for clinical trials of MDM2-and MDMX-targeted single or combination therapies, ultimately improving the prospects for cancer therapy and prevention.

Compared to the progress in the discovery of MDM2 antagonists, the development of selective MDMX inhibitors seems slower, more complicated and challenging. Although the first crystal structure of MDMX-small molecule inhibitor complex was disclosed recently [86], a clinically useful MDMX inhibitor has not yet been reported. However, it is predicated that selective MDMX antagonists or dual inhibitors of MDM2 and MDMX will be of high interest and represent as a promising new approach to fight certain types of cancers that highly express MDMX, such as retinoblastoma [139] or melanoma [17]. Additionally, it has been indicated that systemic inhibition of MDMX is not only feasible as a therapeutic strategy for restoring p53 function in tumors that retain wild-type p53, but also significantly safer than inhibiting MDM2 [18, 22]. Hence, we will certainly not wait for too long to see a promising anti-MDMX small molecule in clinical trials.

A study using an MDM2 RING finger (C462A) mutant knockin mouse model by the Zhang group has shown that the stability of MDM2 is in fact not regulated by autoubiquitination in vivo [140], nor it is capable of blocking p53 activity by binding alone. This finding suggests that the MDM2 RING finger E3 ubiquitin ligase function plays an important role in suppressing p53, although this mutant is also defective in binding to MDMX [141, 142]. It also implies that it is possible to achieve the goal of reactivating p53 in cancers retaining wild type p53 by inhibiting MDM2 E3 ubiquitin ligase activity. However, the development of MDM2 E3 ubiquitin ligase inhibitors has been hampered by the biological complexity of the ubiquitination process. Unlike kinases, which can be inhibited via their ATP binding site, it is not known exactly how RING E3 ubiquitin ligases function, and hence it is unclear how to target their activity. Structural information is urgently needed for better understanding at the atomic and molecular levels how exactly MDM2 functions as an E3 ligase. This information would facilitate structure-based drug design to target MDM2 E3 ubiquitin ligase activity in the future [143].

Another challenge is that although MDM inhibitors mainly target the p53 pathway and kill wild-type p53-containing tumors by inducing p53-dependent apoptosis of cancer cells, more than half of human cancers lack of functional p53 largely because they harbor mutated forms of p53, which either lose wild type function or gain new oncogenic functions. Several approaches for drug discovery have been explored to identify small molecules that target mutant p53, restoring wild type p53 function and/or inhibiting the negative p53 regulators, such as rational design and high throughput screening of chemical libraries [144]. One example is the compound PRIMA-1 that restores wild-type conformation by binding to the core of mutant p53 and induces massive apoptosis in human tumor cells [145]. The PRIMA-1 analog APR-246 is currently in a clinical trial [146–148]. Very recently, another small molecule rescuer of mutant p53 was reported [149]. Interestingly, this small molecule called NSC319726 can convert R175H or R175W mutant p53 into a wild type form by chelating the Zinc ion and altering the redox environment for the p53 protein. Amazingly this compound can inhibit xenograft tumor growth by inducing apoptosis in an R175 mutant-dependent manner. Although much more needs to be investigated, this serves as a lead compound for further development of another mutant p53-targeted cancer therapy. Therefore, systematic and successful development of mutant p53-reactivating anticancer drugs will have a strong impact on the treatment of cancers, particularly those late stage and malignant ones that harbor mutant p53.

As a Chinese saying indicates, "dripping water wears through rock", we strongly believe that the persistent, systematic, intensive and joint efforts from troops of cancer researchers on drug discovery by searching small molecules or agents targeting the p53-MDM2-MDMX pathway will ultimately be translated into a clinically useful anti-cancer therapy or therapies.

Acknowledgements This work was supported in part by NIH-NCI grants CA095441, CA 079721, CA129828, and CA172468, as well as the LLL fund to H.L.

References

1. Brown CJ et al (2009) Awakening guardian angels: drugging the p53 pathway. Nat Rev Cancer 9(12):862–873
2. Kruse JP, Gu W (2009) Modes of p53 regulation. Cell 137(4):609–622
3. Hollstein M et al (1991) p53 mutations in human cancers. Science 253(5015):49–53
4. Vogelstein B, Lane D, Levine AJ (2000) Surfing the p53 network. Nature 408(6810):307–310
5. Vousden KH, Prives C (2009) Blinded by the light: the growing complexity of p53. Cell 137(3):413–431
6. Hong H et al (2009) Suppression of induced pluripotent stem cell generation by the p53-p21 pathway. Nature 460(7259):1132–1135
7. Wu X et al (1993) The p53-mdm-2 autoregulatory feedback loop. Genes Dev 7(7A):1126–1132
8. Momand J et al (1992) The mdm-2 oncogene product forms a complex with the p53 protein and inhibits p53-mediated transactivation. Cell 69(7):1237–1245
9. Barak Y et al (1993) mdm2 expression is induced by wild type p53 activity. EMBO J 12(2):461–468
10. Shvarts A et al (1996) MDMX: a novel p53-binding protein with some functional properties of MDM2. EMBO J 15(19):5349–5357
11. Eischen CM, Lozano G (2009) p53 and MDM2: antagonists or partners in crime? Cancer Cell 15(3):161–162
12. Gu J et al (2002) Mutual dependence of MDM2 and MDMX in their functional inactivation of p53. J Biol Chem 277(22):19251–19254
13. Haupt Y et al (1997) Mdm2 promotes the rapid degradation of p53. Nature 387(6630):296–299
14. Kubbutat MH, Jones SN, Vousden KH (1997) Regulation of p53 stability by Mdm2. Nature 387(6630):299–303
15. Honda R, Tanaka H, Yasuda H (1997) Oncoprotein MDM2 is a ubiquitin ligase E3 for tumor suppressor p53. FEBS Lett 420(1):25–27
16. Wade M, Li YC, Wahl GM (2013) MDM2, MDMX and p53 in oncogenesis and cancer therapy. Nat Rev Cancer 13(2):83–96
17. Gembarska A et al (2012) MDM4 is a key therapeutic target in cutaneous melanoma. Nat Med 18(8):1239–1247

18. Biderman L et al (2012) MdmX is required for p53 interaction with and full induction of the Mdm2 promoter after cellular stress. Mol Cell Biol 32(7):1214–1225
19. Wade M, Wahl GM (2009) Targeting Mdm2 and Mdmx in cancer therapy: better living through medicinal chemistry? Mol Cancer Res 7(1):1–11
20. Zhang Y, Lu H (2009) Signaling to p53: ribosomal proteins find their way. Cancer Cell 16(5):369–377
21. Wade M, Wang YV, Wahl GM (2010) The p53 orchestra: Mdm2 and Mdmx set the tone. Trends Cell Biol 20(5):299–309
22. Garcia D et al (2011) Validation of MdmX as a therapeutic target for reactivating p53 in tumors. Genes Dev 25(16):1746–1757
23. Tang Y et al (2008) Acetylation is indispensable for p53 activation. Cell 133(4):612–626
24. Shieh SY et al (1997) DNA damage-induced phosphorylation of p53 alleviates inhibition by MDM2. Cell 91(3):325–334
25. Banin S et al (1998) Enhanced phosphorylation of p53 by ATM in response to DNA damage. Science 281(5383):1674–1677
26. Maya R et al (2001) ATM-dependent phosphorylation of Mdm2 on serine 395: role in p53 activation by DNA damage. Genes Dev 15(9):1067–1077
27. Kapoor M, Lozano G (1998) Functional activation of p53 via phosphorylation following DNA damage by UV but not gamma radiation. Proc Natl Acad Sci U S A 95(6):2834–2837
28. Zhang Y, Xiong Y, Yarbrough WG (1998) ARF promotes MDM2 degradation and stabilizes p53: ARF-INK4a locus deletion impairs both the Rb and p53 tumor suppression pathways. Cell 92(6):725–734
29. Kobet E et al (2000) MDM2 inhibits p300-mediated p53 acetylation and activation by forming a ternary complex with the two proteins. Proc Natl Acad Sci U S A 97(23):12547–12552
30. Li M et al (2002) Acetylation of p53 inhibits its ubiquitination by Mdm2. J Biol Chem 277(52):50607–50611
31. Ito A et al (2001) p300/CBP-mediated p53 acetylation is commonly induced by p53-activating agents and inhibited by MDM2. EMBO J 20(6):1331–1340
32. Luo J et al (2001) Negative control of p53 by Sir2alpha promotes cell survival under stress. Cell 107(2):137–148
33. Vaziri H et al (2001) hSIR2(SIRT1) functions as an NAD-dependent p53 deacetylase. Cell 107(2):149–159
34. Cheng HL et al (2003) Developmental defects and p53 hyperacetylation in Sir2 homolog (SIRT1)-deficient mice. Proc Natl Acad Sci U S A 100(19):10794–10799
35. Luo J et al (2000) Deacetylation of p53 modulates its effect on cell growth and apoptosis. Nature 408(6810):377–381
36. Onel K, Cordon-Cardo C (2004) MDM2 and prognosis. Mol Cancer Res 2(1):1–8
37. Nosho K et al (2009) SIRT1 histone deacetylase expression is associated with microsatellite instability and CpG island methylator phenotype in colorectal cancer. Mod Pathol 22(7):922–932
38. Jung-Hynes B et al (2009) Role of sirtuin histone deacetylase SIRT1 in prostate cancer. A target for prostate cancer management via its inhibition? J Biol Chem 284(6):3823–3832
39. Ozdag H et al (2006) Differential expression of selected histone modifier genes in human solid cancers. BMC Genomics 7:90
40. Tseng RC et al (2009) Distinct HIC1-SIRT1-p53 loop deregulation in lung squamous carcinoma and adenocarcinoma patients. Neoplasia 11(8):763–770
41. Jones SN et al (1995) Rescue of embryonic lethality in Mdm2-deficient mice by absence of p53. Nature 378(6553):206–208
42. Montes de Oca Luna R, Wagner DS, Lozano G (1995) Rescue of early embryonic lethality in mdm2-deficient mice by deletion of p53. Nature 378(6553):203–206
43. Gannon HS, Jones SN (2012) Using mouse models to explore MDM-p53 signaling in development, cell growth, and tumorigenesis. Genes Cancer 3(3–4):209–218
44. Vassilev LT et al (2004) In vivo activation of the p53 pathway by small-molecule antagonists of MDM2. Science 303(5659):844–848

45. Chargari C et al (2011) Preclinical assessment of JNJ-26854165 (Serdemetan), a novel tryptamine compound with radiosensitizing activity in vitro and in tumor xenografts. Cancer Lett 312(2):209–218
46. Kojima K et al (2010) The novel tryptamine derivative JNJ-26854165 induces wild-type p53- and E2F1-mediated apoptosis in acute myeloid and lymphoid leukemias. Mol Cancer Ther 9(9):2545–2557
47. Yang Y et al (2005) Small molecule inhibitors of HDM2 ubiquitin ligase activity stabilize and activate p53 in cells. Cancer Cell 7(6):547–559
48. Roxburgh P et al (2012) Small molecules that bind the Mdm2 RING stabilize and activate p53. Carcinogenesis 33(4):791–798
49. Herman AG et al (2011) Discovery of Mdm2-MdmX E3 ligase inhibitors using a cell-based ubiquitination assay. Cancer Discov 1(4):312–325
50. Galatin PS, Abraham DJ (2004) A nonpeptidic sulfonamide inhibits the p53-mdm2 interaction and activates p53-dependent transcription in mdm2-overexpressing cells. J Med Chem 47(17):4163–4165
51. Lu Y et al (2006) Discovery of a nanomolar inhibitor of the human murine double minute 2 (MDM2)-p53 interaction through an integrated, virtual database screening strategy. J Med Chem 49(13):3759–3762
52. Yin H et al (2005) Terphenyl-based helical mimetics that disrupt the p53/HDM2 interaction. Angew Chem Int Ed Engl 44(18):2704–2707
53. Hardcastle IR et al (2006) Small-molecule inhibitors of the MDM2-p53 protein-protein interaction based on an isoindolinone scaffold. J Med Chem 49(21):6209–6221
54. Hardcastle IR et al (2011) Isoindolinone inhibitors of the murine double minute 2 (MDM2)-p53 protein-protein interaction: structure-activity studies leading to improved potency. J Med Chem 54(5):1233–1243
55. Koblish HK et al (2006) Benzodiazepinedione inhibitors of the Hdm2:p53 complex suppress human tumor cell proliferation in vitro and sensitize tumors to doxorubicin in vivo. Mol Cancer Ther 5(1):160–169
56. Schilling D et al (2013) Radiosensitization of wildtype p53 cancer cells by the MDM2-inhibitor PXN727 is associated with altered heat shock protein 70 (Hsp70) levels. Cell Stress Chaperones 18(2):183–191
57. Cheok CF et al (2011) Translating p53 into the clinic. Nat Rev Clin Oncol 8(1):25–37
58. Bista M et al (2012) On the mechanism of action of SJ-172550 in inhibiting the interaction of MDM4 and p53. PLoS One 7(6):e37518
59. Reed D et al (2010) Identification and characterization of the first small molecule inhibitor of MDMX. J Biol Chem 285(14):10786–10796
60. Wang H et al (2011) A small-molecule inhibitor of MDMX activates p53 and induces apoptosis. Mol Cancer Ther 10(1):69–79
61. Graves B et al (2012) Activation of the p53 pathway by small-molecule-induced MDM2 and MDMX dimerization. Proc Natl Acad Sci U S A 109(29):11788–11793
62. Qin JJ et al (2012) Natural product MDM2 inhibitors: anticancer activity and mechanisms of action. Curr Med Chem 19(33):5705–5725
63. Zak K et al (2013) Mdm2 and MdmX inhibitors for the treatment of cancer: a patent review (2011-present). Expert Opin Ther Pat 23(4):425–448
64. Kamal A, Mohammed AA, Shaik TB (2012) p53-Mdm2 inhibitors: patent review (2009–2010). Expert Opin Ther Pat 22(2):95–105
65. Weber L (2010) Patented inhibitors of p53-Mdm2 interaction (2006–2008). Expert Opin Ther Pat 20(2):179–191
66. Gu W, Roeder RG (1997) Activation of p53 sequence-specific DNA binding by acetylation of the p53 C-terminal domain. Cell 90(4):595–606
67. Gu W, Shi XL, Roeder RG (1997) Synergistic activation of transcription by CBP and p53. Nature 387(6635):819–823
68. Li L et al (2012) Activation of p53 by SIRT1 inhibition enhances elimination of CML leukemia stem cells in combination with imatinib. Cancer Cell 21(2):266–281

69. Zhang Q et al (2012) A small molecule Inauhzin inhibits SIRT1 activity and suppresses tumour growth through activation of p53. EMBO Mol Med 4(4):298–312
70. Lain S et al (2008) Discovery, in vivo activity, and mechanism of action of a small-molecule p53 activator. Cancer Cell 13(5):454–463
71. Kim WJ et al (2012) The WTX tumor suppressor enhances p53 acetylation by CBP/p300. Mol Cell 45(5):587–597
72. Kussie PH et al (1996) Structure of the MDM2 oncoprotein bound to the p53 tumor suppressor transactivation domain. Science 274(5289):948–953
73. Joseph TL et al (2010) Differential binding of p53 and nutlin to MDM2 and MDMX: computational studies. Cell Cycle 9(6):1167–1181
74. Almerico AM et al (2012) Molecular dynamics studies on Mdm2 complexes: an analysis of the inhibitor influence. Biochem Biophys Res Commun 424(2):341–347
75. Tsuganezawa K et al (2013) A fluorescent-based high-throughput screening assay for small molecules that inhibit the interaction of MdmX with p53. J Biomol Screen 18(2):191–198
76. Grasberger BL et al (2005) Discovery and cocrystal structure of benzodiazepinedione HDM2 antagonists that activate p53 in cells. J Med Chem 48(4):909–912
77. Popowicz GM, Domling A, Holak TA (2011) The structure-based design of Mdm2/Mdmx-p53 inhibitors gets serious. Angew Chem Int Ed Engl 50(12):2680–2688
78. Pazgier M et al (2009) Structural basis for high-affinity peptide inhibition of p53 interactions with MDM2 and MDMX. Proc Natl Acad Sci U S A 106(12):4665–4670
79. Liu M et al (2010) D-peptide inhibitors of the p53-MDM2 interaction for targeted molecular therapy of malignant neoplasms. Proc Natl Acad Sci U S A 107(32):14321–14326
80. Zhan C et al (2012) An ultrahigh affinity d-peptide antagonist Of MDM2. J Med Chem 55(13):6237–6241
81. Sasiela CA et al (2008) Identification of inhibitors for MDM2 ubiquitin ligase activity from natural product extracts by a novel high-throughput electrochemiluminescent screen. J Biomol Screen 13(3):229–237
82. Murray MF et al (2007) A high-throughput screen measuring ubiquitination of p53 by human mdm2. J Biomol Screen 12(8):1050–1058
83. Dudgeon DD et al (2010) Implementation of a 220,000-compound HCS campaign to identify disruptors of the interaction between p53 and hDM2 and characterization of the confirmed hits. J Biomol Screen 15(7):766–782
84. Dudgeon DD et al (2010) Characterization and optimization of a novel protein-protein interaction biosensor high-content screening assay to identify disruptors of the interactions between p53 and hDM2. Assay Drug Dev Technol 8(4):437–458
85. Khoury K et al (2011) The p53-MDM2/MDMX axis – a chemotype perspective. Med Chem Commun 2(4):246–260
86. Popowicz GM et al (2010) Structures of low molecular weight inhibitors bound to MDMX and MDM2 reveal new approaches for p53-MDMX/MDM2 antagonist drug discovery. Cell Cycle 9(6):1104–1111
87. Bernal F et al (2007) Reactivation of the p53 tumor suppressor pathway by a stapled p53 peptide. J Am Chem Soc 129(9):2456–2457
88. Bernal F et al (2010) A stapled p53 helix overcomes HDMX-mediated suppression of p53. Cancer Cell 18(5):411–422
89. Huang Y et al (2010) 1,4-Thienodiazepine-2,5-diones via MCR (I): synthesis, virtual space and p53-Mdm2 activity. Chem Biol Drug Des 76(2):116–129
90. Lee JH et al (2011) Novel pyrrolopyrimidine-based alpha-helix mimetics: cell-permeable inhibitors of protein-protein interactions. J Am Chem Soc 133(4):676–679
91. Czarna A et al (2010) Robust generation of lead compounds for protein-protein interactions by computational and MCR chemistry: p53/Hdm2 antagonists. Angew Chem Int Ed Engl 49(31):5352–5356
92. Fu T et al (2012) Molecular dynamic simulation insights into the normal state and restoration of p53 function. Int J Mol Sci 13(8):9709–9740
93. Ding K et al (2006) Structure-based design of spiro-oxindoles as potent, specific small-molecule inhibitors of the MDM2-p53 interaction. J Med Chem 49(12):3432–3435

94. Shangary S et al (2008) Temporal activation of p53 by a specific MDM2 inhibitor is selectively toxic to tumors and leads to complete tumor growth inhibition. Proc Natl Acad Sci U S A 105(10):3933–3938
95. Yu S et al (2009) Potent and orally active small-molecule inhibitors of the MDM2-p53 interaction. J Med Chem 52(24):7970–7973
96. Rew Y et al (2012) Structure-based design of novel inhibitors of the MDM2-p53 interaction. J Med Chem 55(11):4936–4954
97. Lucas BS et al (2012) An expeditious synthesis of the MDM2-p53 inhibitor AM-8553. J Am Chem Soc 134(30):12855–12860
98. Bowman AL et al (2007) Small molecule inhibitors of the MDM2-p53 interaction discovered by ensemble-based receptor models. J Am Chem Soc 129(42):12809–12814
99. Jacoby E et al (2009) Knowledge-based virtual screening: application to the MDM4/p53 protein-protein interaction. Methods Mol Biol 575:173–194
100. Tovar C et al (2013) MDM2 small-molecule antagonist RG7112 activates p53 signaling and regresses human tumors in preclinical cancer models. Cancer Res 73(8):2587–2597
101. Ray-Coquard I et al (2012) Effect of the MDM2 antagonist RG7112 on the P53 pathway in patients with MDM2-amplified, well-differentiated or dedifferentiated liposarcoma: an exploratory proof-of-mechanism study. Lancet Oncol 13(11):1133–1140
102. Andreeff M et al (2012) Results of the phase 1 trial of RG7112, a small-molecule MDM2 antagonist, in acute leukemia. In: 54 annual meeting American Society of Hematology, Atlanta, 8–11 Dec 2012, p 615, Abstract 675
103. Wang S et al (2012) Targeting the MDM2-p53 protein-protein interaction for new cancer therapeutics. Top Med Chem 8:57–80
104. Mohammad RM et al (2009) An MDM2 antagonist (MI-319) restores p53 functions and increases the life span of orally treated follicular lymphoma bearing animals. Mol Cancer 8:115
105. Menendez S et al (2011) MDM4 downregulates p53 transcriptional activity and response to stress during differentiation. Cell Cycle 10(7):1100–1108
106. Laurie NA, Schin-Shih C, Dyer MA (2007) Targeting MDM2 and MDMX in retinoblastoma. Curr Cancer Drug Targets 7(7):689–695
107. Smith MA et al (2012) Initial testing of JNJ-26854165 (Serdemetan) by the pediatric preclinical testing program. Pediatr Blood Cancer 59(2):329–332
108. Honda R, Yasuda H (1999) Association of p19(ARF) with Mdm2 inhibits ubiquitin ligase activity of Mdm2 for tumor suppressor p53. EMBO J 18(1):22–27
109. Weber JD et al (1999) Nucleolar Arf sequesters Mdm2 and activates p53. Nat Cell Biol 1(1):20–26
110. Kamijo T et al (1998) Functional and physical interactions of the ARF tumor suppressor with p53 and Mdm2. Proc Natl Acad Sci U S A 95(14):8292–8297
111. Bothner B et al (2001) Defining the molecular basis of Arf and Hdm2 interactions. J Mol Biol 314(2):263–277
112. Sivakolundu SG et al (2008) Intrinsically unstructured domains of Arf and Hdm2 form bimolecular oligomeric structures in vitro and in vivo. J Mol Biol 384(1):240–254
113. DiGiammarino EL et al (2001) Solution structure of the p53 regulatory domain of the p19Arf tumor suppressor protein. Biochemistry 40(8):2379–2386
114. Weber JD et al (2000) Cooperative signals governing ARF-mdm2 interaction and nucleolar localization of the complex. Mol Cell Biol 20(7):2517–2528
115. Zhou X et al (2012) Scission of the p53-MDM2 loop by ribosomal proteins. Genes Cancer 3(3–4):298–310
116. Zhang Q et al (2011) Hydrophilic residues are crucial for ribosomal protein L11 (RPL11) interaction with zinc finger domain of MDM2 and p53 protein activation. J Biol Chem 286(44):38264–38274
117. Dai MS et al (2006) Regulation of the MDM2-p53 pathway by ribosomal protein L11 involves a post-ubiquitination mechanism. J Biol Chem 281(34):24304–24313

118. Bhat KP et al (2004) Essential role of ribosomal protein L11 in mediating growth inhibition-induced p53 activation. EMBO J 23(12):2402–2412
119. Zhang Y et al (2003) Ribosomal protein L11 negatively regulates oncoprotein MDM2 and mediates a p53-dependent ribosomal-stress checkpoint pathway. Mol Cell Biol 23(23): 8902–8912
120. Imai S (2007) Is Sirt1 a miracle bullet for longevity? Aging Cell 6(6):735–737
121. Chen D et al (2005) Increase in activity during calorie restriction requires Sirt1. Science 310(5754):1641
122. Stunkel W et al (2007) Function of the SIRT1 protein deacetylase in cancer. Biotechnol J 2(11):1360–1368
123. Inoue Y et al (2011) Suppression of p53 activity through the cooperative action of Ski and histone deacetylase SIRT1. J Biol Chem 286(8):6311–6320
124. Chen WY et al (2005) Tumor suppressor HIC1 directly regulates SIRT1 to modulate p53-dependent DNA-damage responses. Cell 123(3):437–448
125. Blum CA et al (2011) SIRT1 modulation as a novel approach to the treatment of diseases of aging. J Med Chem 54(2):417–432
126. Chakrabarty SP, Balaram H, Chandrasekaran S (2011) Sirtuins: multifaceted drug targets. Curr Mol Med 11(9):709–718
127. Balcerczyk A, Pirola L (2010) Therapeutic potential of activators and inhibitors of sirtuins. Biofactors 36(5):383–393
128. Zhang Y et al (2012) Inauhzin and Nutlin3 synergistically activate p53 and suppress tumor growth. Cancer Biol Ther 13(10):915–924
129. Yiwei Z, Zhang Q, Zeng SX, Qian Hao, Hua Lu (2013) Inauhzin sensitizes p53-dependent cytotoxicity and tumor suppression of chemotherapeutic agents. Neoplasia 15(5):523–534
130. Zhang Q et al (2012) Structure and activity analysis of Inauhzin analogs as novel antitumor compounds that induce p53 and inhibit cell growth. PLoS One 7(10):e46294
131. Liao JM et al (2012) Global effect of inauhzin on human p53-responsive transcriptome. PLoS One 7(12):e52172
132. Azmi AS et al (2011) Network perspectives on HDM2 inhibitor chemotherapy combinations. Curr Pharm Des 17(6):640–652
133. Azmi AS et al (2010) MDM2 inhibitor MI-319 in combination with cisplatin is an effective treatment for pancreatic cancer independent of p53 function. Eur J Cancer 46(6):1122–1131
134. Zheng T et al (2010) Nutlin-3 cooperates with doxorubicin to induce apoptosis of human hepatocellular carcinoma cells through p53 or p73 signaling pathways. J Cancer Res Clin Oncol 136(10):1597–1604
135. Koster R et al (2011) Disruption of the MDM2-p53 interaction strongly potentiates p53-dependent apoptosis in cisplatin-resistant human testicular carcinoma cells via the Fas/FasL pathway. Cell Death Dis 2:e148
136. Michaelis M et al (2011) Adaptation of cancer cells from different entities to the MDM2 inhibitor nutlin-3 results in the emergence of p53-mutated multi-drug-resistant cancer cells. Cell Death Dis 2:e243
137. Muller PA, Vousden KH (2013) p53 mutations in cancer. Nat Cell Biol 15(1):2–8
138. Long J et al (2010) Multiple distinct molecular mechanisms influence sensitivity and resistance to MDM2 inhibitors in adult acute myelogenous leukemia. Blood 116(1):71–80
139. Laurie NA et al (2006) Inactivation of the p53 pathway in retinoblastoma. Nature 444(7115):61–66
140. Clegg HV et al (2012) Mdm2 RING mutation enhances p53 transcriptional activity and p53-p300 interaction. PLoS One 7(5):e38212
141. Huang L et al (2011) The p53 inhibitors MDM2/MDMX complex is required for control of p53 activity in vivo. Proc Natl Acad Sci U S A 108(29):12001–12006
142. Itahana K et al (2007) Targeted inactivation of Mdm2 RING finger E3 ubiquitin ligase activity in the mouse reveals mechanistic insights into p53 regulation. Cancer Cell 12(4):355–366
143. Di J, Zhang Y, Zheng J (2011) Reactivation of p53 by inhibiting Mdm2 E3 ligase: a novel antitumor approach. Curr Cancer Drug Targets 11(8):987–994

144. Maslon MM, Hupp TR (2010) Drug discovery and mutant p53. Trends Cell Biol 20(9):542–555
145. Bykov VJ et al (2002) Restoration of the tumor suppressor function to mutant p53 by a low-molecular-weight compound. Nat Med 8(3):282–288
146. Mohell N et al (2010) Preclinical efficacy and toxicology studies of APR-246, a novel anticancer compound currently in clinical trials for refractory hematological malignancies and prostate cancer. Blood 116(21)
147. Lehmann S et al (2012) Targeting p53 in vivo: a first-in-human study with p53-targeting compound APR-246 in refractory hematologic malignancies and prostate cancer. J Clin Oncol 30(29):3633–3639
148. Shalom-Feuerstein R et al (2013) Impaired epithelial differentiation of induced pluripotent stem cells from ectodermal dysplasia-related patients is rescued by the small compound APR-246/PRIMA-1MET. Proc Natl Acad Sci U S A 110(6):2152–2156
149. Yu X et al (2012) Allele-specific p53 mutant reactivation. Cancer Cell 21(5):614–625
150. Lawrence HR et al (2009) Identification of a disruptor of the MDM2-p53 protein-protein interaction facilitated by high-throughput in silico docking. Bioorg Med Chem Lett 19(14):3756–3759
151. Parks DJ et al (2005) 1,4-Benzodiazepine-2,5-diones as small molecule antagonists of the HDM2-p53 interaction: discovery and SAR. Bioorg Med Chem Lett 15(3):765–770
152. Zhuang C et al (2012) Discovery, synthesis, and biological evaluation of orally active pyrrolidone derivatives as novel inhibitors of p53-MDM2 protein-protein interaction. J Med Chem 55(22):9630–9642
153. Krajewski M et al (2005) NMR indicates that the small molecule RITA does not block p53-MDM2 binding in vitro. Nat Med 11(11):1135–1136; author reply 1136–7
154. Issaeva N et al (2004) Small molecule RITA binds to p53, blocks p53-HDM-2 interaction and activates p53 function in tumors. Nat Med 10(12):1321–1328
155. Michel J et al (2009) In Silico Improvement of beta3-peptide inhibitors of p53 x hDM2 and p53 x hDMX. J Am Chem Soc 131(18):6356–6357
156. Palani CD, Beck JF, Sonnemann J (2012) Histone deacetylase inhibitors enhance the anticancer activity of nutlin-3 and induce p53 hyperacetylation and downregulation of MDM2 and MDM4 gene expression. Invest New Drugs 30(1):25–36
157. Zauli G et al (2012) The sorafenib plus nutlin-3 combination promotes synergistic cytotoxicity in acute myeloid leukemic cells irrespectively of FLT3 and p53 status. Haematologica 97(11):1722–1730
158. Vatsyayan R et al (2013) Nutlin-3 enhances sorafenib efficacy in renal cell carcinoma. Mol Carcinog 52(1):39–48
159. Weisberg E, Sattler M (2012) A novel combination therapy approach for the treatment of acute myeloid leukemia: the multi-kinase inhibitor sorafenib and the HDM2 inhibitor nutlin-3. Haematologica 97(11):1620–1621
160. Vaseva AV et al (2011) Blockade of Hsp90 by 17AAG antagonizes MDMX and synergizes with Nutlin to induce p53-mediated apoptosis in solid tumors. Cell Death Dis 2:e156
161. Zauli G et al (2011) Dasatinib plus Nutlin-3 shows synergistic antileukemic activity in both p53 wild-type and p53 mutated B chronic lymphocytic leukemias by inhibiting the Akt pathway. Clin Cancer Res 17(4):762–770
162. Kurosu T et al (2010) Enhancement of imatinib-induced apoptosis of BCR/ABL-expressing cells by nutlin-3 through synergistic activation of the mitochondrial apoptotic pathway. Apoptosis 15(5):608–620
163. Cheok CF et al (2010) Combination of nutlin-3 and VX-680 selectively targets p53 mutant cells with reversible effects on cells expressing wild-type p53. Cell Death Differ 17(9):1486–1500
164. Azmi AS et al (2011) MI-219-zinc combination: a new paradigm in MDM2 inhibitor-based therapy. Oncogene 30(1):117–126
165. Carter BZ et al (2010) Simultaneous activation of p53 and inhibition of XIAP enhance the activation of apoptosis signaling pathways in AML. Blood 115(2):306–314

166. Ooi MG et al (2009) Interactions of the Hdm2/p53 and proteasome pathways may enhance the antitumor activity of bortezomib. Clin Cancer Res 15(23):7153–7160
167. Tabe Y et al (2009) MDM2 antagonist nutlin-3 displays antiproliferative and proapoptotic activity in mantle cell lymphoma. Clin Cancer Res 15(3):933–942
168. Jin L et al (2010) MDM2 antagonist Nutlin-3 enhances bortezomib-mediated mitochondrial apoptosis in TP53-mutated mantle cell lymphoma. Cancer Lett 299(2):161–170
169. Cross B et al (2011) Inhibition of p53 DNA binding function by the MDM2 protein acidic domain. J Biol Chem 286(18):16018–16029
170. Secchiero P et al (2011) Recent advances in the therapeutic perspectives of Nutlin-3. Curr Pharm Des 17(6):569–577
171. Saha MN et al (2010) MDM2 antagonist nutlin plus proteasome inhibitor velcade combination displays a synergistic anti-myeloma activity. Cancer Biol Ther 9(11):936–944
172. McCormack E et al (2012) Synergistic induction of p53 mediated apoptosis by valproic acid and nutlin-3 in acute myeloid leukemia. Leukemia 26(5):910–917
173. Kojima K et al (2006) Concomitant inhibition of MDM2 and Bcl-2 protein function synergistically induce mitochondrial apoptosis in AML. Cell Cycle 5(23):2778–2786
174. Kojima K et al (2008) Concomitant inhibition of Mdm2-p53 interaction and Aurora kinases activates the p53-dependent postmitotic checkpoints and synergistically induces p53-mediated mitochondrial apoptosis along with reduced endoreduplication in acute myelogenous leukemia. Blood 112(7):2886–2895
175. Kojima K et al (2008) The dual PI3 kinase/mTOR inhibitor PI-103 prevents p53 induction by Mdm2 inhibition but enhances p53-mediated mitochondrial apoptosis in p53 wild-type AML. Leukemia 22(9):1728–1736
176. Ohgami T et al (2010) Low-dose mithramycin exerts its anticancer effect via the p53 signaling pathway and synergizes with nutlin-3 in gynecologic cancers. Cancer Sci 101(6):1387–1395
177. Endo S et al (2011) Potent in vitro and in vivo antitumor effects of MDM2 inhibitor nutlin-3 in gastric cancer cells. Cancer Sci 102(3):605–613
178. Mir R et al (2013) Mdm2 antagonists induce apoptosis and synergize with cisplatin overcoming chemoresistance in TP53 wild-type ovarian cancer cells. Int J Cancer 132(7):1525–1536
179. D'Anneo A et al (2010) Paclitaxel and beta-lapachone synergistically induce apoptosis in human retinoblastoma Y79 cells by downregulating the levels of phospho-Akt. J Cell Physiol 222(2):433–443
180. Cheok CF, Dey A, Lane DP (2007) Cyclin-dependent kinase inhibitors sensitize tumor cells to nutlin-induced apoptosis: a potent drug combination. Mol Cancer Res 5(11):1133–1145
181. Hu B et al (2006) MDMX overexpression prevents p53 activation by the MDM2 inhibitor Nutlin. J Biol Chem 281(44):33030–33035
182. Nadler-Milbauer M et al (2011) Synchronized release of Doxil and Nutlin-3 by remote degradation of polysaccharide matrices and its possible use in the local treatment of colorectal cancer. J Drug Target 19(10):859–873
183. Jones RJ et al (2011) HDM-2 inhibition suppresses expression of ribonucleotide reductase subunit M2, and synergistically enhances gemcitabine-induced cytotoxicity in mantle cell lymphoma. Blood 118(15):4140–4149

Chapter 17
Involvement of p53 in the Repair of DNA Double Strand Breaks: Multifaceted Roles of p53 in Homologous Recombination Repair (HRR) and Non-Homologous End Joining (NHEJ)

Vijay Menon and Lawrence Povirk

Abstract p53 is a tumor suppressor protein that prevents oncogenic transformation and maintains genomic stability by blocking proliferation of cells harboring unrepaired or misrepaired DNA. A wide range of genotoxic stresses such as DNA damaging anti-cancer drugs and ionizing radiation promote nuclear accumulation of p53 and trigger its ability to activate or repress a number of downstream target genes involved in various signaling pathways. This cascade leads to the activation of multiple cell cycle checkpoints and subsequent cell cycle arrest, allowing the cells to either repair the DNA or undergo apoptosis, depending on the intensity of DNA damage. In addition, p53 has many transcription-independent functions, including modulatory roles in DNA repair and recombination. This chapter will focus on the role of p53 in regulating or influencing the repair of DNA double-strand breaks that mainly includes homologous recombination repair (HRR) and non-homologous end joining (NHEJ). Through this discussion, we will try to establish that p53 acts as an important linchpin between upstream DNA damage signaling cues and downstream cellular events that include repair, recombination, and apoptosis.

Keywords DNA damage response • ATM kinase • ATR kinase • Nonhomologous end joining • Homologous recombination repair • Rad51

V. Menon
Goodwin Research Laboratories, Massey Cancer Center,
Virginia Commonwealth University, 401 College Street,
Room No. 380A, Richmond, VA 23298-0035, USA

L. Povirk (✉)
Goodwin Research Laboratories, Massey Cancer Center,
Virginia Commonwealth University, 401 College Street,
Room No. 380A, Richmond, VA 23298-0035, USA

Department of Pharmacology and Toxicology, Virginia Commonwealth University,
401 College Street, Goodwin Research Laboratories, Room No. 380A,
Richmond, VA 23298-0035, USA
e-mail: lpovirk@vcu.edu

S.P. Deb and S. Deb (eds.), *Mutant p53 and MDM2 in Cancer*,
Subcellular Biochemistry 85, DOI 10.1007/978-94-017-9211-0_17,

Introduction

A number of mutagenic and cytotoxic agents pose a major threat to genomic integrity and cellular homeostasis. Different cellular events that occur endogenously such as the reaction of DNA with oxygen and water lead to the formation of a myriad of DNA lesions, primarily involving chemical modifications of bases including oxidation (e.g. 8-oxoguanine) and hydrolysis (e.g. uracil). In addition, replication errors lead to base mismatches, while exogenous agents like ultraviolet light, industrial chemicals, or ionizing radiation produce diverse bulky adducts, alkylated bases and oxidized bases, all of which are potentially cytotoxic and mutagenic.

The tumor suppressor, p53, is a sequence-specific transcription factor, involved in the activation of many signaling molecules. Mutations in the p53 gene are found in ~50 % of tumors, providing a survival advantage to those cells. During normal conditions, the level of wild-type p53 in cells is kept under check by the E3 ubiquitin ligase, HDM2 (human homolog of the mouse double-minute 2 protein), which blocks p53's interaction with other co-activators. Eventually, it ubiquitinates p53 and targets it for proteasomal degradation [21]. As an alternative mechanism, loss of wild-type p53 function has been shown to be due to the dominant-negative effect of mutant p53 [9]. The role of p53 has been viewed as a double-edged sword depending on the severity of damage. Early in the DNA damage response, p53 relays a wide range of pro-survival signals like cell cycle arrest allowing the cells to repair the damage. But if damage continues to accumulate, p53 is seen to shift gears and promote apoptosis or senescence. Over the past decade, p53 has also been shown to positively or negatively regulate autophagy, which is deemed to be another mode of programmed cell death.

The stabilization of p53 is caused by various cellular stresses such as irradiation, exposure to genotoxic chemicals, oncogenic activation, hypoxia, nutrient deprivation, etc. Since most of these processes damage DNA, there are various DNA repair mechanisms to correct the damage incurred and p53 has been shown to play an important role in several of these repair mechanisms including nucleotide excision repair (bulky DNA adducts) [3, 49], base excision repair (base modifications) [56, 100], mismatch repair (base mismatch due to replication errors) [15, 41], homologous recombination repair and non-homologous end joining (DNA double strand breaks). However, this chapter will focus on the role of p53 in the response to DSBs, and its involvement in the repair of DSBs. In this context, p53 primarily plays an "integrator-relayer" function, integrating upstream signaling events and relaying them downstream to activate various cellular events like apoptosis, senescence, or differentiation. Although this role is accomplished primarily by p53-mediated transcriptional activation/repression, p53 has transcription-independent functions in many pathways, including HRR.

DSBs and interstrand crosslinks (ICLs) constitute the most toxic DNA lesions because they involve both the DNA strands. DSBs are mainly induced by ionizing radiation (X-rays and gamma rays) [12, 34, 63] and radiomimetic drugs (bleomycin and neocarzinostatin) [58]. The defective processing of DNA DSBs result in chromosomal translocations, deletions, insertions etc., [18, 74] leading to genomic instability and subsequently malignancy.

In order to circumvent these effects, cells have evolved two major pathways for repairing DNA DSBs; homologous recombination repair (HRR) and non-homologous end joining (NHEJ). In addition to these, a third mechanism, single-strand annealing (SSA), utilizes components from both HRR and NHEJ [8]. The major difference between HRR and NHEJ lies in the fact that HRR is an error-free pathway that plays a pivotal role during meiosis and during S and G2 phases of the cell cycle when sister chromatids are available. On the other hand, NHEJ is an error-prone pathway that occurs throughout the cell cycle and is shown to be important in mitotic cells. Also, it has been shown that in spite of HRR being an error-free pathway of DNA DSB repair, NHEJ competes effectively for the DSBs even when HRR is available. Moreover, it appears that the initial recruitment of repair factors is crucial in selecting one pathway over the other.

Structure of p53

Structural studies of p53 show that it is a tetramer in its active form, containing four identical chains. The N-terminal region of p53 is a disordered, natively unfolded region [7] containing the acidic TAD (transactivation domain), which is a binding site for some of the p53 interacting proteins such as the transcriptional machinery proteins [44], and MDM2 [35]. This is followed by the proline-rich region (PRR) that links the TAD to the DNA binding domain (DBD) in p53 [89]. The central DNA binding domain of p53 is one of the crucial components of p53 wherein most of the cancer-related mutations occur (shown in Fig. 17.1). Most of these mutations are missense mutations [31] wherein not only the wild-type function is lost, but

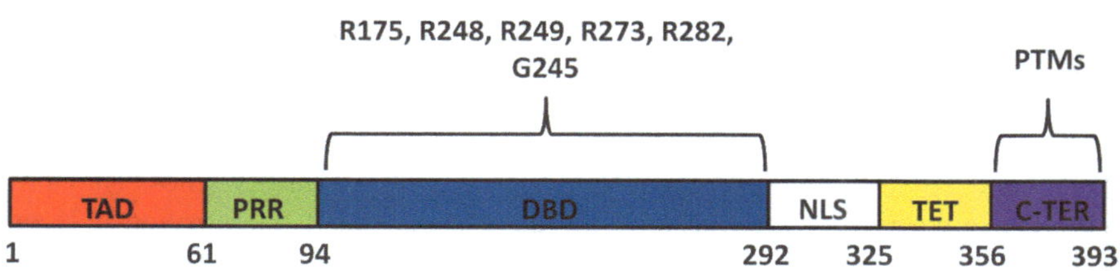

Fig. 17.1 Domain structure of p53. The N-terminal region consists of a transactivation domain followed by a proline-rich region, the central DNA binding domain, the tetramerization domain, and the C-terminal region. The hotspot mutations in the DNA binding domains are indicated. *PTMs* Post-translational modifications (phosphorylation, neddylation, sumoylation, acetylation, methylation etc.)

novel oncogenic functions are acquired along with a dominant-negative phenotype that can inactivate the normal protein functions through heterotetramerization [51], which is elicited via the tetramerization domain that follows the DNA binding domain. The C-terminal domain is also a disordered domain, but undergoes structured transitions on interaction with other proteins [66, 92]. This domain is where most of the post-translational modifications of p53 occur, such as phosphorylation, acetylation, sumoylation, neddylation, etc., which eventually regulate p53 levels and function. In addition to these domains, there is a bipartite nuclear localization signal located between the DBD and TET domain that is required for the nuclear import of p53 [39]. There is also a NES (nuclear export sequence) in the C-terminal region that allows the nuclear export of p53 into the cytoplasm [57].

Activation of p53

In response to various cellular stresses that involve cellular DNA damage, p53 is activated and stabilized. It has been shown to be highly sensitive to small gaps and breaks in DNA that could ultimately lead to an early DNA damage response. However, the primary event in the p53 activation cascade is the disruption of its interaction with its negative regulators, MDM2 or MDM4. One of the mechanisms involved in disrupting this interaction is the activation of p53 by the ataxia-telengiectasia mutated (ATM) and ataxia-telengiectasia and Rad3-related (ATR) protein kinases, which are the major DNA damage sensors mediating the rapid degradation of MDM2 and MDM4 [81]. Both these kinases belong to the phosphatidylinositol-3-kinase-like kinase family and are involved in initiating a myriad of cellular signaling events following different forms of DNA damage.

The activation of p53 is also induced by the tumor suppressor, ARF. The INK/ARF locus produces two proteins, $p16^{Ink4A}$ and $p19^{ARF}$ [73], that are mainly involved in regulating the tumor suppressor functions of the retinoblastoma protein and p53. It has been shown that during oncogenic stimulation or replicative senescence, ARF binds to MDM2 and sequesters it in the nucleolus, thus stabilizing p53 levels [91].

A third mode of p53 activation is via its interaction with CBP (CREB binding protein) and p300, which are important regulators of eukaryotic transcription [27]. They act as co-activators wherein they bind to sequence-specific transcription factors and assist in the initiation of transcription from the target genes. They promote the interaction of the transcription factors with pol II holoenzyme [27] and acetylate the neighboring histones to allow an open configuration of the chromatin [75]. It was shown that there are two docking sites for p53 on p300/CBP; one in the C/H3 and one in the C-terminal domain [4, 29]. Following γ-irradiation, the interaction of p300/CBP with p53 increases due to the phosphorylation of p53 on S15 by the ATM kinase [37]. Furthermore, acetylation of p53 on the C-terminus by p300/CBP and P/CAF histone acetyltransferases (HAT) results in DNA binding activity and transactivation functions of p53 [33].

It is also important to mention that the other members of the p53 family which include p63 and p73 also play an important role in activating p53. Earlier reports using p63- and p73-null MEFs showed that both p63 and p73 are essential for p53-induced apoptosis following DNA damage [19].

DNA Damage Response and p53

ATM is primarily involved in detecting and perhaps binding to DSBs, leading ultimately to the activation of cell cycle checkpoints and modulation of DNA repair pathways. On the other hand, ATR mainly recognizes replication defects or disruption of replication by DNA lesions such as DNA-DNA/DNA-protein crosslinks arising either endogenously (e.g., malondialdehyde from lipid peroxidation) or from exposure to bifunctional DNA-damaging agents such as cisplatin and mitomycin C. During the DNA damage response, ATM and ATR induce a wide variety of post-translational modifications in p53 that promote its activation and stabilization. For instance, it is seen that during DNA damage, ATM phosphorylates the checkpoint kinase Chk2 [46], which in turn phosphorylates serine 20 of p53 [13, 30]. This residue is within the major site for MDM2 attachment [55] and as a result of its phosphorylation, p53-MDM2 interaction is disrupted leading to p53 stabilization. This cascade culminates in the G1 phase checkpoint wherein activated p53 induces the expression of its direct transcriptional target, p21, causing cell cycle arrest. A p53-independent activity of p21 in inducing both G1 and G2 phase cell cycle arrest has also been demonstrated in earlier work. MDM2, like p53, is also subjected to post-translational modifications following DNA damage. The p53 binding domain and the RING finger domains of MDM2 are the hotspots for these modifications. It was shown that following DNA damage, the serine 395 (S395) residue on MDM2 is phosphorylated by ATM both *in vivo* and *in vitro* [47]. This was corroborated recently by the demonstration that ATM-induced MDM2 phosphorylation at S395 increased the interaction of MDM2 with p53 mRNA, leading to increased p53 translation [23]. Furthermore, Gannon et al. [24] showed that ATM-mediated phosphorylation of S394 on MDM2 is important for the increase in p53 activity and subsequent activation of downstream p53 targets. These findings show that ATM plays a key role in regulating the DNA damage response by modifying both p53 and MDM2 in a way that allows activation and stabilization of p53.

In addition to the above regulatory role of ATM via Chk2 and MDM2, ATM directly phosphorylates serine 15 on p53 following exposure to ionizing radiation [36]. This effect was only partially suppressed in AT cells, suggesting that other kinases can also phosphorylate S15 on p53 [5, 77]. The role of ATR in phosphorylating S15 on p53 was shown in γ-irradiated fibroblasts that were transfected with a vector expressing a catalytically inactive mutant of ATR, designated ATR^{ki}. Overexpression of ATR^{ki} abrogated UV-induced p53 S15 phosphorylation [86]. Herein, it was shown that ATR phosphorylates p53 on S37 (also a phosphorylation site for DNA-PK) *in vitro*. Thus, although ATM and ATR are structurally similar, they have both common and unique substrates.

Mutant p53: Gain-of-Function and the DNA Damage Response

The wild-type activity of p53 plays a crucial role in the induction of multiple signaling processes such as cell cycle arrest, apoptosis, senescence, etc. However, in ~50 % of tumors, p53 is mutated [32, 88]. Numerous reports have shown that in addition to losing the tumor suppressive activity, certain p53 mutants acquire a "gain-of-function" (GOF) phenotype leading to oncogenesis and drug resistance. For example, mutant p53 (D281G) activates the expression of MDR1 gene (normally suppressed by wild-type p53) which encodes P-glycoprotein, an energy-dependent drug efflux pump [69, 85].

One of the key mechanisms of mutant p53 GOF is the ability of certain mutant forms of p53 to induce expression of genes that are not induced by wild-type p53. In addition to the MDR1 gene, other genes activated by mutant p53 include VEGFR and EGFR, whose products induce angiogenesis and cellular proliferation, respectively [60]. Mutant p53 also interacts with other proteins, modifying their functions so as to provide a survival advantage to the cells. For example, mutant p53 disrupts the tumor suppressive functions of both p63 and p73 [16, 22, 38] and increases DNA non-homologous recombination by increasing topoisomerase-I activity in cells [2].

However, a more dramatic effect of mutant p53 is seen in the DNA damage response. Earlier work showed an interaction of the R248W and R273H mutant forms of p53 with Mre11, which is a part of the MRN complex (Mre11/Rad50/NBS1) that is involved in the initial sensing of DNA DSBs and subsequent recruitment of ATM [94]. This interaction disrupts the ability of Mre11 to recruit ATM to DSBs, ultimately leading to inter-chromosomal translocations as a result of defective ATM-dependent cell cycle checkpoints [80]. While it is known that ATM-deficient cells are radiosensitive [76, 95], cells harboring mutant p53 usually do not exhibit radiosensitivity. This could be due to the presence of a partially functional ATM in the p53 mutant cells that could still confer some radioresistance despite the lack of interaction with Mre11. Alternatively, mutant p53 could inhibit the function of p73 which is a mediator of apoptosis in the absence of wild-type p53. Also, since mutant p53 has the ability of deregulating the expression of genes involved in cell survival, it could prevent radiosensitivity in ATM null cells [6].

As mentioned, p53 has been shown to play a role in myriad of cellular activities from DNA damage response to gene transcription to differentiation. The following sections will focus on its role specifically in the DSB repair pathways.

Homologous Recombination Repair and p53

Homologous recombination repair (HRR) is an error-free DSB repair pathway, most active during late S and G2 phases of the cell cycle [83], and is primarily governed by the presence of homologous sister chromatids, homologous chromosomes, or

DNA repeats. It is evolutionarily conserved and plays an important role in maintaining genomic integrity. HRR is believed to initiate with ATM sensing and localizing to DSBs, where it phosphorylates H2AX in the surrounding chromatin, which in turn recruits BRCA1 and NBS1, repair proteins that are also phosphorylated by ATM. The DNA is resected in a process that requires the MRN complex [87] and gives rise to single-strand DNA overhangs bound by RPA, which is subsequently replaced by RAD51. Since the MRN complex possesses only $3' \rightarrow 5'$ but not $5' \rightarrow 3'$ exonuclease activity, other nucleases such as CtIP [70] and Exo1 [11] may also be involved in the resection step. The polymerization of RAD51 on the overhangs takes place with the agency of RAD52. RAD51 then searches for DNA homology with the help of RAD54 which binds to RAD51. The ATPase activity of RAD54 helps unwind DNA and facilitates strand invasion [48]. RAD51 then forms a heteroduplex following the acquisition of a homologous duplex. This step is then followed by heteroduplex extension and branch migration. After this, either a non-crossing over event takes place wherein Holliday junctions are disengaged and DNA strands pair followed by gap filling or a crossing-over event takes place resulting from the resolution of Holliday junctions followed by gap filling.

Cells subjected to oxidative stress or to anti-cancer agents undergo a DNA damage response which is characterized by the activation and stabilization of p53, leading to cell cycle arrest in G1 phase, which requires the transactivation ability of p53. However, a plethora of studies have shown that some of the genotoxic agents including the replication elongation inhibitors hydroxyurea and aphidicolin, cause p53 accumulation independent of its role in G1 checkpoint [28, 67]. Likewise, some mutations in p53 (143, 175, 248, 273, and 281) do not affect the G1 phase cell cycle arrest but stimulate both spontaneous and radiation-induced recombination [1, 10, 50, 68].

Although p53 can by itself check the fidelity of homologous recombination by mismatch recognition of heteroduplex intermediates [17, 43], the inhibition of recombination by p53 is primarily mediated by its interaction with RAD51 (Fig. 17.2) and also RAD54. Inhibition of p53 activity promotes spontaneous and radiation-induced homologous recombination between both direct and inverted repeats [68] wherein the latter mainly involves a RAD51-dependent gene conversion process. This interaction of p53 with RAD51 and subsequent inhibition of recombination was further confirmed by overexpressing mutant ^{L186P}RAD51 that prevented p53 binding to RAD51 [43]. A 2–3 fold increase in homologous recombination was seen following the overexpression of this p53 non-binding mutant of RAD51. The interaction of p53 with RAD54 mainly occurs via the extreme C-terminal domain of p53 [43] which is involved in sensing mispaired homologous recombination intermediates. *In vitro* experiments have shown that RAD51 stimulates the $3' \rightarrow 5'$ exonuclease activity of p53 that targets heteroduplexes containing base mismatches [82]. Also, the p53-RAD51 complex inhibits branch migration after the crossing-over or postsynaptic phase of recombination [96]. Further, the regulation of homologous recombination by p53 was found to be biased, with p53 depletion promoting both intra- and extrachromosomal recombination but not homologous DNA integration or gene targeting [97]. This is interesting because

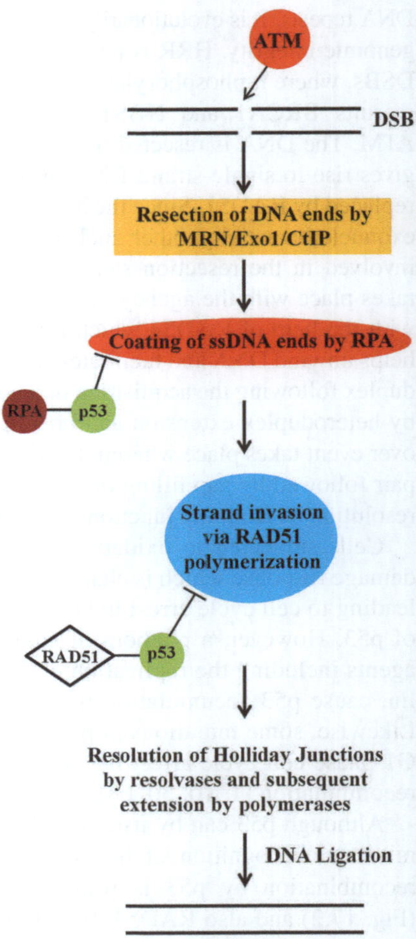

Fig. 17.2 Inhibitory role of p53 in homologous recombination repair

gene targeting or gene disruption could be considered a genome-destabilizing process. Although most of these mechanistic studies involved measurements of spontaneous recombination between ostensibly undamaged loci, other work [1, 10] demonstrates that recombination between a locus containing a site-specific I-SceI-induced DSB and an undamaged homologous sequence, presumably reflecting HRR, is likewise suppressed by both wild-type and transactivation-defective p53.

While all the above studies are in general agreement that HRR suppression by p53 is largely independent of its transcriptional activation ability, this view has recently been challenged [59]. In this study, a wide variety of p53 manipulations were carried out, and in all cases (even those involving transactivation-deficient p53 mutants) the extent of HRR for DSBs induced by the meganuclease *I*-SceI was found to closely correlate with the fraction of cells in S/G2. Thus, these authors argue that most of the reported effects of p53 on HRR can be explained by transactivation-dependent cell cycle perturbations.

During DNA replication, p53 is found at replication sites [93] and also is transported into the nucleus during S-phase [45]. During replicative stress (resulting from treatment with replication inhibitors or DNA crosslinking agents), p53 inhibits homologous recombination and this effect is dependent on the S15 phosphorylation of p53 and the interaction of p53 with the ssDNA binding protein RPA [61]. As discussed earlier, the phosphorylation of p53 on S15 is mediated by both the ATM and ATR kinases. Inhibition of ATM with caffeine or the ATM specific inhibitor KU-55933 still allowed the formation of RAD51 foci, indicative of an active recombination in the absence of ATM. Subsequently, following ATR inhibition using RNA interference, the fraction of RAD51 foci-positive cells was reduced in the presence of either wild-type p53 or the transactivation mutant, p53QS [79]. A recent study [72] has shown that the trio: DNA-PK, ATM, and ATR together downregulate p53-RPA binding. DNA-PK phosphorylates RPA at S46, and ATM/ATR phosphorylates p53 at S37 (preceded by S15 phosphorylation as discussed above) causing the release of p53 and RPA from the p53-RPA complex and thereby allowing RPA to fulfill its normal role in facilitating HRR. DNA-PK is also involved in the other major DNA DSB repair pathway, the non-homologous end joining (NHEJ, discussed in the next section).

Also, an important potential role of p53 in HRR that is not discussed here in detail is its interaction with BLM (Bloom syndrome) and WS (Werner syndrome) proteins that belong to the class of RecQ helicases involved in homologous recombination [78]. It was shown that BLM and p53 show co-localization at sites of stalled replication forks [71]. Although BLM localizes to these sites in a p53-independent manner, it eventually enhances the p53 accumulation at these sites. Sengupta et al. [71] showed that the interaction of BLM with p53 is enhanced by the localization of 53BP1 at these sites, independent of γ-H2AX. This event was dependent on active Chk1 kinase (possibly phosphorylated by ATR), leading to BLM stabilization and ultimately p53 accumulation at stalled replication sites.

To summarize, p53 plays multiple roles in the regulation of the HRR pathway and the mechanism of some of these regulatory roles is still being elucidated.

Non-homologous End Joining and p53

The NHEJ pathway is the predominant pathway for the repair of DNA DSBs, occurring throughout the cell cycle [65]. It is a dynamic process that does not require sequence homology as in HR [64] and has been shown to utilize a wide variety of DNA substrates converting them into joined products. Since this process can lead to the joining of incorrect ends, NHEJ is an error-prone pathway. Radiation-induced DNA damage via the production of oxygen radicals leads to deoxyribose fragmentation, as well as producing modified DNA bases such as 8-oxoguanine and thymine glycols. DNA DSBs induced by deoxyribose oxidation are characterized by both 5′- and 3′-staggered termini with chemically modified ends. Moreover, clusters of localized radicals can produce complex DNA lesions consisting of terminally blocked DSBs flanked by nearby damaged bases. DSBs of widely diverse terminal structure can be repaired via NHEJ irrespective of the sequence or DNA homology.

The first step in NHEJ is the binding of the Ku heterodimers to DSBs [52]. This is followed by DNA-PK binding to Ku and this complex serves as a beacon for nucleases, polymerases, and ligases to bind. However, the formation of a stable complex between Ku and DNA-PK requires conformational changes in Ku which occur only in the presence of DNA ends [40], and promote the interaction of Ku with DNA polymerases μ and λ, and with the XRCC4-DNA ligase IV complex [14]. Interaction between two such complexes on two DNA ends tethers the ends and triggers the kinase activity of DNA-PK which then phosphorylates various repair proteins. Artemis, along with DNA-PK, gains a 3′-endonucleolytic activity that has been shown earlier to act specifically near 3′ DNA termini and resolve noncanonical DNA DSB ends such as 3′-phosphoglycolate moieties. This endonuclease activity is essential for Artemis' role in promoting radioresistance and repair proficiency in mammalian cells [53]. In addition to Artemis and DNA-PK, other proteins such as TDP1 and Metnase have been shown to function in a similar fashion to resolve damaged DNA overhangs. TDP1 removes glycolate residues from 3′ ends followed by additional processing by PNKP to remove the resulting 3′-phosphates [99]. Recently, Metnase was shown to endonucleolytically trim 3′-overhangs greater than 3 bp on DNA duplexes, although its lack of activity when added to NHEJ-competent extracts casts some doubt on such a role *in vivo* [54]. The various enzymes involved in the NHEJ pathway have a high flexibility in binding to DNA lesions, allowing them to interact or bind with a plethora of DNA end structures.

The role of p53 in NHEJ is not clearly understood although p53 has been shown to regulate NHEJ by itself or in association with other NHEJ proteins. Early studies with mice harboring knockouts of the NHEJ factors XRCC4 and Ligase IV have shown that in the absence of NHEJ, DNA DSBs remain unrepaired, and that these eventually trigger apoptosis in a p53-dependent manner [20, 26, 25]. Thus, p53 deficiency can rescue the otherwise embryonic lethal phenotype of Xrcc4–/– or Lig4–/– mice.

In an episomal reactivation assay, p53 was shown to enhance DSB rejoining of transfected linearized plasmids in γ-irradiated cells [84]. This enhancement was found to be dependent on the carboxy terminal domain of p53 (which harbors nonspecific DNA-binding activity), but independent of transcriptional activation ability. Interestingly, only the repair of DNA DSBs with short cohesive ends but not blunt ends was enhanced by p53. In cells harboring an integrated DSB substrate containing tandem sites for meganuclease I-SceI, the p53 inhibitor pifithrin-α reduced precise end-joining while having little or no effect on end joining overall, suggesting a role for p53 in enforcing end-joining fidelity [42]. In a different I-SceI-based NHEJ assay, expression of wild-type p53 inhibited NHEJ events that required trimming of noncomplementary overhangs from the DSB ends. In this case, it was speculated that p53 plays a role in NHEJ either by inhibiting exonucleolytic proofreading or by recognizing heterologies and inhibiting NHEJ [1].

The Artermis endonuclease involved trimming of damaged or noncomplementary ends for NHEJ has been shown to interact with p53 in the suppression of oncogenic *N-Myc* in progenitor B-cells [62]. Other reports have also shown Artemis to be a negative regulator of p53 activity in response to oxidative stress. Artemis knockdown in U2-OS cells induced p53 accumulation, cell cycle arrest, and apoptosis [98]. This reflects on a DNA repair-independent role of Artemis and its subsequent effect on p53 activity.

To summarize, the exact regulatory role of p53 in NHEJ is still poorly understood. Certainly, p53 does have genetic interactions with the players of the NHEJ pathway, eliciting DNA repair-dependent and -independent downstream effects on cell cycle progression and cell survival. However, early suggestions of more direct biochemical effects of p53 on NHEJ itself have been neither refuted nor further elucidated.

Conclusions

Overall, p53 acts as an important link between upstream signaling and activation of downstream signaling cascades depending on the extent of DNA damage. It can activate cell cycle arrest and allow the damage to be repaired or it could transactivate genes involved in the apoptotic machinery.

The role of p53 in regulating cell cycle checkpoints following DNA damage is a pivotal event toward maintaining genomic stability. DNA damage leads to the activation of cell cycle checkpoints in different phases of the cell cycle. The G1/S phase checkpoint that is mainly triggered by DNA DSBs (detected by the presence of γ-H2AX or 53BP1, [90]) involves the ATM kinase that detects DSBs and phosphorylates p53 either directly or indirectly leading to its stabilization. As the repair of some DSBs by NHEJ in G1 requires several hours, this arrest provides a crucial opportunity for the cell to restore the integrity of the genome before it can be replicated. The intra-S-phase checkpoint is mainly activated by stalled replication forks arising as a result of replication defects or DNA damaging agents. It likewise allows for repair of replication-associated DSBs by HRR before new replication forks are initiated. Thus, in either case, cell cycle arrest will serve to enhance genomic stability, and is the primary function of p53 in DSB repair.

More direct regulatory effects of p53 on DSB repair are more difficult to rationalize in terms of genomic stability. It might be expected that the role of p53 in regulating these pathways could be bimodal, either inhibiting these pathways to maintain genomic stability or activating them in response to genotoxic stress or DNA damage cues. While DSBs must be rejoined if genomic integrity is to be preserved, inaccurate repair by the same or very similar mechanisms will lead to rearrangements and instability. It is ostensibly surprising that the primary direct effect of p53 on DSB repair appears to be suppression of HRR, which is usually accurate and certainly more accurate than NHEJ or other alternative DSB repair pathways. Ideally, a genomic surveillance system would evolve so as to specifically detect and suppress events that are likely to be associated with inaccurate repair. There is some suggestion of such a bias in the finding that p53 most strongly inhibits HRR between substrates with a limited extent of homology (<200 bp) – events that could reflect illegitimate recombination between repetitive sequences that could lead to rearrangements [1]. Similarly, p53 reportedly suppresses NHEJ that requires resection of mismatched overhangs (such as might occur in the joining of exchanged ends of two DSBs), while promoting cohesive-end joins [1, 84]. Finally, in cases of extensive damage, the proliferation of cells with unstable genomes might be most efficiently prevented by blocking repair entirely and thereby driving those damaged cells toward apoptosis or senescence.

Acknowledgment The preparation of this review was supported in part by grants CA40615 and CA166264 from the National Cancer Institute, US DHHS.

References

1. Akyuz N, Boehden GS, Susse S, Rimek A, Preuss U, Scheidtmann KH, Wiesmuller L (2002) DNA substrate dependence of p53-mediated regulation of double-strand break repair. Mol Cell Biol 22(17):6306–6317
2. Albor A, Kaku S, Kulesz-Martin M (1998) Wild-type and mutant forms of p53 activate human topoisomerase I: a possible mechanism for gain of function in mutants. Cancer Res 58(10):2091–2094
3. Amundson SA, Patterson A, Do KT, Fornace AJ Jr (2002) A nucleotide excision repair master-switch: p53 regulated coordinate induction of global genomic repair genes. Cancer Biol Ther 1(2):145–149
4. Avantaggiati ML, Ogryzko V, Gardner K, Giordano A, Levine AS, Kelly K (1997) Recruitment of p300/CBP in p53-dependent signal pathways. Cell 89(7):1175–1184
5. Banin S, Moyal L, Shieh S, Taya Y, Anderson CW, Chessa L, Smorodinsky NI, Prives C, Reiss Y, Shiloh Y, Ziv Y (1998) Enhanced phosphorylation of p53 by ATM in response to DNA damage. Science 281(5383):1674–1677
6. Barlow C, Hirotsune S, Paylor R, Liyanage M, Eckhaus M, Collins F, Shiloh Y, Crawley JN, Ried T, Tagle D, Wynshaw-Boris A (1996) Atm-deficient mice: a paradigm of ataxia telangiectasia. Cell 86(1):159–171
7. Bell S, Klein C, Muller L, Hansen S, Buchner J (2002) P53 contains large unstructured regions in its native state. J Mol Biol 322(5):917–927
8. Bennardo N, Cheng A, Huang N, Stark JM (2008) Alternative-NHEJ is a mechanistically distinct pathway of mammalian chromosome break repair. PLoS Genet 4(6):e1000110
9. Blagosklonny MV (2000) P53 from complexity to simplicity: mutant P53 stabilization, gain-of-function, and dominant-negative effect. FASEB J 14(13):1901–1907
10. Boehden GS, Akyuz N, Roemer K, Wiesmuller L (2003) P53 Mutated in the transactivation domain retains regulatory functions in homology-directed double-strand break repair. Oncogene 22(26):4111–4117
11. Bolderson E, Tomimatsu N, Richard DJ, Boucher D, Kumar R, Pandita TK, Burma S, Khanna KK (2010) Phosphorylation of Exo1 modulates homologous recombination repair of DNA double-strand breaks. Nucleic Acids Res 38(6):1821–1831
12. Bradley MO, Kohn KW (1979) X-ray induced DNA double strand break production and repair in mammalian cells as measured by neutral filter elution. Nucleic Acids Res 7(3):793–804
13. Chehab NH, Malikzay A, Stavridi ES, Halazonetis TD (1999) Phosphorylation of Ser-20 mediates stabilization of human p53 in response to DNA damage. Proc Natl Acad Sci U S A 96(24):13777–13782
14. Chen L, Trujillo K, Sung P, Tomkinson AE (2000) Interactions of the DNA ligase IV-XRCC4 complex with DNA ends and the DNA-dependent protein kinase. J Biol Chem 275(34):26196–26205

15. Degtyareva N, Subramanian D, Griffith JD (2001) Analysis of the binding of p53 to DNAs containing mismatched and bulged bases. J Biol Chem 276(12):8778–8784
16. Di Como CJ, Gaiddon C, Prives C (1999) P73 function is inhibited by tumor-derived P53 mutants in Mammalian cells. Mol Cell Biol 19(2):1438–1449
17. Dudenhoffer C, Rohaly G, Will K, Deppert W, Wiesmuller L (1998) Specific mismatch recognition in heteroduplex intermediates by p53 suggests a role in fidelity control of homologous recombination. Mol Cell Biol 18(9):5332–5342
18. Ferguson DO, Alt FW (2001) DNA double strand break repair and chromosomal translocation: lessons from animal models. Oncogene 20(40):5572–5579
19. Flores ER, Tsai KY, Crowley D, Sengupta S, Yang A, McKeon F, Jacks T (2002) p63 and p73 are required for p53-dependent apoptosis in response to DNA damage. Nature 416(6880):560–564
20. Frank KM, Sharpless NE, Gao Y, Sekiguchi JM, Ferguson DO, Zhu C, Manis JP, Horner J, DePinho RA, Alt FW (2000) DNA ligase IV deficiency in mice leads to defective neurogenesis and embryonic lethality via the p53 pathway. Mol Cell 5(6):993–1002
21. Freedman DA, Wu L, Levine AJ (1999) Functions of the MDM2 oncoprotein. Cell Mol Life Sci 55(1):96–107
22. Gaiddon C, Lokshin M, Ahn J, Zhang T, Prives C (2001) A subset of tumor-derived mutant forms of p53 down-regulate p63 and p73 through a direct interaction with the p53 core domain. Mol Cell Biol 21(5):1874–1887
23. Gajjar M, Candeias MM, Malbert-Colas L, Mazars A, Fujita J, Olivares-Illana V, Fahraeus R (2012) The p53 mRNA-Mdm2 interaction controls Mdm2 nuclear trafficking and is required for p53 activation following DNA damage. Cancer Cell 21(1):25–35
24. Gannon HS, Woda BA, Jones SN (2012) ATM phosphorylation of Mdm2 Ser394 regulates the amplitude and duration of the DNA damage response in mice. Cancer Cell 21(5):668–679
25. Gao Y, Ferguson DO, Xie W, Manis JP, Sekiguchi J, Frank KM, Chaudhuri J, Horner J, DePinho RA, Alt FW (2000) Interplay of p53 and DNA-repair protein XRCC4 in tumorigenesis, genomic stability and development. Nature 404(6780):897–900
26. Gao Y, Sun Y, Frank KM, Dikkes P, Fujiwara Y, Seidl KJ, Sekiguchi JM, Rathbun GA, Swat W, Wang J, Bronson RT, Malynn BA, Bryans M, Zhu C, Chaudhuri J, Davidson L, Ferrini R, Stamato T, Orkin SH, Greenberg ME, Alt FW (1998) A critical role for DNA end-joining proteins in both lymphogenesis and neurogenesis. Cell 95(7):891–902
27. Goodman RH, Smolik S (2000) CBP/p300 in cell growth, transformation, and development. Genes Dev 14(13):1553–1577
28. Gottifredi V, Shieh S, Taya Y, Prives C (2001) p53 accumulates but is functionally impaired when DNA synthesis is blocked. Proc Natl Acad Sci U S A 98(3):1036–1041
29. Gu W, Shi XL, Roeder RG (1997) Synergistic activation of transcription by CBP and p53. Nature 387(6635):819–823
30. Hirao A, Kong YY, Matsuoka S, Wakeham A, Ruland J, Yoshida H, Liu D, Elledge SJ, Mak TW (2000) DNA damage-induced activation of p53 by the checkpoint kinase Chk2. Science 287(5459):1824–1827
31. Hollstein M, Rice K, Greenblatt MS, Soussi T, Fuchs R, Sorlie T, Hovig E, Smith-Sorensen B, Montesano R, Harris CC (1994) Database of p53 gene somatic mutations in human tumors and cell lines. Nucleic Acids Res 22(17):3551–3555
32. Hollstein M, Sidransky D, Vogelstein B, Harris CC (1991) P53 mutations in human cancers. Science 253(5015):49–53
33. Hupp TR, Meek DW, Midgley CA, Lane DP (1992) Regulation of the specific DNA binding function of p53. Cell 71(5):875–886
34. Kraxenberger F, Weber KJ, Friedl AA, Eckardt-Schupp F, Flentje M, Quicken P, Kellerer AM (1998) DNA double-strand breaks in mammalian cells exposed to gamma-rays and very heavy ions. Fragment-size distributions determined by pulsed-field gel electrophoresis. Radiat Environ Biophys 37(2):107–115
35. Kussie PH, Gorina S, Marechal V, Elenbaas B, Moreau J, Levine AJ, Pavletich NP (1996) Structure of the MDM2 oncoprotein bound to the p53 tumor suppressor transactivation domain. Science 274(5289):948–953

36. Lakin ND, Jackson SP (1999) Regulation of p53 in response to DNA damage. Oncogene 18(53):7644–7655
37. Lambert PF, Kashanchi F, Radonovich MF, Shiekhattar R, Brady JN (1998) Phosphorylation of p53 serine 15 increases interaction with CBP. J Biol Chem 273(49):33048–33053
38. Li Y, Prives C (2007) Are interactions with p63 and p73 involved in mutant p53 gain of oncogenic function? Oncogene 26(15):2220–2225
39. Liang SH, Clarke MF (1999) A bipartite nuclear localization signal is required for p53 nuclear import regulated by a carboxyl-terminal domain. J Biol Chem 274(46):32699–32703
40. Lieber MR (2008) The mechanism of human nonhomologous DNA end joining. J Biol Chem 283(1):1–5
41. Lin X, Ramamurthi K, Mishima M, Kondo A, Howell SB (2000) p53 interacts with the DNA mismatch repair system to modulate the cytotoxicity and mutagenicity of hydrogen peroxide. Mol Pharmacol 58(6):1222–1229
42. Lin Y, Waldman BC, Waldman AS (2003) Suppression of high-fidelity double-strand break repair in mammalian chromosomes by pifithrin-alpha, a chemical inhibitor of p53. DNA Repair 2(1):1–11
43. Linke SP, Sengupta S, Khabie N, Jeffries BA, Buchhop S, Miska S, Henning W, Pedeux R, Wang XW, Hofseth LJ, Yang Q, Garfield SH, Sturzbecher HW, Harris CC (2003) p53 interacts with hRAD51 and hRAD54, and directly modulates homologous recombination. Cancer Res 63(10):2596–2605
44. Lu H, Levine AJ (1995) Human TAFII31 protein is a transcriptional coactivator of the p53 protein. Proc Natl Acad Sci U S A 92(11):5154–5158
45. Martinez J, Georgoff I, Martinez J, Levine AJ (1991) Cellular localization and cell cycle regulation by a temperature-sensitive p53 protein. Genes Dev 5(2):151–159
46. Matsuoka S, Rotman G, Ogawa A, Shiloh Y, Tamai K, Elledge SJ (2000) Ataxia telangiectasia-mutated phosphorylates Chk2 in vivo and in vitro. Proc Natl Acad Sci U S A 97(19): 10389–10394
47. Maya R, Balass M, Kim ST, Shkedy D, Leal JF, Shifman O, Moas M, Buschmann T, Ronai Z, Shiloh Y, Kastan MB, Katzir E, Oren M (2001) ATM-dependent phosphorylation of Mdm2 on serine 395: role in p53 activation by DNA damage. Genes Dev 15(9):1067–1077
48. Mazin AV, Alexeev AA, Kowalczykowski SC (2003) A novel function of Rad54 protein. Stabilization of the Rad51 nucleoprotein filament. J Biol Chem 278(16):14029–14036
49. McKay BC, Ljungman M, Rainbow AJ (1999) Potential roles for p53 in nucleotide excision repair. Carcinogenesis 20(8):1389–1396
50. Mekeel KL, Tang W, Kachnic LA, Luo CM, DeFrank JS, Powell SN (1997) Inactivation of p53 results in high rates of homologous recombination. Oncogene 14(15):1847–1857
51. Milner J, Medcalf EA (1991) Cotranslation of activated mutant p53 with wild type drives the wild-type p53 protein into the mutant conformation. Cell 65(5):765–774
52. Mimori T, Hardin JA (1986) Mechanism of interaction between Ku protein and DNA. J Biol Chem 261(22):10375–10379
53. Mohapatra S, Kawahara M, Khan IS, Yannone SM, Povirk LF (2011) Restoration of G1 chemo/radioresistance and double-strand-break repair proficiency by wild-type but not endonuclease-deficient Artemis. Nucleic Acids Res 39(15):6500–6510
54. Mohapatra S, Yannone SM, Lee SH, Hromas RA, Akopiants K, Menon V, Ramsden DA, Povirk LF (2013) Trimming of damaged 3′ overhangs of DNA double-strand breaks by the Metnase and Artemis endonucleases. DNA Repair 12(6):422–432
55. Moll UM, Petrenko O (2003) The MDM2-p53 interaction. Mol Cancer Res 1(14):1001–1008
56. Offer H, Wolkowicz R, Matas D, Blumenstein S, Livneh Z, Rotter V (1999) Direct involvement of p53 in the base excision repair pathway of the DNA repair machinery. FEBS Lett 450(3):197–204
57. O'Keefe K, Li H, Zhang Y (2003) Nucleocytoplasmic shuttling of p53 is essential for MDM2-mediated cytoplasmic degradation but not ubiquitination. Mol Cell Biol 23(18):6396–6405
58. Povirk LF (1996) DNA damage and mutagenesis by radiomimetic DNA-cleaving agents: bleomycin, neocarzinostatin and other enediynes. Mutat Res 355(1–2):71–89

59. Rieckmann T, Kriegs M, Nitsch L, Hoffer K, Rohaly G, Kocher S, Petersen C, Dikomey E, Dornreiter I, Dahm-Daphi J (2013) p53 modulates homologous recombination at I-SceI-induced double-strand breaks through cell-cycle regulation. Oncogene 32(8):968–975
60. Roemer K (1999) Mutant p53: gain-of-function oncoproteins and wild-type p53 inactivators. Biol Chem 380(7–8):879–887
61. Romanova LY, Willers H, Blagosklonny MV, Powell SN (2004) The interaction of p53 with replication protein A mediates suppression of homologous recombination. Oncogene 23(56):9025–9033
62. Rooney S, Sekiguchi J, Whitlow S, Eckersdorff M, Manis JP, Lee C, Ferguson DO, Alt FW (2004) Artemis and p53 cooperate to suppress oncogenic N-myc amplification in progenitor B cells. Proc Natl Acad Sci U S A 101(8):2410–2415
63. Roots R, Kraft G, Gosschalk E (1985) The formation of radiation-induced DNA breaks: the ratio of double-strand breaks to single-strand breaks. Int J Radiat Oncol Biol Phys 11(2):259–265
64. Roth DB, Porter TN, Wilson JH (1985) Mechanisms of nonhomologous recombination in mammalian cells. Mol Cell Biol 5(10):2599–2607
65. Rothkamm K, Kruger I, Thompson LH, Lobrich M (2003) Pathways of DNA double-strand break repair during the mammalian cell cycle. Mol Cell Biol 23(16):5706–5715
66. Rustandi RR, Baldisseri DM, Weber DJ (2000) Structure of the negative regulatory domain of p53 bound to S100B(betabeta). Nat Struct Biol 7(7):570–574
67. Saintigny Y, Lopez BS (2002) Homologous recombination induced by replication inhibition, is stimulated by expression of mutant p53. Oncogene 21(3):488–492
68. Saintigny Y, Rouillard D, Chaput B, Soussi T, Lopez BS (1999) Mutant p53 proteins stimulate spontaneous and radiation-induced intrachromosomal homologous recombination independently of the alteration of the transactivation activity and of the G1 checkpoint. Oncogene 18(24):3553–3563
69. Sampath J, Sun D, Kidd VJ, Grenet J, Gandhi A, Shapiro LH, Wang Q, Zambetti GP, Schuetz JD (2001) Mutant p53 cooperates with ETS and selectively up-regulates human MDR1 not MRP1. J Biol Chem 276(42):39359–39367
70. Sartori AA, Lukas C, Coates J, Mistrik M, Fu S, Bartek J, Baer R, Lukas J, Jackson SP (2007) Human CtIP promotes DNA end resection. Nature 450(7169):509–514
71. Sengupta S, Linke SP, Pedeux R, Yang Q, Farnsworth J, Garfield SH, Valerie K, Shay JW, Ellis NA, Wasylyk B, Harris CC (2003) BLM helicase-dependent transport of p53 to sites of stalled DNA replication forks modulates homologous recombination. EMBO J 22(5):1210–1222
72. Serrano MA, Li Z, Dangeti M, Musich PR, Patrick S, Roginskaya M, Cartwright B, Zou Y (2013) DNA-PK, ATM and ATR collaboratively regulate p53-RPA interaction to facilitate homologous recombination DNA repair. Oncogene 32(19):2452–2462
73. Sharpless NE, DePinho RA (1999) The INK4A/ARF locus and its two gene products. Curr Opin Genet Dev 9(1):22–30
74. Sharpless NE, Ferguson DO, O'Hagan RC, Castrillon DH, Lee C, Farazi PA, Alson S, Fleming J, Morton CC, Frank K, Chin L, Alt FW, DePinho RA (2001) Impaired nonhomologous end-joining provokes soft tissue sarcomas harboring chromosomal translocations, amplifications, and deletions. Mol Cell 8(6):1187–1196
75. Shiama N (1997) The p300/CBP family: integrating signals with transcription factors and chromatin. Trends Cell Biol 7(6):230–236
76. Shiloh Y, Kastan MB (2001) ATM: genome stability, neuronal development, and cancer cross paths. Adv Cancer Res 83:209–254
77. Siliciano JD, Canman CE, Taya Y, Sakaguchi K, Appella E, Kastan MB (1997) DNA damage induces phosphorylation of the amino terminus of p53. Genes Dev 11(24):3471–3481
78. Singh DK, Ahn B, Bohr VA (2009) Roles of RECQ helicases in recombination based DNA repair, genomic stability and aging. Biogerontology 10(3):235–252
79. Sirbu BM, Lachmayer SJ, Wulfing V, Marten LM, Clarkson KE, Lee LW, Gheorghiu L, Zou L, Powell SN, Dahm-Daphi J, Willers H (2011) ATR-p53 restricts homologous recombination in response to replicative stress but does not limit DNA interstrand crosslink repair in lung cancer cells. PLoS One 6(8):e23053

80. Song H, Hollstein M, Xu Y (2007) p53 gain-of-function cancer mutants induce genetic instability by inactivating ATM. Nat Cell Biol 9(5):573–580
81. Stommel JM, Wahl GM (2004) Accelerated MDM2 auto-degradation induced by DNA-damage kinases is required for p53 activation. EMBO J 23(7):1547–1556
82. Susse S, Janz C, Janus F, Deppert W, Wiesmuller L (2000) Role of heteroduplex joints in the functional interactions between human Rad51 and wild-type p53. Oncogene 19(39):4500–4512
83. Takata M, Sasaki MS, Sonoda E, Morrison C, Hashimoto M, Utsumi H, Yamaguchi-Iwai Y, Shinohara A, Takeda S (1998) Homologous recombination and non-homologous end-joining pathways of DNA double-strand break repair have overlapping roles in the maintenance of chromosomal integrity in vertebrate cells. EMBO J 17(18):5497–5508
84. Tang W, Willers H, Powell SN (1999) p53 directly enhances rejoining of DNA double-strand breaks with cohesive ends in gamma-irradiated mouse fibroblasts. Cancer Res 59(11):2562–2565
85. Thottassery JV, Zambetti GP, Arimori K, Schuetz EG, Schuetz JD (1997) p53-dependent regulation of MDR1 gene expression causes selective resistance to chemotherapeutic agents. Proc Natl Acad Sci U S A 94(20):11037–11042
86. Tibbetts RS, Brumbaugh KM, Williams JM, Sarkaria JN, Cliby WA, Shieh SY, Taya Y, Prives C, Abraham RT (1999) A role for ATR in the DNA damage-induced phosphorylation of p53. Genes Dev 13(2):152–157
87. van den Bosch M, Bree RT, Lowndes NF (2003) The MRN complex: coordinating and mediating the response to broken chromosomes. EMBO Rep 4(9):844–849
88. Vogelstein B, Lane D, Levine AJ (2000) Surfing the p53 network. Nature 408(6810):307–310
89. Walker KK, Levine AJ (1996) Identification of a novel p53 functional domain that is necessary for efficient growth suppression. Proc Natl Acad Sci U S A 93(26):15335–15340
90. Ward IM, Minn K, Jorda KG, Chen J (2003) Accumulation of checkpoint protein 53BP1 at DNA breaks involves its binding to phosphorylated histone H2AX. J Biol Chem 278(22):19579–19582
91. Weber JD, Taylor LJ, Roussel MF, Sherr CJ, Bar-Sagi D (1999) Nucleolar Arf sequesters Mdm2 and activates p53. Nat Cell Biol 1(1):20–26
92. Weinberg RL, Freund SM, Veprintsev DB, Bycroft M, Fersht AR (2004) Regulation of DNA binding of p53 by its C-terminal domain. J Mol Biol 342(3):801–811
93. Wilcock D, Lane DP (1991) Localization of p53, retinoblastoma and host replication proteins at sites of viral replication in herpes-infected cells. Nature 349(6308):429–431
94. Xu Y (2006) DNA damage: a trigger of innate immunity but a requirement for adaptive immune homeostasis. Nat Rev Immunol 6(4):261–270
95. Xu Y, Baltimore D (1996) Dual roles of ATM in the cellular response to radiation and in cell growth control. Genes Dev 10(19):2401–2410
96. Yoon D, Wang Y, Stapleford K, Wiesmuller L, Chen J (2004) P53 inhibits strand exchange and replication fork regression promoted by human Rad51. J Mol Biol 336(3):639–654
97. Yun S, Lie-A-Cheong C, Porter AC (2004) Discriminatory suppression of homologous recombination by p53. Nucleic Acids Res 32(22):6479–6489
98. Zhang X, Zhu Y, Geng L, Wang H, Legerski RJ (2009) Artemis is a negative regulator of p53 in response to oxidative stress. Oncogene 28(22):2196–2204
99. Zhou T, Akopiants K, Mohapatra S, Lin PS, Valerie K, Ramsden DA, Lees-Miller SP, Povirk LF (2009) Tyrosyl-DNA phosphodiesterase and the repair of 3′-phosphoglycolate-terminated DNA double-strand breaks. DNA Repair 8(8):901–911
100. Zurer I, Hofseth LJ, Cohen Y, Xu-Welliver M, Hussain SP, Harris CC, Rotter V (2004) The role of p53 in base excision repair following genotoxic stress. Carcinogenesis 25(1):11–19

Chapter 18
The Role of Tumor Suppressor p53 in the Antioxidant Defense and Metabolism

Andrei V. Budanov

Abstract Tumor suppressor p53 is inactivated in most cancers and the critical role of p53 in the suppression of carcinogenesis has been confirmed in many mouse models. The protein product of the tumor suppressor p53 gene works as a transcriptional regulator, activating expression of numerous genes involved in cell death, cell cycle arrest, senescence, DNA-repair and many other processes. In spite of the multiple efforts to characterize the functions of p53, the mechanisms of tumor suppression by p53 are still elusive. Recently, new activities of p53 such as regulation of reactive oxygen species (ROS) and metabolism have been described and the p53-regulated genes responsible for these functions have been identified. Metabolic derangements and accumulation of ROS are features of carcinogenesis, supporting the idea that many tumor suppressive effects of p53 can be mediated by regulation of metabolism and/or ROS. Mutations in the p53 gene can not only inactivate wild type function of p53 but also endow p53 with new functions such as activation of new metabolic pathways contributing to carcinogenesis. Understanding the metabolic and antioxidant functions of p53 allows us to develop approaches to restore p53 function in cancers, where p53 is inactivated, in other to ensure the best outcome of anti-cancer treatment.

Keywords p53 • Stress response • ROS • Metabolism • mTOR

A.V. Budanov (✉)
Department of Neurosurgery & Biochemistry and Molecular Biology, Massey Cancer Center, Virginia Commonwealth University, Richmond, VA, USA
e-mail: abudanov@vcu.edu

S.P. Deb and S. Deb (eds.), *Mutant p53 and MDM2 in Cancer*,
Subcellular Biochemistry 85, DOI 10.1007/978-94-017-9211-0_18,

Abbreviations

4EBP1	eIF-4E binding protein 1
Atg1-17	autophagy gene 1–17
AIF	apoptosis-inducing factor
AMPK	AMP activated protein kinase
Arf1	alternative reading frame 1
ATP	adenosine triphosphate
ATM	ataxia-telangiectasia mutated kinase
ATR	ATM related kinase
Chk1/Chk2	checkpoint kinase 1/2
COP1	constitutively photomorphogenic 1
eIF-4E	eukaryotic translation initiation factor 4E
FPP	farnesyl pyrophosphate
FoxO	forkhead box O transcription factors
GAMT	guanidinoacetate methyltransferase
GGPP	geranylgeranyl pyrophosphate
GLS2	glutaminase 2
GPX1	glutathione peroxidase 1
GSH	glutathione
GSSG	oxidized form of glutathione
HIF1	hypoxia-inducible factor 1
HK2	mitochondrial hexokinase 2
Keap1	Kelch-like ECH-associated protein 1
MnSOD	manganese superoxide dismutase
Mdm2	mouse double minute
NAC	N-acetyl cysteine
NCF2/p67phox	neutrophil cytosol factor 2
NRF2	nuclear factor (erythroid-derived 2)-like 2
mTOR	mammalian target of rapamycin kinase
mTORC1/2	mTOR complex 1/2
p53INP1	p53-inducible nuclear protein 1
PET	paired-end ditag
PGC1α	PPARγ coactivator 1α
PGM	phosphoglycerate mutase
PFK-2/FBPase-2	6-phosphofructo-2-kinase/fructose-2,6-biphosphatase
PIG1-13	p53-inducible genes 1–13
PIRH2	p53-induced protein with a RING (Really Interesting New Gene)-H2 domain
PPARα/γ	peroxisome proliferator-activated receptor α/γ
RagA,B,C,D	Ras-related GTP-binding protein A,B,C,D
Rheb	Ras homolog enriched in brain
ROS	reactive oxygen species
SCO2	synthesis of cytochrome C oxidase 2
SREBP1	sterol-regulatory element binding protein 1

TAp73	transcriptionally active p73
TIGAR	p53-induced glycolysis and apoptotic regulator
TSC1/2	tuberoses sclerosis complex protein 1/2

Introduction

p53 was discovered more than three decades ago as a protein interacting with large T-antigen of human polyoma virus SV40, which is known to be responsible for malignant transformation of different human and mouse cells. This discovery paved the way for intensive studies of the role of p53 in transformation and carcinogenesis [61]. A few years later, the tumor suppressor p53 (TSp53) gene was cloned by several groups and it was demonstrated that TSp53 is mutated in a vast majority of human cancers [61]. Beyond point mutations, p53 can be inactivated via many other mechanisms including chromosome deletions, amplification of its natural negative regulator Mdm2 (mouse double minute 2) or expression of viral oncogenes such as papillomavirus E6 or adenovirus E1B proteins [61]. The following studies found inactivation of one p53 allele in families with Li-Fraumeni syndrome, the disease characterized by the predisposition to many types of cancer at an early age [63, 64]. The ultimate evidence that p53 is a critical tumor suppressor came with the advent of gene knockout technology. p53 knockout mice were normal, indicating that p53 does not play a role in embryogenesis and differentiation, but they developed cancers, mostly lymphomas and sarcomas with 100 % penetrance and died from cancer by the age of 6 months [27]. Surprisingly, the pattern of cancers observed in the p53-knockout animals differed from the majority of human cancers, which are mostly carcinomas. This may be because human cancers mostly bear a mutant form of the proteins rather than total inactivation of p53 and the most common mutations in p53, called "hot-spot" mutations, give the mutant form of p53 an ability to stimulate invasiveness or cancer-associated metabolism in the process called "gain-of-function" [27, 33, 61]. Accordingly, two mouse p53 knockin strains with mutations R172H and R270H (where arginines (R) are replaced to histidines (H), corresponding to hot-spot human p53 mutations in codons 175 and 273) were generated, and it was demonstrated that mice developed mostly highly-invasive carcinomas [53, 81].

As it has been believed for many years, p53 protects from carcinogenesis via "crime and punishment" mechanisms by restricting proliferation of genetically damaged cells via activation of cell death or permanent cell cycle arrest or senescence [102]. This dogma has been significantly re-visited during recent years when it was shown that inactivation of major targets of p53 involved in regulation of cell death – Noxa and Puma, or senescence/cell cycle arrest gene p21 does not recapitulate the cancer-prone phenotype of p53-deficient mice [25, 42, 100]. In following studies it was also established that p53 triple mutant knockin mice ($p53^{3KR}$), where 3 lysines (117, 161 and 162), subjects of p53 acetylation, were replaced with arginines, lost the ability to activate cell death, cell cycle arrest or senescence but still maintained the ability to suppress carcinogenesis [65]. This data opens a new venue for the studies of the mechanisms of "good maintenance" applied by p53 to prevent accumulation of damage via regulation of reactive oxygen species and metabolism leaving the "punishment" as the last resort for the cells which go awry.

Tumor Suppressor Gene p53 Encodes Stress-Responsive Transcriptional Factor

p53 works as a transcription factor involved in transcriptional regulation of multiple genes. p53 functions in the form of homo-tetramers, recognizing p53-responsive elements composed of two decamers separated by short spacer 5′-RRRCWWGYYYN$_{0-14}$RRRCWWGYYY-3′ where R – purine, Y – pyrimidine, C – cytosine, G – guanine, W – adenine or thymine and N – nucleotide and activates or suppresses different promoters dependent on context [33]. It is well established that p53 is induced by DNA-damage via consequent activation of ATM/ATR and Chk1/Chk2 kinases which phosphorylate p53 on N-terminus, causing p53 stabilization and activation. As a result, p53 induces cell cycle arrest, stimulates DNA-repair and protects genomic stability earning p53 the name "guardian of the genome" [52]. Other stimuli, including oxidative stress, hypoxia, nutrient deprivation or activation of oncogenes, induce p53 via mechanisms yet to be established and stimulate expression of p53-dependent genes, facilitating stress consequences [101]. Nevertheless, if the stress insult is too intense, p53 can stimulate pathways that lead to the elimination of cells via induction of cell death or senescence. The choice between a pro-survival or pro-death outcome is determined by the sensitivity of different promoters to p53 [101]. In response to weak or moderate stress insults, p53 stimulates expression of pro-survival genes which protect cells from damage. The promoters of these genes are very sensitive to even low p53 activation and are usually activated very quickly after the stress is applied. On the contrary, pro-apoptotic genes are activated in response to intense stress and with a significant delay as compared to pro-survival genes [90, 102] (Fig. 18.1).

Among the targets of p53 are its own negative regulators such as Mdm2, PIRH2 (p53-induced protein with a RING (Really Interesting New Gene)-H2 domain) and COP1 (constitutively photomorphogenic 1). They work as E3-ubiquitine ligases which stimulate p53 ubiquitination and degradation by proteasomes [28, 37, 60]. Mdm2 plays a major role in regulation of p53 stability and activity via constant control of p53 expression [10, 51]. Inactivation of Mdm2 in mice leads to permanent p53 activation, causing embryonic lethality and the lethal phenotype was completely rescued by p53 inactivation supporting the critical role of Mdm2 in p53 regulation [44]. Moreover, single nucleotide polymorphisms in SP1 site in regulatory area of the human Mdm2 gene causes elevated Mdm2 expression, which increases susceptibility of the organism to carcinogenesis [9].

Although p53 protein is suppressed by Mdm2 under non-stressed conditions and has a lifespan around 20 min, it can be easily activated by multiple stress insults [61]. Mild stress can be induced by mistakes in DNA-replication, ROS accumulation or decrease in ATP levels. It causes modification of p53 via

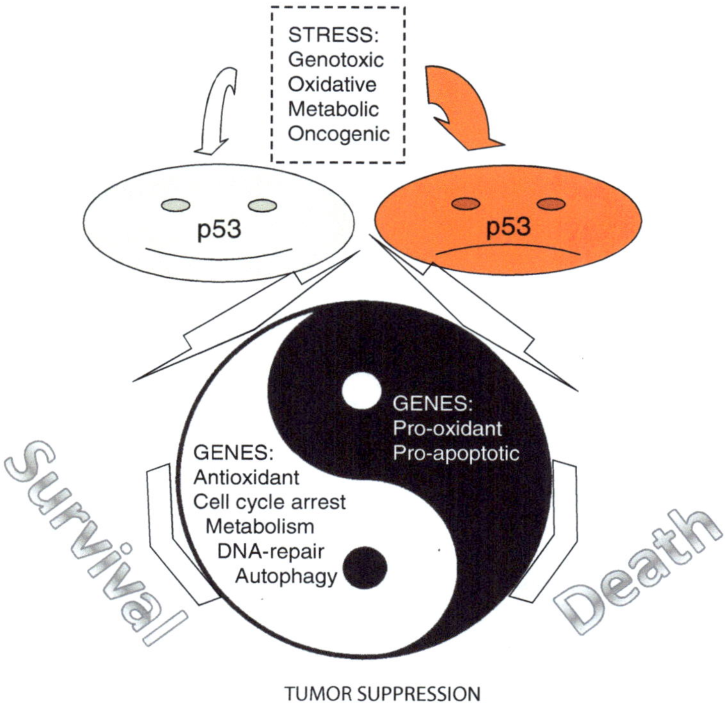

Fig. 18.1 Yin and Yang activities of p53 in tumor suppression. Many stress insults induce p53 which activates expression of multiple genes via interaction with p53-responsive elements on their promoters. The outcome of p53 activation can be cell protection or cell death depending on the nature and intensity of the stress. Both processes can be required to reach maximal protection of the organism against carcinogenesis

phosphorylation and some other mechanisms such as acetylation, methylation, ubiquitylation, neddylation or summoylation [101]. p53 can be also activated via redox-dependent mechanisms involving redox factor Ref1 [95]. As a result, it leads to stimulation of p53-dependent genes involved in suppression of ROS and tuning up the metabolism. In contrast, activation of p53 by severe stress leads to cell death or permanent cell cycle arrest [12, 101].

p53 Regulates Redox Balance in Cells

Pro-oxidant Function of p53

p53 is the major regulator of programmed cell death or apoptosis, and for many years this activity was considered to be the major mechanism of p53-controlled tumor suppression [92]. In an attempt to explain the mechanisms of the pro-apoptotic activity of p53, several targets named PIG1-13 (p53-inducible genes 1-13) with strong

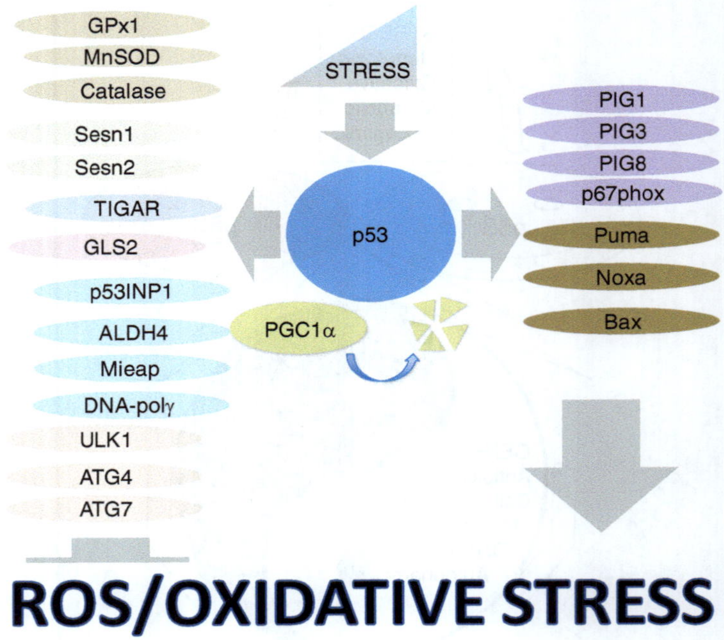

Fig. 18.2 Regulation of anti-oxidant and pro-oxidant genes by p53. Low-intensities of stress stimulate expression of highly sensitive p53-dependent pro-survival genes involved in ROS suppression, metabolism, mitochondrial function and autophagy, which protect cell viability. Highly intense stress insults activate cell death via the induction of pro-apoptotic and pro-oxidant genes. In response to low stress, PGC1**α** binds p53 and stimulates expression of pro-survival genes, although its degradation induced by chronic prolonged stress may be responsible for the activation of genes responsible for cell death

pro-oxidant properties were described [85]. Some examples include: PIG1 – member of galectin family involved in superoxide production; PIG3 – homolog of NADPH–quinone oxidoreductase, a potent ROS generator; PIG8 – human homolog of mouse E-24 gene, a quinone known to regulate ROS [85]. Accordingly, activation of p53 in different cancer cell lines via overexpression or strong genotoxic stress causes accumulation of reactive oxygen species (ROS) and oxidative stress contributing to induction of cell death [85, 90]. Overexpression of BH3-only protein Puma, a critical mediator of p53-activated cell death, also stimulates ROS production which most likely is associated with mitochondria disintegration [90]. In accordance with these data, overexpression of p53 stimulates ROS production mostly in cells with intact mitochondria, which are susceptible to p53-induced cell death [90]. Besides its effects on mitochondria, the parallel mechanism of ROS production by p53 can involve transcriptional activation of a component of NADPH oxidase, the critical enzyme responsible for O_2^- production, NCF2/p67phox (neutrophil cytosol factor 2) [41]. Although pro-oxidant activity of p53 can be beneficial for elimination of cancer cells, it might have some disadvantages for the organism under conditions of ischemia, neurodegeneration or aging [102]; consequently tight p53 regulation is critical for proper control of many physiological processes associated with p53 activation (Fig. 18.2).

Antioxidant Function of p53

The paradigm that p53 is a bona fide pro-oxidant factor changed since it had been found that under physiological or low stress conditions p53 suppresses ROS accumulation [90]. ROS cause oxidative DNA-damage which increase the rate of mutagenesis and chromosomal instability. Being widely recognized as the protector of genomic stability, p53 inhibits DNA-oxidation and mutagenesis via suppression of ROS [12]. ROS are also involved in the activation of signaling pathways controlling cell growth, proliferation, viability and transformation such as PI3K-AKT, JAK-STAT, PLC-PKC, MAPK cascade or IKK-NF-κB [71]. They also play an important role in stimulation of angiogenesis and epithelial-mesenchymal transition, both critical steps of cancer progression [22].

Thus, ROS accumulation, often observed in cancer cells [5], can be responsible for the high rate of mutagenesis in oncogenes and tumor suppressor genes, stimulation of cell proliferation and malignant transformation, and eventually for angiogenesis, invasiveness and metastasis [5, 71].

These considerations are supported by animal studies. The lifespans of the p53-deficient mice, characterized by increased ROS levels, are shortened by fivefold to sixfold due to accelerated carcinogenesis [27, 90]. Increased levels of oxidative stress is also observed in patients with Li-Fraumeni cancer predisposition syndrome characterized by inactivation of one copy of the p53 gene [103]. Inactivation of ATM, an upstream p53 regulator, also leads to oxidative stress, DNA oxidation, mutagenesis and carcinogenesis [88, 107]. Tumor phenotypes of p53 and ATM knockout mice were significantly suppressed in animals supplemented with an antioxidant N-acetyl-cysteine (NAC) [90, 91]. Additionally xenograft studies on lung adenocarcinoma cells demonstrated that p53 silencing accelerates tumor growth, while these effects were strongly suppressed by NAC treatment [90].

In another set of experiments, a mouse strain carrying extra-genomic copies of p53 and its upstream regulator Arf1 (alternative reading frame 1) was generated [72]. These mice, designated s-p53/Arf1 (super-p53/Arf1) are characterized by slightly elevated p53 expression and a continuous activation of p53-dependent antioxidant genes. s-p53/Arf1 mice have an extended lifespan as compared to wild type control and are highly resistant to carcinogenesis. Interestingly, the cells isolated from the s-p53/Arf1 mice are also well protected against transformation by cooperating oncogenes, providing a protecting mechanism against carcinogenesis [72].

Since characterization of the antioxidant function of p53, several groups of p53-inducible antioxidant proteins were identified.

Antioxidant Enzymes

MnSOD (Manganese superoxide dismutase) is an enzyme responsible for decomposition of superoxide (O_2^-) converting it to less toxic H_2O_2 form in the reaction $2O_2^- + 2H^+ = H_2O_2 + O_2$ [4]. Superoxide is produced as a by-product of mitochondrial oxidative phosphorylation or via activation of NADPH oxidases. Mitochondria are the major source of superoxide and other ROS, and as estimated approximately 2 % of oxygen consumed by healthy mitochondria is converted into the O_2^- form [31]. Moreover, mitochondrial dysfunction induced by stress or improper control of mitochondrial integrity causes increased leakage of electrons from the respiratory chain. Superoxide is extremely reactive and, consequently, very unstable. Nevertheless, it can damage different macromolecules in mitochondria including lipids, proteins and DNA, affecting mitochondrial function and stimulating extensive electron leakage and ROS production [4, 31]. MnSOD, residing in the mitochondrial matrix, plays a critical role in the detoxification of O_2^-, producing less reactive H_2O_2. p53 activates expression of the MnSOD via direct recognition of MnSOD promoter in -2032 - -2009 position of the human gene [39].

Another critical antioxidant enzyme regulated by p53 is GPx1 (glutathione peroxidase 1), an enzyme responsible for decomposition of H_2O_2 via the reaction $2GSH + H_2O_2 = GS\text{-}SG + H_2O$ [4], where GSH and GS-SG are reduced and oxidized forms of glutathione, a major cellular antioxidant [31]. Selenoprotein GPx1 plays a major role in decomposition of H_2O_2, a highly diffusible molecule which reacts with different macromolecules including membrane lipids, proteins and nuclear DNA. p53 activates expression of the GPx1 gene through binding to the -694 - -720 region of its promoter, stimulating an antioxidant response [39]. In particular cell types, such as developing rat retina and retinal ganglion cells, p53 can stimulate expression of catalase, another enzyme involved in H_2O_2 decomposition [80]. It was also reported that p53 regulates activity of catalase via direct protein-protein interactions [46]. Interestingly, p53-responsive p53R2 (p53-inducible ribonucleotide reductase) was found in a p53-catalase complex under physiological conditions and stimulates catalase activity, while another p53 target PIG3 was found in the p53-catalase complex under genotoxic stress conditions and inhibits catalase activity [46].

Sestrins

Another group of antioxidant genes, Sestrins, play a major role in regulation of ROS in response to p53 activation [15]. Sestrins are a family of genes highly conserved in evolution from protists to mammals [14]. The mammalian Sestrin gene family is composed of three members, Sesn1, Sesn2 and Sesn3, while the invertebrate genome contains only one Sestrin gene [14, 56]. The first family member, Sesn1, was identified as a p53-inducible gene activated by DNA-damage [99]. Several potential p53-responsive elements were described within intron 1 and 2 [99, 104]. Sesn1 gene is transcribed in three mRNA forms using alternative promoters and produces three different proteins with molecular weights of 46, 55 and 68 kDa [99]. Among these proteins, only the 46 and 55 kDa forms are regulated by p53, while the 68 kDa form is constantly expressed [99].

The second member of the family, Sesn2, was isolated as hypoxia inducible gene [16]. In following studies it was established that the Sesn2 gene is also activated by many other stimuli including oxidative stress, nutrient deficiency and

DNA-damage [6, 15, 16]. p53 is responsible for Sesn2 activation by DNA-damage, and contributes to Sesn2 activation by oxidative stress [15, 16]. The p53-responsive element was identified 9.7 kb downstream of the Sesn2 gene by PET (paired-end ditag) sequencing [104]. In different study, another p53-responsive element was found in the 1st exon of the Sesn2 gene [59]. The Sesn2 gene encodes a protein with a molecular weight of 60 kDa showing close similarity with 55 kDa protein product of the Sesn1 gene [16].

Sesn3, the third family member, was identified *in silico* as a close homolog of both Sesn1 and Sesn2 genes [16, 83]. Interestingly, while two of three members of the Sestrin family – Sesn1 and Sesn2 – are direct targets of p53, Sesn1 and Sesn3 are activated by the FoxO family of transcription factors [14, 79]. Similar to p53, FoxO controls both pro-survival and pro-apoptotic processes via activation of different sets of genes involved in regulation of cell death and cell cycle arrest as well as metabolism and ROS [36]. Regulation of Sestrins by p53 and FoxO is very conserved in evolution, as demonstrated by the *Drosophila* Sestrin gene [57].

Although Sestrins do not have similarity with other proteins, detailed sequence analysis has shown that part of Sestrin protein (corresponding to 100-175 aa of Sesn2) shares homology with the *Mycobacterium Tuberculosis* antioxidant protein AhpD, which is a critical regulator of bacterial thiol peroxidase AhpC [11, 15]. During the catalytic cycle AhpC is oxidized and then restored by AhpD. Mammalian peroxiredoxins are AhpC homologs, which work via similar mechanisms [23]. In contrast to bacterial proteins, mammalian peroxiredoxins can easily be inactivated via overoxidation of catalytical cysteine [105]. Sestrins contribute to regeneration of peroxiredoxins potentially through activation of sulfinilreductase for mammalian peroxiredoxins – Srx (sulfiredoxin) [14]. Sestrins activate expression of Srx and some other antioxidant genes stimulating autophagic degradation of protein Keap1 (Kelch-like ECH-associated protein 1), an inhibitor of antioxidant transcription factor NRF2 (nuclear factor (erythroid-derived 2)-like 2) [3]. The regulation of antioxidant response by Sestrins via regeneration of peroxiredoxins can play an important role in protection of neurons and macrophages from oxidative stress [29, 82]. Sestrin-mediated autophagy also plays a critical role in support of integrity of the mitochondria and other organelles responsible for ROS production [13, 57, 58]. Inactivation of Sestrins via p53 mutation or other mechanism might play a critical role in carcinogenesis, facilitating cell transformation. Accordingly, Sesn2-deficient cells are much more susceptible to transformation than their wild-type counterparts [13]. Moreover, cell transformation by Ras oncogenes requires ROS production and Ras stimulates ROS via suppression of Sesn1 and Sesn3 gene expression [49].

TIGAR

TIGAR (p53-induced glycolysis and apoptotic regulator) was discovered as a p53-inducible gene by microarray analysis. It is a direct p53 target with kinetics of activation similar with other genes involved in pro-survival activities, such as ROS regulation, cell cycle control and DNA-repair, which are activated quickly after p53 induction [8]. Two p53-responsive elements were found in the promoter and first intron of the human TIGAR gene [8]. TIGAR shares similarity with proteins of the PGM (phosphoglycerate mutase) family, but the highest degree homology is observed with the biphosphatase domain of PFK-2/FBPase-2 (6-phosphofructo-2-kinase/fructose-2,6-biphosphatase) [8]. Ectopic expression of TIGAR decreases the levels of Fru-2,6-P_2 (fructose-2,6-biphosphate). As expected, it leads to a decrease in glycolytic rates, and silencing TIGAR stimulates Fru-2,6-P_2 production and glycolysis. Being an inhibitor of glycolysis, TIGAR can re-direct glycolytic intermediates into the pentose phosphate pathway (PPP), which is involved in the production of the reducing agent NADPH. NADPH is used by glutathione reductase to convert GSSG to GSH. The GSH/GSSG ratio is the major indicator of redox balance in the cell, which affects cell metabolism, viability and antioxidant defense. Accordingly, activation of TIGAR in response to p53 decreases the levels of ROS and suppresses p53-induced cell death [8]. TIGAR protein was also found in mitochondria where it interacts with mitochondrial HK2 (hexokinase 2) and stimulates its activity. HK2 arguably helps to couple glycolysis with oxidative phosphorylation and limits ROS production in mitochondria via control of ADP levels [19].

GLS2

The GLS2 (Glutaminase 2) gene encodes a mitochondrial glutaminase which catalyzes the hydrolysis of glutamine to glutamate. Activation of GLS2 by p53 leads to increased production of glutamate and α-ketoglutarate, last one is the critical substrate for the tricarboxylic acid (TCA) cycle. As a result, GLS2 stimulates mitochondrial respiration and ATP production. Moreover, GLS2 induction causes GSH accumulation and ROS downregulation, providing protection against oxidative stress [38]. Thus, GLS2 regulates ROS supporting mitochondrial function and preventing ROS generation by dysfunctional mitochondria. Activation of GLS2 also protects cells against oxidative DNA-damage and supports genomic stability [96].

Other p53 Targets Involved in Redox Regulation

Several other p53-inducible proteins are involved in redox regulation, although the mechanisms are yet to be established. p53INP1 (p53-inducible nuclear protein 1) when overexpressed facilitates cell death. However, p53INP1-deficient MEFs and splenocytes accumulate ROS, and p53INP1-null thymocytes are highly susceptible to cell death linked with enhanced ROS production [18, 78]. p53INP1-null cells are characterized by decreased levels of autophagy which might be responsible for p53INP1-regulated antioxidant protection [78]. Another p53 target, ALDH4, is a mitochondrial-matrix NAD^+ -dependent enzyme which catalyzes the second step in the proline degradation pathway. ALDH4 downregulates ROS levels via a yet to be defined mechanism and its silencing enhances cell death in response to p53 overexpression [108]. Many effects of p53 on ROS can be explained via support of mitochondrial function. p53 is involved in the maintenance of mitochondrial DNA copy number and mitochondrial mass, and p53 inactivation can lead to improper control of mitochondrial integrity and an increased ROS production [55]. p53 can also support stability of mitochondrial DNA interacting with mitochondrial DNA-polymerase γ, ensuring integrity and proper functioning of mitochondria [1]. As an additional mechanism of control of mitochondrial function, p53 activates expression of Mieap which induces the formation of a lysosome-like vacuole structure in mitochondria and is involved in mitochondrial quality control, suppressing ROS production [48] (Fig. 18.2).

p53 and Autophagy

Macroautophagy (therein in the text – autophagy) is the mechanism of two-membrane encapsulation and lysosomal degradation of cellular constituents such as organelles, protein aggregates and bulk of cytoplasm [50]. Autophagy is responsible for the control of the integrity of organelles, such as mitochondria and peroxisomes, which, when damaged, produce excessive amounts of ROS. Thus, impaired autophagy can be responsible for oxidative stress and genomic instability [73]. p53 regulates autophagy in a positive and negative manner dependent on context [30, 97]. The nuclear form of p53 transcriptionally activates expression of multiple genes involved in autophagy such as DRAM, Sesn2, ULK1, ULK2, Atg4a, Atg4c, Atg7, and Uvrag in response to DNA-damage or oncogene activation [21, 30, 34, 47, 70]. In contrast, the cytoplasmic form of p53 inhibits autophagy via an interaction with autophagy proteins Atg17 and RB1CC1/FLP200 [76, 97]. This promiscuous effect of p53 on autophagy regulation can be explained by the link of autophagy with cell death. Autophagy can suppress cell death in response to stress supporting integrity and function of organelles, suppressing ROS production and providing ATP for the repair processes [50]. When p53 functions in a pro-survival fashion, it might support autophagy via transcriptional activation of autophagy genes. On the contrary, when cell death is the more desirable scenario p53 can suppress autophagy via protein-protein interactions and ensure completion of cell death program. Interestingly, non-nuclear p53 plays an important role in activation of cell death via an interaction with the proteins of the Bcl2-family, so these p53 forms can also be responsible for regulation of autophagy [35] (Fig. 18.2).

Role p53 in Regulation of Metabolism

AMPK-mTOR Pathway

Carcinogenesis involves metabolic changes required for tumor growth and adaptation to the environment. Constant growth and proliferation of tumor cells requires constant synthesis of proteins, lipids and nucleotides and all of these processes are regulated by mTOR (mammalian target of rapamycin) kinase [7, 84, 106]. mTOR is a highly conserved protein kinase which belongs to the PIKK (phosphatidylinositol kinase related kinase) family found in all eukaryotic organisms [106]. mTOR forms two complexes mTORC1 (mTOR complex 1) and mTORC2 (mTOR complex 2) [106]. mTORC1 regulates protein synthesis via phosphorylation of p70S6K (p70S6 kinase) and 4EBP1 (eIF4E binding protein 1). While p70S6K phosphorylates ribosomal S6 protein potentially modulating its activity, 4EBP1 binds and inactivates a factor of initiation of Cap-dependent transcription eIF4E. In its hypophosphorylated forms, 4EBP1 has high avidity to eIF4E causing inhibition of translation of many mRNA with structured 5'UTR, encoding proteins critical for cell growth and metabolism. Phosphorylation of 4EBP1 on multiple sites by mTORC1 leads to disintegration of the 4EBP1-eIF4E protein complex, leading to eIF4E release and the activation of translation [106]. mTORC1 is activated by Rheb and RagA,B,C, and D small GTPases in a cooperative manner. While Rag proteins are activated by amino acids, Rheb is tightly controlled by growth factors or stress insults via tuberoses scleroses 1 and 2 protein complex (TSC1:TSC2) where TSC2 is GAP (GTP activating protein) for Rheb [54]. Being activated by upstream signals, the TSC1:TSC2 complex causes conversion of Rheb bound GTP to GDP form leading to Rheb inhibition. Insulin and growth factors inhibit TSC1:TSC2 complex via phosphorylation by AKT kinase, causing activation of Rheb-mTORC1 axis and stimulation of cell growth and proliferation. Among the negative regulators of mTORC1, AMPK (AMP-activated protein kinase) plays a major role due to its susceptibility to many stress factors, including nutrient/energy deficiency, accumulation of Ca^{2+}, oxidative stress and DNA-damage. AMPK directly phosphorylates TSC2 and Raptor proteins, which leads to inhibition of TORC1 [54, 75].

p53 inhibits mTORC1 in response to genotoxic stress and some other stimuli via activation of expression of mTORC1 inhibitors such as IGF-BP3, TSC2, PTEN and AMPKβ1 [12, 62]. p53 also stimulates phosphorylation of AMPKα subunits on T172, required for mTORC1 inhibition [30]. p53-activated Sesn1 and Sesn2 play a major role in activation of AMPK phosphorylation in response to p53 [13]. As a result, mTORC1 inhibition leads to suppression of protein synthesis and activation of autophagy [12, 67], preventing accumulation of protein aggregates and damaged mitochondria, which is detrimental for cell viability. Inhibition of protein synthesis can also re-direct ATP from energy-consuming anabolism toward repair processes to protect cell homeostasis. In parallel, autophagy supplies cells with demanded ATP via digestion of cellular constituents [50]. Inactivation of p53 in cancer cells has long-lasting consequences on metabolism and angiogenesis instigated by hyperactivation of mTORC1. One of the critical targets of mTORC1 is the transcription factor hypoxia-inducible factor 1 (HIF1) [54]. HIF1 is composed from stable HIF1β and inducible HIF1α subunits [93]. mTORC1 stimulates the translation of HIF1α, activating expression of many genes involved in adaptation to metabolic derangements and hypoxia such as glycolytic enzymes and angiogenic factors supporting the viability of cancer cells in conditions of nutrient and oxygen deprivation [93]. mTORC1 is also a critical activator of transcriptional factor SREBP (sterol regulatory element-binding protein), that induces genes of lipid biosynthesis required for the growth of cancer cells [110] (Fig. 18.3).

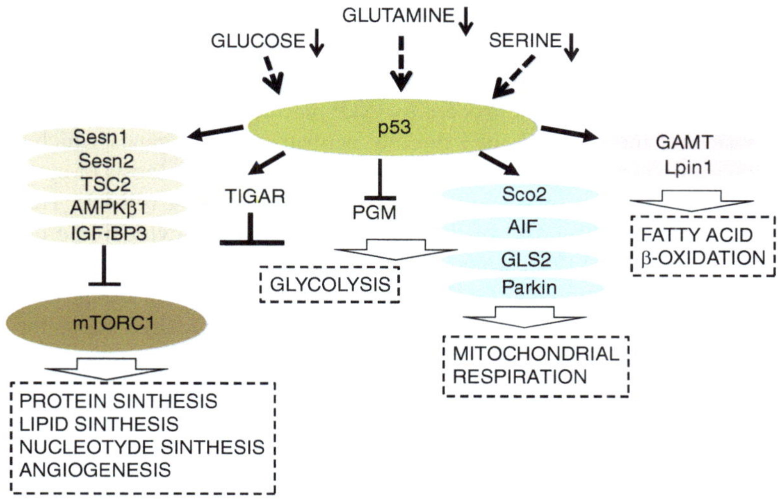

Fig. 18.3 p53 plays an important role in adaptation to metabolic derangements. A shortage in metabolites such as glucose, glutamine and serine activate p53 which tunes up metabolism to suppress undesired anabolic processes via the inhibition of mTORC1, stimulates ATP production via mitochondrial respiration rather than glycolysis, and use alternative fatty acid oxidation as other sources of ATP production. Dysregulation of these processes due to p53 inactivation in cancer cells can be beneficial for tumor growth

Regulation of Glycolysis and Oxidative Phosphorylation by p53

Glycolysis

Cancer cells have higher glycolytic rates and rely more on glycolysis for ATP production than normal cells [24]. This observation has been made by Otto Warburg in the 1930s and was called the "Warburg effect". Although the glycolytic pathway is much less efficient for ATP production, under abundant glucose supply, ATP production through glycolysis is faster than via oxidative phosphorylation. Moreover, glycolysis allows cells to generate many intermediates of biosynthetic pathways involved in the synthesis of carbohydrates, DNA, proteins and lipids [24]. p53 inhibits the Warburg effect by suppressing glycolysis and stimulating mitochondrial respiration [69]. As discussed earlier, p53-inducible protein TIGAR acts as a PFK-2/FBPase-2 biphosphatase [8]. Consequently, via induction of TIGAR, p53 re-directs glucose derivatives to the pentose phosphate pathway (PPP) [8]. This leads to the production of NADPH, the important component required for the biosynthesis of ribose-6-phosphate [69]. p53 also inhibits glycolysis via ubiquitination and degradation of the PGM protein [66]. Besides its direct effects on glycolytic enzymes, p53 also suppresses glycolysis by restricting glucose transport by inhibiting glucose transporters such as GLUT1, GLUT4 and GLUT3 [66].

p53 and Mitochondrial Respiration

While having a negative impact on glycolysis, p53 also stimulates mitochondrial respiration, the most efficient process for ATP production [24]. It can be regulated through several mechanisms. p53 directly activates several genes which support oxidative phosphorylation in mitochondria such as SCO2 (synthesis of cytochrome C oxidase 2), apoptosis-inducing factor (AIF), glutaminase 2 (GLS2), Parkin and p53R2 [12, 66]. SCO2 is involved in the formation of the cytochrome c oxidase complex (complex IV) [74], while AIF supports the stability of the mitochondrial complex I [98]. Additionally, GLS2, the enzyme that controls the production of glutamate and α-ketoglutarate, supports mitochondrial oxidative phosphorylation and ATP production [38]. Moreover, glutamate is a precursor of antioxidant GSH, thus GLS2 stimulates GSH production and the antioxidant response [96]. Parkin also supports mitochondrial oxidative phosphorylation [109]. The potential mechanism involves stimulation of the expression of PDHA1 (pyruvate dehydrogenase E1α1), a component of the pyruvate dehydrogenase complex (PDH), which converts pyruvate into acetyl-CoA, the primary substrate for the TCA cycle. In parallel, p53 stimulates PDH through the repression of PDK2 (pyruvate dehydrogenase kinase 2), which phosphorylates and inhibits PDHA1. As a result p53 stimulates acetyl-CoA production and mitochondrial respiration, preventing the conversion of pyruvate to lactate [109]. p53 can also enhance mitochondrial respiration through the control of mitochondrial integrity via regulation of autophagy, DNA-stability and antioxidant protection (Fig. 18.3).

p53 Controls of Cell Viability in Response to Nutrient Deprivation

Metabolic stress induced by glucose deprivation leads to a significant decrease in ATP production and induces p53 via activation of AMPK and ATM. Both kinases directly and indirectly phosphorylate human p53 on Ser15, leading to the activation of inhibitor Cyclin-CDK complexes p21 and reversible cell cycle arrest in G1 [2, 43]. As a result, p53 inhibits cell proliferation to protect cells against cell death induced by glucose deprivation [43]. Interestingly, pharmacological activators of AMPK such as metformin and Aicar suppress growth of p53-deficient tumors, but has no effect on p53-positive tumors in xenograft based studies, indicating the important role of p53 in protection of cell viability in response to AMPK activation. Consequently, metformin stimulates cell death in p53-deficient, but not in p53-proficient cells deprived of glucose [17]. The protective effect of p53 against cell death can be explained by several mechanisms including the activation of autophagy and fatty acid β-oxidation, which supply cells with ATP. Besides glucose deprivation, serine starvation is another type of metabolic stress involved in p53 activation. Similar to its effects on the viability of glucose-starved cells, p53 activates p21 and induces cell cycle arrest preventing cell death in response to serine withdrawal.

Moreover, p53-proficient cells show suppressed glycolysis and higher rates of oxidative phosphorylation stimulating higher outcome of ATP production. Serine starved p53-deficient cells fail to recover from serine depletion, resulting in oxidative stress and cell death [68]. Depletion of another amino acid – glutamine also activates p53, that supports cell viability under glutamine-depleted conditions [87]. Glutamine induces p53 through transcriptional upregulation of B55α, a regulatory subunit of PP2A (protein phosphatase 2A). PP2A activates p53 via dephosphorylation and inactivation of EDD (E3 identified by differential display), HECT domain-containing E3 ubiquitin-ligase which negatively regulates p53 [87]. Nevertheless, in the conditions of genotoxic stress, metabolic regulation by p53 can play pro-apoptotic role. In response to DNA-damage p53 directly activates expression of GAMT (guanidinoacetate methyltransferase), an enzyme involved in creatine synthesis [40]. Creatine metabolism helps to maintain the proper levels of ATP, which is critical for apoptosis. GAMT also stimulates fatty acid β-oxidation, the source of energy in nutrient-depleted conditions. Accordingly GAMT silencing resulted in a significant decrease of ATP production and suppression of p53-induced cell death [40]. Lpin1 is another factor that stimulates fatty acid β-oxidation in a cell-type specific manner in response to p53 activation. It interacts with PPARα (peroxisome proliferator-activated receptor α) and PGC1α (PPARγ coactivator 1α) stimulating transcription of genes involved in fatty acid β-oxidation. As a result, it potentiates the effects of p53 on the regulation of metabolism, ROS and cell viability [2] (Fig. 18.3).

Among the factors, which modulate pro-survival or pro-apoptotic functions of p53 via the activation of pro-survival genes involved in regulation of ROS, metabolism, and cell cycle arrest, a critical role is played by p53-co-activator PGC1α. Under starvation conditions PGC1α binds p53 and stimulates activation of pro-survival genes including TIGAR, GADD45, SCO2, Sesn2 and p21, but not pro-apoptotic Puma and Bax [94]. Prolonged starvation causes PGC1α degradation via ubiquitin-proteosomal pathway. This process is controlled by binding of PGC1α with RNF2, a polycomb group protein that possesses E3 ubiquitin ligase activity. Activation of RNF2 by prolonged stress causes inactivation of PGC1α and a transcriptional switch toward p53-regulated proapoptotic genes [94].

Mutant p53 and Regulation of Metabolism

p53 is inactivated in many human tumors through point mutations, deletions, overexpression of Mdm2 or some viral proteins leading to the loss-of-function phenotype [61]. Inactivation of p53 leads to metabolic derangements, uncontrolled ROS production and genomic instability [12]. In addition, many tumors express a mutant form of p53 which, although unable to normally regulate the expression of many p53-regulated genes, still may have residual activity of the wild-type protein or regulate expression of a new set of genes via the gain-of-function mechanism. It was demonstrated that mutant p53 forms contribute to viability, invasiveness, migration, and metastasis of cancer cells [77]. Mutant p53 can also stimulate persistent inflammation by activating the NF-κB transcription factor. Inflammation is a critical promoter of carcinogenesis, involved in the production of ROS and RNS, which can fuel mutagenesis [20].

In contrast to the wild-type form, a hot-spot p53 R273H mutant stimulates production of reactive oxygen species via inhibition of the NRF2 transcription factor causing a decrease in expression of antioxidant and detoxifying enzymes such as NQO1 and HO1 [45]. Mutant p53 can also inhibit p73, another member of the p53 family, via direct protein-protein interactions [26]. As reported, the transcriptionally active form of p73 (TAp73) regulates mitochondrial activity and prevents ROS accumulation via transcriptional activation of the mitochondrial complex IV cytochrome C oxidase subunit 4 (Cox4i1). Accordingly, deficiency in TAp73 suppresses ATP production, mitochondrial complex IV activity, and oxygen consumption causing accumulation of ROS [89].

Additionally, mutant p53 stimulates the expression of genes involved in biosynthesis of sterols (mevalonate pathway) and other lipids [32]. The mevalonate pathway is responsible for the synthesis of cholesterol, a critical component of the cell membrane. Lipids are a necessary component required for cell growth and proliferation. Moreover mevalonate pathway is critical for production of farnesyl pyrophosphate (FPP) and geranylgeranyl pyrophosphate (GGPP), the indispensable components of posttranslational modification of Ras or RhoA, correspondingly. Both proteins play a major role in carcinogenesis, supporting proliferation, invasiveness and metastatic ability of cancer cells. Mutant p53 regulates the mevalonate pathway via interaction and activation of transcriptional factor SREBP [32].

Mutant p53 can also affect pyrimidine metabolism providing resistance against anticancer fluoropyrimidine drugs such as 5FU (5-fluorouracil) [86]. The p53H175 hot-spot mutant activates the expression of the enzyme dUTPase, which exerts conversion of dUTP to dUMP. The enzyme thymidylate synthase uses dUMP for synthesis of dTMP. Uracil can be mistakenly incorporated into DNA and its mis-incorporation is responsible for the cytotoxic activity of 5FU; consequently, inhibitors of thymidylate synthase cause increase in dUTP and increased uracil incorporation into DNA. As a result, it leads to massive DNA-damage, which cause DNA-strain breaks and cell death. On the contrary, dUTPase diminishes dUTP levels preventing uracil mis-incorporation and protects the cell from DNA-damage induced by fluoropyrimidine drugs [86] (Fig. 18.4).

Conclusion

While in previous years the p53 research was focused on its role in restraining and eliminating cancer cells as a major mechanism of tumor suppression, now the paradigm is being revisited. The mechanism of "good maintenance" involved in the control of metabolism and ROS seems to be indispensable for the tumor suppressive function of p53 and their inefficiency and perversion can cause numerous defects leading to the attainment of carcinogenic properties. Interestingly, some hot-spot

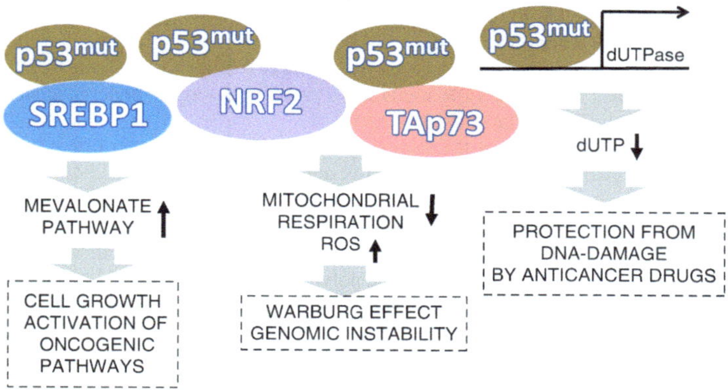

Fig. 18.4 Mutant p53 can contribute to carcinogenesis via gain-of-function mechanism. Hot-spot p53 mutants can regulate metabolism, suppress mitochondrial respiration, stimulate ROS production and protect against anti-cancer therapy. These activities are mediated by an interaction with different transcription factors such as: SREBP, critical for activation of mevalonate pathway, as well as NRF2 and TAp73, involved in control of mitochondrial respiration and antioxidant defense. Mutant p53 can also activate the promoter of the dUTPase gene, inducing the expression of enzyme dUTPase which suppresses therapeutic effects of fluoropyrimidine anticancer drugs

p53 mutants can contribute to carcinogenesis through activation of metabolic pathways not exerted by wild-type p53, which play a role in cell growth and proliferation. Finally, mutant p53 can also decrease sensitivity to anticancer drugs through metabolic modulation, which can make tumors more resistant to anticancer therapy. Thus, restoration of the wild-type function of p53 or inactivation of the mutant p53 form is a highly desirable approach for the future of anticancer treatments.

Acknowledgements This work is supported by NIH NCI 1RO1CA172660 to Andrei Budanov. The author thanks Brittany Binion for the help in preparation of the manuscript. The author is especially grateful to Nadulya Pryadilova for everyday support.

References

1. Achanta G, Sasaki R, Feng L, Carew JS, Lu W, Pelicano H, Keating MJ, Huang P (2005) Novel role of p53 in maintaining mitochondrial genetic stability through interaction with DNA Pol gamma. EMBO J 24:3482–3492
2. Assaily W, Rubinger DA, Wheaton K, Lin Y, Ma W, Xuan W, Brown-Endres L, Tsuchihara K, Mak TW, Benchimol S (2011) ROS-mediated p53 induction of Lpin1 regulates fatty acid oxidation in response to nutritional stress. Mol Cell 44:491–501
3. Bae SH, Sung SH, Oh SY, Lim JM, Lee SK, Park YN, Lee HE, Kang D, Rhee SG (2013) Sestrins activate Nrf2 by promoting p62-dependent autophagic degradation of Keap1 and prevent oxidative liver damage. Cell Metab 17:73–84
4. Balaban RS, Nemoto S, Finkel T (2005) Mitochondria, oxidants, and aging. Cell 120:483–495

5. Behrend L, Henderson G, Zwacka RM (2003) Reactive oxygen species in oncogenic transformation. Biochem Soc Trans 31:1441–1444
6. Ben-Sahra I, Dirat B, Laurent K, Puissant A, Auberger P, Budanov A, Tanti JF, Bost F (2013) Sestrin2 integrates Akt and mTOR signaling to protect cells against energetic stress-induced death. Cell Death Differ 20:611–619
7. Ben-Sahra I, Howell JJ, Asara JM, Manning BD (2013) Stimulation of de novo pyrimidine synthesis by growth signaling through mTOR and S6K1. Science 339:1323–1328
8. Bensaad K, Tsuruta A, Selak MA, Vidal MN, Nakano K, Bartrons R, Gottlieb E, Vousden KH (2006) TIGAR, a p53-inducible regulator of glycolysis and apoptosis. Cell 126:107–120
9. Bond GL, Hu W, Bond EE, Robins H, Lutzker SG, Arva NC, Bargonetti J, Bartel F, Taubert H, Wuerl P et al (2004) A single nucleotide polymorphism in the MDM2 promoter attenuates the p53 tumor suppressor pathway and accelerates tumor formation in humans. Cell 119:591–602
10. Bond GL, Hu W, Levine AJ (2005) MDM2 is a central node in the p53 pathway: 12 years and counting. Curr Cancer Drug Targets 5:3–8
11. Bryk R, Lima CD, Erdjument-Bromage H, Tempst P, Nathan C (2002) Metabolic enzymes of mycobacteria linked to antioxidant defense by a thioredoxin-like protein. Science 295:1073–1077
12. Budanov AV (2011) Stress-responsive sestrins link p53 with redox regulation and mammalian target of rapamycin signaling. Antioxid Redox Signal 15:1679–1690
13. Budanov AV, Karin M (2008) p53 target genes sestrin1 and sestrin2 connect genotoxic stress and mTOR signaling. Cell 134:451–460
14. Budanov AV, Lee JH, Karin M (2010) Stressin' Sestrins take an aging fight. EMBO Mol Med 2:388–400
15. Budanov AV, Sablina AA, Feinstein E, Koonin EV, Chumakov PM (2004) Regeneration of peroxiredoxins by p53-regulated sestrins, homologs of bacterial AhpD. Science 304:596–600
16. Budanov AV, Shoshani T, Faerman A, Zelin E, Kamer I, Kalinski H, Gorodin S, Fishman A, Chajut A, Einat P et al (2002) Identification of a novel stress-responsive gene Hi95 involved in regulation of cell viability. Oncogene 21:6017–6031
17. Buzzai M, Jones RG, Amaravadi RK, Lum JJ, DeBerardinis RJ, Zhao F, Viollet B, Thompson CB (2007) Systemic treatment with the antidiabetic drug metformin selectively impairs p53-deficient tumor cell growth. Cancer Res 67:6745–6752
18. Cano CE, Gommeaux J, Pietri S, Culcasi M, Garcia S, Seux M, Barelier S, Vasseur S, Spoto RP, Pebusque MJ et al (2009) Tumor protein 53-induced nuclear protein 1 is a major mediator of p53 antioxidant function. Cancer Res 69:219–226
19. Cheung EC, Ludwig RL, Vousden KH (2012) Mitochondrial localization of TIGAR under hypoxia stimulates HK2 and lowers ROS and cell death. Proc Natl Acad Sci U S A 109:20491–20496
20. Cooks T, Pateras IS, Tarcic O, Solomon H, Schetter AJ, Wilder S, Lozano G, Pikarsky E, Forshew T, Rozenfeld N et al (2013) Mutant p53 prolongs NF-kappaB activation and promotes chronic inflammation and inflammation-associated colorectal cancer. Cancer Cell 23:634–646
21. Crighton D, Wilkinson S, O'Prey J, Syed N, Smith P, Harrison PR, Gasco M, Garrone O, Crook T, Ryan KM (2006) DRAM, a p53-induced modulator of autophagy, is critical for apoptosis. Cell 126:121–134
22. Cui X (2012) Reactive oxygen species: the achilles' heel of cancer cells? Antioxid Redox Signal 16:1212–1214
23. D'Autreaux B, Toledano MB (2007) ROS as signalling molecules: mechanisms that generate specificity in ROS homeostasis. Nat Rev Mol Cell Biol 8:813–824
24. DeBerardinis RJ, Thompson CB (2012) Cellular metabolism and disease: what do metabolic outliers teach us? Cell 148:1132–1144
25. Deng C, Zhang P, Harper JW, Elledge SJ, Leder P (1995) Mice lacking p21CIP1/WAF1 undergo normal development, but are defective in G1 checkpoint control. Cell 82:675–684

26. Di Como CJ, Gaiddon C, Prives C (1999) p73 function is inhibited by tumor-derived p53 mutants in mammalian cells. Mol Cell Biol 19:1438–1449
27. Donehower LA, Harvey M, Slagle BL, McArthur MJ, Montgomery CA Jr, Butel JS, Bradley A (1992) Mice deficient for p53 are developmentally normal but susceptible to spontaneous tumours. Nature 356:215–221
28. Dornan D, Wertz I, Shimizu H, Arnott D, Frantz GD, Dowd P, O'Rourke K, Koeppen H, Dixit VM (2004) The ubiquitin ligase COP1 is a critical negative regulator of p53. Nature 429:86–92
29. Essler S, Dehne N, Brune B (2009) Role of sestrin2 in peroxide signaling in macrophages. FEBS Lett 583:3531–3535
30. Feng Z, Zhang H, Levine AJ, Jin S (2005) The coordinate regulation of the p53 and mTOR pathways in cells. Proc Natl Acad Sci U S A 102:8204–8209
31. Finkel T, Holbrook NJ (2000) Oxidants, oxidative stress and the biology of ageing. Nature 408:239–247
32. Freed-Pastor WA, Mizuno H, Zhao X, Langerod A, Moon SH, Rodriguez-Barrueco R, Barsotti A, Chicas A, Li W, Polotskaia A et al (2012) Mutant p53 disrupts mammary tissue architecture via the mevalonate pathway. Cell 148:244–258
33. Freed-Pastor WA, Prives C (2012) Mutant p53: one name, many proteins. Genes Dev 26:1268–1286
34. Gao W, Shen Z, Shang L, Wang X (2011) Upregulation of human autophagy-initiation kinase ULK1 by tumor suppressor p53 contributes to DNA-damage-induced cell death. Cell Death Differ 18:1598–1607
35. Green DR, Kroemer G (2009) Cytoplasmic functions of the tumour suppressor p53. Nature 458:1127–1130
36. Hagenbuchner J, Kuznetsov A, Hermann M, Hausott B, Obexer P, Ausserlechner MJ (2012) FOXO3-induced reactive oxygen species are regulated by BCL2L11 (Bim) and SESN3. J Cell Sci 125(Pt 5):1191–1203
37. Honda R, Tanaka H, Yasuda H (1997) Oncoprotein MDM2 is a ubiquitin ligase E3 for tumor suppressor p53. FEBS Lett 420:25–27
38. Hu W, Zhang C, Wu R, Sun Y, Levine A, Feng Z (2010) Glutaminase 2, a novel p53 target gene regulating energy metabolism and antioxidant function. Proc Natl Acad Sci U S A 107:7455–7460
39. Hussain SP, Amstad P, He P, Robles A, Lupold S, Kaneko I, Ichimiya M, Sengupta S, Mechanic L, Okamura S et al (2004) p53-induced up-regulation of MnSOD and GPx but not catalase increases oxidative stress and apoptosis. Cancer Res 64:2350–2356
40. Ide T, Brown-Endres L, Chu K, Ongusaha PP, Ohtsuka T, El-Deiry WS, Aaronson SA, Lee SW (2009) GAMT, a p53-inducible modulator of apoptosis, is critical for the adaptive response to nutrient stress. Mol Cell 36:379–392
41. Italiano D, Lena AM, Melino G, Candi E (2012) Identification of NCF2/p67phox as a novel p53 target gene. Cell Cycle 11:4589–4596
42. Jeffers JR, Parganas E, Lee Y, Yang C, Wang J, Brennan J, MacLean KH, Han J, Chittenden T, Ihle JN et al (2003) Puma is an essential mediator of p53-dependent and -independent apoptotic pathways. Cancer Cell 4:321–328
43. Jones RG, Plas DR, Kubek S, Buzzai M, Mu J, Xu Y, Birnbaum MJ, Thompson CB (2005) AMP-activated protein kinase induces a p53-dependent metabolic checkpoint. Mol Cell 18:283–293
44. Jones SN, Roe AE, Donehower LA, Bradley A (1995) Rescue of embryonic lethality in Mdm2-deficient mice by absence of p53. Nature 378:206–208
45. Kalo E, Kogan-Sakin I, Solomon H, Bar-Nathan E, Shay M, Shetzer Y, Dekel E, Goldfinger N, Buganim Y, Stambolsky P et al (2012) Mutant p53R273H attenuates the expression of phase 2 detoxifying enzymes and promotes the survival of cells with high levels of reactive oxygen species. J Cell Sci 125:5578–5586
46. Kang MY, Kim HB, Piao C, Lee KH, Hyun JW, Chang IY, You HJ (2013) The critical role of catalase in prooxidant and antioxidant function of p53. Cell Death Differ 20:117–129

47. Kenzelmann Broz D, Spano Mello S, Bieging KT, Jiang D, Dusek RL, Brady CA, Sidow A, Attardi LD (2013) Global genomic profiling reveals an extensive p53-regulated autophagy program contributing to key p53 responses. Genes Dev 27:1016–1031
48. Kitamura N, Nakamura Y, Miyamoto Y, Miyamoto T, Kabu K, Yoshida M, Futamura M, Ichinose S, Arakawa H (2011) Mieap, a p53-inducible protein, controls mitochondrial quality by repairing or eliminating unhealthy mitochondria. PLoS One 6:e16060
49. Kopnin PB, Agapova LS, Kopnin BP, Chumakov PM (2007) Repression of sestrin family genes contributes to oncogenic Ras-induced reactive oxygen species up-regulation and genetic instability. Cancer Res 67:4671–4678
50. Kroemer G, Marino G, Levine B (2010) Autophagy and the integrated stress response. Mol Cell 40:280–293
51. Lahav G, Rosenfeld N, Sigal A, Geva-Zatorsky N, Levine AJ, Elowitz MB, Alon U (2004) Dynamics of the p53-Mdm2 feedback loop in individual cells. Nat Genet 36:147–150
52. Lane DP (1992) Cancer. p53, guardian of the genome. Nature 358:15–16
53. Lang GA, Iwakuma T, Suh YA, Liu G, Rao VA, Parant JM, Valentin-Vega YA, Terzian T, Caldwell LC, Strong LC et al (2004) Gain of function of a p53 hot spot mutation in a mouse model of Li-Fraumeni syndrome. Cell 119:861–872
54. Laplante M, Sabatini DM (2012) mTOR signaling in growth control and disease. Cell 149:274–293
55. Lebedeva MA, Eaton JS, Shadel GS (2009) Loss of p53 causes mitochondrial DNA depletion and altered mitochondrial reactive oxygen species homeostasis. Biochim Biophys Acta 1787:328–334
56. Lee H, Cho JS, Lambacher N, Lee J, Lee SJ, Lee TH, Gartner A, Koo HS (2008) The Caenorhabditis elegans AMP-activated protein kinase AAK-2 is phosphorylated by LKB1 and is required for resistance to oxidative stress and for normal motility and foraging behavior. J Biol Chem 283:14988–14993
57. Lee JH, Budanov AV, Park EJ, Birse R, Kim TE, Perkins GA, Ocorr K, Ellisman MH, Bodmer R, Bier E et al (2009) Sestrin as a feedback inhibitor of TOR that prevents age-related pathologies. Science 327:1223–1228
58. Lee JH, Budanov AV, Talukdar S, Park EJ, Park HL, Park HW, Bandyopadhyay G, Li N, Aghajan M, Jang I et al (2012) Maintenance of metabolic homeostasis by sestrin2 and sestrin3. Cell Metab 16:311–321
59. Lee SO, Andey T, Jin UH, Kim K, Sachdeva M, Safe S (2011) The nuclear receptor TR3 regulates mTORC1 signaling in lung cancer cells expressing wild-type p53. Oncogene 31(27):3265–3276
60. Leng RP, Lin Y, Ma W, Wu H, Lemmers B, Chung S, Parant JM, Lozano G, Hakem R, Benchimol S (2003) Pirh2, a p53-induced ubiquitin-protein ligase, promotes p53 degradation. Cell 112:779–791
61. Levine AJ (1997) p53, the cellular gatekeeper for growth and division. Cell 88:323–331
62. Levine AJ, Feng Z, Mak TW, You H, Jin S (2006) Coordination and communication between the p53 and IGF-1-AKT-TOR signal transduction pathways. Genes Dev 20:267–275
63. Li FP, Fraumeni JF Jr (1969) Rhabdomyosarcoma in children: epidemiologic study and identification of a familial cancer syndrome. J Natl Cancer Inst 43:1365–1373
64. Li FP, Fraumeni JF Jr (1969) Soft-tissue sarcomas, breast cancer, and other neoplasms. A familial syndrome? Ann Intern Med 71:747–752
65. Li T, Kon N, Jiang L, Tan M, Ludwig T, Zhao Y, Baer R, Gu W (2012) Tumor suppression in the absence of p53-mediated cell-cycle arrest, apoptosis, and senescence. Cell 149:1269–1283
66. Liang Y, Liu J, Feng Z (2013) The regulation of cellular metabolism by tumor suppressor p53. Cell Biosci 3:9
67. Loayza-Puch F, Drost J, Rooijers K, Lopes R, Elkon R, Agami R (2013) p53 induces transcriptional and translational programs to suppress cell proliferation and growth. Genome Biol 14:R32
68. Maddocks OD, Berkers CR, Mason SM, Zheng L, Blyth K, Gottlieb E, Vousden KH (2013) Serine starvation induces stress and p53-dependent metabolic remodelling in cancer cells. Nature 493:542–546

69. Maddocks OD, Vousden KH (2011) Metabolic regulation by p53. J Mol Med (Berl) 89:237–245
70. Maiuri MC, Malik SA, Morselli E, Kepp O, Criollo A, Mouchel PL, Carnuccio R, Kroemer G (2009) Stimulation of autophagy by the p53 target gene Sestrin2. Cell Cycle 8:1571–1576
71. Martindale JL, Holbrook NJ (2002) Cellular response to oxidative stress: signaling for suicide and survival. J Cell Physiol 192:1–15
72. Matheu A, Maraver A, Klatt P, Flores I, Garcia-Cao I, Borras C, Flores JM, Vina J, Blasco MA, Serrano M (2007) Delayed ageing through damage protection by the Arf/p53 pathway. Nature 448:375–379
73. Mathew R, Karantza-Wadsworth V, White E (2007) Role of autophagy in cancer. Nat Rev Cancer 7:961–967
74. Matoba S, Kang JG, Patino WD, Wragg A, Boehm M, Gavrilova O, Hurley PJ, Bunz F, Hwang PM (2006) p53 regulates mitochondrial respiration. Science 312:1650–1653
75. Mihaylova MM, Shaw RJ (2011) The AMPK signalling pathway coordinates cell growth, autophagy and metabolism. Nat Cell Biol 13:1016–1023
76. Morselli E, Shen S, Ruckenstuhl C, Bauer MA, Marino G, Galluzzi L, Criollo A, Michaud M, Maiuri MC, Chano T et al (2011) p53 inhibits autophagy by interacting with the human ortholog of yeast Atg17, RB1CC1/FIP200. Cell Cycle 10:2763–2769
77. Muller PA, Vousden KH (2013) p53 mutations in cancer. Nat Cell Biol 15:2–8
78. N'Guessan P, Pouyet L, Gosset G, Hamlaoui S, Seillier M, Cano CE, Seux M, Stocker P, Culcasi M, Iovanna JL et al (2011) Absence of tumor suppressor tumor protein 53-induced nuclear protein 1 (TP53INP1) sensitizes mouse thymocytes and embryonic fibroblasts to redox-driven apoptosis. Antioxid Redox Signal 15:1639–1653
79. Nogueira V, Park Y, Chen CC, Xu PZ, Chen ML, Tonic I, Unterman T, Hay N (2008) Akt determines replicative senescence and oxidative or oncogenic premature senescence and sensitizes cells to oxidative apoptosis. Cancer Cell 14:458–470
80. O'Connor JC, Wallace DM, O'Brien CJ, Cotter TG (2008) A novel antioxidant function for the tumor-suppressor gene p53 in the retinal ganglion cell. Invest Ophthalmol Vis Sci 49:4237–4244
81. Olive KP, Tuveson DA, Ruhe ZC, Yin B, Willis NA, Bronson RT, Crowley D, Jacks T (2004) Mutant p53 gain of function in two mouse models of Li-Fraumeni syndrome. Cell 119:847–860
82. Papadia S, Soriano FX, Leveille F, Martel MA, Dakin KA, Hansen HH, Kaindl A, Sifringer M, Fowler J, Stefovska V et al (2008) Synaptic NMDA receptor activity boosts intrinsic antioxidant defenses. Nat Neurosci 11:476–487
83. Peeters H, Debeer P, Bairoch A, Wilquet V, Huysmans C, Parthoens E, Fryns JP, Gewillig M, Nakamura Y, Niikawa N et al (2003) PA26 is a candidate gene for heterotaxia in humans: identification of a novel PA26-related gene family in human and mouse. Hum Genet 112:573–580
84. Peterson TR, Sengupta SS, Harris TE, Carmack AE, Kang SA, Balderas E, Guertin DA, Madden KL, Carpenter AE, Finck BN et al (2011) mTOR complex 1 regulates lipin 1 localization to control the SREBP pathway. Cell 146:408–420
85. Polyak K, Xia Y, Zweier JL, Kinzler KW, Vogelstein B (1997) A model for p53-induced apoptosis. Nature 389:300–305
86. Pugacheva EN, Ivanov AV, Kravchenko JE, Kopnin BP, Levine AJ, Chumakov PM (2002) Novel gain of function activity of p53 mutants: activation of the dUTPase gene expression leading to resistance to 5-fluorouracil. Oncogene 21:4595–4600
87. Reid MA, Wang WI, Rosales KR, Welliver MX, Pan M, Kong M (2013) The B55alpha subunit of PP2A drives a p53-dependent metabolic adaptation to glutamine deprivation. Mol Cell 50:200–211
88. Reliene R, Fischer E, Schiestl RH (2004) Effect of N-acetyl cysteine on oxidative DNA damage and the frequency of DNA deletions in atm-deficient mice. Cancer Res 64:5148–5153
89. Rufini A, Niklison-Chirou MV, Inoue S, Tomasini R, Harris IS, Marino A, Federici M, Dinsdale D, Knight RA, Melino G et al (2012) TAp73 depletion accelerates aging through metabolic dysregulation. Genes Dev 26:2009–2014

90. Sablina AA, Budanov AV, Ilyinskaya GV, Agapova LS, Kravchenko JE, Chumakov PM (2005) The antioxidant function of the p53 tumor suppressor. Nat Med 11:1306–1313
91. Schubert R, Erker L, Barlow C, Yakushiji H, Larson D, Russo A, Mitchell JB, Wynshaw-Boris A (2004) Cancer chemoprevention by the antioxidant tempol in Atm-deficient mice. Hum Mol Genet 13:1793–1802
92. Selivanova G (2004) p53: fighting cancer. Curr Cancer Drug Targets 4:385–402
93. Semenza GL (2012) Hypoxia-inducible factors in physiology and medicine. Cell 148:399–408
94. Sen N, Satija YK, Das S (2011) PGC-1alpha, a key modulator of p53, promotes cell survival upon metabolic stress. Mol Cell 44:621–634
95. Seo YR, Kelley MR, Smith ML (2002) Selenomethionine regulation of p53 by a ref1-dependent redox mechanism. Proc Natl Acad Sci U S A 99:14548–14553
96. Suzuki S, Tanaka T, Poyurovsky MV, Nagano H, Mayama T, Ohkubo S, Lokshin M, Hosokawa H, Nakayama T, Suzuki Y et al (2010) Phosphate-activated glutaminase (GLS2), a p53-inducible regulator of glutamine metabolism and reactive oxygen species. Proc Natl Acad Sci U S A 107:7461–7466
97. Tasdemir E, Maiuri MC, Galluzzi L, Vitale I, Djavaheri-Mergny M, D'Amelio M, Criollo A, Morselli E, Zhu C, Harper F et al (2008) Regulation of autophagy by cytoplasmic p53. Nat Cell Biol 10:676–687
98. Vahsen N, Cande C, Briere JJ, Benit P, Joza N, Larochette N, Mastroberardino PG, Pequignot MO, Casares N, Lazar V et al (2004) AIF deficiency compromises oxidative phosphorylation. EMBO J 23:4679–4689
99. Velasco-Miguel S, Buckbinder L, Jean P, Gelbert L, Talbott R, Laidlaw J, Seizinger B, Kley N (1999) PA26, a novel target of the p53 tumor suppressor and member of the GADD family of DNA damage and growth arrest inducible genes. Oncogene 18:127–137
100. Villunger A, Michalak EM, Coultas L, Mullauer F, Bock G, Ausserlechner MJ, Adams JM, Strasser A (2003) p53- and drug-induced apoptotic responses mediated by BH3-only proteins puma and noxa. Science 302:1036–1038
101. Vousden KH, Lane DP (2007) p53 in health and disease. Nat Rev Mol Cell Biol 8:275–283
102. Vousden KH, Prives C (2009) Blinded by the light: the growing complexity of p53. Cell 137:413–431
103. Wang PY, Ma W, Park JY, Celi FS, Arena R, Choi JW, Ali QA, Tripodi DJ, Zhuang J, Lago CU et al (2013) Increased oxidative metabolism in the Li-Fraumeni syndrome. N Engl J Med 368:1027–1032
104. Wei CL, Wu Q, Vega VB, Chiu KP, Ng P, Zhang T, Shahab A, Yong HC, Fu Y, Weng Z et al (2006) A global map of p53 transcription-factor binding sites in the human genome. Cell 124:207–219
105. Wood ZA, Poole LB, Karplus PA (2003) Peroxiredoxin evolution and the regulation of hydrogen peroxide signaling. Science 300:650–653
106. Wullschleger S, Loewith R, Hall MN (2006) TOR signaling in growth and metabolism. Cell 124:471–484
107. Xu Y, Ashley T, Brainerd EE, Bronson RT, Meyn MS, Baltimore D (1996) Targeted disruption of ATM leads to growth retardation, chromosomal fragmentation during meiosis, immune defects, and thymic lymphoma. Genes Dev 10:2411–2422
108. Yoon KA, Nakamura Y, Arakawa H (2004) Identification of ALDH4 as a p53-inducible gene and its protective role in cellular stresses. J Hum Genet 49:134–140
109. Zhang C, Lin M, Wu R, Wang X, Yang B, Levine AJ, Hu W, Feng Z (2011) Parkin, a p53 target gene, mediates the role of p53 in glucose metabolism and the Warburg effect. Proc Natl Acad Sci U S A 108:16259–16264
110. Zoncu R, Efeyan A, Sabatini DM (2011) mTOR: from growth signal integration to cancer, diabetes and ageing. Nat Rev Mol Cell Biol 12:21–35

Chapter 19
Lung Cancer Stem Cells, p53 Mutations and MDM2

Venkat Sundar Gadepalli, Swati Palit Deb, Sumitra Deb, and Raj R. Rao

Abstract Over the past few decades, advances in cancer research have enabled us to understand the different mechanisms that contribute to the aberrant proliferation of normal cells into abnormal cells that result in tumors. In the pursuit to find cures, researchers have primarily focused on various molecular level changes that are unique to cancerous cells. In humans, about 50 % or more cancers have a mutated tumor suppressor p53 gene thereby resulting in accumulation of p53 protein and losing its function to activate the target genes that regulate cell cycle and apoptosis. Extensive research conducted in murine cancer models with activated p53, loss of p53, or p53 missense mutations have facilitated researchers to understand the role of this key protein. Despite the identification of numerous triggers that causes lung cancer specific cure still remain elusive. One of the primary reasons attributed to this is due to the fact that the tumor tissue is heterogeneous and contains numerous sub-populations of cells. Studies have shown that a specific sub-population of cells termed as cancer stem cells (CSCs) drive the recurrence of cancer in response to standard chemotherapy. These CSCs are mutated cells with core properties similar to those of adult stem cells. They reside in a microenvironment within the tumor tissue that supports their growth and make them less susceptible to drug treatment. These cells possess properties of symmetric self-renewal and migration thus driving tumor formation and metastasis. Therefore, research specifically targeting these

V.S. Gadepalli
Integrated Life Sciences Program, Virginia Commonwealth University, Richmond, VA, USA

S.P. Deb • S. Deb
Department of Biochemistry and Molecular Biology, and the Massey Cancer Center, Virginia Commonwealth University, Richmond, VA 23298, USA

R.R. Rao (✉)
Chemical and Life Science Engineering, Virginia Commonwealth University, Richmond, VA, USA
e-mail: rrrao@vcu.edu

S.P. Deb and S. Deb (eds.), *Mutant p53 and MDM2 in Cancer*,
Subcellular Biochemistry 85, DOI 10.1007/978-94-017-9211-0_19,

cells has gained prominence towards developing new therapeutic agents against cancer. This chapter focuses on lung cancer stem cells, p53 mutations noted in these cells, and importance of MDM2 interactions. Further, research approaches for better understanding of molecular mechanisms that drive CSC function and developing appropriate therapies are discussed.

Keywords Cancer stem cells • Lung cancer • p53 mutations • MDM2

Lung Cancer

One of the leading causes of cancer-related deaths is cancer of the lung, of which 80–90 % being attributed to tobacco smoking [1, 2]. More than 50 different histological variants of lung cancer have been recognized by the World Health Organization (WHO) and classified based on their phenotype, their zone of origin in the lung, or by tumors arising from functionally diverse lung cells [3]. This diversity characterizes the neoplasms in lung as heterogeneous with different histological subtypes. Almost 98 % of lung cancers are carcinomas (tumors arising from epithelial cells) and based on the size of a cancer cell, categorized as small cell lung cancer (SCLSC) and non-small cell lung cancer (NSCLC) carcinomas. SCLSCs and NSCLCs constitute approximately 10–15 % and 85–90 % of lung cancers respectively [4, 5]. The SCLSCs are malignant small epithelial cells with scanty cytoplasm, while the NSCLCs are further classified based on their size and shape as large cell carcinoma, squamous carcinoma, and adenocarcinoma. NSCLCs are relatively larger in size and contain a high nucleus to cytoplasmic ratio [3, 4, 6]. Additionally, other rare subtypes of lung cancer include bronchioloalveolar carcinoma, carcinoid, glandular, and neuroendocrine tumors.

According to NCI PDQ®, of all the subtypes in lung cancer, the incidence of squamous and adenocarcinoma are considered to be the highest [7]. Studies comparing the major four subtypes of lung cancer recognized that the rate of development of adenocarcinoma is more common and constitute approximately 40 % of these lung cancer subtypes in humans, with the cause strongly associated with tobacco smoking [3, 7, 8]. Various other factors such as asbestos, arsenic, radon (radioactive gas formed as a result of breakdown of uranium in soil), and environmental air pollution also pose a risk of lung cancer [9]. Epidemiological studies and molecular biology studies have indicated a high risk of at least 20 carcinogens in tobacco smoke that can cause lung cancer [10]. Research has shown that tobacco carcinogens such as polycyclic aromatic hydrocarbons (benzo[a]pyren) target hot spots in codon regions of TP53 by forming DNA adduct, thus forming sites for mutations in cancer (reviewed and summarized in [11]). The systematic analysis based on different lung cancer research data uploaded at International Agency for Research on Cancer (IARC) indicates a moderate relationship between smoking exposure and mutation pattern in codon regions (157,158, 175,245, 248, 249 and

273) of TP53, suggesting that mutational pattern in cancers arising in smokers is not specific to a single codon. Studies also confirmed a high frequency in G: C to T: A transversions in TP53 coding region smokers (16 %) than in non-smokers (5.8 %). One key observation in their analysis is change of mutational spectrum based on gender with G:C to T:A transversions found to be higher in female smokers (36 %) than male smokers (27 %) [11, 12]. The advances at molecular level in understanding the cause of cancer and research studies targeting identification of cancer stem cells hold a promise for development of novel approaches in diagnosis and treatment of lung cancer.

Stem Cells in the Lung

The lung is considered to be a highly heterogeneous organ with a variety of cells located in distinct regions of the tissue. Functionally distinct putative stem cells were shown to reside in different anatomical regions of the respiratory system, which play a key role in repopulating the cells in their local area [13–17]. Studies have reviewed the role played by the local stem cells found in trachea (basal, mucous secretory), bronchus (basal, mucous secretory), bronchiole (Clara), and alveolus (type II pneumocyte) and have shown that they primarily contribute to regeneration of lost cells/tissues in response to injury [18]. Identification of resident multipotent lung stem cells that can regenerate any lung cell is still an area of active research. Two major types of stem cells, namely epithelial and mesenchymal stem cells have been reported in the lung so far [13, 16, 19]. These cells were isolated and characterized based on specific cell surface markers that are unique to certain cell lineages. In line with the heterogeneous nature of the lung, the cells that reside in different regions of the lung exhibit differential expression of various cell surface markers. A classic stem cell marker used in the identification of hematopoietic stem cells, Stem cell antigen (Sca-1), has also been found to be expressed in some cells of mesenchymal origin [20]. Studies have demonstrated that lung cells expressing Sca-1 were predominantly found in distal regions of lungs and were shown to possess a temporal emergence, indicated by enrichment of Sca-1 expressing cells in adult mouse lungs when compared to neonatal lungs [21]. Sca-1^{pos} cells have been shown to emerge in postnatal lung during the branching of the airways/lung vasculature and increase exponentially in adult lungs. Sca-1 [22]. Based on the expression of Sca-1 and other markers, various studies have identified unique sub-populations in lung tissue with stem cell characteristics. Thus Sca-1 emerged as a representative cells surface marker to identify the lung stem cells.

Bronchioalveolar stem cells (BASCs) isolated, from bronchioalveolar duct junction in adult mouse lungs, based on expression of Sca-1 and CD 34 (Epithelial and hematopoietic markers) were shown to exhibit self-renewal and multipotent capabilities. In *in vivo* studies, the BASCs were shown to participate in lung epithelial cell renewal and maintain bronchiolar, clara and alveolar cell populations in the distal lung [13]. Gene expression analysis on Sca-1^{neg}, $CD45^{neg}$, $CD31^{neg}$ lung populations and corresponding Sca-1^{pos} cell lines were shown to possess epithelial and mesenchymal gene expression profiles respectively, signifying the presence of Sca-1pos cells with mesenchymal characteristics [16, 21]. Moreover, the Sca-1^{pos}, CD45^{neg}, CD31^{neg} were enriched with mesenchymal progenitor cells in culture as shown by their spindle shaped morphology and expression of mesenchymal markers (CD 104a, Vimentin). In contrast, Sca-1^{neg}, CD45^{neg}, CD31^{neg} cells were shown to possess cobblestone epithelial cell morphology and epithelial marker expression (E-cadherin, cytokeratins 5 and 14, and proSP-C) [21]. From a functional standpoint, the isolated and characterized stem cells in the lung are believed to play an important role in maintaining lung homeostasis. Bronchiolar stem cells have been functionally defined by their expression of clara cell secretory protein (CCSP), pro-surfactant protein C and they belong to airway epithelium [23, 24]. Tiesanu et al., in 2009 have identified bronchiolar stem cells as CD45^{neg}, CD31^{neg}, CD34^{neg}, Sca-l^{low} and AF^{low} as opposed to Sca-1^{pos}, CD45^{pos}, CD45^{neg}, CD31^{neg} reported by Kim et al. [13]. Their transgenic mice models study associated with stem cell expansion, ablation, and lineage tracing, demonstrated CD34^{pos} does not belong to air way epithelium and CCSP expressing cells are found in CD34^{neg}, Sca-l^{low} and AF^{low} [25]. It is useful to note that evidence of different subpopulations in lung cells with potential stem cell properties has been attributed to the method of isolation, culturing conditions, and choice of markers [22]. These studies indicate the complex nature of lung and presence of one or more putative stem cells in the pool Sca-1^{pos}CD 45^{neg}CD31^{neg}, details of isolation and characterization of these cells has been detailed in recent articles published by our group and others [16, 26].

In humans, a class of somatic lung stem cells with self-renewing, clonogenic, and multipotent (in vitro and in vivo) properties were shown to exist using c-kit as the stem cell marker [17]. The c-kit^{pos} cells were negative for hematopoietic and mesenchymal markers and interestingly demonstrated positive expression of key markers associated with pluripotency: *OCT4*, *NANOG*, *KLF4*, and *SOX2*. However, a key defining feature of somatic stem cells that differentiates them from pluripotent stem cells is that they undergo asymmetric division that results in the generation of a heterogeneous population of stem cells and progenitor cells. This study has been received with some skepticism putting forward several questions and need for independent studies to ascertain existence of somatic lung stem cells in human [27].

The fact that lung tissue is composed of a variety of cells with distinct phenotype and functions complicates our understanding of lung regeneration. This is evident from multiple research studies where lung cells characterized by different markers were shown to possess core stem cell properties of self-renewal, clonality, and multipotent characteristics [13, 16, 22]. While these research studies promise a step ahead in identifying putative lung stem cells, there are challenges that need to be addressed. One such challenge is to define a unique set of cells that play a crucial role in lung regeneration. This is expected to strengthen our attempts to develop focused therapeutic strategies in the context of wound healing but also in identifying and targeting putative cancer stem cells in the lung. Some of the studies carried out in identifying lung stem cells in mouse and their characteristic features described by experimental evidence are summarized in Table 19.1.

Table 19.1 Summary of putative lung stem cells sorted by cell surface markers

Sorted based on	Sca-1 expression anatomical location	Morphology	Comments	Reference
Sca-1^{pos}, CD 34^{pos}, CD 45^{neg}, CD31^{neg}	Bronchi alveolar duct junction	Epithelial	These cells were positive for Clara cell marker and Surfactant protein marker.	BASCcs [13]
Hoechst ^{low}, CD 45^{neg}, CD31^{neg}	Sca-1 expression is predominantly found in Distal Lung, restricted to Endothelial and Perivascular Cells	Mesenchymal	Similar to other mesenchymal progenitors, these cells express Sca-1, CD106, CD 140a (PDFGR-a) and CD44	[19]
Sca-1^{pos}, CD 45^{neg}, CD31^{neg}	–	Mesenchymal	Differentiated into endothelial and lung epithelial (alveolar type I, II, and Clara)	[16]
Sca-1^{low}, AF ^{low}CD34^{neg}, CD 45^{neg}, CD31^{neg}	Endothelium and proximal airway epithelium	Epithelial (Bronchiolar)	Sca-1^{low}, CD4^{5neg}, CD31^{neg}, CD34^{neg} include both clara cells & bronchiolar stem cells. Bronchiolar stem cells are distinguished from clara cells by low autofluorescence	[25]
Sca-1^{pos}, CD 45^{neg}, CD31^{neg}	Endothelium and distal parenchymal	Mesenchymal	Cells were also positive for CD 34 and Thy1. Capable of differentiating into lipofibroblastic, osteogenic and chondrogenic cell lineages	[21]

Cancer Stem Cells in the Lung

Numerous studies have demonstrated the presence of subpopulations of cells in tumors that play a critical role in initiating a tumor during post chemotherapy or radiation treatments [15, 28]. These cells are present as a small population within the tumor and appear to be more potent in initiating the tumor than other subpopulations and were classified as cancer stem cells (CSCs). These CSCs were characterized by independent research studies and were shown to sustain their malignant phenotype against drugs targeting cancer [28–30]. Interestingly, these CSC subpopulations were found to possess stem like properties of self-renewal and differentiation similar to those exhibited by somatic stem cells. Certain signaling pathways such as Hedgehog, Notch, and WNT that are important for maintenance of embryonic stem cells were also shown to have a role in putative CSCs found in the lung [14, 31–34]. Thus the discovery of CSCs has opened a new area of research in cancer that focuses on understanding and targeting the cells that drive the recurrence of tumor and metastasis.

CSCs share similarities with the resident somatic stem cells in their respective tissues of origin. Somatic stem cells are characterized by their oligopotent property, where they continuously renew themselves as well as differentiate into distinct descendants that are specific to a tissue. Somatic stem cells are found along with specialized cells of an adult tissue or organ as rare side populations. For prolonged periods of time they reside in quiescence (G-0/resting) phase of the cell cycle, a stage that is an actively controlled phase involving various epigenetic, transcriptional, and signaling pathways [35]. In response to injury or stimuli, these somatic stem cells enter mitosis and give rise to a stem cell and a progenitor cell by the process of asymmetric cell division. The stem cell resides back in quiescence stage until the next signal to re-enter the cell cycle, while the progenitor cells undergo a series of amplifications that give rise to post mitotic differentiated cells in respective tissues or organs of an animal (Fig. 19.1).

This characteristic asymmetric division not only plays a role in maintaining homeostasis in adult tissues by replacing the dead or aging cells, but also avoids repetitive entry of stem cells into the cell cycle, which may increase the chance of DNA damage. This similar kind of hierarchy is observed in CSCs, where a side population of cells forms the backbone to drive relapse of tumor and metastasis. However, unlike normal somatic stem cells these CSCs possess abnormal characteristics which are currently being explored in the context of understanding their role in specific cancers. A broad perspective and future directions in identifying cancer stem cells, *in vitro and in vivo* assays to characterize them, and developing drug screening strategies have been critically discussed [36].

Identification and isolation of these CSCs from the bulk of tumors have been reported based on presence of specific markers that differ from those from used to identify adult lung stem cells reviewed in [15]. The phenotypic characterization of CSCs include the activity of cytoplasmic enzyme aldehyde dehydrogenase (ALDH), expression of cell surface markers CD 133 and CD 44, or capacity of cells to efflux

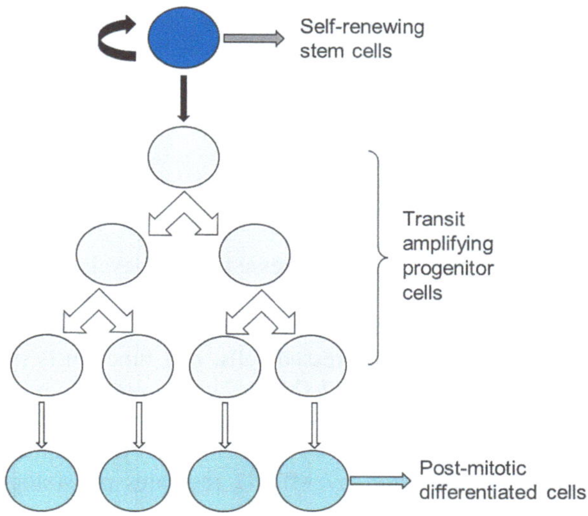

Fig. 19.1 Asymmetric division of somatic stem cells. Asymmetric self-renewal properties of somatic stem cells results in a stem cell and a progenitor cell. The progenitor cells divides repeatedly and give rise to post-mitotic terminally differentiated cells, thus maintaining homeostasis and stem cell pool of a tissue

membrane permeable dyes such as Hoechst 33342 dye and existing as a side population (SP) in bulk of tumor cells [15]. CSCs expressing CD133 cell surface markers were identified to be a putative marker for NSCLC and SCLC, while CD44 is found to be enriched only in NSCLC and not in SCLC. Similarly, NSCLC demonstrate positive activity for ALDH. Based on these studies, identifying a panel of universal markers to classify CSCs is an active area of research.

Signaling pathways such as Hedgehog (Hh), Notch, and WNT are important in the maintenance of stem cells and tissue homeostasis found in CSCs. It is believed that dysregulation of these pathways in CSCs could drive their tumorigenic activities with several reports focused on developing therapeutic strategies to target these pathways [15]. For example, inhibiting Hh signaling pathway in lung cancer cell lines resulted in loss of side population cells, while targeting Notch and Wnt signaling resulted in reduction of ALDH positive tumor cells or induction of apoptosis or growth inhibition in NSCLC [15]. These clinical trials provide some novel developments in treating lung cancer but further trials are needed to demonstrate efficacy. Despite these encouraging clinical results in treating cancer, other mechanisms that are being discovered in CSCs still need to be further researched to develop feasible therapies. One such key mechanism is the Epithelial-mesenchymal transition (EMT) in CSCs, first reported in breast cancer stem cells [37], where the CSCs were shown to exploit this EMT mechanism that is normally observed during developmental process of the mesoderm. It involves a process by which the epithelial cells lose their morphology and gain migratory and invasive properties to become mesenchymal cells. This EMT mechanism was found to be activated during cancer invasion and metastasis and results in the generation of mesenchymal cells that express the stem cell marker CD44 and form tumors effectively in mammary epithelial cancer cells [37] and believed to be associated with drug resistance and cancer progression. Our current understanding is limited on signaling pathways and transcriptional factors that takes place in these CSCs. Expression of EMT associated genes is also being assessed in the context of lung cancer, but the role of EMT in progression of lung cancer is yet to be established [38].

P53 and MDM2 in Lung Cancer

In normal cells, p53 is expressed at low levels but as a result of stress or cellular damage, it activates a host of different proteins that are involved in cell cycle, apoptosis, and senescence, thereby prevents proliferation of cells that carry mutations or DNA damage. In unstressed cells, p53 function is regulated by its specific target murine double minute 2 (MDM2) by a process of ubiquitination. MDM2, an E3 ubiquitin-protein ligase, binds N-terminal transactivation domain of p53, thus mediating p53 degradation by nuclear and cytoplasmic proteasomes. This constant mono-ubiquitination by MDM2 regulates physiological levels and functions of p53 in normal cells [39].

Apart from known functions of p53, recent evidence suggests that p53 plays a crucial role in regulating stem cell homeostasis [40]. Studies involving re-programming of differentiated cells into induced pluripotent stem cells have noted that inhibition or loss of p53 increases the re-programming efficiency by 3–10 fold [41–43]. These studies indicate that p53 has a pivotal role in restricting the reprogramming process. Other studies involving adult mammary stem cells derived from p53–/– mice were shown to possess immortal behavior by increased self-renewal and symmetric division as opposed to limited self-renewal and asymmetric division observed in their wild type counterparts [44]. Similarly in haematopoietic stem cells (HSC), expression of p53 has been found to be critical for regulation of several aspects of HSC behavior. Deletion of p53 in mice was also shown to contribute to increased HSC self-renewal and as well as an increase in the HSC pool [45]. In HSCs, p53 was also found to regulate cellular response to oncogene expression in progenitor cells, where absence of p53 and expression of proto-oncogene KRAs was found to promote acute myeloid leukemia [46]. A recent study in hematopoietic stem cells and mammary stem cells has noted that DNA damage by irradiation induces up regulation of p21, a known p53 inhibitor [47]. These studies indicated that elevated levels of p21 prevent p53 activation and its basal activity, thus preventing stem cells from apoptosis, and allowing them to enter cell cycle. This study identified a unique mode of p21-dependent response to DNA damage in stem cells wherein p21 activates DNA repair, minimizing DNA damage accumulation, and exhausting the stem cells to divide symmetrically as opposed to less stressful asymmetric division [47]. In summary, these studies suggest that, apart from its normal functional role in tumor suppression and cell cycle regulation, p53 is able to restrain adult stem cell self-renewal, and impose asymmetric mode of cell division.

The loss of tumor suppressor function of p53 is either impaired by deletion of Tp53 gene or expression of mutated p53 protein. Alternatively in some human cancers even though wild type p53 is active, its function is diminished by its primary cellular inhibitor, MDM2. These functional disparities are the most commonly observed causes of cancers in humans. MDM2 has a dual function towards p53, by acting as a positive regulator of p53 by interacting with p53 mRNA when the ATM (Ataxia telangiectasia mutated) pathway is active. As the ATM activity ceases, MDM2 acts as a negative regulator to suppress the p53 protein activity by mediating its degradation under normal conditions [48].

The incidence of p53 missense mutations (70 %) is highest in lung cancer compared to all cancer types. Approximately 90 % of these p53 mutations are missense mutations that result in accumulation of mutant p53 protein [49, 50]. These impact molecular activity of cells and cause novel tumors not commonly observed in p53−/− (p53 null) cancerous mice. These missense mutation effects have been explained based on two primary models: dominant negative (DN) activity or oncogenic gain-of-function (GOF) [51]. In the first model, it is proposed that the mutant protein forms a hetero tetramer with wild type p53 and exerts a dominant negative effect on wild type p53 function. In the second model, it is projected that the mutant allele confers oncogenic progression irrespective of the wild type p53 allele counterpart. Studies carried out to understand the GOF mutations by transforming p53 null mice with mutant p53 constructs have supported the effects based on the GOF model. Phenotypic characteristics ascribed to GOF activity of mutant p53 include increased tumorigenicity, growth rate, motility, metastasis, invasiveness and decreased sensitivity to chemotherapeutic drugs [51]. Understanding the changes in these cancer lines with GOF activities is basically proposed as an important area of research for drug targeting. Towards use in these studies, many researchers have generated p53 mutant mouse models [51] and human cell line models based on mutations reported by research studies complied in IARC TP53 database (http://www-p53.iarc.fr/). Recent studies have also highlighted the use of lentiviral approaches (endogenous expression) or transfections (transient expression) to express tumor-derived mutant p53 in cells [52]. Engineered lung cancer cell lines by lentiviral approaches, were shown to possess differential up-regulation of genes between the p53 mutants and differences in their GOF activities [53, 54]. It is observed that these mutant derived p53 show enhanced expression of NF-kappaB2 and receptor tyrosine kinase AXL [53, 55] that could be potential targets for therapies. However, mechanisms underlying these differential expression are unclear and yet to be defined.

Summary

Cancer stem cells are found to conserve many properties of normal somatic stem cells that relate to self-renewal and differentiation. However, they develop resistance to the action of drugs by activating new molecular mechanisms to protect themselves from apoptosis. Some of these mechanisms, specifically related to self-renewal and differentiation are defining features of somatic stem cells. Along with stem-cell like characteristics, CSCs also carry dysregulated activities in of p53, MDM2 and pathways dependent on these proteins. It is thus important to understand pro-tumorigenic effects of missense mutations and their impact on cancer progression. Defining the mechanisms underlying the molecular level changes observed in cancer as it relates to CSCs will enable us to develop effective therapeutic strategies.

References

1. Ferlay J, Soerjomataram I, Ervik M, Dikshit R, Eser S, Mathers C, Rebelo M, Parkin DM, Forman D, Bray F (2013) GLOBOCAN 2012 v1.0, Cancer incidence and mortality worldwide: IARC CancerBase No. 11 [Internet]. International Agency for Research on Cancer, Lyon. Available from: http://globocan.iarc.fr. Accessed 7 Aug 2014.
2. Jemal A, Bray F, Center MM, Ferlay J, Ward E et al (2011) Global cancer statistics. CA Cancer J Clin 61:69–90.
3. Travis WD (2004) Pathology and genetics of tumours of the lung, pleura, thymus, and heart. IARC Press, Lyon, 344 p
4. Travis WD (2002) Pathology of lung cancer. Clin Chest Med 23:65–81, viii
5. Travis WD (2011) Pathology of lung cancer. Clin Chest Med 32:669–692
6. Travis WD, Travis LB, Devesa SS (1995) Lung cancer. Cancer 75:191–202
7. NCI (2014) PDQ® Non-small cell lung cancer treatment. National Cancer Institute, Bethesda
8. Kenfield SA, Wei EK, Stampfer MJ, Rosner BA, Colditz GA (2008) Comparison of aspects of smoking among the four histological types of lung cancer. Tob Control 17:198–204
9. Subramanian J, Govindan R (2007) Lung cancer in never smokers: a review. J Clin Oncol 25:561–570
10. Hecht SS (1999) Tobacco smoke carcinogens and lung cancer. J Natl Cancer Inst 91:1194–1210
11. Toyooka S, Tsuda T, Gazdar AF (2003) The TP53 gene, tobacco exposure, and lung cancer. Hum Mutat 21:229–239
12. Ezzati M, Henley SJ, Lopez AD, Thun MJ (2005) Role of smoking in global and regional cancer epidemiology: current patterns and data needs. Int J Cancer 116:963–971
13. Kim CF, Jackson EL, Woolfenden AE, Lawrence S, Babar I et al (2005) Identification of bronchioalveolar stem cells in normal lung and lung cancer. Cell 121:823–835
14. Singh S, Trevino J, Bora-Singhal N, Coppola D, Haura E et al (2012) EGFR/Src/Akt signaling modulates Sox2 expression and self-renewal of stem-like side-population cells in non-small cell lung cancer. Mol Cancer 11:73
15. Alamgeer M, Peacock CD, Matsui W, Ganju V, Watkins DN (2013) Cancer stem cells in lung cancer: evidence and controversies. Respirology 18:757–764
16. Hegab AE, Kubo H, Fujino N, Suzuki T, He M et al (2010) Isolation and characterization of murine multipotent lung stem cells. Stem Cells Dev 19:523–536
17. Kajstura J, Rota M, Hall SR, Hosoda T, D'Amario D et al (2011) Evidence for human lung stem cells. N Engl J Med 364:1795–1806
18. Otto WR (2002) Lung epithelial stem cells. J Pathol 197:527–535
19. Summer R, Fitzsimmons K, Dwyer D, Murphy J, Fine A (2007) Isolation of an adult mouse lung mesenchymal progenitor cell population. Am J Respir Cell Mol Biol 37:152–159
20. Holmes C, Stanford WL (2007) Concise review: stem cell antigen-1: expression, function, and enigma. Stem Cells 25:1339–1347
21. McQualter JL, Brouard N, Williams B, Baird BN, Sims-Lucas S et al (2009) Endogenous fibroblastic progenitor cells in the adult mouse lung are highly enriched in the sca-1 positive cell fraction. Stem Cells 27:623–633
22. Raiser DM, Kim CF (2009) Commentary: Sca-1 and Cells of the Lung: a matter of different sorts. Stem Cells 27:606–611

23. Hong KU, Reynolds SD, Giangreco A, Hurley CM, Stripp BR (2001) Clara cell secretory protein-expressing cells of the airway neuroepithelial body microenvironment include a label-retaining subset and are critical for epithelial renewal after progenitor cell depletion. Am J Respir Cell Mol Biol 24:671–681
24. Giangreco A, Reynolds SD, Stripp BR (2002) Terminal bronchioles harbor a unique airway stem cell population that localizes to the bronchoalveolar duct junction. Am J Pathol 161:173–182
25. Teisanu RM, Lagasse E, Whitesides JF, Stripp BR (2009) Prospective isolation of bronchiolar stem cells based upon immunophenotypic and autofluorescence characteristics. Stem Cells 27:612–622
26. Gadepalli VS, Vaughan C, Rao RR (2013) Isolation and characterization of murine multipotent lung stem cells. Methods Mol Biol 962:183–191
27. Stripp BR, BHaVT (2011) Lung stem cells: looking beyond the hype. Nat Med 17:788–789
28. Perona R, Lopez-Ayllon BD, de Castro Carpeno J, Belda-Iniesta C (2011) A role for cancer stem cells in drug resistance and metastasis in non-small-cell lung cancer. Clin Transl Oncol 13:289–293
29. Berns A (2005) Stem cells for lung cancer? Cell 121:811–813
30. Kratz JR, Yagui-Beltran A, Jablons DM (2010) Cancer stem cells in lung tumorigenesis. Ann Thorac Surg 89:S2090–S2095
31. Hassan KA, Wang L, Korkaya H, Chen G, Maillard I et al (2013) Notch pathway activity identifies cells with cancer stem cell-like properties and correlates with worse survival in lung adenocarcinoma. Clin Cancer Res 19:1972–1980
32. He B, Barg RN, You L, Xu Z, Reguart N et al (2005) Wnt signaling in stem cells and non-small-cell lung cancer. Clin Lung Cancer 7:54–60
33. Zhang S, Wang Y, Mao JH, Hsieh D, Kim IJ et al (2012) Inhibition of CK2alpha down-regulates Hedgehog/Gli signaling leading to a reduction of a stem-like side population in human lung cancer cells. PLoS One 7:e38996
34. Zhang S, Long H, Yang YL, Wang Y, Hsieh D et al (2013) Inhibition of CK2alpha down-regulates Notch1 signalling in lung cancer cells. J Cell Mol Med 17(7):854–862. http://www.ncbi.nlm.nih.gov/pubmed/23651443.
35. Cheung TH, Rando TA (2013) Molecular regulation of stem cell quiescence. Nat Rev Mol Cell Biol 14:329–340
36. Clarke MF, Dick JE, Dirks PB, Eaves CJ, Jamieson CH et al (2006) Cancer stem cells – perspectives on current status and future directions: AACR Workshop on cancer stem cells. Cancer Res 66:9339–9344
37. Mani SA, Guo W, Liao MJ, Eaton EN, Ayyanan A et al (2008) The epithelial-mesenchymal transition generates cells with properties of stem cells. Cell 133:704–715
38. Xiao D, He J (2010) Epithelial mesenchymal transition and lung cancer. J Thorac Dis 2:154–159
39. Moll UM, Petrenko O (2003) The MDM2-p53 interaction. Mol Cancer Res 1:1001–1008
40. Bonizzi G, Cicalese A, Insinga A, Pelicci PG (2012) The emerging role of p53 in stem cells. Trends Mol Med 18:6–12
41. Kawamura T, Suzuki J, Wang YV, Menendez S, Morera LB et al (2009) Linking the p53 tumour suppressor pathway to somatic cell reprogramming. Nature 460:1140–1144
42. Marion RM, Strati K, Li H, Murga M, Blanco R et al (2009) A p53-mediated DNA damage response limits reprogramming to ensure iPS cell genomic integrity. Nature 460:1149–1153
43. Li H, Collado M, Villasante A, Strati K, Ortega S et al (2009) The Ink4/Arf locus is a barrier for iPS cell reprogramming. Nature 460:1136–1139
44. Cicalese A, Bonizzi G, Pasi CE, Faretta M, Ronzoni S et al (2009) The tumor suppressor p53 regulates polarity of self-renewing divisions in mammary stem cells. Cell 138:1083–1095
45. Liu Y, Elf SE, Miyata Y, Sashida G, Huang G et al (2009) p53 regulates hematopoietic stem cell quiescence. Cell Stem Cell 4:37–48

46. Zhao Z, Zuber J, Diaz-Flores E, Lintault L, Kogan SC et al (2010) p53 loss promotes acute myeloid leukemia by enabling aberrant self-renewal. Genes Dev 24:1389–1402
47. Insinga A, Cicalese A, Faretta M, Gallo B, Albano L et al (2013) DNA damage in stem cells activates p21, inhibits p53, and induces symmetric self-renewing divisions. Proc Natl Acad Sci U S A 110:3931–3936
48. Gajjar M, Candeias MM, Malbert-Colas L, Mazars A, Fujita J et al (2012) The p53 mRNA-Mdm2 interaction controls Mdm2 nuclear trafficking and is required for p53 activation following DNA damage. Cancer Cell 21:25–35
49. Govindan R, Weber J (2014) TP53 mutations and lung cancer not all mutations are created equal. Clinical Cancer Res 2014 Jun 10. pii: clincanres.0899.2014. http://www.ncbi.nlm.nih.gov/pubmed/24916693 (Epub ahead of print)
50. Govindan R, Ding L, Griffith M, Subramanian J, Dees ND et al (2012) Genomic landscape of non-small cell lung cancer in smokers and never-smokers. Cell 150:1121–1134
51. Olive KP, Tuveson DA, Ruhe ZC, Yin B, Willis NA et al (2004) Mutant p53 gain of function in two mouse models of Li-Fraumeni syndrome. Cell 119:847–860
52. Vaughan C, Deb SP, Deb S (2013) Generation of p53 knock-down cell lines. Methods Mol Biol 962:193–199
53. Vaughan CA, Singh S, Windle B, Yeudall WA, Frum R et al (2012) Gain-of-function activity of mutant p53 in lung cancer through up-regulation of receptor protein tyrosine kinase Axl. Genes Cancer 3:491–502
54. Vaughan CA, Frum R, Pearsall I, Singh S, Windle B et al (2012) Allele specific gain-of-function activity of p53 mutants in lung cancer cells. Biochem Biophys Res Commun 428:6–10
55. Scian MJ, Stagliano KE, Anderson MA, Hassan S, Bowman M et al (2005) Tumor-derived p53 mutants induce NF-kappaB2 gene expression. Mol Cell Biol 25:10097–10110

Index

A

Akt signaling, 55, 223–224
Alternative splicing, 19–21, 32, 109, 237, 247, 248, 251, 254–255, 273, 275
Angiogenesis, 77, 105–107, 109–113, 206, 209–210, 282, 326, 343, 348
Apoptosis, 2, 8–10, 12, 18, 21, 24, 26, 33, 43, 45, 52–54, 56, 57, 60, 74, 75, 80, 92, 93, 96, 97, 109, 134, 138–142, 144, 162, 163, 165, 168, 170–173, 175, 176, 188, 189, 201, 207, 209, 218, 220, 235, 236, 238, 239, 255, 263, 267, 271, 282, 283, 293, 295–297, 299, 303–306, 311, 312, 322, 325, 326, 330–332, 338, 341, 350, 351, 365–367
Ataxia telangiectasia mutant (ATM), 26, 92–94, 166–169, 189, 225, 324–327, 329, 331, 343, 350, 367
ATM kinase, 94, 324, 331
ATR kinase, 225, 329

C

c-Abl, 168, 169, 173
Cancer, 1, 17, 41, 72, 92, 106, 119, 134, 162, 187, 201, 216, 236, 247, 263, 281, 323, 339, 359
Cancer-associated mutations, 41–64
Cancer genetics, 2, 5
Cancer predisposition, 10, 343
Cancer stem cells, 359–368
Cancer therapy, 80, 81, 96, 134, 145, 147, 148, 206, 281–312, 353
Cell cycle, 2, 10, 12, 18, 26, 33, 51, 53, 54, 74, 80, 92–94, 97, 106, 120, 123, 127, 128, 134, 142, 147, 168, 174, 188, 189, 195, 207, 217–221, 223, 225, 228, 235, 237–239, 263, 267, 270, 283, 297, 303, 322, 323, 325–329, 331, 339, 341, 345, 350, 351, 364, 366, 367
Cell cycle control, 142, 168, 201, 345
Cellular homeostasis, 106, 322
Chemokine, 56, 96, 107, 110, 111
Chemoresistance, 50, 56, 96, 98, 143, 144
Chemotherapy, 26, 28, 32, 48, 55, 59, 63, 94, 98, 133–153, 275, 307, 364
Chk, 167, 168, 171, 176, 225, 325, 329, 340
Crystallography, 121, 292

D

Degradation, 8, 45–48, 52, 107, 112, 163–169, 171, 172, 176, 177, 188–189, 191–195, 200, 202–206, 208, 210, 218, 224, 226, 236, 238–241, 254, 267, 269, 282–284, 289, 295, 296, 305–308, 322, 324, 340, 345–347, 349, 351
DNA damage, 18, 24, 26, 28, 57, 92–94, 120, 162, 167–169, 171, 173, 174, 176, 177, 188–191, 193, 195, 208, 210, 225, 226, 236, 240–242, 263, 265, 269, 270, 283, 322, 324, 325, 329, 331, 344–348, 351, 352, 364, 366
DNA damage response, 73, 92, 171, 191, 194, 229, 236, 240, 322, 324–327

S.P. Deb and S. Deb (eds.), *Mutant p53 and MDM2 in Cancer*, Subcellular Biochemistry 85, DOI 10.1007/978-94-017-9211-0,

Drug
design, 123, 124, 289, 290, 299–301, 311
development, 299
discovery, 289–291, 304, 306, 307, 311, 312

E
E3 Ligase, 164, 165, 173, 191, 240, 265, 269, 276, 287, 296, 311
Epithelial-mesenchymal transition (EMT), 96, 106, 108, 112, 343, 365, 366
ET-1, 205
Extracellular matrix (ECM), 55, 105, 107

G
Gain of function (GOF), 11, 42, 44, 45, 58, 59, 61, 75–80, 92–93, 95, 96, 108–111, 120, 134, 135, 138–140, 142–144, 188, 190, 192, 194, 326, 339, 351, 353, 367
Gene amplification, 27, 28, 77, 215, 216, 242, 269, 272–275, 311
Genomics, 6, 9, 10, 18, 19, 21, 23, 25–28, 44, 60, 64, 72, 77, 140, 215, 220, 228, 240, 264, 266, 267, 270–272, 322, 323, 327, 331, 340, 343, 346, 347, 351
GLUT, 209, 349
Glycolysis, 140, 200, 209, 345, 346, 349–351
GOF. *See* Gain of function (GOF)
G-protein, 108
Growth-factor response, 210

H
HIF1α, 107, 199–211, 348
Homologous recombination repair (HRR), 321–332
Hypoxia, 18, 47, 162, 199–211, 283, 322, 340, 344, 348
Hypoxia-inducible factor 1 (HIF1), 107, 202, 348

K
Knockout, 2, 10, 18, 22–26, 77, 94, 216, 267, 268, 270, 271, 330, 339, 343

L
Li-Fraumeni syndrome (LFS), 2–12, 22, 72, 75, 77, 339
Lung cancer, 3, 28–31, 33, 80, 108, 110, 111, 144, 149, 150, 226, 274, 306, 307, 359–368

M
Mammalian target of rapamycin kinase (mTOR), 62, 310, 347, 348
MDM2. *See* Mouse double minute 2 (MDM2)
Mdm2 (Hdm2), 162
MdmX, 18, 23, 163, 164, 166, 167, 169, 239, 263–276, 281–312
Metabolism, 2, 73, 206, 209–210, 263, 282, 303, 337–353
Metastasis, 24, 27, 29, 31, 44, 46, 52, 53, 59, 63, 77, 105–113, 134, 142, 238, 282, 343, 351, 364, 365, 367
MicroRNA, 96, 112, 113, 140, 143, 207, 210
Motility, 29, 77, 78, 105, 107, 108, 111, 112, 144, 223, 238, 387
Mouse, 2, 10–12, 18, 19, 21, 22, 24–26, 43, 44, 46, 52, 76, 77, 93, 107, 108, 110, 138, 163, 167, 168, 170, 173, 174, 190, 192, 208, 215–217, 219–221, 236, 252, 255, 256, 264, 266–268, 270–273, 283, 295, 311, 322, 339, 342, 343, 361, 362, 367
Mouse double minute 2 (MDM2), 8, 18, 45, 77, 96, 137, 162, 188, 189, 191–194, 199, 215, 235, 247, 263, 282, 323, 339, 359
mTOR. *See* Mammalian target of rapamycin kinase (mTOR)
Mutant p53, 2, 31, 41, 71, 91, 105, 119, 133, 165, 187, 217, 238, 264, 290, 322, 339, 359
Mutation, 1, 22, 41–64, 92, 105–111, 118, 134, 162, 188, 204, 215, 234, 263, 282, 322, 339, 359–368

N
Nonhomologous end joining (NHEJ), 321–332
Nuclear export, 165, 168, 175, 268, 324

O
Oncogenesis, 72, 77, 80, 91–98, 134, 142, 216, 225, 227, 241, 242, 247–257, 326
Oncogenic signaling, 45, 48, 63, 236
Overexpression, 18, 21, 24, 25, 27–33, 48, 52, 56, 57, 59, 61–63, 76, 137, 138, 143, 176, 203, 204, 207, 210, 215–228, 236–238, 240, 242, 252, 265, 266, 270–276, 325, 327, 342, 346, 351

P
P53, 1, 18, 41, 71, 91, 105, 133, 161, 187, 199, 215, 235, 251, 263, 281, 321, 337, 359
p63, 7, 17–33, 58, 109–112, 127, 135, 325, 326
p73, 7, 17–33, 52, 56–58, 78–80, 92–94, 97, 98, 109, 121, 123, 127–129, 135, 204, 325, 326, 352
p53 activation, 49, 74, 120, 166–168, 172, 267, 269, 282, 283, 305–308, 324, 340–342, 344, 350, 351, 366
Phosphorylation, 26, 45, 49–53, 57, 63, 72, 77, 94, 143, 166–177, 189, 190, 193, 201, 205, 206, 218, 219, 224, 225, 236, 239, 268, 282, 323–325, 329, 341, 344, 346, 348–351
Plasminogen activator inhibitor-1 (PAI-1), 54, 107, 205
p53 missense mutations, 52, 59, 60, 62, 63, 92, 367
PML nuclear bodies, 175–177
p53 mutations, 2, 31, 41, 71, 91, 105, 119, 133, 165, 187, 217, 238, 264, 290, 322, 339, 359
PNUTS, 203
Post-translational modifications (PTMs), 49, 54, 72, 92, 162, 166, 174, 175, 188–190, 236, 323–325
Prognostic factor, 52, 134
Prolyl-hydroxylases (PHD1-3), 200
Proteasome, 188, 191, 195, 200, 204, 238, 254, 269, 276, 284, 287, 295, 306, 309, 340, 366
Protein degradation, 164
Protein-protein interactions, 18, 20, 21, 78, 79, 162, 172–173, 202, 282, 299, 344, 347, 352
p53 structure, 119–129
PTMs. *See* Post-translational modifications (PTMs)
p53 transactivation, 75, 218, 236, 239, 265, 269, 298
pVHL, 200, 202, 204, 208

R
Rad51, 327, 329
Radiation, 46, 49, 77, 133–153, 174, 189, 208, 218, 220, 227, 255, 272, 322, 325, 327, 329, 364
Radioresistance, 137–140, 150, 326, 330
Reactive oxygen species (ROS), 22, 47, 49, 142, 189–190, 195, 340–347, 351–353

S
Splicing, 6, 11, 18–33, 109, 237, 247, 248, 251, 253–256, 273, 275
Stability, 10, 12, 23, 25, 26, 42, 45–47, 51, 54, 61, 73, 123, 125, 127, 168, 170, 171, 176, 188–195, 202, 204, 205, 208, 239, 241, 269, 282, 293, 294, 306, 307, 311, 331, 340, 343, 346, 347, 350
Stress response, 340–341
Suppression, 2, 12, 18, 24, 25, 29, 30, 52, 53, 58, 75, 107, 110, 112, 120, 125–127, 162, 168, 171, 174, 221, 225, 242, 275, 328, 330, 331, 341–343, 345, 348, 351, 352, 366

T
Target, 21, 46, 74, 92, 106, 121, 135, 162, 188, 200, 219, 236, 254, 266, 284, 322, 339, 360
Target genes, 21, 22, 24, 26, 27, 29, 30, 50, 53, 57, 58, 63, 74, 78,–80, 92–97, 110, 121, 123, 127, 128, 143, 162, 163, 165, 170, 171, 175, 193–195, 201, 202, 206–208, 216, 239, 297, 305, 324
Transactivation, 6, 19–22, 29, 30, 50, 73–75, 77–80, 108–110, 120–123, 125–129, 142, 143, 163–165, 173, 176, 189, 191, 193–194, 208, 218, 231, 236, 239, 265, 269, 298, 305, 323, 324, 327–329, 366
Transcription, 2, 18, 45, 71, 91, 109, 120, 138, 162, 193, 200, 216, 236, 247, 263, 282, 322, 340, 363
Transforming growth factor beta (TGFbeta), 110
Tumor suppressor, 2, 3, 11, 12, 18, 20, 21, 24, 26, 28, 30, 33, 42, 48, 58, 63, 72, 75, 76, 107, 109, 188, 190, 235, 238, 254, 263, 265, 282, 306, 322, 324, 337–353, 366

U
Ubiquitin, 8, 23, 46, 163, 188, 200, 217, 236, 265, 282, 322, 340, 366
Ubiquitination, 49, 50, 92, 164–169, 171, 177, 188, 191, 192, 195, 204, 218, 226, 238, 241, 269, 283, 289, 295–296, 311, 349, 366
Unscheduled DNA replication, 225

V
Vascular endothelial growth factor (VEGF), 76, 107, 204–207, 210

P

p53, 1, 18, 41, 71, 91, 105, 133, 161, [illegible]
213, 235, 250, 263, 281, [illegible]
p63, 7, 17, 35, 58, 108, 112, 127, 132, [illegible]
327, 328
p73, 7, 17, 33, 52, 56–58, 78–80, [illegible]
96, 100, 121, 123, 127–129, [illegible]
325, 326, 352
p53 activation, 49, 7[illegible], 120, 166–168, [illegible]
26[illegible], 282, 284, 30[illegible]–308, [illegible]
340–342, [illegible]
Phosphorylation, 26, 45, 49, 53, [illegible]
94, 143, 166–177, 180, 190, [illegible]
205, 206, 218, 219, 224, 225, [illegible]
268, 282, [illegible], 315, 320, 341, [illegible]
348, 351
Plasminogen activator inhibitor-1 (PAI-1), [illegible]
54, 107, 205
p53 missense mutations, 62, 59, 60, [illegible]
63, 92, [illegible]
PML nuclear bodies, 175, 177
p53 mutations, 2, [illegible], 41, 71, 91, 105, [illegible]
129, 162, 187, 217, 238, 264, 290, [illegible]
322, 330, [illegible]
PNUTS, 203
Post-translational modifications (PTMs), [illegible]
49, 54, 72, 92, 115, 116, 139, [illegible]
188–190, 226, 323, 325
Prognostic factor, 57, 234
Prolyl hydroxylase (PHD), 197, 200
Proteasome, 188, 191–193, 200, 204, [illegible]
254, 257, 276, 284, 287, 292, [illegible]
309, 340, 356
Protein degradation, 164
Protein-protein interactions, 18, 20, [illegible]
94, 162, 172, 174, 202, 282, 292, [illegible]
347, 352
p53 structure, 119, 129
PTMs. See Post-translational modifications (PTMs)
p53 transactivation, 7[illegible], 178, 236, 238, [illegible]
280, 29[illegible]
pVHL, 200, 202, 204, 205

R

Rad51, 327, 329
Radiation, 46, 49, 72, 155, 157, 170, 171, [illegible]
218, 220, 227, 233, 237, [illegible]
32[illegible], 364
Radioresistance, 137–140, [illegible]
[illegible]
Reactive oxygen species (ROS), [illegible]
183, 188–190, 195, 210, 217, [illegible]
31[illegible]–313

S

Splicing, [illegible]
Stability, [illegible]
Stress response, [illegible]
Suppress[illegible]

T

Target[illegible]
TATA[illegible]
Transactivation domain, [illegible]
Transcription[illegible]
Tumor growth[illegible]
Tumorigenesis, [illegible]

U

Ubiquitin[illegible]
Ubiquitination, [illegible]

V

Vaccination, [illegible]

MIX
Papier aus verantwortungsvollen Quellen
Paper from responsible sources
FSC® C105338

If you have any concerns about our products,
you can contact us on
ProductSafety@springernature.com

In case Publisher is established outside the EU,
the EU authorized representative is:
Springer Nature Customer Service Center GmbH
Europaplatz 3, 69115 Heidelberg, Germany

Printed by Libri Plureos GmbH
in Hamburg, Germany